Lecture Notes in Networks and Systems

Volume 109

The series "Lecture Notes in Networks and Systems" publishes the latest developments in Networks and Systems—quickly, informally and with high quality. Original research reported in proceedings and post-proceedings represents the core of LNNS.

Volumes published in LNNS embrace all aspects and subfields of, as well as new challenges in, Networks and Systems.

The series contains proceedings and edited volumes in systems and networks, spanning the areas of Cyber-Physical Systems, Autonomous Systems, Sensor Networks, Control Systems, Energy Systems, Automotive Systems, Biological Systems, Vehicular Networking and Connected Vehicles, Aerospace Systems, Automation, Manufacturing, Smart Grids, Nonlinear Systems, Power Systems, Robotics, Social Systems, Economic Systems and other. Of particular value to both the contributors and the readership are the short publication timeframe and the world-wide distribution and exposure which enable both a wide and rapid dissemination of research output.

The series covers the theory, applications, and perspectives on the state of the art and future developments relevant to systems and networks, decision making, control, complex processes and related areas, as embedded in the fields of interdisciplinary and applied sciences, engineering, computer science, physics, economics, social, and life sciences, as well as the paradigms and methodologies behind them.

**** Indexing: The books of this series are submitted to ISI Proceedings, SCOPUS, Google Scholar and Springerlink ****

More information about this series at http://www.springer.com/series/15179

Mihir Narayan Mohanty ·
Swagatam Das
Editors

Advances in Intelligent Computing and Communication

Proceedings of ICAC 2019

Springer

Editors
Mihir Narayan Mohanty
Department of Electronics and Communication Engineering, Institute of Technical Education and Research
Siksha 'O' Anusandhan Deemed to be University
Bhubaneswar, Odisha, India

Swagatam Das
Electronics and Communication Sciences Unit
Indian Statistical Institute
Kolkata, West Bengal, India

ISSN 2367-3370 ISSN 2367-3389 (electronic)
Lecture Notes in Networks and Systems
ISBN 978-981-15-2773-9 ISBN 978-981-15-2774-6 (eBook)
https://doi.org/10.1007/978-981-15-2774-6

This Springer imprint is published by the registered company Springer Nature Singapore Pte Ltd.
The registered company address is: 152 Beach Road, #21-01/04 Gateway East, Singapore 189721, Singapore

Preface

This issue of Lecture Notes in Networks and Systems is dedicated to the International Conference on Intelligent Computing and Advances in Communication (ICAC) held at the campus of Institute of Technical Education and Research (Faculty of Engineering and Technology), Siksha 'O' Anusandhan Deemed to be University, Bhubaneswar, from November 15 to 17, 2019. This conference was organized by the Department Electronics and Communication Engineering of Institute of Technical Education and Research (Faculty of Engineering and Technology). This conference had three tracks, namely advances in communication systems, intelligent systems, and signal processing. From 165 papers received from three countries, 63 were selected for inclusion into the conference proceedings. Each paper was peer-reviewed by at least two reviewers.

The objective of the conference is to bring together experts from academic institutions, industries, research organizations, and professional engineers for sharing of knowledge, expertise, and experience in emerging trends related to the computer, communication, and electrical topics. The aim of this international conference is to cover all the issues on a single platform and provide international forum for researchers to discuss the real-time problems and solutions to exchange their valuable ideas and showcase the ongoing works which may lead to path-breaking foundation of the futuristic engineering. This conference mainly aims at advanced communication protocol, database security and privacy, advanced computing system, energy saving, etc., on several updated techniques. The conference offers a platform to focus on the inventive information and computing toward the investigation of cognitive mechanisms and processes of human information processing, and the development of the next-generation engineering and advanced technological systems.

This conference was held in the Bansuri Guru Auditorium, Institute of Technical Education and Research (Faculty of Engineering and Technology), Siksha 'O' Anusandhan Deemed to be University, Bhubaneswar, Odisha, India. Siksha 'O' Anusandhan is a Deemed University located in Bhubaneswar, Odisha. It was originally founded as Institute of Technical Education and Research (ITER) in the year 1996. The university has been at the forefront nourishing a learning ambience, encouraging academic, research, and innovations since its inception in 2007. The university is composed of nine degree-granting schools with 10,000 students. Many of SOAU's programs are nationally accredited for meeting high standards of academic quality, including engineering, medicine, pharmacy, business, nursing, biotechnology, humanities, environment, nanotechnology, agriculture, and law. The university was ranked 16th by National Institutional Ranking Framework (NIRF) under the aegis of Ministry of Human Resource Development, Government of India, and has been awarded 'A' grade by NAAC. It has established ten research centers and 23 research laboratories to fulfill the need of faculties and students.

Faculty of Engineering and Technology is a constituent of SOAU having 13 departments with more than 400 faculty members. The Department of ECE is continually working to provide quality research outputs in the areas of signal and image processing, communication engineering, and microelectronics devices.

The editors thank the authors for extending their fullest cooperation in preparing the manuscripts to the Springer Lecture Notes guidelines, taking aboard the additional review comments.

The editors would also like to convey their heartfelt thanks to Prof. M. R. Nayak, President, Siksha 'O' Anusandhan Deemed to be University; Prof. Amit Bannerjee, Vice-Chancellor, Siksha 'O' Anusandhan Deemed to be University; Prof. M. R. Mallick, Director, ITER (FET), Siksha 'O' Anusandhan Deemed to be University; Prof. P. K. Nanda, Dean (R&D), Siksha 'O' Anusandhan Deemed to be University; and Prof. P. K. Sahoo, Dean, ITER (FET), Siksha 'O' Anusandhan Deemed to be University, for their constant inspiration and motivation in all stages of the conference.

Sincerely,

Bhubaneswar, India
Kolkata, India

Mihir Narayan Mohanty
Swagatam Das
Guest Editors

Contents

Editors and Contributors

About the Editors

Dr. Mihir Narayan Mohanty received his M.Tech. in Communication System Engineering from Sambalpur University, Odisha, and his Ph.D. in Applied Signal Processing from Biju Patnaik University of Technology, Odisha. He is currently serving as a Professor at the Department of Electronics and Communication Engineering, Institute of Technical Education and Research (FET), Siksha 'O' Anusandhan (Deemed University), Bhubaneswar, Odisha. With 24 years of teaching experience, his research interests include intelligent signal and image processing, digital signal/image processing, biomedical signal processing, microwave communication engineering, and antennas. He has published over 300 research papers in national and international journals and conference proceedings, and has received various national and international awards for his contributions. He is an active member of numerous professional societies, e.g. the IEEE, IET, IETE, ISTE, EMC & EMI Engineers India, etc. Dr. Mohanty is also an active reviewer for the IEEE, Elsevier, Springer, and several international conferences.

Swagatam Das received his B. E. Tel. E., M. E. Tel. E (Control Engineering specialization) and Ph. D. degrees, all from Jadavpur University, India, in 2003, 2005, and 2009 respectively. Currently serving as an Associate Professor at the Electronics and Communication Sciences Unit of the Indian Statistical Institute, Kolkata, India, his research interests include evolutionary computing, pattern recognition, multi-agent systems, and wireless communication. Dr. Das has published more than 300 research articles in peer-reviewed journals and international conference proceedings. He is the founding co-editor-in-chief of Swarm and Evolutionary Computation, an international journal from Elsevier. He has also served as or is serving as an associate editor of various journals, including Pattern Recognition, Neurocomputing, Information Sciences, IEEE Access, and so on.

He is an editorial board member of numerous other journals, including Progress in Artificial Intelligence, Applied Soft Computing, and Artificial Intelligence Review.

Dr. Das is the recipient of the 2012 Young Engineer Award from the Indian National Academy of Engineering (INAE), and of the 2015 Thomson Reuters Research Excellence India Citation Award for the highest cited researcher from India in the Engineering and Computer Science category for the period 2010 to 2014.

Contributors

Ramachandra Agrawal Department of CSIT, ITER, S‘O’A University, Deemed to be University, Bhubaneswar, India;
EEE Department, ITER, S‘O’A (Deemed to be University), Bhubaneswar, India

Sunanda Alamuru SreeNidhi Institute of Science and Technology, Hyderabad, India

Bhavya Alankar Department of Computer Science and Engineering, School of Engineering Sciences and Technology, New Delhi, India

Bhukya Arun Kumar School of Electronics and Electrical Engineering, Lovely Professional University, Phagwara, India

Mohd. Atif Department of Computer Science and Engineering, School of Engineering Sciences and Technology, JamiaHamdard, New Delhi, India

Prasanta Kumar Bal Department of Computer Science & Application, Utkal University, Bhubaneswar, India

Shekharesh Barik CSE Department, DRIEMS (Autonomous), Cuttack, India

Bhagyalaxmi Behera ITER, Siksha ‘O’ Anusandhan (Deemed to be University), Bhubaneswar, India

Debasis Behera C. V. Raman College of Engineering, Bhubaneswar, Odisha, India

Suchismita Behera Department of ECE, Siksha, ‘O’ Anusandhan (Deemed to be University), Bhubaneswar, Odisha, India

Satyanarayan Bhuyan Department of Electronics and Communication Engineering, SOA Deemed to be University, Bhubaneswar, Odisha, India

B. N. Biswal Department of Computer Science & Engineering, Bhubaneswar Engineering Colleges, Bhubaneswar, Odisha, India

Bedanta Kumar Biswal Electronics and Communication Engineering, Faculty of Engineering, Siksha ‘O’Anusandhan (Deemed to be University), Bhubaneswar, India

Kumar Biswal School of Electronics Engineering, Kalinga Institute of Industrial Technology (DU), Bhubaneswar, India

Pradyut Kumar Biswal Department of Electronics Engineering, IIIT Bhubaneswar, Bhubaneswar, India

Pravat Kumar Biswal Department of EEE, KIIT, Bhubaneswar, Odisha, India

S. Chakravarty Department of Computer Science & Engineering, Centurion University of Technology & Management, Bhubaneswar, Odisha, India

Ritu Chauhan Centre for Computational Biology and Bioinformatics, Amity University, Noida, U.P, India

Satish Choudhury Department of Electrical Engineering, ITER Siksha 'O' Anusandhan (Deemed to be University), Bhubaneswar, India

Abhisek Das ITER, Siksha 'O' Anusandhan (Deemed to be University), Bhubaneswar, India

Abhishek Das ITER, Siksha 'O' Anusandhan (Deemed to be University), Bhubaneswar, India

Kunal Kumar Das Electronics and Communication Engineering, Institute of Technical Education and Research, Siksha 'O' Anusandhan Deemed University, Bhubaneswar, Odisha, India

Manoranjan Das Institute of Technical Education and Research, SOA (Deemed to be University), Bhubaneswar, India

Nilima R. Das Faculty of Engineering & Technology, Siksha 'O' Anusandhan Deemed to Be University, Bhubaneswar, Odisha, India

Niva Das Department of ECE, Siksha, 'O' Anusandhan (Deemed to be University), Bhubaneswar, Odisha, India

P. S. Das Department of Electronics and Communication Engineering, GITA, Bhubaneswar, India

Pradipta Das Veer Surendra Sai University of Technology, Burla, Odisha, India

Pranati Das Department of Electrical Engineering, Indira Gandhi Institute of Technology, Sarang, Odisha, India

Radhakrishna Das Department of Electrical Engineering, GIET, Bhubaneswar, India

Rajendra Kumar Das Department of ENTC, DRIEMS Autonomous Engineering College, Tangi, Cuttack, Odisha, India

S. R. Das Department of Electronics and Communication Engginerring, GITA, Bhubaneswar, India

Smruti Rekha Das Department of Computer Science and Engineering, Gandhi Institute for Education and Technology, Bhubaneswar, Odisha, India

Lipsa Dash New Horizon College of Engineering, Bengaluru, India

Meera Dash Department of Electronics and Communication Engineering, SOA (Deemed to be) University, Bhubaneswar, India

Sachikanta Dash Department of CSE, DRIEMS Autonomous Engineering College, Tangi, Cuttack, Odisha, India

Shibanee Dash Department of ECE, FET, ITER, S'O'A Deemed to be University, Bhubaneswar, India

Prajna Paramita Debata International Institute of Information Technology, Bhubaneswar, Odisha, India

Dinesh Kumar Garg Department of Computer Science and Engineering, SSCET, Pathankot, India

Debalina Ghosh School of Electrical Sciences, IIT Bhubaneswar, Bhubaneswar, India

Nagendra Kumar Goel GoVivace Inc., McLean, VA, USA

Sonali Goel Department of Electrical Engineering, ITER Siksha 'O' Anusandhan (Deemed to be University), Bhubaneswar, India

Vishal Goyal Department of Electronics and Communication Engineering, GLA University, Mathura, India

Bibekananda Jena Department of Electrical Engineering, ITER Siksha 'O' Anusandhan (Deemed to be University), Bhubaneswar, India

Lambodar Jena Department of Computer Science and Engineering, Siksha O Anusandhan (Deemed to Be University), Bhubaneswar, India

Narendra Kumar Jena Department of Electrical Engineering, Siksha 'O' Anusandhan Deemed to be University, Bhubaneswar, Odisha, India;
Department of EEE, ITER, S'O'A University, Bhubaneswar, India

Durga P. Kar Department of Electronics and Communication Engineering, SOA Deemed to be University, Bhubaneswar, Odisha, India

Nihar Kar Department of ECE, Faculty of Engineering, Siksha 'O' Anusandhan, Bhubaneswar, Odisha, India

Sanjeeb Kar ITER, SoA, Deemed to be University, Bhubaneswar, Odisha, India

Harleen Kaur Department of Computer Science and Engineering, School of Engineering Sciences and Technology, JamiaHamdard, New Delhi, India

Mehreen Kirmani Department of Computer Science and Engineering, SSCET, Pathankot, India

Sonagara Abhishek M Department of Electronics and Communication Engineering, ITER, Siksha 'O' Anusandhan (Deemed to be University), Bhubaneswar, Odisha, India

Shubham Mahajan School of Electronics & Communication, Shri Mata Vaishno Devi University, Katra, Reasi, J&K, India

Shaktijeet Mahapatra Department of ECE, ITER, Siksha 'O' Anusandhan (Deemed to be University), Bhubaneswar, India

Suman Malik School of Electrical Science, Indian Institute of Technology Bhubaneswar, Bhubaneswar, Odisha, India

Alok Kumar Mishra Department of EEE, ITER, S'O'A University, Deemed to be University, Bhubaneswar, India

Debahuti Mishra Department of Computer Science and Engineering, Siksha 'O'Anusandhan Deemed to be University, Bhubaneswar, Odisha, India

Guru Prasad Mishra Department of ECE, FET, ITER, S'O'A Deemed to be University, Bhubaneswar, India

Laxmi Prasad Mishra Department of Electronics and Communication Engineering, Institute of Technical Education and Research, Siksha 'O' Anusandhan (Deemed to be University), Bhubaneswar, Odisha, India

Puneet Mishra Department of Electrical and Electronics Engineering, Birla Institute of Technology and Science, Pilani, Jhunjhunu, Rajasthan, India

Sivkumar Mishra Department of EE, CAPGS, BPUT, Rourkela, Odisha, India; Department of EEE, CAPGS, BPUT, Rourkela, Odisha, India

Bibhuprasad Mohanty Department of Electronics and Communication Engineering, ITER, Siksha 'O' Anusandhan University (Deemed to be University), Bhubaneswar, India

Kanungo B. Mohanty Department of Electrical Engineering, NIT, Rourkela, Rourkela, Odisha, India

Mihir Narayan Mohanty Department of ECE, ITER, Siksha 'O' Anusandhan (Deemed to be University), Bhubaneswar, India

Monalisa Mohanty Department of Electronics and Communication, Institute of Technical Education & Research, SOA Deemed to be University, Bhubaneswar, India

Sangram Kishore Mohanty Department of ECE, ICE, GITA, Bhubaneswar, India;
Department of ECE, ICE, Bhubaneswar, India

Saumendra Kumar Mohanty Department of ECE, FET, ITER, S'O'A Deemed to be University, Bhubaneswar, India

Saumendra Kumar Mohapatra ITER, Siksha 'O' Anusandhan (Deemed to be University), Bhubaneswar, India

Subhasish Mohapatra Department of Computerscience Engineering, Adamas University, Kolkata, India

Laxmipriya Moharana Electronics and Communication Engineering, Faculty of Engineering, Siksha 'O'Anusandhan (Deemed to be University), Bhubaneswar, India

Ashish Kumar Mourya Department of Computer Science and Engineering, School of Engineering Sciences and Technology, JamiaHamdard, New Delhi, India

Manoj Kumar Naik Department of ECE, ITER, Siksha O Anusandhan, Bhubaneswar, Odisha, India

Suraj Naik Electronics and Communication Engineering, Faculty of Engineering, Siksha 'O'Anusandhan (Deemed to be University), Bhubaneswar, India

Amar Bijaya Nanda Department of Electrical Engineering, Siksha 'O' Anusandhan Deemed to be University, Bhubaneswar, Odisha, India

Anuja Nanda Department of EEE, ITER, S'O'A University, Deemed to be University, Bhubaneswar, India

Pradipta Kumar Nanda Image & Video Analysis Laboratory, Department of ECE, S O A, Deemed to be University, Bhubaneswar, India

Ajit Nayak Faculty of Engineering & Technology, Siksha 'O' Anusandhan Deemed to Be University, Bhubaneswar, Odisha, India

Soumen Nayak Department of Computer Science and Engineering, Siksha O Anusandhan (Deemed to Be University), Bhubaneswar, India;
Department of Computer Science and Engineering, Indian Institute of Technology (ISM), Dhanbad, India

Suryakanta Nayak Department of Computer Science and Application, Utkal University, Bhubaneswar, India

Praveen P. Nayk Department of Electronics and Communication Engineering, SOA Deemed to be University, Bhubaneswar, Odisha, India

B. K. Padhi Department of Computer Science & Engineering, Centurion University of Technology & Management, Bhubaneswar, Odisha, India

Jagadish C. Padhi Department of Electronics and Communication Engineering, SOA Deemed to be University, Bhubaneswar, Odisha, India

G. Palai Department of Electronics and Communication Engineering, GITA, Bhubaneswar, India

Hemanta Kumar Palo Siksha O Anusandhan (Deemed to Be University), Bhubaneswar, Odisha, India

Leeza Panda Department of Electronics and Communication Engineering, ITER, Siksha 'O' Anusandhan University (Deemed to be University), Bhubaneswar, India

Mohit Ranjan Panda School of Computer Engineering, KIIT University, Bhubaneswar, Odisha, India

Mrutyunjaya Panda Department of Computer Science and Application, Utkal University, Bhubaneswar, India

Prakash Kumar Panda Department of ECE, ITER, SOA Deemed to be University, Bhubaneswar, India

Susmita Panda Institute of Technical Education and Research, SOA Deemed to be University, Bhubaneswar, Odisha, India

Amit Kant Pandit School of Electronics & Communication, Shri Mata Vaishno Devi University, Katra, Reasi, J&K, India

Sasmita Pani Department of Computer Science and Engineering, Government College of Engineering, Keonjhar, Odisha, India

Santisudha Panigrahi Department of CSE, ITER, S'O'A University, Bhubaneswar, India

Trilochan Panigrahi Department of Electronics and Communication Engineering, SOA (Deemed to be) University, Bhubaneswar, India

Pournamasi Parhi Department of Computer Science and Engineering, Siksha 'O'Anusandhan Deemed to be University, Bhubaneswar, Odisha, India

Smita Parija Department of Eletronics, CV Raman College of Engineering, Odisha, India

Dipti Patnaik Department of Electronics and Communication Engineering, SOA Deemed to be University, Bhubaneswar, Odisha, India

Abhishek Patra Department of CSIT, ITER, S'O'A University, Deemed to be University, Bhubaneswar, India

Akshaya Kumar Patra Department of EEE, ITER, S'O'A University, Deemed to be University, Bhubaneswar, India

Binod Kumar Pattanayak Department of Computer Science and Engineering, Institute of Technical Education and Research, Siksha 'O' Anusandhan Deemed to be University, Bhubaneswar, Odisha, India

Omkar Pattnaik Department of Computer Science and Engineering, Government College of Engineering, Keonjhar, Odisha, India

Sateesh Kumar Pradhan Department of Computer Science & Application, Utkal University, Bhubaneswar, India

Sovit Kumar Pradhan Department of Electrical Engineering, NIT, Silchar, India

S. S. P. B. K. Prasad Department of Electrical & Electronics Engineering, GIET, Bhubaneswar, India

Rojalina Priyadarshini School of Computer Engineering, KIIT University, Bhubaneswar, Odisha, India

Pankaj Prusty Electronics and Communication Engineering, Faculty of Engineering, ITER, Siksha 'O' Anusandhan (Deemed to be University), Bhubaneswar, India

Vedanta Prusty Department of ETC, IIIT, Bhubaneswar, Odisha, India

Satyananda C. Rai Department of IT, Silicon Institute of Technology, Bhubaneswar, Odisha, India

Raman Raj Electronics and Communication Engineering, Faculty of Engineering, Siksha 'O'Anusandhan (Deemed to be University), Bhubaneswar, India

Rashmirekha Ram ITER, Siksha 'O' Anusandhan (Deemed to be University), Bhubaneswar, India

Abhisek Rath Department of ETC, IIIT, Bhubaneswar, Odisha, India

Debswarup Rath Department of Electrical & Electronics Engineering, GIET, Bhubaneswar, India

Bidyadhar Rout Department of EEE, Veer Surendra Sai University of Technology, Burla, India

Bikash Chandra Rout Siksha O Anusandhan (Deemed to Be University), Bhubaneswar, Odisha, India

Kshirod Kumar Rout Department of EE, IIIT, Bhubaneswar, Odisha, India; Department of EEE, IIIT, Bhubaneswar, Odisha, India

Pravat Kumar Rout Department EEE, ITER, Siksha 'O' Anusandhan (Deemed to be University), Bhubaneswar, India

Jibendu Sekhar Roy School of Electronics Engineering, Kalinga Institute of Industrial Technology (KIIT), Deemed to be University, Bhubaneswar, Odisha, India

Sukanta Sabut School of Electronics Engineering, KIIT Deemed to beUniversity, Bhubaneswar, Odisha, India

Arabinda Sahoo Department of ECE, ITER, Siksha 'O' Anusandhan Deemed to be University, Bhubaneswar, Odisha, India

Santanu Sahoo Department of ECE, Faculty of Engineering, Siksha 'O' Anusandhan, Bhubaneswar, Odisha, India

Subhadra Sahoo Department of Electrical Engineering, Siksha 'O' Anusandhan Deemed to be University, Bhubaneswar, Odisha, India

Subhaluxmi Sahoo Image & Video Analysis Laboratory, Department of ECE, S O A, Deemed to be University, Bhubaneswar, India

Swarnaprava Sahoo Department of Electronics and Communication Engineering, Institute of Technical Education and Research, Siksha 'O' Anusandhan (Deemed to be University), Bhubaneswar, Odisha, India

Tamanna Sahoo Department of Electronics and Communication Engineering, ITER, Siksha 'O' Anusandhan University (Deemed to be University), Bhubaneswar, India

Tapasmini Sahoo Electronics and Communication Engineering, Institute of Technical Education and Research, Siksha 'O' Anusandhan Deemed University, Bhubaneswar, India

Badrinarayan Sahu Department of ECE, Faculty of Engineering, ITER, Siksha 'O' Anusandhan (Deemed to be University), Bhubaneswar, Odisha, India

Benudhar Sahu Institute of Technical Education and Research, SOA (Deemed to be University), Bhubaneswar, India

Binod Kumar Sahu Department of Electrical Engineering, Siksha 'O' Anusandhan Deemed to be University, Bhubaneswar, Odisha, India

Prasant Kumar Sahu Indian Institute of Technology Bhubaneswar, Argul, Odisha, India;
School of Electrical Science, Indian Institute of Technology Bhubaneswar, Bhubaneswar, Odisha, India

Sanjay Kumar Sahu School of Electronics and Electrical Engineering, Lovely Professional University, Phagwara, India

Mohammad Salik Department EEE, ITER, Siksha 'O' Anusandhan (Deemed to be University), Bhubaneswar, India

Debashisa Samal Department of ECE, ITER, S'O'A Deemed to be University, Bhubaneswar, India

Laxmipriya Samal Department of ECE, ITER, S'O'A Deemed to be University, Bhubaneswar, India

Sunita Samant Image & Video Analysis Laboratory, Department of ECE, S O A, Deemed to be University, Bhubaneswar, India

Barsa Samantaray ECE Department, Institute of Technical Education and Research, Siksha 'O' Anusandhan Deemed to be University, Bhubaneswar, Odisha, India

Kandarpa Kumar Sarma Department of Electronics and Communication Engineering, Gauhati University, Guwahati, India

Mousmita Sarma Department of Electronics and Communication Engineering, Gauhati University, Guwahati, India

Lalit Mohan Satapathy EEE Department, ITER, S'O'A (Deemed to be University), Bhubaneswar, India

Samarjeet Satapathy EEE Department, ITER, S'O'A (Deemed to be University), Bhubaneswar, India

Sarmistha Satrusallya Department of ECE, ITER, Siksha 'O' Anusandhan (Deemed to be University), Bhubaneswar, India

Kaibalya Kumar Sethi Department of Electronics and Communication Engineering, GITA, Bhubaneswar, India

ShafqatUlAhsaan Department of Computer Science and Engineering, School of Engineering Sciences and Technology, JamiaHamdard, New Delhi, India

Renu Sharma Department of Electrical Engineering, Institute of Technical Education and Research, SOA (Deemed to be) University, Bhubaneswar, India

Aasheesh Shukla Department of Electronics and Communication Engineering, GLA University, Mathura, India

Asit Kumar Subudhi Department of Electronics and Communication, Institute of Technical Education & Research, SOA Deemed to be University, Bhubaneswar, India

Dillip Kumar Subudhi Department of CSIT, ITER, S'O'A University, Deemed to be University, Bhubaneswar, India

A. Susmitha New Horizon College of Engineering, Bengaluru, India

Biswaranjan Swain Department of Electronics and Communication Engineering, SOA Deemed to be University, Bhubaneswar, Odisha, India

Jnana Ranjan Swain EEE Department, ITER, S'O'A (Deemed to be University), Bhubaneswar, India

K. P. Swain Department of Electronics and Communication Engineering, GITA, Bhubaneswar, India

Ramakrushna Swain Department of Computer Science and Engineering, Silicon Institute of Technology, Bhubaneswar, India

Ramji Tangudu Indian Institute of Technology Bhubaneswar, Argul, Odisha, India

Madhab Chandra Tripathy Department of Instrumentation & Electronics Engineering, CET, BPUT, Bhubaneswar, India

Shuvendra Kumar Tripathy Trident Academy of Technology, Bhubaneswar, India

S. K. Varshney Department of Electronics and Electrical Communication Engineering, IIT Kharagpur, Kharagpur, India

Rahul Vishwakarma Department of Electronics and Communication Engineering, ITER, Siksha 'O' Anusandhan (Deemed to be University), Bhubaneswar, Odisha, India

Aneesh Wunnava Department of ECE, ITER, Siksha O Anusandhan, Bhubaneswar, Odisha, India

Children's Age and Gender Recognition from Raw Speech Waveform Using DNN

Mousmita Sarma[1(✉)], Kandarpa Kumar Sarma[1], and Nagendra Kumar Goel[2]

[1] Department of Electronics and Communication Engineering, Gauhati University, Guwahati, India
{mousmita.s, kandarpaks}@gauhati.ac.in
[2] GoVivace Inc., McLean, VA, USA
nagendra.goel@govivace.com

1 Introduction

Speaker's age and gender are two most important para-linguistic characteristics extracted from speech signal, and there has been extensive research conducted in this direction. Extraction of age and gender like information from children speech (below 14 years) is considered as a difficult task, and it has not been very well studied. Most of the previous works are focused mainly on adult's speech. A few works considered children as a separate age group along with other age groups like young, adult, senior, etc. for the design of classifiers [1, 2]. The reason for children's speech not getting much attention may be because greater extent of variability is involved in this age group. This is mainly due to children's shorter vocal tracts, higher fundamental frequencies, developing cognitive issues relating to language acquisition before reaching the adult stage, physiological variations in growth, etc. [3] and [4].

This work is focused on the design of end-to-end DNN for automatic extraction of age and gender information from speech signal of children within the age group of 4–14 years. Initially, we design separate DNN for both the tasks, using age or gender target in the output layer. However, age and gender are speaker characteristics which may have shared interdependencies. Therefore, such tasks may better fit in a multi-task learning setup and help to improve recognition accuracy. To explore such possibility, we also design multi-task learning DNN, where we use speaker's age and gender as target in two separate output layers, where shared hidden representation is used to learn both the objectives. The proposed multi-task learning DNN system leverages speaker's age information to improve gender identification accuracy and vice versa.

The DNN models developed in this work uses data-driven approach of filter learning from time-delay waveform, which is previously described in [5, 6]. It uses convolutional layer and network-in-network (NIN) layer for DNN front end, and the learned feature representation is used by temporal context modeling. Thus, the DNN learns features within the network and jointly optimizes age and gender identification objective with the rest part of the network. Such networks are able to learn evidence of features during training time. For temporal context modeling in the DNN, we use interleaving time-delay neural network (TDNN) with unidirectional long short-term

M. N. Mohanty and S. Das (eds.), *Advances in Intelligent Computing and Communication*, Lecture Notes in Networks and Systems 109,
https://doi.org/10.1007/978-981-15-2774-6_1

memory (LSTM) cell-based recurrent layer (TDNN-LSTM)-based configuration. The configuration also uses time-restricted attention mechanisms at the end to make the network attentive toward relevant features of speech. The proposed DNN is trained and evaluated on OGI Kids' corpus [7]. We provide a description of the proposed multi-task learning DNN for gender recognition in Sect. 2. Section 3 explains experimental setup which includes details of train and test datasets, network design consideration and result and analysis. In Sect. 4, we summarize the findings of this work and subsequently conclude the description.

2 System Description

In this work, we design single-task and multi-task DNN models for age and gender recognition from children's speech as shown in Fig. 1a and b. The DNN has a front end consisting of convolutional layer and NIN layers. Convolutional filter learns band-pass filter like structure within DNN and generates time-delay features. In our previous work on the domain of emotion recognition [6], we reported that this time-delay raw waveform front end layer learns emotion-specific cues better than MFCC features. This raw waveform front end is a four-layer block jointly trained with rest part of the network. A $1-d$ time convolution layer consisted of 100 filters of 31.25 ms dimension operates on 40 ms raw signal with step size 1.25 ms. Absolute logarithm of filter outputs are computed using a logarithmic layer, and next filter outputs are aggregated using two trainable NIN nonlinearity layers. More details of the raw waveform front end are given in [5, 6].

Another crucial issue in speech-based application is temporal context modeling of speech. Particularly for para-linguistic information extraction, it is important to preserve long temporal context since such information perhaps mostly lies in a longer span of time. In this work, we explore temporal convolution in the form of TDNN layers along with LSTM layers for long temporal context modeling. Interleaving of temporal convolution with unidirectional LSTM is reported to outperform bidirectional LSTM to preserve future context. We use higher frame rate at lower layers of LSTM and TDNN layer in the network, and we decrease layer frame rate with layer depth. The dimensions of projection and the recurrence in LSTM layers are one-quarter of the cell dimension. We found 128 cell dimension optimal for the current gender identification task with the recurrence of dimension 32 and the LSTM output of dimension 64. The LSTMs operate with a recurrence that spans 3 time steps. Time-restricted attention layer is used as the last layer along with TDNN and unidirectional LSTM layer. Use of an attention layer also helps the network to put less weight in non-speech or non-relevant regions of the signal. Removing non-speech frames using fixed speech activity detection (SAD) module beforehand may create a discontinuity in the temporal content. However, DNN setups with an attention layer become more attentive toward the salient regions of speech ignoring non-speech regions. The attention layer used has 12 heads, a context of [−5, 2], a key dimension of 40 and value dimension of 60.

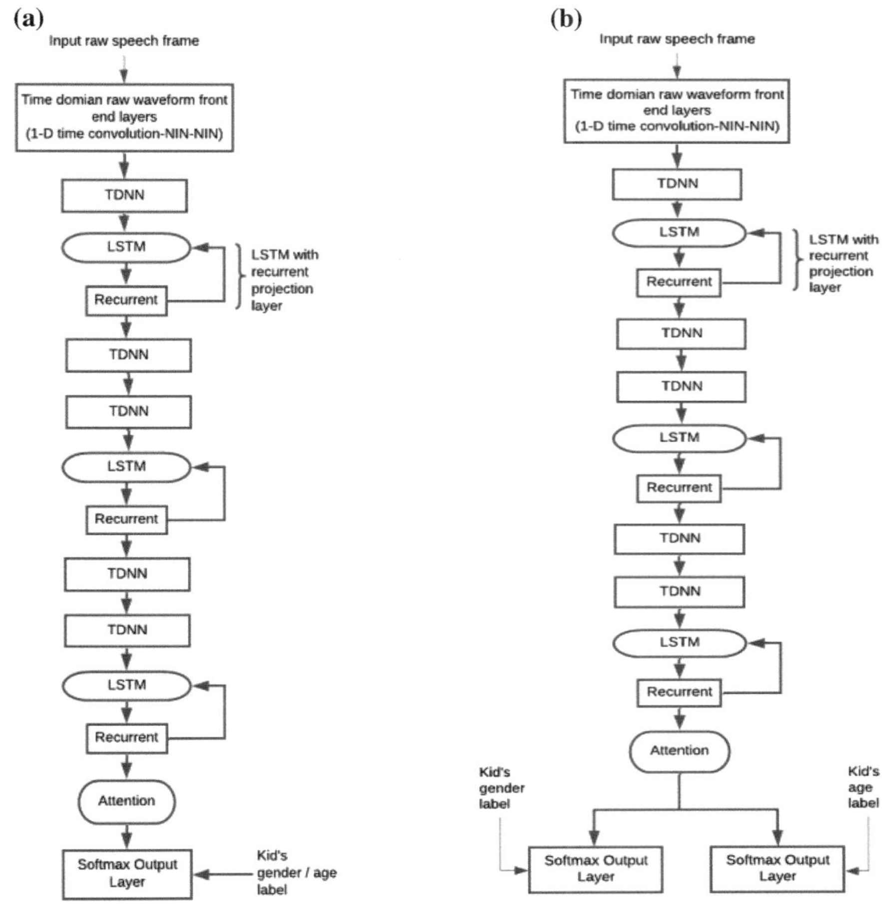

Fig. 1. **a** Single-task learning model. **b** Multi-task learning model

The output of attention layer is used as input to the output layers of the DNN. In the single-task learning model (Fig. 1a), age or gender label are used as target. The age task can be considered either as a regression or a classification problem. However, this work models the 11 different age (4–14 years) of OGI Kids' corpus as 11 different classes and trains the network as a classification problem. Male and female gender is modeled as separate output classes in a different output layer. Thus, the two output layers of the DNN are trained using softmax layer with cross-entropy loss function. This gives the network freedom to model any distribution over output. We train the network using per-frame objective method, where the label of the class is repeated for each frame of the speech chunk. On the other hand, we use speaker's age and gender both as targets in two separate output layers in the multi-task learning DNN model. It has three distinct parts: feature extraction front end layers, temporal modeling layers and output layers. However, all layers are jointly optimized with two separate output layers with gender and age target, respectively. Therefore, the multi-task objective loss sums the loss of each task.

3 Experimental Details

This section describes the experimental details and results. All our experiments are done using Kaldi toolkit. Following here is a detail description of experimental considerations.

The proposed work is trained and tested on OGI Kids' corpus [7]. We create train and test partition of 1093 children for tenfold training. The entire dataset has children from 4 years to 14 years age, i.e., total 11 different ages. We create ten test sessions where every session contains on an average 100 non-overlapping speakers and against every test session there is a train session made up of the speakers from the rest of the test sessions.

To further increase the amount of data in the training set, we perform data augmentation by means of amplitude and speed perturbation. Amplitude perturbation also mitigates the effect of low and high amplitude variations of speech over ages. Amplitude perturbation is applied by modulating the amplitude of the raw speech signal. Each recording in the training data has been scaled with a random variable drawn from a uniform distribution over [0.125, 2]. Speed perturbation is performed on the amplitude modulated signals with speed factors 0.9, 1.0 and 1.1.

The performance of the gender identification tasks is reported using two parameters, weighted accuracy (WA) which is the overall classification accuracy and unweighted accuracy (UA) which is the average recall over the classes. On the other hand, the performance of the age estimation is measured by computing the mean absolute error (MAE) and root mean square error (RMSE) for every utterance of evaluation datasets. The lower the MAE and RMSE, the better is the performance.

The MAE and RMSE is given by,

$$\text{MAM} = \frac{1}{M}\sum_{i=1}^{M} |(\text{EstimatedAge} - \text{TrueAge})| \quad (1)$$

$$\text{RMSE} = \sqrt{\frac{1}{M}\sum_{i=1}^{M} (\text{EstimatedAge} - \text{TrueAge})^2} \quad (2)$$

where M is the total number of examples in the dataset.

3.1 DNN Design Consideration

One of the challenging issues in training speech-based task is to mitigate the effect of variable lengths of training and test examples. It is important to minimize network sensitivity to varying speech duration. One solution is to train the network on chunks of different duration and making mini-batch size a function of chunk length. For example, for the present work the length of training chunks is varied from 1 to 6 s, and the entire utterance is used as an example if it is shorter than 6 s. Corresponding to it mini-batch sizes used for training examples with length 100, 200, 300 are 128, 64, 32, respectively, and so on. This way the total number of frames is almost equal in different mini-batches. We call this training approach variable chunk and variable mini-batch. However, we compare this variable chunk and variable mini-batch approach with fixed chunk and fixed mini-batch approach as shown in Table 1. It can be observed that using variable chunk and variable mini-batch network performs better.

Table 1. Effect of chunk length of training examples (test session 1 is used to derive this comparison)

Training chunk length	Age		Gender	
	MAE	RMSE	WA	UA
Fixed (100 frames)	1.61	2.42	75	74.24
Variable (100–600 frames)	1.51	2.33	82.69	82.30

We train DNNs with per-frame objective, i.e., label is repeated for every frame of speech chunk. The learning rate of the training is set to gradually decrease from $6 \times 10 - 4$ to $6 \times 10 - 5$ over the course of 30 epochs. We use per-frame dropout using the dropout schedule method, where the entire vector is forced to be zero or one. The dropout schedule is expressed as a piece-wise linear function on the interval [0, 1], where f (0) gives the dropout proportion at the start of training and f (1) gives the dropout proportion after seeing all the data. A dropout schedule of the form 0, 0@0.20, p@0.5, 0@0.75, 0 is used in this setup, where p is 0.3 in the results reported here. Thus, the dropout probability is 0 at f (0), 0 at f (0.2), 0.3 at f (0.5), 0 at f (0.75) and 0 at f (1). In DNNs trained with per-frame objective, we give some extra left context at the time of decoding which provides flexibility to the network regarding number of frames it sees in addition to what was provided during training and we evaluate the model several times to tune this length of decode left time context. However, we also compare the effect of a fixed length decode time left context with that where we provide an effectively unlimited left context during decode time. Unlimited means the network is allowed to reuse hidden state activations from previously computed chunk. As can be seen from Table 2 unlimited left context does not help in the present multi-task learning set up and finding an effective fixed length left context provides better performance.

Table 2. Effect of decode time left context (test session 1 is used to derive this comparison)

Training chunk length	Age		Gender	
	MAE	RMSE	WA	UA
20 frames	1.51	2.33	82.69	82.30
Unlimited left context	1.67	2.37	76.92	76.17

The speech segments are converted to 8000 Hz sample frequency and quantized at 16 bit per sample. Mean and variance normalization is done at the segment level and low energy dither is added to the windowed signal to avoid runs of digital zeros. The pre-processed raw speech utterances are passed to the input layer of the DNN.

Label smoothing regularizes our model based on a softmax with k output values by replacing the hard 0 and 1 classification targets (one hot encoding) with targets of $\varepsilon/k-1$ and $1-\varepsilon$, respectively, where ε is small constant.

3.2 Result and Analysis

Results of age and gender identification have been evaluated on ten test sessions of OGI Kids' corpus for both single-task and multi-task training. To evaluate every test session, we use the model which is trained with rest of the test sessions data. The single-task training of DNN follows the same process as shown in Fig. 1, which uses only one output layer for the specific task. On the other hand, multi-task DNN uses an auxiliary task trained in an extra output layer. The auxiliary task is gender for age estimation and age for gender recognition task. The results are shown for files longer than 30 s and files shorter than 30 s separately. It is observed that except test session 2, all other test sessions achieve less MAE and RMSE using single-task training of DNN for age estimation. In contrary to age estimation, we observe that for all test sessions, multi-task training provides better WA and UA.

We summarize the results of age and gender recognition in Table 3, where average performance has been shown. We achieved best age estimation MAE-RMSE pair of 0.91 and 1.43 using single-task DNN training and 1.13 and 1.72 using multi-task DNN training for test set longer than 10 s. The results show that gender information is not much helpful while estimating age. On the other hand, we obtain best gender recognition WA-UA pair 55.63 and 50.97 using single-task DNN training and 72.67 and 70.55% using multi-task DNN training for test set longer than 10 s. The result clearly tells that for gender recognition age information is highly useful which increases the overall accuracy by 17.04%. Previous works by Safavi et al. 2014 and 2018 are also focused on age and gender recognition from children's speech and reported to obtain 78% accuracy on age independent gender recognition. However, we perform a tenfold testing and obtain 82.69% highest accuracy among ten different test sessions and an average 72.67% accuracy. On the other hand, in case of age recognition we attempt to estimate exact age of the child which is a more difficult task compared to the work by Safavi et al. 2014 and 2018 which have performed an age group classification. Table 3 also shows how the performance is affected by the duration of speech. In an obvious manner, better performance is observed using longer duration of speech for both age and gender recognition.

Table 3. Summary of results (averaged over all test sessions)

Model	Duration of speech (s)	Age		Gender	
		MAE	RMSE	WA	UA
Single-task	≥ 30	0.91	1.43	82.69	82.30
	<30	1.13	1.72	59.31	50.33
Multi-task	≥ 30	1.39	2.04	72.67	70.55
	<30	1.86	2.57	64.86	59.64

Table 4 reflects the gender recognition performance across the 11 age classes of children using the multi-task DNN model. Although the observation here is not very conclusive, it can be observed that (at least from the <30 s files results) for younger children below 9 years accuracy is less. It clearly tells that gender discrimination in case of younger children is more difficult. Table 5 shows the age estimation MAE and RMSE across the two gender classes using the single-task DNN model. The MAE for female and male is 0.85 and 0.89 for ≥ 30 s duration of speech and 1.21 and 1.61 for <30 s speech duration. The observation here tells that, although age estimation performance is not heavily biased toward one particular gender, MAE of male speakers is little less. Perhaps, this is because the kids' voice most of the time is like a female voice and it is difficult even for human to discriminate male and female for kids younger than 7–8 years. Therefore, it is not very surprising that machine's age estimation error is more in case of male children.

Table 4. Result of gender identification across all ages for test session 1(multi-task learning model)

Speaker's age	≥ 30 s				<30 s			
	WA	UA	%F	%M	WA	UA	%F	%M
4	90	90	80	100	60.2	60.2	60	60.41
5	80	80	80	80	58.52	58.87	65.13	51.91
6	87.5	90	80	100	58.18	59.08	48.1	70.06
7	66.66	65	80	50	49.45	49.72	53.33	46.12
8	80	80	100	60	73.3	73.36	73.5	73.22
9	70	70	80	60	56.07	56.11	62.06	50.15
10	80	80	100	60	70.67	70.83	77.91	63.74
11	77.77	75	100	50	71.54	72.02	60.49	83.55
12	100	100	100	100	76.83	76.66	61.07	92.26
13	77.77	77.5	80	75	75.89	75.9	71.51	80.29
14	100	100	100	100	89.41	89.38	84.98	93.78

Table 5. Result of age identification across all genders for test session 1 (single-task learning model)

Speaker's gender	≥ 30 s		<30 s	
	MAE	RMSE	WA	UA
Female	0.85	1.28	1.21	1.88
Male	0.89	1.34	1.168	1.83

4 Summary of Findings and Conclusion

We described single-task and multi-task DNN performance for age and gender recognition task. For single-task model, we train age and gender estimation objectives separately with softmax output layer. On the other hand, for multi-task training we train single DNN model with two different output layers for age and gender and jointly optimize the objectives along with feature extraction and temporal modeling layers. In order to exploit the feature learning and discriminating ability of multi-task learning process of DNN, we used raw speech waveform in this work. A1 − d time convolutional and NIN layer-based raw waveform front end have been used in the proposed DNN, and these layers are jointly trained with the speech temporal modeling parts of the DNN to optimize gender classification objectives. We use interleaving TDNN-LSTM-based learning to preserve long temporal context of speech, and it uses time-restricted self-attention mechanism to attend age- and gender-specific features of speech. The results obtained throughout this study show that although children's gender recognition process is very much benefited by age information provided during multi-task learning process, the vice versa is not true. We achieve less MAE in age estimation task when we train the network with one output layer compared to the network trained with gender auxiliary task. Therefore, further investigation of age estimation process in multi-task learning DNN is required. We are not aware of any previous works where multi-task learning process of DNN is utilized for para-linguistic information extraction from children's speech. The proposed work shall serve as a baseline for DNN-based recognition system design from children's speech.

References

1. Kumar N, Nasir M, Georgiou P, Narayanan SS (2016) Robust multichannel gender classification from speech in movie audio. In: Proceeding INERSPEECH 2016, The 17th annual conference of the international speech communication association. SanFrancisco, USA, 8–12 September, 2016
2. Kabil SH, Muckenhirn H, Doss MM (2018) on learning to identify genders from raw speech signal using CNNs. In: Proceeding INERSPEECH 2018, The 19th annual conference of the international speech communication association. Hyderabad, India, 2–6 September, 2018
3. Safavi S, Russell M, Jancovic P (2014) Identification of age-group from childrens speech by computers and humans. In: Proceeding INTERSPEECH 2014, The 15th annual conference of the international speech communication association. Singapore, 14–18 September, 2014
4. Safavi S, Russell M, Jancovic P (2018) Automatic speaker, age-group and gender identification from childrens speech. Comput Speech Lang 50:141–156
5. Ghahremani P, Manohar V, Povey D, Khudanpur S (2016) Acoustic modellingfrom the signal domain using CNNs. In: Proceeding INTERSPEECH 2016, 17th annual conference of the international speech communication association. San Francisco, CA, USA, 8–12 September 2016

6. Sarma M, Ghahremani P, Povey D, Goel NK, Sarma KK, Dehak N (2018) Emotion identification from raw speech signals using DNNs. In: Proceeding INERSPEECH2018, The 19th annual conference of the international speech communication association. Hyderabad, India, 2–6 September, 2018
7. Shobaki K, Hosom J, Cole (2000) The OGI kids speech corpus and recognizers. In: Proceeding INTERSPEECH 2000, The 6th international conference on spoken language processing, ICSLP 2000/INTERSPEECH 2000, Beijing, China, 16–20 October, 2000

Inter-vehicular Communication in Urban Traffic Scenario

Suman Malik(✉) and Prasant Kumar Sahu

School of Electrical Science, Indian Institute of Technology Bhubaneswar, Bhubaneswar, Odisha 752050, India
sm38@iitbbs.ac.in

1 Introduction

Vehicular ad hoc network (VANET) aims at enhancing the transportation efficiency and road traffic safety through the application of advance information processing, communication and control techniques [1]. A key to implement the intelligent transportation system (ITS) is traffic management and the road traffic state monitoring. VANET is the most suitable solution to this regard. The unique features offered by VANETs are higher mobility and higher density of mobile nodes [2]. VANET deals with V2V and V2I communication link using wireless local area network (WLAN) technologies [3]. Routing of data packet and selection of suitable wireless channel is a major scientific challenging task in VANET because of higher mobility and dense mobile nodes in urban environment [4]. Ad hoc on demand distance vector (AODV), dynamic source routing (DSR) protocol and optimized link state routing (OLSR) are generally employed in VANET to set up the communication link between V2V and V2I in an ad hoc network scenario [5]. The wireless channel characteristics do play a major role within the VANET disseminating of information between vehicles and infrastructures. The channel impairments [6] especially path loss and fading effect also affect the signal power during the propagation through free space channel. Many fading channel models, like Rayleigh, Nakagami and Rician, were proposed as a wireless channel corresponds to VANET [7].

With respect to the previously published paper, the main contribution of this paper is outlined as follows;

- Primarily, ideal wireless channels are taken into consideration, and thereafter, Rayleigh and Nakagami fading channel with higher network size in terms of larger number of vehicles and random mobility is employed to VANET scenario for performance analysis.
- We have also investigated the performance of the VANET, parameters like packet loss ratio and average delay with different number of vehicles under higher mobility condition.

M. N. Mohanty and S. Das (eds.), *Advances in Intelligent Computing and Communication*, Lecture Notes in Networks and Systems 109,
https://doi.org/10.1007/978-981-15-2774-6_2

2 Related Work

Several investigations were reported in the scientific literature for studying and comparing the routing protocols to enhance the performance of VANETs [8–11]. The significant characteristics of a routing protocol are to determine the effective way between the source and destination for information dissemination in a reliable manner.

In the published literature, the performance analysis has been carried out using metrics like PDR, delay, route length and overhead packets of VANET and by using position-based routing approach in dense vehicular environment [8]. The overall performance of VANET was studied using different routing protocols where wireless channel and higher network size are not considered [9]. In [10], review and performance analysis has been studied for VANETs' protocols. However, the effect of channel model and vehicular mobility on VANET was not considered in their studies. Similarly, in the work of Chandel and Sharma et al. [11], cluster-based routing protocols were used to study the performance metric of VANET such as PDR, throughput, jitter and average delay with respect to network size without considering the effect of free space channel model, dense traffic and vehicular mobility.

3 VANET Model, Methodology and Simulation Setup

VANET testbed deployment is very expensive, and to avoid complicated and expensive setup, software simulation platforms are in wide use by research community. We have created and simulated the proposed model by using the NS2 and NetSim which are event-oriented simulator which is used to carry out the simulation of the wireless network and SUMO which is a road traffic simulator and is used to analyze the realistic simulations of traffic mobility [12] (see Fig. 1). The main reason for selecting the proposíed simulation environment is its capability to handle the distribution of vehicles as well as the support for channel model.

3.1 Simulation Using NS2 and NetSim with SUMO Tools

The VANET model is simulated using NS2 using mobility trace that is generated using SUMO. SUMO gains rapid research interest to the transportation engineer due to its capability of multi-dimensional traffic simulation, traffic lights and for time schedule generation. Here, we have considered ideal wireless channel model for VANET. The step required to create and simulate the proposed scenario is shown in Fig. 2. NetSim along with SUMO tool be used to perform the simulation of the proposed model where realistic mobility-based traffic platforms are considered. It permits different vehicle mobility modeling and simulation with different types of vehicle, traffic lights and multi-lane roads with lane alternating features. The step required to create and simulate the proposed scenario is shown in Fig. 3.

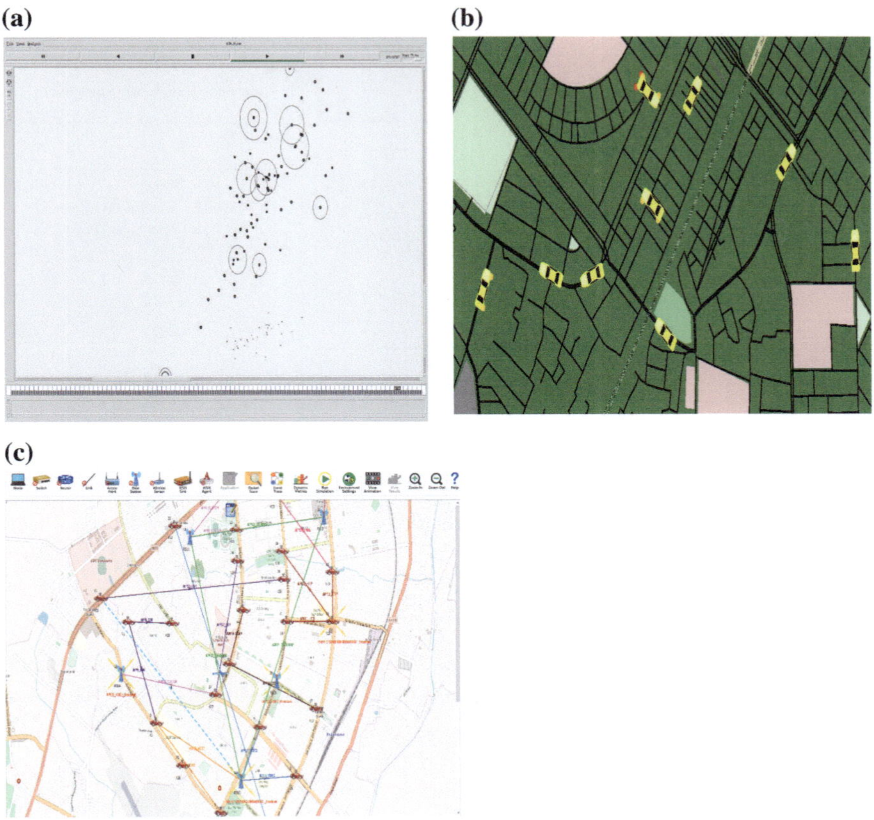

Fig. 1. VANET simulation scenario **a** mobility model in NS2, **b** mobility model in NetSim, **c** VANET scenario in NetSim

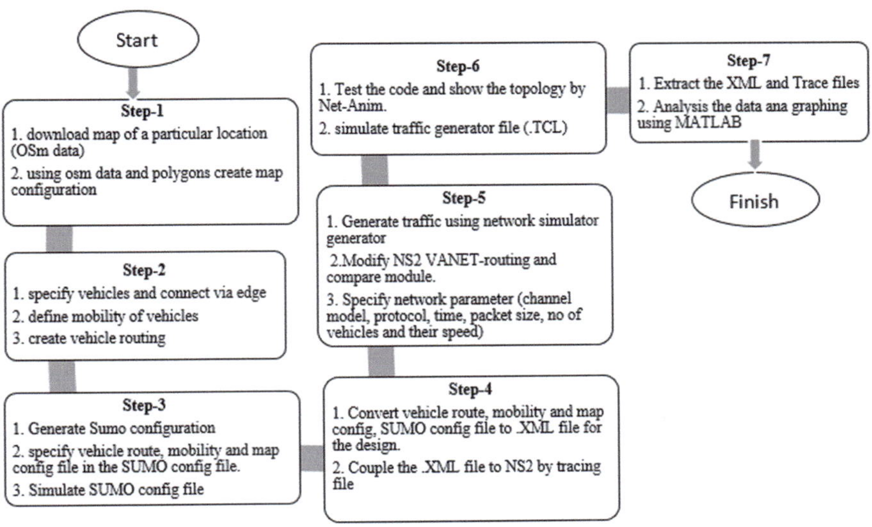

Fig. 2. Methodology flowchart for SUMO and NS2

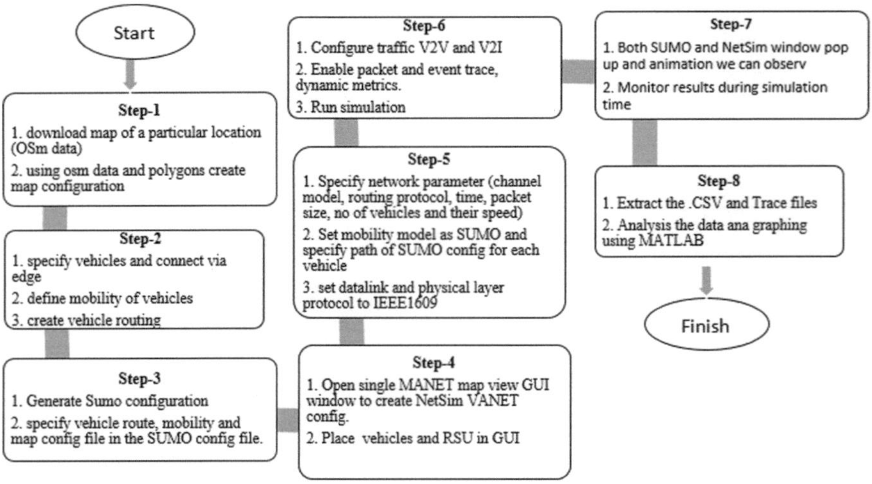

Fig. 3. Methodology flowchart for SUMO and NetSim

3.2 Packet Loss Ratio

Packet loss ratio is the percentage of the MAC collisions packets loss or packet dropped during total packet dissemination. Dense network increases the channel utilization which is the cause of packet loss. It is expressed as [13]:

$$\text{PLR} = \text{Total packet dropped/Total packet send} \tag{1}$$

3.3 Average Delay

Delay is the average time consumed by the node when the data packets delivered successfully to the targeted vehicles. Smaller average delay should be maintained for better network performance. Propagation, processing and transmission delay are the main factor for average delay in VANET. Average delay is expressed as follows [14]:

$$\text{Average Delay} = \text{Time consumed to transfer packets successfully/}\\ \text{Total no of packets delivered} \tag{2}$$

4 Simulation Results

Figure 4 presents the PLR for the specified routing protocols and for different network sizes. Here, network simulator NS2 tool is used for simulation using an ideal wireless channel (see Table 1). We can observe that PLR is increasing with network size and is changing with routing protocols. It is also noticed that packet drop is lower when less vehicles are present in the VANET. DSR routing protocol provides the lower PLR with respect to AODV and OLSR for same vehicles.

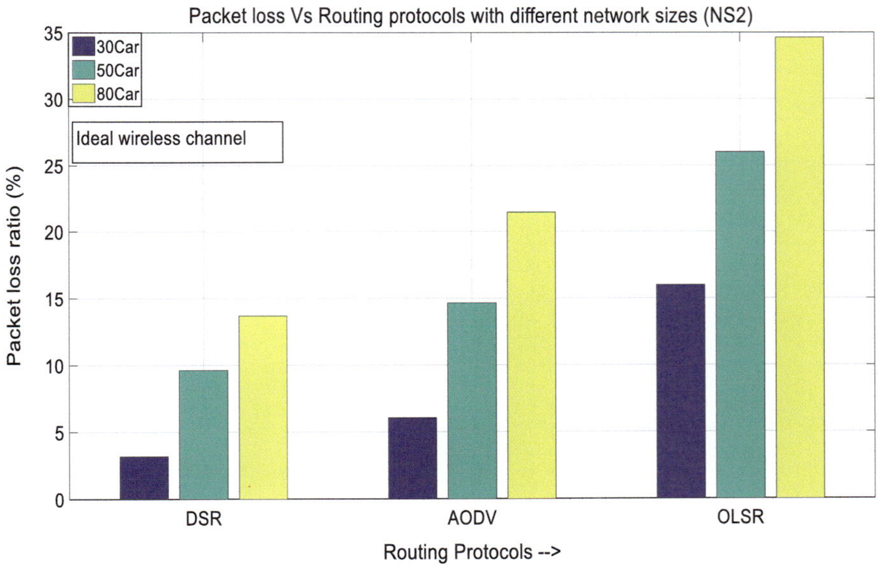

Fig. 4. Packet loss ratio versus routing protocol with different number of vehicles using NS2

Table 1. Simulation parameters

Parameters	Values	Parameters	Values
Simulation tool	NS2, NetSim 10.2	Bit rate	27 Mbps
Mobility model	SUMO	Link mode	Half duplex
Simulation area	Bhubaneswar city	Transmitter power	40 mw
No of vehicles	30, 50, 80	link range	1 km
Simulation time	400 s	Message type	Unicast, broadcast
Fading models	Rayleigh, Nakagami	Packet size	1420 bytes
Shadowing	Lognormal model	Frequency band	5.9 GHz
Vehicle speed	40–120 kmph	Routing protocols	DSR, AODV, OLSR
Performance metric	PLR, average delay	IEEE standard protocol	IEEE-1609, IEEE-802.11p

In a similar way, Fig. 5a, b presents the same performance metric with respect to total number of vehicles in VANET. Here, we have considered two fading channel model, i.e., Rayleigh and Nakagami fading channel and NetSim-based simulation platform. As shown in Figs. 4 and 5, packet loss ratio is significantly increased when we changed the ideal wireless scenario to wireless fading channel model (Rayleigh and Nakagami), due to more number of packets drop because of fading channel. This observation shows that DSR could be a suitable routing protocol with Rayleigh fading channel for application where lower packet loss is required over a large network.

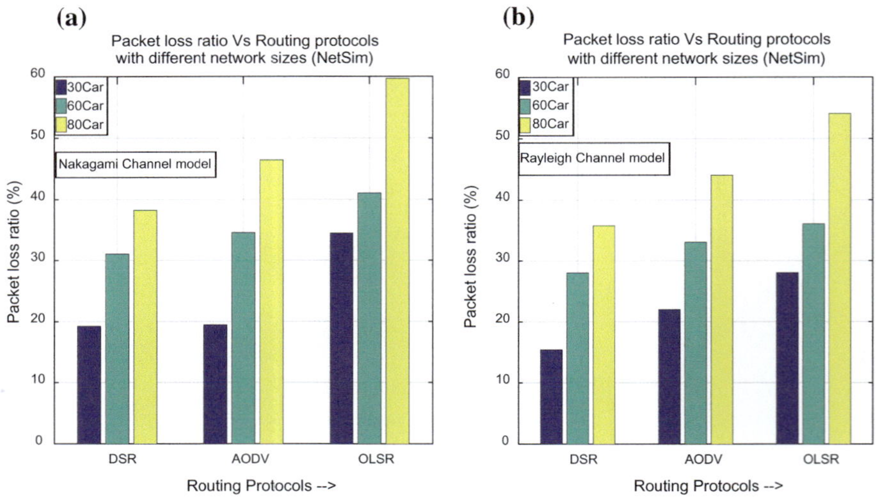

Fig. 5. Packet loss ratio versus routing protocol with different number of vehicles using NetSim **a** Rayleigh channel model, **b** Nakagami channel model

Figure 6 demonstrates the relation between packet loss ratio and vehicular speed with specified wireless channel model and routing protocol. Here, Rayleigh and Nakagami fading channel model is considered with DSR routing protocol along with variable vehicular mobility ranging from 40 kmph to 120 kmph to analyze the PLR in VANET. We can observe that the PLR reached maximum of 50% at 120 kmph of vehicle speed with the combination of DSR routing and Rayleigh fading channel, whereas it becomes 54.2% with Nakagami fading channel (see Table 2). Hence, we can conclude that PLR is increased with increase in vehicle speed for a particular number of vehicles because of higher mobility of vehicles cause a fast alternate of topology and packet drop in the network.

Figure 7 demonstrates the delay for three mentioned routing protocols under different number of vehicles. Two appropriate fading models, like Rayleigh and Nakagami fading channel, are introduced in VANET for simulation. It can be seen that the average delay is stick to 0.8–1 ms and the maximum it can reach up to 1.3 ms. From Fig. 7, we can determine that DSR is taking less time as compared to AODV and

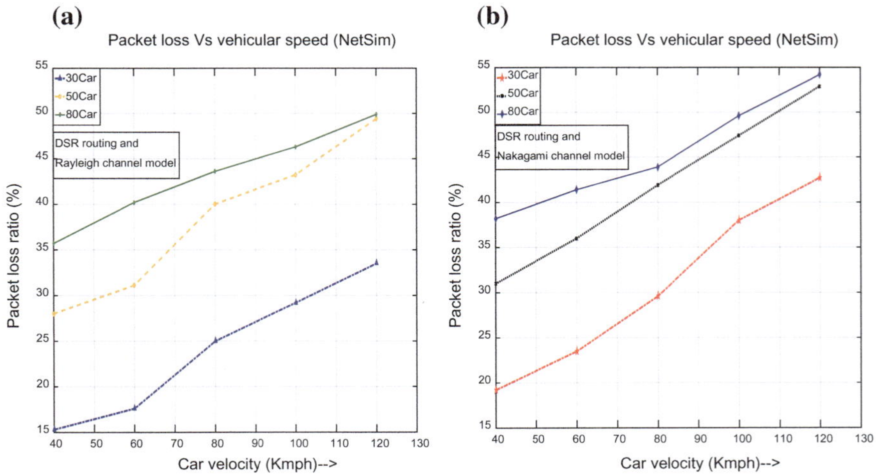

Fig. 6. PLR versus vehicular speed **a** Rayleigh channel model **b** Nakagami channel mode

Table 2. Comparative analysis of PLR

Parameter		DSR	AODV	OLSR	Vehicles
PLR (%) (NetSim)	Rayleigh	15.3	22	28	30
		28	33	36	50
		35.7	44	54	80
	Nakagami	19.2	19.4	34.4	30
		31	34.5	41	50
		38.2	46.4	59.6	80
PLR (%) (NS2)		3.2	6	16	30
		9.6	14.62	26	50
		13.65	21.4	34.6	80

OLSR, irrespective of the vehicle's number. This can be possible when the packet transmits through shortest path by suitable routing protocol.

Figure 8a, b presents the average delay with respect to the vehicle speed under the Rayleigh and Nakagami fading channel with DSR routing, respectively. It can be seen that slow-moving vehicle exhibits a slight increase in average delay, and it slowly decays with increase in vehicle speed. Further, it is also noticed that as the density of the traffic increases (30–80 vehicles), delay also increases for both type of fading channel. Under dense traffic scenario (around 80 vehicles) and slow-moving vehicles (40 kmph), network suffers for 1 ms average delay under Rayleigh fading channel, whereas for network utilizing Nakagami fading channel, the delay is of the order of 1.1 ms delay. Therefore, we can conclude that traffic density, vehicle speed and fading channel have detrimental impact on VANET performance.

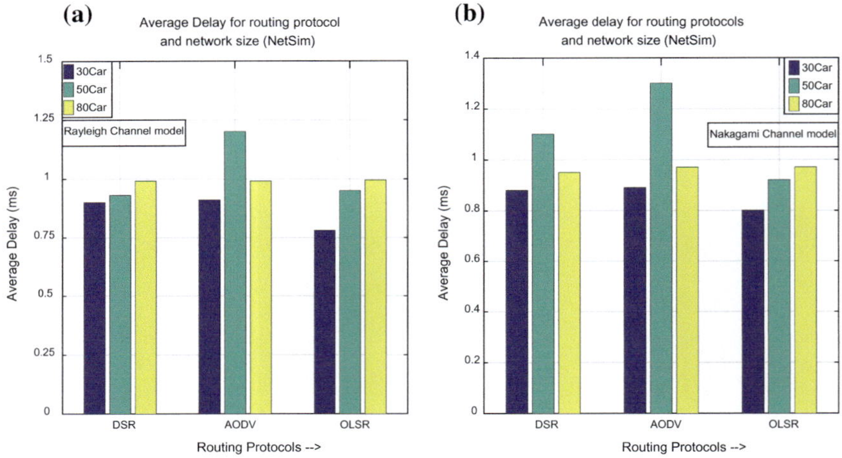

Fig. 7. Average delay versus routing protocol with different number of vehicles using NetSim **a** Rayleigh channel model, **b** Nakagami channel mode

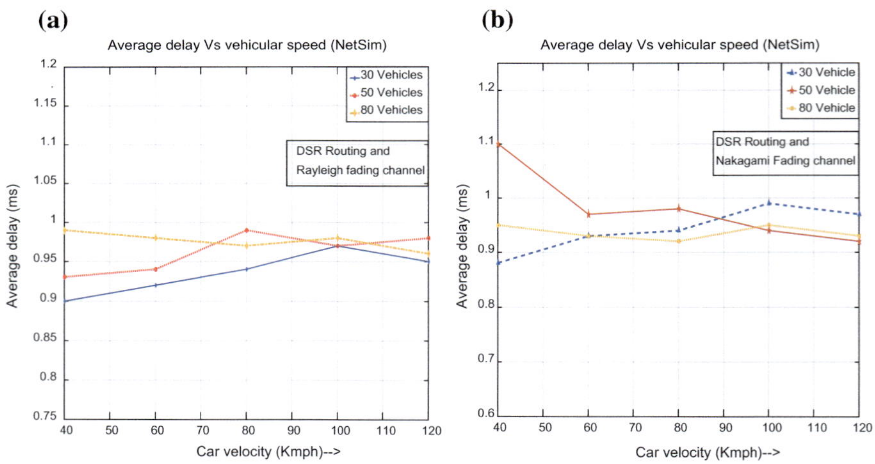

Fig. 8. Average delay versus vehicular speed **a** DSR routing and Rayleigh fading channel, **b** DSR routing and Nakagami fading channel

5 Conclusion

In this work, the performance of vehicular ad hoc network using three existing topology-based routing protocols was presented and analyzed. Simulation results show that DSR performs better than AODV and OLSR routing protocol over ideal wireless as well as for fading channel models. DSR has lower packet loss ratio and approximately minimum average delay compared to other protocols. The DSR also exhibits reduced data delivery latency. Though, VANET would provide better platform and effective communication between vehicles, but under higher vehicle density and higher mobility condition, the network QoS parameter suffers due to network congestion and bandwidth wastage.

Acknowledgements. This work was carried out in the FIST supported Optical Engineering Laboratory of SES, IIT Bhubaneswar.

References

1. Cuomo F, Rubin I, Baiocchi A, Salvo P (2014) Enhanced VANET broadcast throughput capacity via a dynamic backbone architecture. Ad-hoc Netw 21:42–59
2. Omar H, Zhuang W, Li L (2013) Ve-MAC: a TDMA-based MAC protocol for reliable broadcast in VANETs. IEEE Trans Mob Comput 12(9):1724–1736
3. Lin YY, Rubin I (2015) Vehicular and messaging throughput tradeoffs in autonomous highway systems. IEEE Global Commun Conf (GLOBECOM) 1–6
4. Al-Bayatti AH, Zedan H (2014) A comprehensive survey on vehicular Ad-Hoc network. J Netw Comput Appl 37:380–392
5. Awang A, Husain K, Kamel N, Aissa S (2017) Routing in vehicular Ad-Hoc networks: a survey on single- and cross-layer design techniques, and perspectives. IEEE Access 5:9497–9517
6. Uzcátegui R, Marum GA (2009) WAVE: a tutorial. IEEE Commun Mag 47(5):126–133
7. Sandalidis HG, Tsiftsis TA, Karagiannidis GK (2009) Optical wireless communications with heterodyne detection over turbulence channels with pointing errors. J Lightw Technol 27 (20):4440–4445
8. La D, Fuente MG, Labiod H (2007) Performance analysis of position-based routing approaches in VANETs. In: 9th IFIP international conference on mobile wireless Communications Networks, pp 16–20
9. Paul B, Ibrahim M, Bikas MAN (2011) Experimental analysis of AODV & DSR over TCP and CBR connections with varying speed and node density in VANET. Int J Comp App 24(4)
10. Shakya M, Markam K (2016) Comparison of AODV and DSR routing protocols by using group mobility model with varying pause time in vehicular Ad-Hoc network. Int J Hybrid Info Technol 9(2):177–182
11. Chandel N, Gupta V (2014) Comparative analysis of AODV, DSR and DSDV routing protocols for VANET City Scenario. Int J Recent and Innovation Trends in Comp Commun 2(6):1380–1384
12. Piorkowski M, Raya M, Lugo AL, Papadimitratos P, Grossglauser M, Hubaux JP (2008) TraNS: realistic joint traffic and network simulator for VANETs. ACM Mob Comput Commun Rev 12(1):31–33

13. Oliveria R, Montez C, Boukerche A, Wangham MS (2017) Reliable data dissemination protocol for vanet traffic safety. Ad-hoc Netw 63:30–44
14. Malik S, Sahu PK (2019) A comparative study on routing protocols for VANETs. Heliyon, 5(8)

Distributed Estimation of IIR System's Parameters in Sensor Network by Multihop Diffusion LMS Algorithm

Meera Dash[1](✉), Trilochan Panigrahi[1], and Renu Sharma[2]

[1] Department of Electronics and Communication Engineering, SOA (Deemed to be) University, Bhubaneswar, India
meeradash@soa.ac.in

[2] Department of Electrical Engineering, Institute of Technical Education and Research, SOA (Deemed to be) University, Bhubaneswar, India
renusharma@soa.ac.in

1 Introduction

Wireless sensor networks (WSNs) are comprised of thousands of tiny wireless sensor nodes which are distributed in an area for specific remote sensing activities [1]. Every node then estimates the desired parameters out of the observed data of its own and also depends on the data from all the wireless sensor nodes of the network. For this, each sensor node after sensing the environment processes the data and then sends the data to its predefined neighbors. To improve the desired parameter estimation at each node, distributed algorithm plays an important role with information exchange between them [2]. But, in less connected or sparse wireless sensor networks, there is a difficulty in accessing information from the sensor network for estimating the parameters [3, 4].

FIR system's estimation of the parameter using a distributed strategy is discussed in literature [5, 6]. Mostly infinite impulse response (IIR) modeling is applied in cases of naturally occurring systems in which the unknown system has feedback. Thus, having reasonable length systems using FIR filter may not have an acceptable model of the feedback system [7]. Therefore, an IIR adaptive filter is used to model an unknown system.

But, the disadvantage of the IIR filter is, it is prone to have non-quadratic and multi-modal error surfaces with regard to the coefficients of the filter [8]. However, the general IIR LMS solution has been proved useful, and the local minima problem is less severe and the IIR LMS algorithm extensively improves after incorporating themselves in the censor network. Thus, the objective of the paper is to approximate the parameters of the IIR system in sparse sensor network by employing diffusion cooperation. Further, for sparse network where the nodes are connected to less number of neighbors, multihop diffusion is proposed in literature [4, 9]. The number of hops to and fro between the nodes impacts the overall network performance [10]. The algorithm is simulated in MATLAB, and results are compared with and without multihop strategy for standard IIR system.

M. N. Mohanty and S. Das (eds.), *Advances in Intelligent Computing and Communication*, Lecture Notes in Networks and Systems 109,
https://doi.org/10.1007/978-981-15-2774-6_3

2 Formulation of the Estimation Problem

A sensor node with k nodes is distributed in the area of interest to estimate system parameters. The network topology is assumed as an undirected graph.

Let $S = \{s_1, s_2, \ldots, s_K\}$ be the number of nodes in the network. The node s_k is denoted as k for compact notation. Let an undirected graph G be defined by a set of sensor nodes as vertices V and a set of edges E. Sensor nodes j and k are called neighbors if they are connected by an edge. The links are assumed to be bidirectional. The neighborhood of k is defined by

$$N_k = \{j|(j,k) \in E\} \tag{1}$$

Each node is modeled as IIR system. Figure 1 shows the block diagram of the system identification at each node and every node, $k \in S$ is in association with an output $d_k(i)$, which is noisy, to $x_k(i)$, which is the input at time instant i.

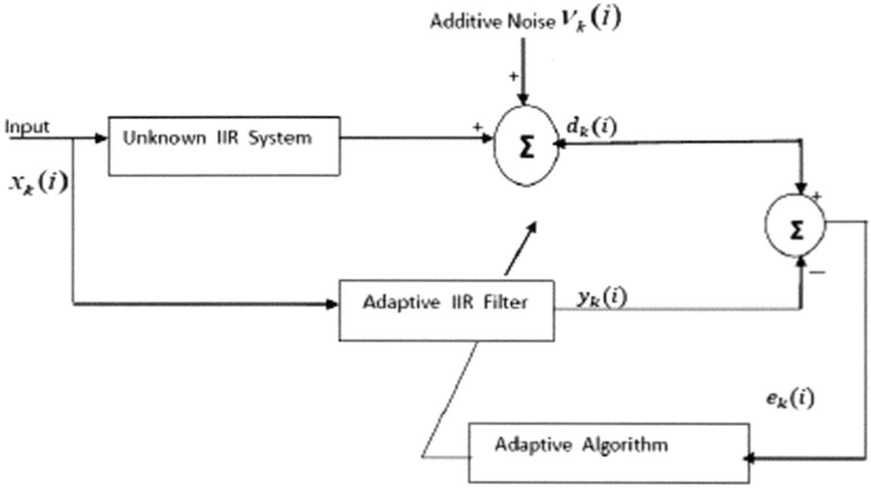

Fig. 1. Block diagram of IIR system at kth node in a WSN

Noisy observations are assumed to be independent to space and time and to input and output observations too. Each sensor k has a set of measured data $d_k(i)$ to its corresponding $x_k(i)$, i.e., the input vector at time instant i [11]. The IIR filter system is modeled having N number of feedback and M number of feed-forward parameters, respectively. The relationship between the output and input measured values are given by equation as [8]

$$d_k(i) = \sum_{n=1}^{N} a_n d_k(i-n) + \sum_{m=0}^{M} b_m x_k(i-m) + v_k(i) \tag{2}$$

where $\{a_n\}_{n=1}^{N}$ and $\{b_m\}_{m=0}^{M}$ are the poles and zeroes of the IIR system, and the noise measurement is denoted by $v_k(i)$ at the ith instant for the kth sensor with variance $\sigma_{v,k}^2$ and mean as zero. Noise in the system is supposed as white Gaussian which is not correlated to the input data $x_k(i)$. In compact form, the equation can be written as

$$d_k(i) = a^{\mathrm{T}} d_k^{i-1} + b^{\mathrm{T}} x_k^i + v_k(i) = \left[d_k^{i-1^{\mathrm{T}}} x_k^{i^{\mathrm{T}}} \right] w \quad (3)$$

where

$$a = [a_1, a_2, \ldots, a_N]^{\mathrm{T}},$$
$$b = [b_0, b_1, \ldots, b_M]^{\mathrm{T}}, d_k^{i-1} = [d_k(i-1), d_k(i-2), \ldots d_k(i-N)]^{\mathrm{T}}$$
and $w^{\mathrm{T}} = \left[a^{\mathrm{T}} b^{\mathrm{T}} \right]$.

Modeling of the adaptive system is done as IIR type with an equal number of parameters, respectively. Each sensor node estimates the output of the adaptive system given by as

$$y_k(i) = \sum_{n=1}^{N} \hat{a}_{k,n}(i) y_k(i-n) + \sum_{m=0}^{M} \hat{b}_{k,m}(i) x_k(i-m) \quad (4)$$

where IIR system output of the IIR adaptive system is represented by $y_k(i)$ at the kth node. The current estimate of the output is dependent on N outputs previously estimated. The estimated coefficients $\left\{ \hat{a}_{k,n}(i) \right\}_{n=1}^{N}$ and $\left\{ \hat{b}_{k,m}(i) \right\}_{m=0}^{M}$ are at each sensor node k, such that during training they are adjusted to achieve steady-state value after the training is done. Equation (4) in compact form is given as

$$y_k(i) = \hat{a}_k^{\mathrm{T}} y_k^{i-1} + \hat{b}_k^{\mathrm{T}} x_k^i = w_k^{\mathrm{T}} u_k^i \quad (5)$$

where $\hat{a}_k = \left[\hat{a}_{k,1}, \hat{a}_{k,2}, \ldots, \hat{a}_{k,N} \right]^{\mathrm{T}}$, $\hat{b}_k = \left[\hat{b}_{k,0}, \hat{b}_{k,1}, \ldots, \hat{b}_{k,M} \right]^{\mathrm{T}}$, and $w_k^{\mathrm{T}} = \left[\hat{a}_k^{\mathrm{T}} \right]$.

The input data vector is defined as

$$u_k^i = [y_k(i-1), \ldots, y_k(i-N), x_k(i), x_k(i-1), \ldots, x_k(i-M)] \quad (6)$$

Global parameter vector w is estimated using data $\left\{ d_k(i), x_k^i \right\}$ available at every node using the diffusion mode of cooperation method [12] which is the objective of this paper.

Diffusion LMS (DLMS) algorithm is used for approximation of the IIR coefficients. By optimizing the following global cost function w can be estimated [13],

$$J(w) = \sum_{k=1}^{K} E \left\| d_k(i) - u_k^i \right\|^2 \quad (7)$$

where $E\|\cdot\|$ indicates the expectation operator and $J_k(w) = E\|d_k(i) - u_k^i w\|^2$ indicates the local cost function.

The optimal value is calculated by solving Eq. (7), [11].

$$\hat{w} = \arg\min_{w} \sum_{k=1}^{K} E\|d_k(i) - u_k^i w\|^2 \tag{8}$$

Desired response, $d_k(i)$, is approximated by using LMS algorithm [7]. To resolve Eq. (7), the standard steepest-descent recursive algorithm given in (9) is used. To resolve the difficulty, the equivalent implementation is introduced by using the adapt then combine (ATC) strategy as in [5]. Therefore, the modified diffusion algorithm is

$$\begin{aligned} i = 1, 2, \ldots; k &= 1, 2, \ldots, K; \hat{w}_k^0 = w_{\text{init}} \\ e_k(i) &= d_k(i) - u_k^i w_k^{i-1} \\ \hat{w}_k^i &= \hat{w}_k^{i-1} + \mu_k e_k(i) u_k^{iT} \\ \hat{w}_k^i &= \sum_{l \in N_k}^{k} c_{lk} \hat{w}_l^i \end{aligned} \tag{9}$$

By following the diffusion cooperation algorithm, each sensor node cooperates among each other (9) to find the IIR system parameter.

3 Distributed Estimation Using Multihop Diffusion Strategy

To upgrade the diffusion cooperation strategy, many authors have proposed different methods in the literature. Mostly, many of the authors have tried to minimize the communication overhead in the network because more energy is required for communication than computation [6, 14, 15]. But these methods failed to provide the desired performance in a less connected sparse network. By adopting a partial diffusion [14], the number of communications is reduced. In [15], energy-aware distributed diffusion strategy is discussed. These methods are also not suitable for IIR system parameter estimation problems. It is because the poles play an important role in IIR system and slight deviation in the value during the learning process may cause instability.

The multihop DLMS is proposed in [3, 9] with constrained energy. As per the energy of the nodes, the 1-hop neighbors and multihop neighbors are specified. Data aggregation is a difficult process and takes more time as each sensor node has to find the information of energy of the neighbors.

Here, we propose a multihop diffusion adapt then combine (ATC) strategy without constraints. In comparison to combine then adapt (CTA), the ATC is more adaptive. From each h-hop neighbors, every node can obtain the information. The h-hop sensor nodes are defined as a set of nodes available at h-hop distance from itself. The 1-hop network topology and 2-hop network topology are described in [3].

The update equations for h-hop ATC (hATC) [3] are following the same steps as that of one-hop strategy as in (9),

$$\begin{aligned} & i = 1,2,\ldots; k = 1,2,\ldots,K; \hat{w}_k^0 = w_{\text{init}} \\ & e_k(i) = d_k(i) - u_k^i w_k^{i-1} \hat{w}_k^i = \hat{w}_k^{i-1} + \mu_k e_k(i) u_k^{i^T} \\ & \hat{w}_k^i = \sum_{l \in N_k^h}^{k} c_{lk} \hat{w}_l^i \end{aligned} \tag{10}$$

where μ_k denotes the step size at sensor node k, and c_{lk} is combination weights satisfying

$$c_{lk} \geq 0, \sum_{l \in N_k^h} c_{kl} = 1, \quad \text{and}\, c_{lk} = 0 \text{ if } l \text{ not belonging to } N_k^h \tag{11}$$

Conventional ATC and hATC varies with each other with respect to the combination step [6, 16]. But, in distributed signal processing in sensor network, each sensor is assumed to communicate only with 1-hop neighbors only. So, it is difficult to combine the estimates from multihop neighbors in. This can be calculated by repeating 1-hop combination times in h-hop diffusion strategy [4]. The multihop algorithm can be implemented with same communication protocol used in conventional diffusion LMS.

The algorithm also does not require additional storage capacity. But the algorithm needs $(|N_k| - 1)h$ communication cost in each iteration. This is times more than the communication cost for DLMS $(|N_k| - 1)$. The additional communication requirement can be reduced by employing block LMS method in multihop diffusion LMS algorithm.

4 Results Obtained by Simulation

Here, the performance analysis of diffusion LMS algorithm is done with and without multihop for parameter estimation of IIR system with simulations. For sparse sensor network, multihop diffusion LMS (hDLMS-IIR) is used. The simulations are done using shift registers and regressor input. According to the model in (3), the desired data are generated. For simulation purpose, the standard IIR 2nd, 3rd, and 4th order transfer function of IIR system systems are considered [17, 18].

$$H_2(z) = \frac{1.25 - 0.25z^{-1}}{1 + 0.32z^{-1} - 0.412z^{-2}} \tag{12}$$

$$H_3(z) = \frac{0.6 - 0.4z^{-1}}{1 + 0.2z^{-1} - 0.51z^{-1} - 0.1z^{-3}} \tag{13}$$

And

$$H_4(z) = \frac{1 - 0.9z^{-1} + 0.812z^{-2} - 0.73z^{-3}}{1 + 0.042z^{-1} + 0.28z^{-2} - 0.21z^{-3} + 0.1z^{-4}} \tag{14}$$

Here, we have used IIR adaptive filter of 2nd, 3rd, and 4th order at each sensor node to model the corresponding orders of IIR system, respectively. The transfer function for the adaptive IIR system models to estimate their parameters is

$$\hat{H}_{2k}(z) = \frac{\hat{b}_{k0} + \hat{b}_{k1}z^{-1}}{1 - \hat{a}_{k1}z^{-1} - \hat{a}_{k2}z^{-2}} \tag{15}$$

$$\hat{H}_{2k}(z) = \frac{\hat{b}_{k0} + \hat{b}_{k1}z^{-1} + \hat{b}_{k2}z^{-2}}{1 - \hat{a}_{k1}z^{-1} - \hat{a}_{k2}z^{-2} - \hat{a}_{k3}z^{-3}} \tag{16}$$

$$\hat{H}_{4k}(z) = \frac{\hat{b}_{k0} + \hat{b}_{k1}z^{-1} + \hat{b}_{k2}z^{-2} + \hat{b}_{k3}z^{-3}}{1 - \hat{a}_{k1}z^{-1} - \hat{a}_{k2}z^{-2} - \hat{a}_{k3}z^{-3} - \hat{a}_{k4}z^{-4}} \tag{17}$$

The average over 500 independent experiments is plotted. Performance of diffusion LMS algorithm in less connected network is discussed here. Let a sparse sensor network with 20 sensor nodes. Network topology is given in Fig. 2. From the figure, it is evident that the ILMS algorithm is not possible here to estimate the systems' parameter using distributed approach, whereas diffusion LMS may apply for the same. But the edge nodes may not get sufficient information from the network because of less connectivity. Therefore, multihop diffusion LMS algorithm is used here for IIR systems' parameter estimation (hDLMS-IIR).

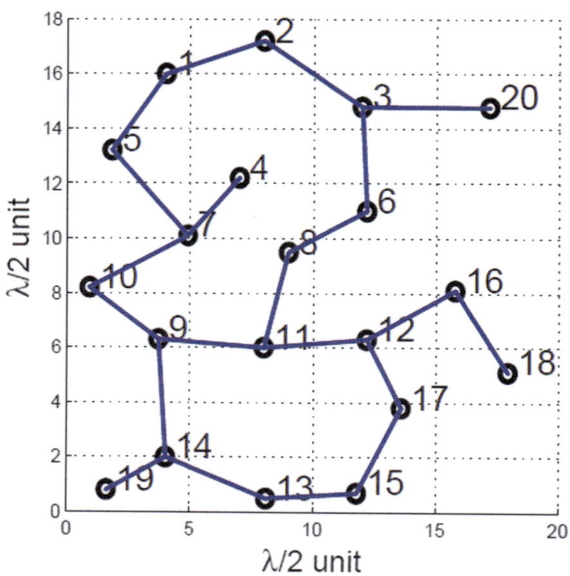

Fig. 2. Network topology used for sparse sensor network

The parameters of each sensor node of the sensor network are given in Fig. 3. Number of iterations to simulate for each system is taken 3000. The block size is chosen as length of the optimum vector that is the sum of the feed-forward and feedback parameters. For the simulations, block length L has been assumed to be 4, 5, 8 for 2nd, 3rd, and 4th order IIR systems, respectively. The parameter μ (step size) is chosen as 0.075 for DLMS-IIR and non-cooperative algorithms, whereas μ is chosen as $0.075 * L$ for BDLMS-IIR and BhDLMS-IIR algorithms to get similar performance. This is based on the theory for block adaptive algorithm and also valid for distributed block adaptive algorithms [6, 19].

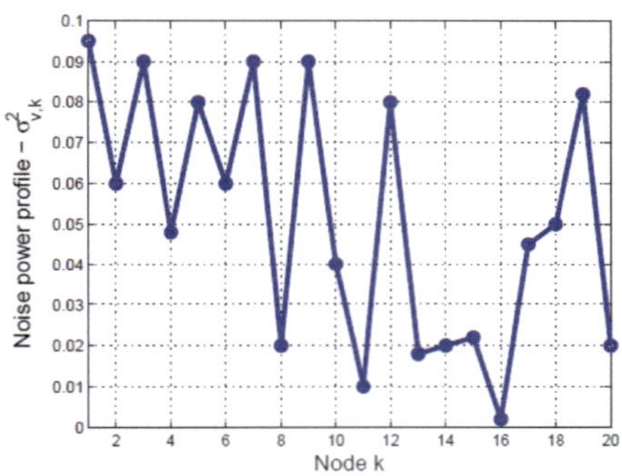

(a) Noise Power Profile

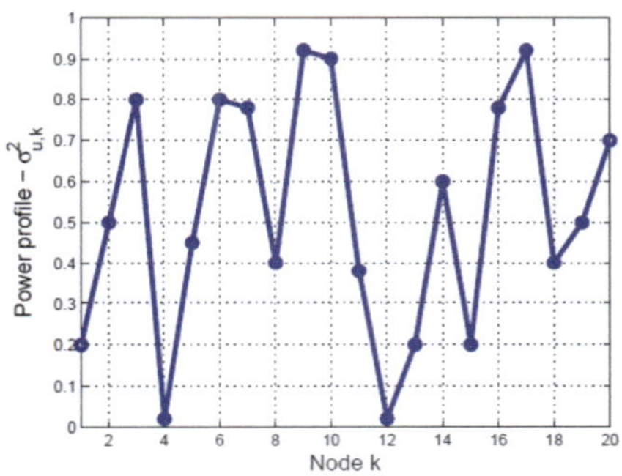

(b) Signal Power at Each Node

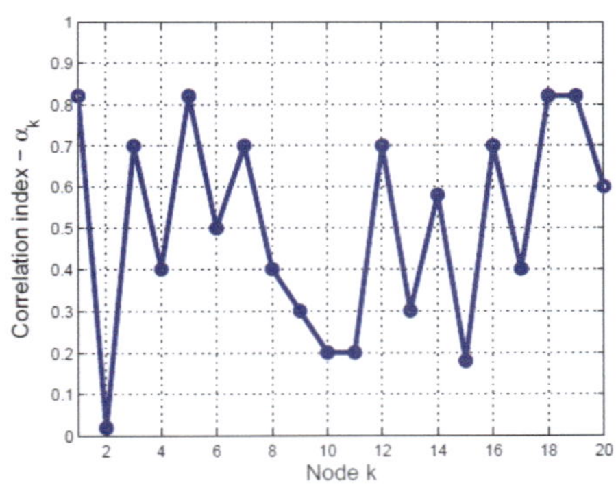

(c) Correlation Index α_k at Each Node

Fig. 3. Sensor node's parameters for less connected wireless sensor network

In literature, the performance of distributed block LMS and distributed LMS is nearly the same if the step size of LMS is chosen L time and the step size for LMS algorithm is without batch processing [6]. The MSE and MSD curves are plotted by averaging 500 Monte Carlo simulations for 3000 iterations for 1-hop DLMS (DLMS-IIR), 2-hop DLMS (hDLMS-IIR) and non-cooperation LMS algorithms, and then block version of all the algorithms. The simulation results show that the performance of

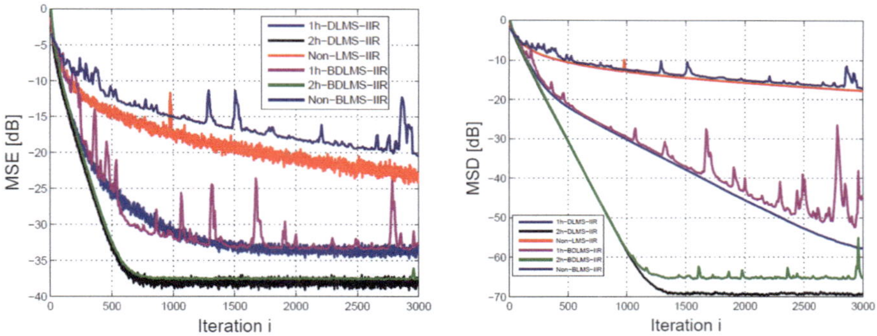

(a) MSE Performance (b) MSD Performance

Fig. 6. Average MSE and MSD performances by 1-hop DLMS-IIR, 2-hop hDLMS-IIR, non-cooperative LMS-IIR, 1-hop BDLMS-IIR, 2-hop BhDLMS-IIR, and non-cooperative block LMS-IIR algorithms for 4th order IIR system

But the 2-hop BhDLMS algorithm gives similar to that of non-block algorithm. There is an improvement of 5 and 10 dB in MSE and MSD performance of 2-hop hDLMS and BhDLMS algorithm over 1-hop algorithms, respectively. We suppress IIR in the name of the algorithm for compactness.

5 Conclusion

Distributed IIR systems' parameter estimation in sparse WSNs using multihop DLMS algorithm has been studied in this paper. In the wireless sensor network, every node is assumed to be an adaptive IIR filter, and each and every node in the network collaborates among themselves by using diffusion cooperation method. MSE and MSD are computed through simulation and compared the non-cooperative LMS algorithm for benchmark IIR systems. Further to reduce the increase in communication overhead in the network, block distributed multihop diffusion LMS algorithm is proposed where each node shares data after processing a block of data. By doing this, the multihop block diffusion LMS provides the best MSE and MSD performance with minimum number of iterations.

References

1. Li Y, Thai MTE (2008) Wireless sensor networks and applications. Springer
2. Sayed AH (2014) Adaptive networks. Proc IEEE 102(4):460–497
3. Nayak M, Panigrahi T, Sharma R (2015) Distributed estimation using multi-hop adaptive diffusion in sparse wireless sensor networks. In: International Conference on Microwave, Optical and Communication Engineering (ICMOCE), Dec 2015, pp 318–321

4. Kong J-T, Lee J-W, Kim S-E, Shin S, Song W-J (2017) Diffusion LMS algorithms with multi combination for distributed estimation: formulation and performance analysis. Dig Signal Proc 71(Supplement C), 117–130
5. Cattivelli FS, Sayed AH (2010) Diffusion LMS strategies for distributed estimation. IEEE Trans Signal Process 58(3):1035–1048
6. Panigrahi T, Pradhan PM, Panda G, Mulgrew B (2012) Block least-mean square algorithm over distributed wireless sensor network. J Comput Netw Commun., vol. 2012, Hindawi Publishing Corporation
7. Shynk JJ (1989) Adaptive IIR filtering. IEEE ASSP Mag 6(2):4–21
8. Agrawal N, Kumar A, Bajaj V, Singh GK (2017) High order stable infinite impulse response filter design using cuckoo search algorithm. Int J Autom Comput 14(5):589–602
9. Hu W, Tay WP (2015) Multi-hop diffusion LMS for energy-constrained distributed estimation. IEEE Trans Signal Process 63(15):4022–4036
10. Jerew O, Blackmore K (2014) Estimation of hop count in multi-hop wireless sensor networks with arbitrary node densitym. Int J Wireless Mob Comput (IJWMC) 7(3):207–216
11. Schizas I, Mateos G, Giannakis G (2009) Distributed LMS for consensus-based in-network adaptive processing. IEEE Trans Signal Process 57(6):2365–2382
12. Nedic A, Ozdaglar A (2009) Distributed subgradient methods for multi-agent optimization. IEEE Trans Autom Control 54(1):48–61
13. Khalili, Rastegarnia A (2016) Tracking analysis of augmented complex least mean square algorithm. Int J Adapt Control Signal Proc 30(1):106–114
14. Arablouei R, Werner S, Huang Y-F, Dogancay K (2014) Distributed least mean-square estimation with partial diffusion. IEEE Trans Signal Process 62(2):472–484
15. Gharehshiran ON, Krishnamurthy V, Yin G (2013) Distributed energy-aware diffusion least mean squares: game-theoretic learning. IEEE J Select Top Signal Proc 7(5):821–836
16. Xiao L, Boyd S, Lall S (2005) A scheme for robust distributed sensor fusion based on average consensus. In: Proceedings of 4th international symposium on information processing in sensor networks, Loss Angles, CA, pp. 63–70, Apr 2005
17. Scarpiniti M, Comminiello D, Parisi R, Uncini A (2015) Nonlinear system identification using IIR spline adaptive filters. Signal Process 108:30–35
18. Majhi B, Panda G (2013) Distributed and robust parameter estimation of IIR systems using incremental particle swarm optimization. Dig Signal Proc 23(4):1303–1313
19. Panda G, Mulgrew B, Cowan CFN (1986) A self-orthogonalizing efficient block adaptive filter. IEEE Trans Acoust Speech, Signal Process 34(6): 1573–1582, Dec 1986

Implementation of Raspberry Pi for Fault Detection in Optic Fibre Line

K. P. Swain[1(✉)], S. R. Das[1], Sangram Kishore Mohanty[2], and G. Palai[1]

[1] Department of Electronics and Communication Enginerring, GITA, Bhubaneswar, India
[2] Department of ECE, ICE, GITA, Bhubaneswar, India

1 Introduction

Optical fibres are long cables made of glass with two layers called core and cladding which works on the principle of total internal reflection where the light is propagated through the length of the cable, reflecting back in the core by enabling the light to pass through the cross-section area even in the bend regions. But, the signal is attenuated while travelling for a certain distance which causes weakening of signal strength and introduces noise in the signal. So repeaters are generally used to amplify the optical signal and transmit to further stages. But, when there is a fault in the optical fibre, it is usually not known to the concerned person about the fault details.

There are more than a few works are discussed in the literature for finding the fault in the optic fibre line. One foremost technology like distributed temperature system (DTS) is used in [1, 2] for real-time monitoring system. In [1], a cable monitoring system for underground power system is discussed by DTS technology in real-time basis and online. Here, DTS technology is applied by comparing the temperature of conductor and cables in real time, and also a test-bed is erected to ensure the applicability of the system towards the underground system. By this technology, the operators can forecast the fault so that the issue can be resolved within a very short span of time. This technology is also used in [2] for real-time temperature monitoring system which is applied to many fields like power cable monitoring, underground power system monitoring, power cable monitoring in tunnel, etc.

Optical time-domain reflectometer (OTDR) is used in [3] to identify different kinds of fault like bend, crack, splice, etc. It also conversed a MATLAB-based GUI to indicate the fault condition with time. OTDR is interfaced with computer screen to show every measurement which can be further analysis. OTDR is worked on the principle of Rayleigh backscattering (RBS) which detects an abrupt shift known as point defects. In [4], limited perimeter vector matching (LVM) protocol is used to locate single link failure in the optical network. Here, the optimization technique is implemented to minimize the fault detection time in a static optical network. A photon-counting OTDR with linearized Breyman iteration algorithm-based processing unit is discoursed in [5] to find an optical fault. But in [6], minimum length probe cycle cover algorithm is realized for fault detection and in which m2-cycle mechanism is found superior among others. A single-mode fault detection is discussed in [7] by using

M. N. Mohanty and S. Das (eds.), *Advances in Intelligent Computing and Communication*, Lecture Notes in Networks and Systems 109,
https://doi.org/10.1007/978-981-15-2774-6_4

approximation algorithm and GIS-based fibre optic monitoring system is illustrated in [8], where alarms are used to indicate in case of any fault or degradation of optical signal is occurred. Microcontroller is used in [9, 10], to observe the received power from an optical fibre line by using a sensor circuit mostly a photodiode. In [9], message is sent through GSM technology if any abrupt change is occurred in received voltage, whereas in [10], a database is maintained by using MySQL to keep record of the received voltage. Arduino Uno is used along with a LDR to sense the received voltage in real-time basis and to display the fault status continuously through a LCD. Here, a database is created by using Apache Web server, perl, MySQL and PHP which are very popular package for Web application and XAMPP is also used for hosting the server locally.

In the previous work, on optical fault detection system [10], Arduino was used to acquire data from the optical fibres and processing was done in a different computer to detect fault and update the data to the server. But sending data to the server requires an Internet connection and setting up a computer altogether is a difficult task and sometimes not feasible. But in this work, we propose a system using Raspberry Pi to implement the fault detection by constantly monitoring the received power through a photodiode. Raspberry Pi used in this work is a mini-computer having multithreading and multitasking capability which makes it as a superior processing power than Arduino and also it can directly host Web server by eliminating the need of separate computer for maintaining the database.

2 System Overview

See Fig. 1.

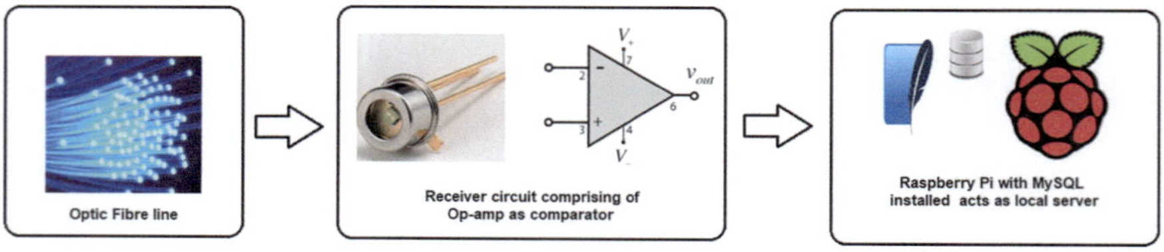

Fig. 1. Graphical representation of system overview

2.1 Sensor

As shown in Fig. 2, the sensor circuit consists of a photodiode along with an op-amp (LM358) which acts as a comparator. At the non-inverting end of the op-amp, a reference voltage is set by the voltage divider circuit comprising of R1 and R2 of value 20 K and 30 K, respectively, whereas capacitive coupled voltage is available at the inverting end generated by C2 (100 nF) which is coming from the fibre line through photodiode. Then, these voltages are compared to give an output to GPIO pin of

Raspberry Pi. A Zener diode is also connected at the output to maintain 3.3 V for protection of Pi as the GPIO pins are operated at 3.3 V.

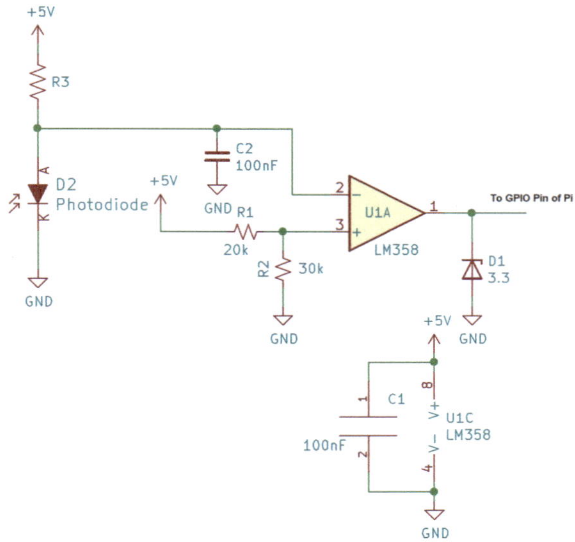

Fig. 2. Circuit diagram of the sensor unit

2.2 Signal Processing Through Raspberry Pi

Raspberry Pi acts as a brain of the system where GPIO pins are exposed for direct interfacing of variety of sensors and electronic systems. With higher voltages, there is a high chance of damaging the processor as it works on 3.3v logic level. MySQL database is installed on the Raspberry Pi to store real-time data. It features Wi-Fi and Bluetooth which enables to connect with existing Wi-Fi networks with ease. The interfacing and processing logic is written in python which gives us more computing power and control to handle and store data simultaneously without any lag. The complete circuit diagram of the proposed system is shown in Fig. 3 where GPIO 2 pin is connected to the output pin of op-amp which is used to push the data in case of any fault is raised.

2.3 Web Server (My SQL Server)

MySQL is database management software which stores data in the form of tables and has the capability to form relations within tables. It was created by Oracle Corporation which now powers applications right from low-end to high-end data science. It is a relational database, which means tables within a database can refer to each other, using foreign key concept. MySQL database provides users to query data from the tables through commands known as structured query language or SQL for short. In this optical fault detection system, MySQL is used to store the fault data for monitoring

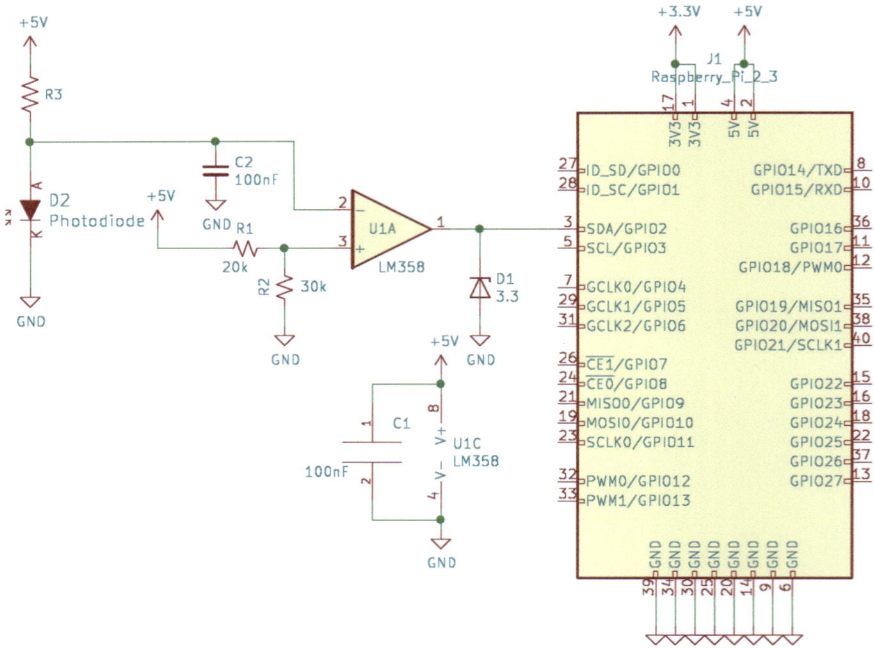

Fig. 3. Complete circuit diagram of the proposed system

which is installed on Raspberry Pi. Data from the digital inputs is detected by the GPIO of the Raspberry Pi and are pushed to MySQL database through python script. The python script acts as a software interface between the Raspberry Pi GPIOs and database. The script constantly listens for change in signal at the GPIO and as soon as the signal change is detected, it stores the data along with the timestamp in the database.

3 Result and Discussion

The fault detection system includes a sensing unit, a comparator unit and data processing unit, where signal is coming from the optical fibre and fell on the sensing unit, which converts the optical signal to the electrical form. The electric signals are in the form of voltages which represent the strength of a signal. To detect a fault, the average of a signal is taken over a time period and its corresponding voltage is generated. For benchmarking, samples of the signals with typical strength is recorded and stored for comparison. The comparator circuit is developed using an op-amp (LM358) with capacitive coupled input on the inverting terminal of it in which average voltage required for comparison is provided by capacitor (C2 of 100 nF). The non-inverting terminal of the comparator is given with a voltage reference through a resistor arrangement (R1 and R2 of 20 K and 30 K, respectively). When the sensed signal strength falls below the setpoint, the op-amp sends a high signal which represents a fault in the system. During this period, timestamp of the event occurred and the fault status is updated by Raspberry Pi. This data is recorded in the database and notification is generated for the operator for inspection.

3.1 Physical Realization of the System

The physical implementation is realized in the laboratory by using two optical demo kits of Scientech 2506 model with a two metre poly methyl methacrylate also known as acrylic glass (PMMA) fibre cable. Here, the two kits are used to demonstrate optical sender and receiver through the fibre cable. A sinusoidal signal is generated using a function generator of Scientech 4061 (3 MHz) model which drives the sender kit for converting the electrical signals into the optical signals and transmitted through the optical fibre. The receiver has optical transducer which converts optical signal back to electrical signals, and both the output of the receiver and sender are shown in the oscilloscope as shown in Fig. 4. The receiver output is connected to the op-amp which is configured as a comparator. The output of the op-amp is connected to the GPIO of the Raspberry Pi where the Pi constantly listens for the interrupt signal from the op-amp. As soon as a signal is generated, Raspberry Pi stores the signal information along with the timestamp. The monitor in the image shows MYSQL database with the records inserted during the fault. To test the system, the fault signals are generated by reducing the gain of the amplifier at the sending end simulating an attenuated signal.

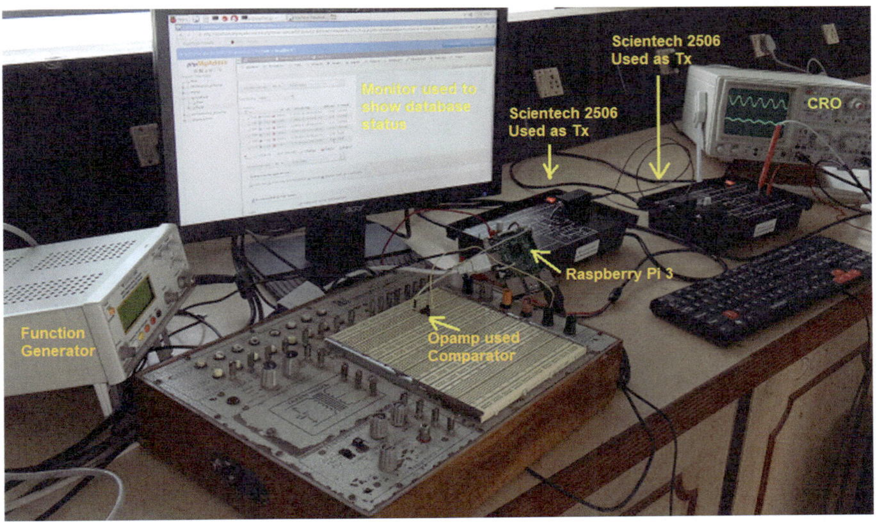

Fig. 4. Realization of the system in laboratory using Scientech 2506 kit

3.2 Database Apprising

The table shown in Fig. 5 is used to store the status of the signal strength coming from the fibre line where the columns include timestamp, signal strength and its corresponding remark. The values in the signal strength represent the fault present or absent. The lowest value 1 represents the signal has fault and value 3 represents the system is normally operational condition. To support this, remark column shows the same and the timestamp column represents the data insertion time.

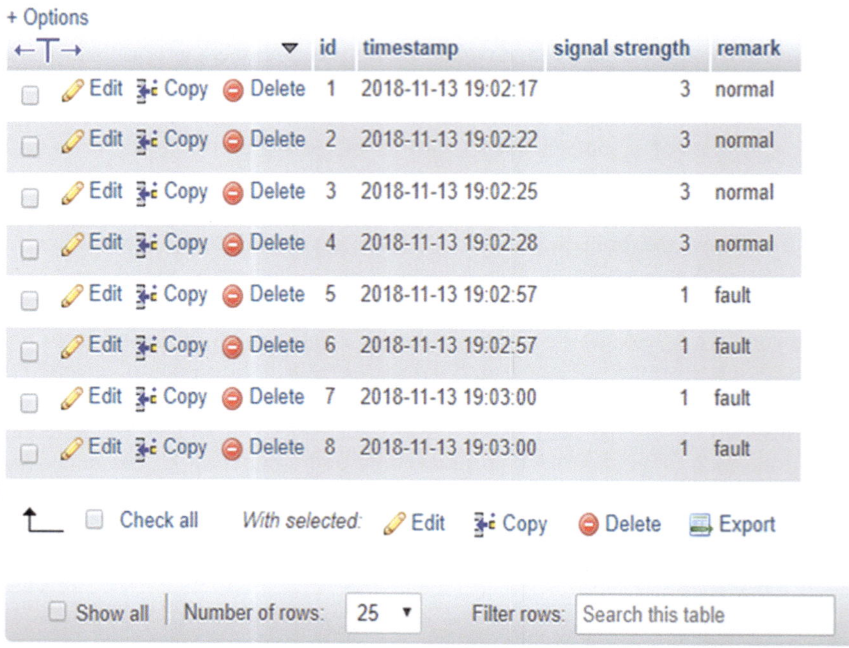

Fig. 5. Update of database by experimental method

4 Conclusion

An optical fault detecting system using Raspberry Pi is realized successfully in this paper, which is used to signpost the fault status ensued in the fibre line through a local server. Here, MySQL is installed in the Raspberry Pi which acts as a local server to update the fault info in real-time basis by the aid of a comparator containing an op-amp. Here, op-amp with a photodiode used together to compare the incoming line voltage through photodiode available at inverting terminal with a reference voltage maintained at non-inverting input of it. To end with, an experiment is also conducted successfully in the laboratory by using two optical transceiver kit of model Scientech 2506 to verify the result in the database by varying the voltage manually at the transmitting end.

References

1. Cho J, Kim J, Lee H, Kim J, Song I, Choi J (2014) Development and improvement of an intelligent cable monitoring system for underground distribution networks using distributed temperature sensing. Energies 7:1076–1094
2. Downe J, Leung HY (2004) Distributed temperature sensing worldwide power circuit monitoring applications. IEEE ICED 1804–1809
3. Ab-Rahman MS, Chuan NB, Premadi A, Jumari K (2008) A new approach for identifying fiber fault and detecting failure location. IEEE ICED 1–4
4. Khair MG, Kantarci B, Zheng J, Mouftah HT (2008) Optimization for minimizing fault localization time in all-optical networks. IEEE ICTON 63–66
5. Lunglmayr M, Amaral GC, Linearized B (2018) Iteration for automatic optical fault analysis. IEEE Trans Instrum Meas 1–13
6. Wang N, Wang R, Huang S (2009) A low cost fault detection mechanism based on cycle cover algorithm for optical burst switching networks. IEEE ICCMC 545–549
7. Nayek P, Pal S, Choudhury B, Mukherjee A, Saha D, Nasipuri M (2005) Optimal monitor placement scheme for single fault detection in optical network. IEEE ICTON 433–436
8. Akdemir OK, Dursun T, Arslan S, Benzer R, Akcayol MA (2016) A GIS-based novel active monitoring system for fiber networks. Turk. J. Electr. Eng. Comput. Sci. 24:247–261
9. Baskaran G, Seethalakshmi R (2012) Intelligent fault detecting system in an optical fibre. J. Theor. Appl. Inf. Technol. 39:178–187
10. Swain KP, Sahoo J, Prasad MVSV, Palai G (2015) Fault detection system in an optical fiber using arduino. Int. J. Appl. Eng. Res. 10:31745–31749

Distributed Optical Fiber Sensing System Performance Improvement Using Signal Processing Techniques

Ramji Tangudu(✉) and Prasant Kumar Sahu

Indian Institute of Technology Bhubaneswar, Argul, Odisha, India
{rt10, pks}@iitbbs.ac.in

1 Introduction

In this paper, the detailed study of Brillouin optical time-domain reflectometry (BOTDR)-based distributed optical fiber sensing (DOFS) system is reported. Nowadays, optical fiber is being utilized for both communication and sensing purposes [1]. DOFS systems are more effective, cost-effective, and popular, due to the feature of replacing thousands of point sensors and the associated complexity by using single long-range fiber sensors typically of the orders of few kilometers [1]. The DOFS system operation is based on backscattered signals, namely Rayleigh, Brillouin, and Raman. Raman backscattered signal-based DOFS system is useful for temperature sensing only, whereas Brillouin backscattered signal-based DOFS system is useful for both temperature and strain sensing. Due to the inherent advantages of Brillouin sensing, this system has higher potential applications in oil and gas pipeline leakage detection, fire detection, high power cable monitoring, structural health monitoring (bridges, dams, tunnels, and roads), as well as in defense and military applications [1]. The major advantages of fiber-based sensing system include the electromagnetic interference immunity, lightweight, design simplicity, and more survivability under harsh environmental conditions [1]. The variation in temperature along the fiber will result in fluctuations in frequency of Brillouin backscattered signal. From these parametric fluctuations, we can estimate the actual temperature along the fiber. Typically, the value of Brillouin frequency shift (BFS) for 1550 nm pumping wavelength is $\sim$11 GHz [1].

Brillouin backscattered signal-based DOFS system can further be classified as distributed temperature sensing (DTS) system and distributed strain sensing (DSS) systems. In this paper, we have focused on DTS system only. According to literature survey, Brillouin backscattered signal-based DTS (BDTS) system is broadly classified into two categories, such as spontaneous Brillouin scattering (SpBS) and stimulated Brillouin scattering (SBS)-based BDTS system. Under SpBS type, BOTDR, Brillouin optical correlation domain reflectometry (BOCDR) and Brillouin optical frequency-domain reflectometry (BOFDR)-based DTS systems are grouped together [2–4], whereas under SBS type, Brillouin optical time-domain analysis (BOTDA), Brillouin optical correlation domain analysis (BOCDA) and Brillouin optical frequency-domain analysis (BOFDA)-based DTS systems are grouped together [5–7]. In all analysis based DTS systems which requires the double ended accesses (probe and

M. N. Mohanty and S. Das (eds.), *Advances in Intelligent Computing and Communication*, Lecture Notes in Networks and Systems 109,
https://doi.org/10.1007/978-981-15-2774-6_5

pump lights) and stimulated Brillouin scattering are complex to design and costly [5–7]. BOCDR-based DTS system is useful for short-range applications typically for meter-long sensing only, whereas BOFDR-based DTS system is having high cost, due to usage of electrical spectrum analyzer (ESA) and vector network analyzer (VNA) [4].

Song and Zhao [8] proposed a BOTDR system using Hilbert transform method for sensing the temperature for a range up to 25 km with an input optical launch power of 22 dBm of peak power. Bernini et al. [9] proposed a combined scheme of BOTDR technique and harmonic approach for obtaining 40 cm spatial resolution in temperature sensing. The major feature of their proposed system is as follows: 1 mW of laser source power over a sensing range of 35 m only. Soto and Sahu [2] have designed a distributed temperature sensing system using BOTDR and 127-bit simplex coding scheme for 21 km of sensing range and the reported temperature resolution was of 3.1 K. However, this method suffers from the limitations of design complexity and bandwidth requirement when the coding length increases. In [10], combination of BOTDA technique and optimized pulse using 127-bit simplex coding is proposed for getting 2.6 °C temperature resolution as well as Raman amplification being used for getting 120 km of sensing range in the presence of more relative intensity noise (RIN). However, since the system is designed on the analysis-based technique, coding scheme and amplification process, complexity, bandwidth requirement, and cost will be more [10].

Minardo et al. [7] presented BOFDA technique-based system for temperature sensing using polymer optical fiber (POF). However, the sensing ranging is limited to 20 m only. Pradhan and Sahu [11] proposed a BOTDR system using Fourier wavelet regularized deconvolution (FourWaRD) scheme for achieving a 1.85 K temperature resolution and 30 m spatial resolution using 400 ns laser pulse width. The system was designed using evolutionary optimization schemes for achieving better SNR value. Though the designed system offered a sensing range up to 50 km of fiber length, but the spatial resolution was 30 m, which needs improvement. Li et al. [5] proposed a differential pulse pair (DPP)-BOTDA technique for achieving 2 m of spatial resolution. Here, they have used two pulse signals, the respective pulse widths are 80 and 100 ns, respectively, and they have achieved a 40 km of fiber sensing length. This scheme was based on SBS waveform. The major drawbacks of this scheme include higher degree of system complexity and suffers from nonlinearity effects. Shlomi et al. [6] proposed a system for temperature sensing operation with DPP-BOCDA technique, and the achieved spatial resolution was 2.7 cm with a 1.9 MHz of BFS error. However, the major drawbacks of this technique include higher system complexity and lower sensing range. The sensing range was 43 m only.

Bao et al. [12] have presented a work on analysis of temperature and strain sensing. The temperature and strain sensing was carried out from 8 °C to 50 °C (with Panda fiber), from 0 °C to 43 °C (with Bow-Tie fiber), from 0 $\mu\varepsilon$ to 408 $\mu\varepsilon$ (with Panda fiber), and from 0 $\mu\varepsilon$ to 1021 $\mu\varepsilon$ (with Tiger7 fiber), respectively. Because of using polarization maintaining fiber (PMF), the polarization mode dispersion (PMD) effect was removed. The system features a 20 cm spatial resolution with 2 ns pulse width and PMF. The system offered achieved 15 cm of spatial resolution with pulse width of 1.5 ns and photonic crystal fiber (PCF). Employing PCF, they have obtained temperature and strain resolutions of 1.3 °C and 15 $\mu\varepsilon$, respectively. From PMF, they have obtained temperature and strain resolutions varied from 2 °C to 38 °C and varied from 39 με to 598 με, respectively. These resolutions are obtained for 2 m of sensing fiber only [12].

Soto et al. [13] presented a BOTDR system using 127-bit simplex coding technique. The system was designed for temperature and strain sensing over 53 km of measurement range with 10 mW of laser source power. The reported SNR enhancement was 7.4 dB with 127-bit simplex coding technique. In this work, the temperature resolution is achieved as 8.8 K and strain resolution is achieved as 220 με. They have achieved spatial resolution as 32 m using 320 ns of laser source pulse width [13]. Zou et al. [14] proposed BOCDR technique with PMF. The system offered spatial resolution of 5 mm, 5 cm, and 16 cm using 18.525 MHz, 1.785 MHz, and 528.968 kHz of modulation frequency. However, the sensing range was limited to 200, 56, and 5 m only. The BFS error reported was ±3 MHz for 200 m and ±5 MHz for 56 m of sensing range. For 200 m sensing range, temperature and strain resolutions are ±3 °C and με, respectively. For 56 m sensing range, temperature and strain resolutions are ±5 °C and με [14].

Bernini et al. [15] proposed a BOFDA system, using peak search, inverse Fourier transform (IFT), and iterative procedures. This system has been offered a 29 mm of spatial resolution and has low spurious effects on the collected data (with iterative algorithm). The FUT length was up to ~5.5 m with 40 mW of laser input power. Yamashita et al. [16] proposed a combination of BOCDA and temporal gating technique. They have obtained 5 cm of spatial resolution and almost zero error in strain sensing. Due to the usage of temporal gating technique, their FUT length was obtained as 500 m (here fiber is PMF). The amount of error in temperature sensing was 10 °C.

Hayashi et al. [3] proposed BOCDR system for sensing temperature and strain. The system was designed based on per fluorinated graded index—polymer optical fiber (PFGI-POF) and using heterodyne detection scheme. For 2 m of POF length, 34 cm spatial resolution was obtained with 11.698 MHz modulation frequency, and for 1.3 m POF length, 7.4 cm spatial resolution was obtained with 53.451 MHz modulation frequency. The system offered a 1.3 m of FUT length for temperature sensing and 2 m of FUT length for strain sensing. Since PFGI-POF fiber has 250 dB/km of attenuation loss, the system required 27 dBm of laser input power. The achieved temperature and strain resolution were ±3 °C and ±0.09%, respectively [3]. Wada et al. [17] proposed a system based on OFDR technique and using PMF-PANDA. The system has been offered 0.9 °C of temperature resolution and 7.5 με of strain resolution. Since the spatial resolution was 1 mm, the length of FUT is restricted to 10 m only [17].

In order to minimize the specified drawbacks of BOTDR-based DTS system, we have proposed two signal processing techniques, such as empirical mode decomposition (EMD) and lifting wavelet transform-optimum modified differential threshold (LWT-OMDT). The purposes behind the usage of proposed (EMD) [18] signal processing technique are: It decomposes any non-stationary and nonlinear signals. It gives fast response. It contains an adaptive nature. This technique operates at the scale of one oscillation; it means high locality is available. It offers multiresolution property [18]. The purposes behind the usage of proposed (LWT-OMDT) signal processing technique are: Lifting scheme [19] does not require any complex mathematical calculation. This technique is very easy to implement; it gives the quick response, and it does not depend on Fourier transform. The modified differential thresholding (MDT) function produces differentiable and continuous segmentation at all points of a signal. In order to get optimum solution, we have optimized the MDT function using the modified particle swarm optimization (MPSO) [19]. The MPSO algorithm majorly depends on a velocity parameter only. MPSO algorithm contains very few number of control parameters, better convergence, and simplest implementation for linear and nonlinear signals. Unlike PSO algorithm, it consists of adaptive nature and rejects local optimum, due to the weight factor variation for every iteration [19].

2 Theoretical Background

The considered BOTDR setup is illustrated in Fig. 1. In Fig. 1, the laser diode generates a series of optical pulses and these pulses are injected into an optical fiber. These pulses are passed through a 3-port circulator as shown in the figure below. This circulator's second port is connected with 50 km of FUT or sensing fiber. When the optical signal propagates through the FUT, the spontaneous Brillouin backscattered signal will be generated along an entire FUT due to an interaction between the incident optical pulses and the acoustic phonons in the sensing fiber. The frequency of this Brillouin backscattered signal is different from an incident optical signal's frequency. This frequency difference is called as BFS.

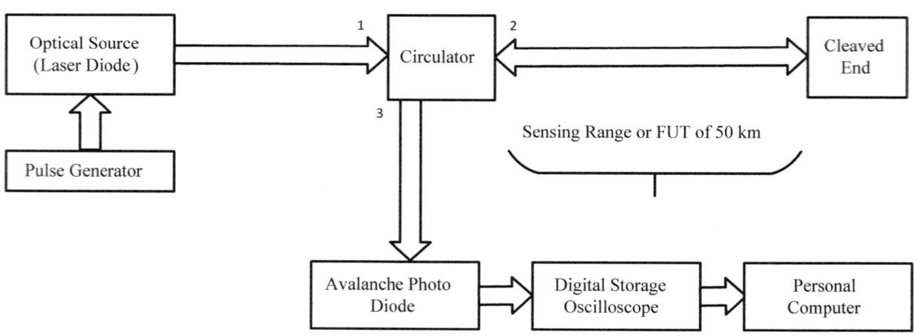

Fig. 1. BOTDR schematic setup

The Brillouin backscattered signal from the FUT will be transmitted toward the circulator, which then directs the backscattered signal toward the avalanche photodiode (APD) equipped with an amplifier. Since the APD device is used to convert an optical signal into an electrical signal with gain, this electrical signal is displayed in time domain with the help of a digital storage oscilloscope (DSO) and finally it will be processed using a personal computer. In Fig. 1, the FUT under consideration is subjected to temperature variation. The structure of FUT for temperature sensing is shown in Fig. 2.

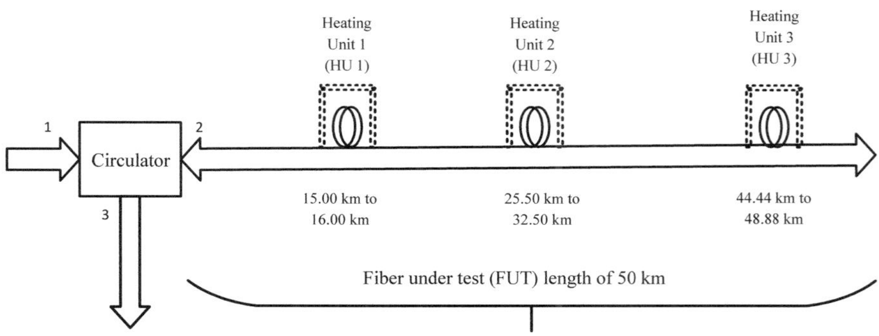

Fig. 2. Structure of silica optical fiber (SOF) under test for temperature

In Fig. 1, the FUT under consideration is subjected to temperature variations. The structure of FUT for temperature sensing is shown in Fig. 2. Brillouin frequency shift $\vartheta_B(T)$ is produced by the applied temperature on the sensing fiber or FUT. It can be written as [1]:

$$\vartheta_B(T) = \vartheta_B(T_r) + K_T^{\vartheta}(T - T_r) \tag{1}$$

K_T^{ϑ} parameter denotes temperature coefficient of BFS in a silica single mode optical fiber. T and T_r parameters denote non-room and room temperatures. For silica single mode optical fiber, the value of K_T^{ϑ} is 1.07 MHz/°C.

The mathematical background for the proposed (EMD) signal processing technique is available in [18]. The mathematical background for the proposed (LWT-OMDT) signal processing technique is available in our previously contributed work [19]. In the proposed (EMD) signal processing technique, the total number of levels, 0 = 4. It means the four number of empirical mode decompositions are required. Thus, the four number of an intrinsic mode functions (IMF) are required. In the proposed (LWT-OMDT) signal processing technique, the MPSO evolutionary algorithm is applied to each level of LWT scheme. Here, the total number of levels, 0 = 3. It means the three number of lifting wavelet decompositions are required.

3 Simulation Results and Discussions

In this paper, we have presented a BOTDR-based DTS system. The system response is simulated using MATLAB. These results are presented for the proposed (EMD and LWT-OMDT) and conventional (discrete wavelet transform (DWT)-soft threshold and DWT-hard threshold) technique-based systems.

The simulation parameters of these systems are: optical launch power of 10 mW, sensing fiber length 50 km, laser source pulse width of 400 ns, refractive index of silica fiber core as 1.48, and pumping wavelength as 1550 nm. We have considered three different temperatures or events generated by different heating units (HU) on the sensing fiber for BOTDR-based DTS system. The locations of these heating units are shown in Fig. 2. The temperatures from these HUs are 37.77 °C, 48.88 °C, and 66.66 °C, respectively, and the rest part of the sensing fiber is under room temperature.

In order to minimize the temperature error in BOTDR-based DTS system, we have implemented conventional (DWT-hard threshold and DWT-soft threshold) [20] and our proposed (EMD and LWT-OMDT) signal processing techniques.

By the conventional DWT-soft threshold technique, we have obtained an improvement in temperature extraction over other conventional DWT-hard threshold [20] technique as reported in Figs. 3 and 4. Though the system response has improved, scope is there for further improvement.

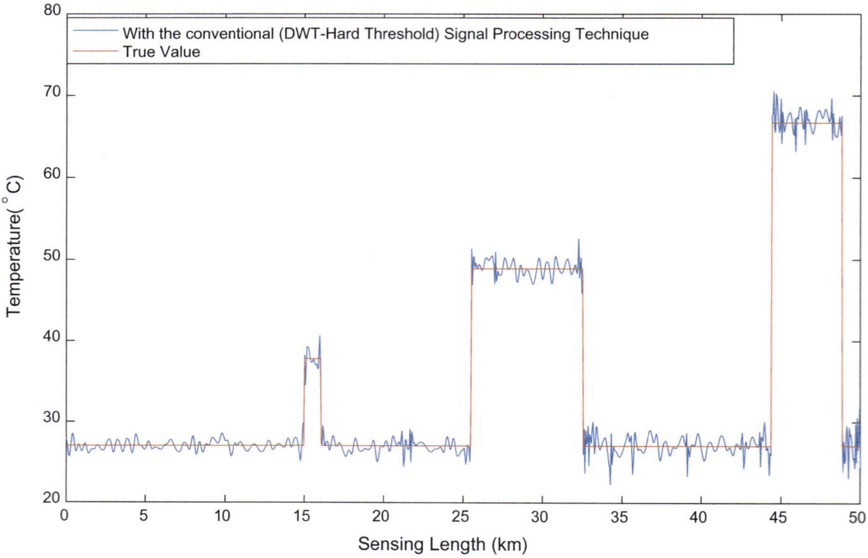

Fig. 3. Extraction of temperature from BOTDR-DWT-hard threshold-based DTS system

In our proposed EMD signal processing technique, IMF has played a major role to produce minimum error in extraction of temperature. From Fig. 5, it is clear that the proposed BOTDR-EMD-based DTS system offers higher degree of temperature resolution than the conventional BOTDR-DWT-hard threshold and BOTDR-DWT-soft threshold based DTS systems. The EMD-based technique resulted in ~1.14 °C of an average temperature resolution value for BOTDR-based DTS system over a 50 km of sensing range. However, the proposed LWT-OMDT signal processing technique produced minimum error in temperature extraction using an OMDT function and the observed an average temperature resolution was of ~0.80 °C as shown in Fig. 6. The OMDT function value was obtained using MPSO algorithm.

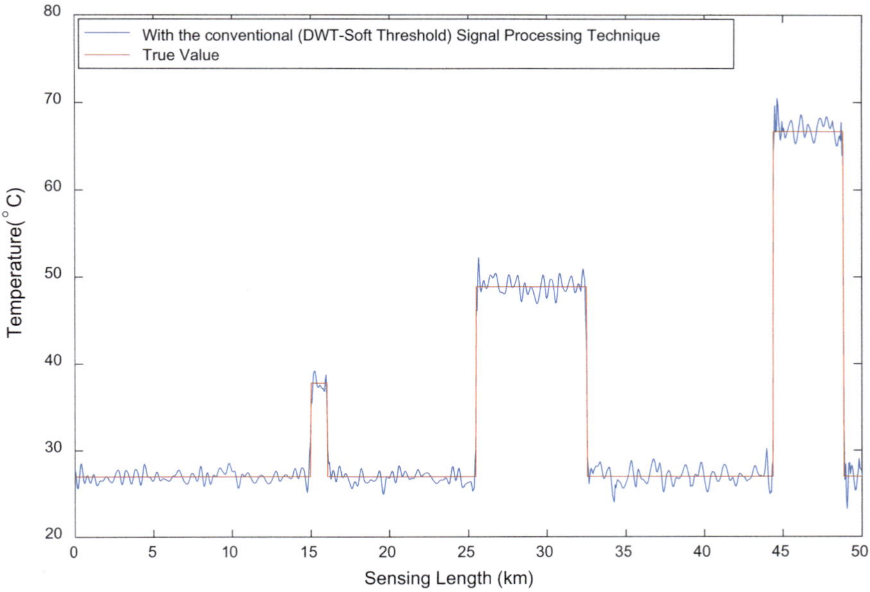

Fig. 4. Extraction of temperature from BOTDR-DWT-soft threshold-based DTS system

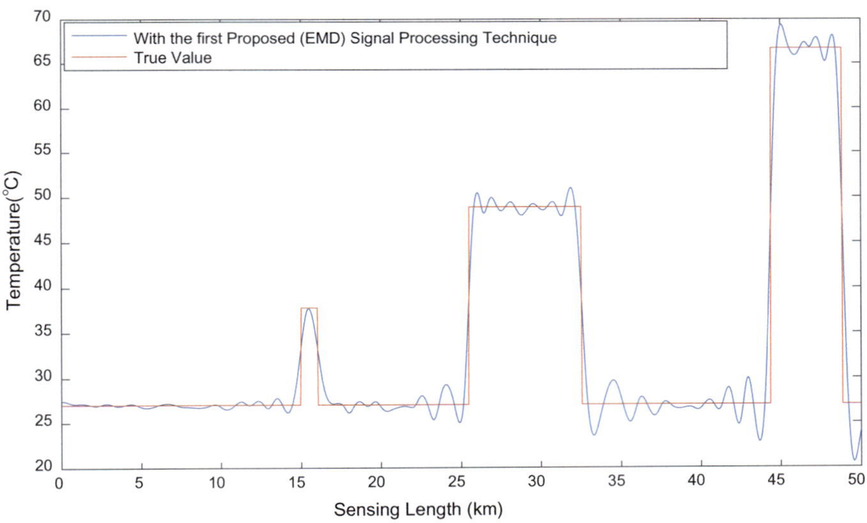

Fig. 5. Extraction of temperature from our proposed (BOTDR-EMD) DTS system

the hDLMS-IIR provides best performance compared to the other algorithms. The MSE and MSD simulation results for 2nd, 3rd, and 4th order IIR system are plotted in Figs. 4, 5, and 6, respectively. From the figure, it is evident that the 2-hop hDLMS algorithm performs well by providing less MSD and MSE performance with faster convergence compared to that of 1-hop DLMS algorithm in a sparse WSN. The block version of both 1-hop and 2-hop algorithms perform similar to that of non-block algorithms for 2nd order IIR system with less communication overhead, Whereas in the case of 3rd and 4th order system, 1-hop BDLMS algorithms' performance fluctuates, especially the MSD performance.

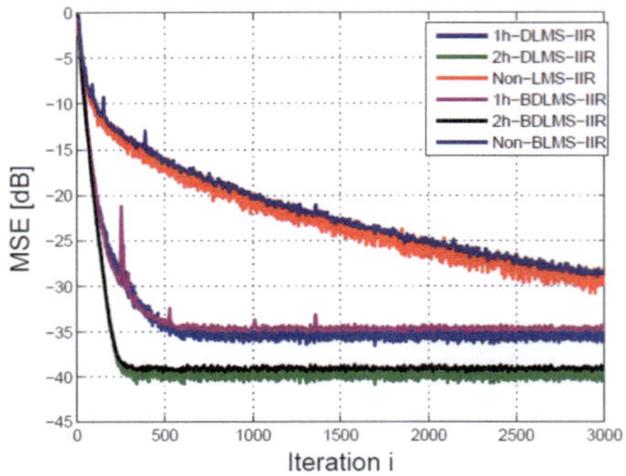

(a) MSE Performance

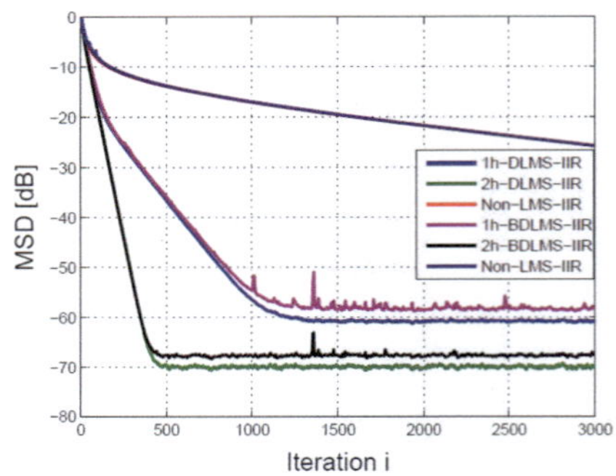

(b) MSD Performance

Fig. 4. Average MSE and MSD performances by 1-hop DLMS-IIR, 2-hop hDLMS-IIR, non-cooperative LMS-IIR, 1-hop BDLMS-IIR, 2-hop BhDLMS-IIR, and non-cooperative block LMS-IIR for 2nd order IIR system

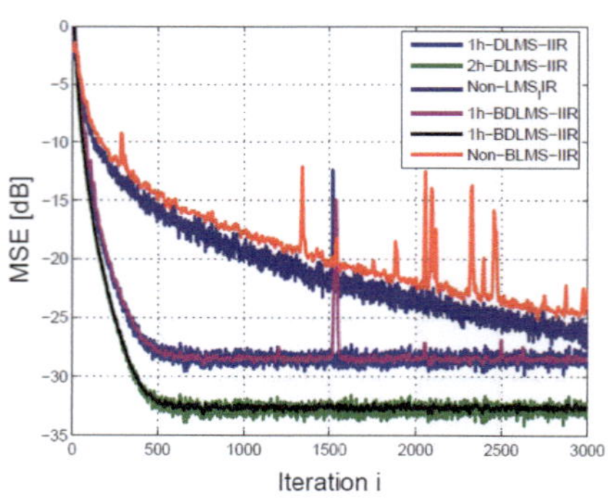

(a) MSE Performance

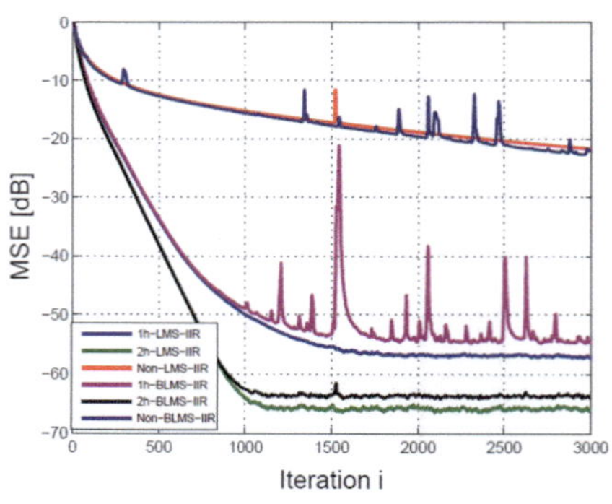

(b) MSD Performance

Fig. 5. Average MSE and MSD performances by 1-hop DLMS-IIR, 2-hop hDLMS-IIR, non-cooperative LMS-IIR, 1-hop BDLMS-IIR, 2-hop BhDLMS-IIR, and non-cooperative block LMS-IIR algorithms for 3rd order IIR system

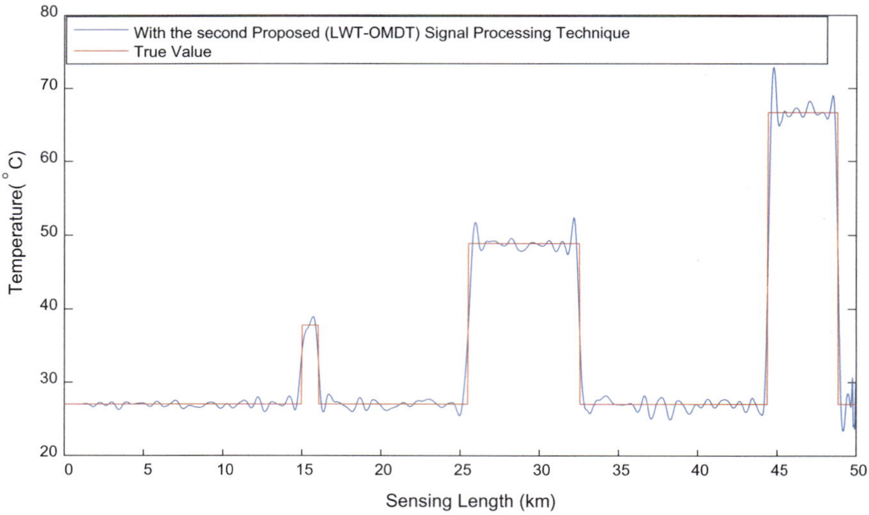

Fig. 6. Extraction of temperature from our proposed (BOTDR-LWT-OMDT) DTS system

4 Conclusion

In conclusion, the proposed DOFS system has many potential applications ranging from biomedical engineering to oil and gas pipeline monitoring, high power cable monitoring, fire detection, to defense and aviation industries. The proposed BOTDR-EMD-based DTS system offers an average temperature resolution of ~ 1.14 °C, whereas the BOTDR-LWT-OMDT-based DTS system offered a temperature resolution of ~ 0.80 °C.

Acknowledgements. This work was carried out in the FIST supported optical engineering laboratory of School of Electrical Sciences (SES), Indian Institute of Technology Bhubaneswar.

References

1. Lee B (2003) Review of the present status of optical fiber sensors. J. Opti Fiber Technol 9:57–79
2. Soto MA, Sahu PK, Bolognini G, Pasquale FD (2008) Brillouin-based distributed temperature sensor employing pulse coding. J Sens 8:225–226
3. Hayashi N, Mizuno Y, Nakamura K (2014) Distributed Brillouin sensing with centimeter-order spatial resolution in Polymer optical fibers. J Lightw Technol 32:399–4003
4. Lombera RR, Minardo A, Bernini R, Mirapeix J, Higuera JML, Zeni L (2017) Experimental demonstration of a Brillouin optical frequency-domain reflectometry (BOFDR) sensor. In: Proceedings of the SPIE International conference on optical fiber sensors (SPIE OFS '25), pp 1–4
5. Li Z, Yan L, Shao L, Pan W, Luo B, Liang J, He H (2016) Coherent BOTDA sensor with single- sideband modulated probe light. J Photon 8:13804–13810

6. Shlomi O, Preter E, Dexin B, London Y, Antman Y, Zadok A (2016) Double-pulse pair Brillouin optical correlation domain analysis. J Op Exp 24:26867–26876
7. Minard OA, Bernini R, Zeni L (2014) Distributed temperature sensing in polymer optical fiber by BOFDA. Photon Technol Lett 26:387–390
8. Song MP, Zhao B (2005) Accuracy enhancement in Brillouin scattering distributed temperature sensor based on Hilbert transform. J Opt Commun 250:252–257
9. Bernini R, Minardo A, Zeni L (2006) An accurate high-resolution technique for distributed sensing based on frequency-domain Brillouin scattering. Photon Technol Lett 18:280–282
10. Taki M, Soto MA, Pasquale FD, Bolognini G (2011) Long-range BOTDA sensing using optical pulse coding and single source bi-directional distributed Raman amplification. In: Proceedings of the IEEE Sens (IEEE Sensors '10). 6127160 (2011) 382–385
11. Pradhan HS, Sahu PK (2014) High-performance Brillouin distributed temperature sensor using Fourier wavelet regularized deconvolution algorithm. J Optoelectron 8:203–209
12. Bao X, Zou L, Yu Q, Chen L (2005) Development and applications of the distributed temperature and strain sensors based on Brillouin scattering. In: Proceedings of the IEEE Sens. (IEEE Sensors '04) 1210–1213
13. Soto MA, Bolognini G, Pasquale FD (2007) Enhanced simultaneous distributed strain and temperature fiber sensor employing spontaneous Brillouin scattering and optical pulse coding. Photon Technol Lett 21:450–452
14. Zou W, He Z, Hotate K (2010) Single-end-access correlation-domain distributed fiber-optic sensor based on stimulated Brillouin scattering. J Lightw Technol 28:2736–2742
15. Belal M, Newson TP (2010) Enhanced performance of a temperature-compensated sub meter spatial resolution distributed strain sensor. Photon Technol Lett 22:1705–1707
16. Bernini R, Minardo A, Zeni L (2012) Distributed sensing at centimeter-scale spatial resolution by BOFDA: measurements and signal processing. J Photon 4:48–56
17. Yamashita RK, Zou W, He Z, Hotate K (2012) Measurement range elongation based on temporal gating in Brillouin optical correlation domain distributed simultaneous sensing of strain and temperature. Photon Technol Lett 24:1006–1008
18. Maheshwari S, Kumar A (2014) Empirical mode decomposition: theory and applications. J Electron and Elect Eng 7:873–878
19. Tangudu R, Sahu PK (2018) Strain resolution enhancement in Rayleigh-OTDR based DSS system using LWT-MPSO scheme. J Optik 176:102–119
20. Huimin C, Ruimei Z, Yanli H (2012) Improved threshold denoising method based on wavelet transform. In: Proceedings of the Elsevier international conference on medical physics and biomedical engineering (ICMPBE), pp 1354–1359

A Brief Understanding of IOT Health Care Service Model Over Remotely Cloud Connected Environment

Subhasish Mohapatra[1] and Smita Parija[2](✉)

[1] Department of Computerscience Engineering, Adamas University, Kolkata, India
[2] Department of Eletronics, CV Raman College of Engineering, Odisha, India

1 Introduction

Human body depicts a temple so we must take care of it properly. It is the vital backbone for any developing nation like India. The most hopeful and awaited smart city project health care is the vital aspect so the Government can't ignore this cutting edge application for underprivileged person. There are patient go in a parallel way. It can provide tangible benefit to patient by accurate, reliable, and in a timely manner, consequently it imparts healthcare service provider in a mode of reliability many application-based technique are available for citizen health access parameter [1].

Here, the patient-specific information is passed by a sensor to gateway which is connected over cloud. In this model, we can take advantage of public cloud which can

Send patient data to cloud data store after accessing the internet. Here web is adhering to hoisting the health care application project where a patient can fill out the necessary credential. A doctor can verify patient data securely and give his valuable feedback over cloud. Here a citizen can collaborate access his claim from the company database and bank for transferring a bill payment. The propound model in this paper provide abstraction of citizen information and flexibility of cloud data processing Sensed by IOT mechanism [2, 3].

Section 2 gives IOT application for hospitals. Section 3 discuss the motivation of the work; Sect. 4 gives proposed System architecture, Sect. 5 explains the working model and Sect. 6 presents the conclusion and future scope,

2 IOT for Hospitals

Here citizens can swipe the card which is supplied by unique Id authority the feature of Unique Id for this cloud-based access is given in Fig. 2 Sect. 3 of this paper. Subsequently, an OTP is generated for the registered mobile no of the person credential is send over the public kiosk of the hospital and a person can check in his detail. The sensed data is transmitted over ZigBee network to intrahospital-based cloud. In the

M. N. Mohanty and S. Das (eds.), *Advances in Intelligent Computing and Communication*, Lecture Notes in Networks and Systems 109,
https://doi.org/10.1007/978-981-15-2774-6_6

proposed method, ZigBee network is used as it is a low power, low cost, and long battery life device [4–7].

For remote accessing, the sensed data can be viewed to web module. The benefit of cloud-based model enhances documentation and quick dissemination of valuable information to citizens, and secure service-centric data delivery as such claim settlement with the bank. It also helps research group to study the patient history data for further research in health sector without going pillar to post. For grasping the sweet of this effort, every citizen can go for this ID card and every hospital including CHC, PHC present at nook and corner of the country must supply with this sensed bar code reader and ZigBee network to hoist application over cloud so this ambitious model will be successful in future [8, 9].

3 Motivation

Now, IOT is transforming the future of every technology and has created a lot of envision in a wide coverage, it enables more smart services that provide huge amount of on-demand access to a shared pool of configurable resources, e.g., computer, cloud, device, software, etc. So in this paper, we illustrate barcode sensing in healthcare industry to hold massive amount of particular data at any place around us in a user-friendly manner.

4 Proposed System Architecture

The underlying backbone of this model is cloud connection and sensing of data by barcode reader and sending it over public kiosk cloud is depicted as below Fig. 1. The most promising application of this model is it can provide data level security at entry point and passing through firewall for hoisting the data over cloud sub-module data store [10, 11]. Before to this validation of user credential is done by OTP (one time password)-based mechanism is projected ass below Fig. 1.

5 Working Model

Figure 1 elaborates the cascading flow of this model how a citizen avail services through Unique id with barcode, which will serve as an elementary IOT-based facilitation of quick service delivery over cloud [12]. Electronic services like checking of claim status, banking transaction for payment of the bill, Research group can check the medical history database, and viewing of an item to transact over cloud to citizen. Here, the Fig. 1 diagram depicts Citizen2Health interface2cloud is illustrated as below:

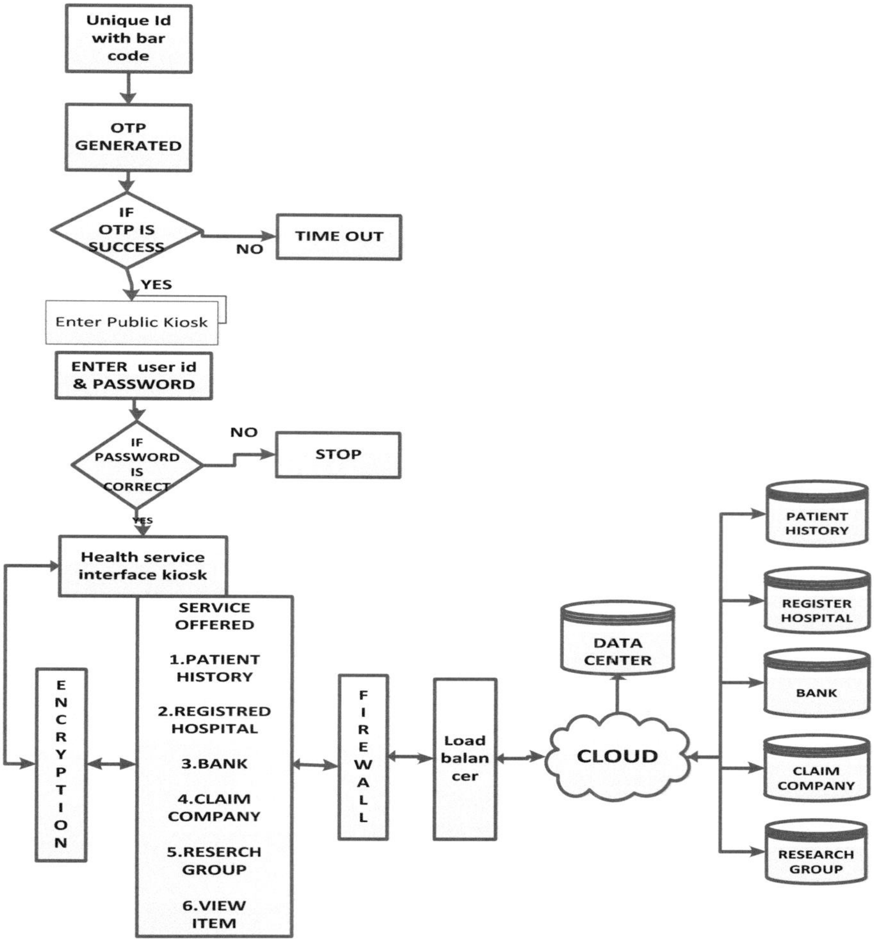

Fig. 1. Barcode sensing healthcare service application model for cloud-connected infrastructure

Service side of Citizen

Step-1-citizen trigger out the transaction by using an ID with barcode supplied by certain unique identification authority.

Step1.1-citizen provide a unique set of attributes to Health interface for a sign of verification. This has been discussed below.

Step1.2-Here citizen credential is checked by health interface and further channelization of the service to C-Health interface.

Service side of Public kiosk

Step 2-Public kiosk verifies the user input for authentication which is the Current environment of the hospital sector connected to cloud over a gateway.

Scenario1-For the successful validation the transaction proceeds further to step3.

Scenario2-In case of unsuccessful validation the transaction is aborted. The system timeout or negations of validation success, i.e., negative acknowledgement give a status of failure transaction to citizen.

Step3-Transaction of further service request is processed for health interface that deliberate the vita aspect of this paper here we have channelized the connected service for the ease of citizen service. In this step, Citizen2Health interface2cloud service delivery notifies the citizen and doctor about this model.

Figure 2 project the Unique ID card issued by unique identification authority to citizen for accessing medical service and other vital aspects from the hospital sector.

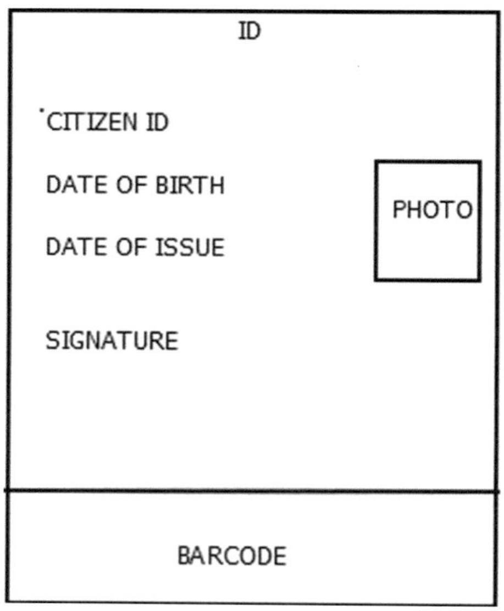

Fig. 2. Pictorial representation of Unique id with barcode sensing feature

- Citizen ID-combination of character and integer special character type
- Name of citizen
- Date of birth of a citizen
- Date of issue
- Photo and signature

This thing will be shown as soon as citizen touch this Id card in a barcode reader machine that will be connected to cloud over gateway as soon as the credential is checked it redirected to cloud by a gateway and OTP is generated for the request for a secure transaction purpose.

And a person can login to health interface service kiosk.

After the citizen chooses the menu for service he desired for and views the transaction over the cloud.

For security here we inculcate two-way data security paradigm as patient data grab high attention in black market so input side citizen data-centric encryption and service filtration on output side is the beauty of this model which enhances further data security.

One submodule research group is added over for further research upon electronic health record data that is stored in submodule patient history module.

The proposed system also has taken the optimistic approach of load balancing to further protect the cloud data store and segregate the load among the submodule of cloud as per user request selected in View item section in service offered menu in Fig. 2.

6 Conclusion and Future Scope

In this paper, we propose an effective cloud-based model for better solutions over the tedious traditional methods. The advantage of cloud can better serve especially for hospital where many patients lacking the claim settlement over bank. Accessibility of patient data over any hospital in the country will demarcate as a tremendous revolution. Furthermore, every citizen must take advantage of the process of digitization incurred automatic remote access of data and service. This model is adding future implementation scope of healthcare industry by transferring and storing messages over cloud. It captures user authentication by barcode sensor to process and capture user authentication. This model implements a device to device (D2D) communication and device to cloud (D2C) communication. The D2D exhibits communication by bluetooth or ZigBee for establishing direct communication. Here, in this model, in Fig. 1, we gather knowledge of implementation for D2C. Here IoT devices share with internet cloud directly like an application service provider to exchange data and control message over network. This model takes the advantage of Wi-Fi network for establishing a connection between device and IP network that will absolutely connect to cloud service. We as an author must not forget to check vulnerabilities of all cloud interface and API for authentication and security. IOT to cloud communication is going through this paradigm as given in Fig. 1.

Device to gateway and then to cloud infrastructure. A device is a thing that interacts with the real-life world directly in this paper we introduce ID card of a patient that can be read by barcode reader and data is sensibly passed to gateway for propagating it to cloud infrastructure. Where the device information of patient should be captured in this format.{Citizen ID: “dhsl888ja”, “Date of birth”: “Date of issue”, “Signature”: “Photo”, Barcode}. This IoT-based model uses the hardware interface to exchange data serially over the network we can general purpose input–output serial interface for building this mechanism. IoT-based barcode scanner can be used by medical for dissemination of patient information to cloud.

References

1. Tochukwu C, Abraham I, Somtoochukwu I (2015) Cloud based health care information system for rural clinics in Nigeria, Dec 2015
2. Abayomi-Alli AA, Ikuomola AJ, Robert IS, Abayomi-Alli OO (2014) An enterprise cloud-based electronic health records system
3. Boyinbode O, Toriola G (2015) CloudeMR: a cloud based electronic medical record system. Abouelmehdi K, Beni-Hessane A, Khaloufi H (2018) Big healthcare data: preserving security and privacy. J Big Data 5(1). https://doi.org/10.1186/s40537-017-0110-7
4. Chiyan Z (2014) Cloud-based hospital information system as a service for grassroots healthcare institutions. Singh M, Srivastava VM (2018) Multiple regression based cloud adoption factors for online firms. In: 2018 International Conference on Advances in Computing and Communication Engineering (ICACCE). https://doi.org/10.1109/icacce.2018.8457722
5. Hitachi (2012) How to improve healthcare with cloud computing, Hitachi Data Systems, White Paper, May, 2012, 1–20
6. Kavitha R, Kannan E, Kotteswaran S (2016) Implementation of cloud based Electronic Health Record (EHR) for Indian Healthcare Needs, Jan 2016
7. Hu NZ, Lee CY, Hou MC et al (2013) A cloud system for mobile medical services of traditional Chinese Medicine. J Med Syst 37(6):1–13
8. Karthikeyan N, Sukanesh R (2012) Cloud based emergency health care information service in India. J Med Syst 36(6):4031–4036. https://doi.org/10.1007/s10916-012-9875-6
9. Xi AB, Qin BQ, Huang CL (2013) The regional medical business process optimization based on cloud computing medical resources sharing environment. Cybern Inf Technol 13 (Special Issue):18–29
10. Muthaiyan M, Goel N, Prakash DS (2012) Virtual E-Medic: a cloud based medical aid. In: Proceedings of World Academy of Science, Engineering and Technology (WASET), issue 71, p 1344
11. Min C, Tao L, Le Z (2015) Explore the development of medical and health tourism industry in Wuzhishan City Hainan Province. Consume Guide 1:45–47
12. Zhang J, Gu Z, Zheng C (2010) The progress on cloud computing research

Fusion-Based Multi-biometrics Authentication System with Intrusion Detection and Prevention Anamoly

Mehreen Kirmani and Dinesh Kumar Garg(✉)

Department of Computer Science and Engineering, SSCET, Pathankot, India

1 Introduction

The "Biometrics" combines the two Greek words "bio" that represents "the life" and the word "metrics" that denotes "measuring something". The biometric devices are easily available from the last few decades on a very large scale that is mainly because of the tremendous elevations in the field of computer science and technology. Biometrics recognizes every person distinctly, depending upon one or more than one inborn physical trait or the behavioural traits. Therefore, the biometric system is used to make sure that the security and integrity of the personal or the business assets is well-protected. The output of the verification step in biometric systems produces a matching a score that is equated and then compared with a decision step. When the result of the decision step is true, the user is considered as a genuine and a legitimate user; otherwise, the user is treated as illegitimate and hence, rejected as an attacker [1]. Adding the extra preventive steps helps to reduce the chances of the spoofing attacks, and our system gets secured and it becomes less susceptible to attacks [2]. Every biometric trait has got its pre-defined merits and demerits. Face is treated as the common form of biometrics, due to its ease of use, and the user-friendliness [3]. On the other hand, Iris identification is the most dependable and the precise form of the biometrics, available nowadays [4]. Biometric determiners, therefore, is the easy way to secure those system, which needs to be protected. The entire biometric community is going through the difficult tasks of developing an efficient, and secure way, against such threats [5]. Over a unimodal-biometric system, a multi-biometric system has an advantage of making authentication systems trustworthy and difficult to attack, also offering the advantage of user-choice and last not the least, enhancing the authentication-performance. Although when it comes to implementing the design of multi-biometric systems, it can give rise to some of the issues [6]. The security of the network is the challenging and one of the most obvious applications of the biometrics. The most concerning element of the network security is the authentication of the users. The authenticity of the user is fused with the identification, and these two together, comprises of the whole purpose of biometrics technology. The network which requires the users to identify and authenticate themselves as legitimate users, therefore, is the main prospect for biometrics [7]. In the current society, demand for such a network is very high, and also, the use of such a network, is growing at an exponential rate [8].

M. N. Mohanty and S. Das (eds.), *Advances in Intelligent Computing and Communication*, Lecture Notes in Networks and Systems 109,
https://doi.org/10.1007/978-981-15-2774-6_7

1.1 Architecture of Biometrics System

The below block diagram illustrates the architecture of biometrics. It starts from extracting the images from the sensors, moving towards extracting the features from the specific biometric traits and moving towards matching and verifying the biometrics, that is stored in the form of images, and therefore, deciding whether the user is a legitimate or an illegitimate one [9]. The working of the whole biometrics system is divided into two phases.

First is enrolling the individual's phase, in which the system captures and processes the biometric traits (one or more than one trait) and stores the unique values for every individual in the database, on the basis of which access is granted to the individuals.

The block diagram shows the enrolling process (Fig. 1).

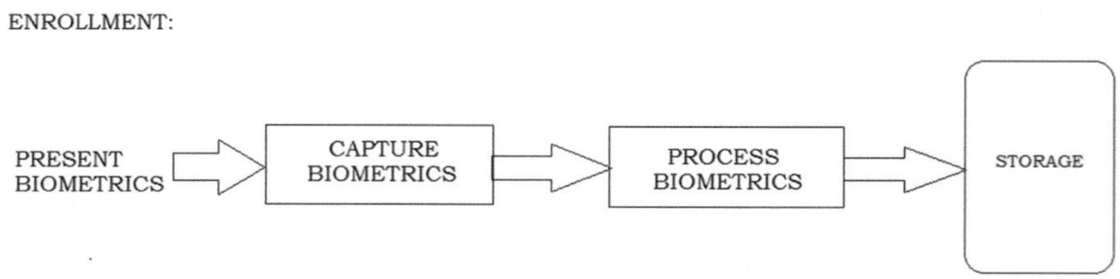

Fig. 1. Enrolling the biometric traits

Second phase, which is the verification phase, is a phase during which the security system collects and processes the biometrics to match with the biometric templates which are stored during the enrolment phase to grant or deny access to the system [8]. The given block diagram depicts the working of the verification phase (Fig. 2).

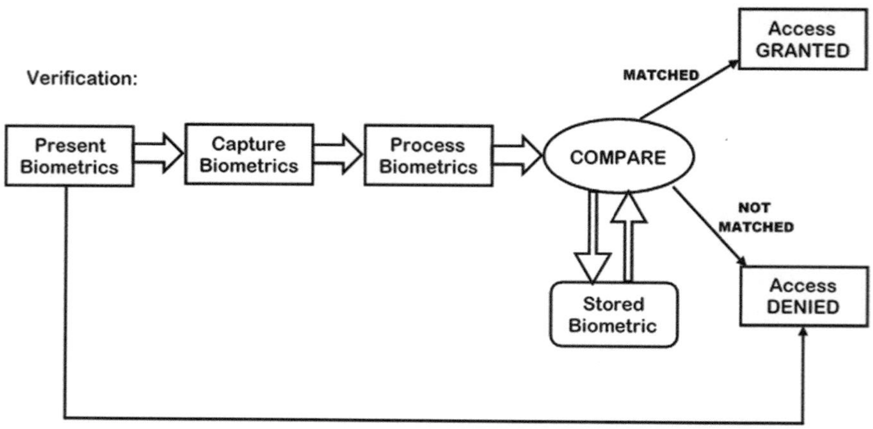

Fig. 2. Verifying biometric traits

1.2 Advantages of Biometric System Over Traditional Security Systems

Some the advantages due to which the biometric recognition systems are considered over traditional security systems are as follows:

1. Security is high:

The traditional security viz. passwords comprise of the numbers, the alphabets and the symbols, which are becoming easy to hack every day. There are Zillions of hacking episodes that are taking place every year and because of this reason, people are losing a lot of money. However, with the advent of the latest Biometric technologies, different solutions are put forward, which are almost impossible to hack unlike passwords.

2. Accuracy is better:

The most usual traditional security systems such as the passwords or personal identification numbers (PINs), when hacked costs some amount of money and time, as well as other resources and usually aren't accurate. However, the biometric security systems work by considering the individuals' biometric traits, may that be physical or behavioural, such as fingerprints, palm pattern, vein pattern, retina, etc. which always behaves accurately at any time.

3. Accountability is required:

In other verification methods, apart from the biometric traits, anyone can make use of the password or security PIN number to get the personal information of an individual, which is considered highly risky. However, if we take the case of the biometric security system, it always requires the interactions of particular individuals to login or passes as the legitimate user in the security system that is employed, which will allow 100% accountability for any sort of the activities.

4. Convenient or User-friendly:

It requires knowledge and is somewhat tiring to remember or pen-down the passwords. People also are more prone to forget the passwords at some point in future due to the uneasy situations in one's life. In such cases, the biometrics, therefore, stands out to be the most convenient solutions.

5. High Scalability:

The biometric security solutions are the scalable solutions for any kind of projects and it can be used in almost every organization that is the banking, workforce-management systems, or semi-government and government projects.

2 Problem Formulation and Methodology

2.1 Problem Formulation

The Spoofing attack is usually defined as an illegal-way by which a person(s) tries to masquerade as an individual and, therefore, makes an attempt to get the access of a system.

2.2 Existing Methodology

The exiting methodology presented an idea of fusing the biometric traits (iris and FKP) together. A matching score is obtained by fusion, which is used to decide the authenticity of an individual(s). The focus was laid on the detecting intrusions, while the prevention of the intrusions was neglected. Many of the biometric systems undergo the spoofing attacks that are mostly because they ignore verifying all the parameters. The existing method monitors for the intrusion detections by getting through the log-analysis of the system. The extra user-interface and steps enhance the security. The Database is stored with existing collected featured images of iris and FKP of the user(s). After the first matching-step, the framework performs the image acquisitions and afterward, a matching-procedure is performed. In this research work, the biometric traits which were studied in detail were iris and Fingerprint.

3 Proposed Work

The main aim of the proposed work is to use image-processing techniques for biometric acquisition and identification process, for as many of the biometric determiners like the knuckle-print, iris, and the face that are represented by the images. Depending upon the data for the pre-processing, the algorithms have been used. To enhance the working of the biometric system using the integrated or multi-modal approach, techniques are used and are presented in this work. This research work is based on the concept that each individual has to pass the various biometrics checkpoints in order to get access to the system. Multi-modal Biometrics recognition systems include making use of more than one biometric trait to login into the system, which in turn makes the environment more reliable, and therefore, less prone to the spoofing attacks. Using the combination of the face, iris and the finger-knuckle print, which are normally registered at the time of the registration and then verified at the time of the login, followed by generating unique random OTP's to the email of individual guarantees the security to the greater extent.

Proposed Algorithm:

Step 1. Multi-Biometric trait acquisition.

Step 2. Extracting the features from the acquired images.

Step 3. Matching the biometric images.

Step 4. Generate the random OTP's to the Email of the authorized and registered user for logging in the system.

Step 5. Verification of the generated unique OTP's, for detecting and preventing Spoofing attacks.

Step 6. Encryption/decryption of the data files done using the AES algorithm.

Step 7. Transfer of the encrypted data files via an intermediary node, making use of Dijkstra's shortest path algorithm.

Step 8. Preventing and detecting the intrusions performed by intermediary node.

3.1 Unimodal Biometric System Versus Multi-modal Biometrics System

The mentioned table compares and gives a contrast between the unimodal security and the multi-modal biometrics-security system (Table 1).

Table 1. Unimodal-biometric versus multi-modal biometric authentication system

S. no	Unimodal biometric system	Multi-modal biometric system
1.	Unimodal biometric systems make use of only a single biometric trait, that may be a physiological trait, such as the iris, fingerprint, face, etc. or a behavioural trait such as typing rhythm of an individual, signature, etc.	Multi-modal biometric systems integrate or make fusion of more than one biometric trait of an individual that can be a combination of both the types of biometric traits
2.	A single biometric trait can be stolen as nowadays individual's information such as face images are easily available on social networking sites	It is tough to steal the multiple biometric traits of an individual and also impossible to steal the internal organs of an individual, such as retina, iris, etc.
3.	Security is less than expected by the organization	It offers more security compared to the Unimodal biometric system
4.	Cannot be trusted	Can be trusted as compared to the former authentication systems
5.	Because only one biometric trait is included, it is less complex and costly	It is costly, but it can be a better option for organizations that needs high security and integrity

4 Results

Security, nowadays, is on the top of the list of every individual or an organization that has some data to protect. This research is mainly themed on resolving the security concerns related to individual(s) or organization. Every biometric trait of the user is captured, matched, verified and only after then, the system generates a unique random OTP's for allowing access to the system (Fig. 3).

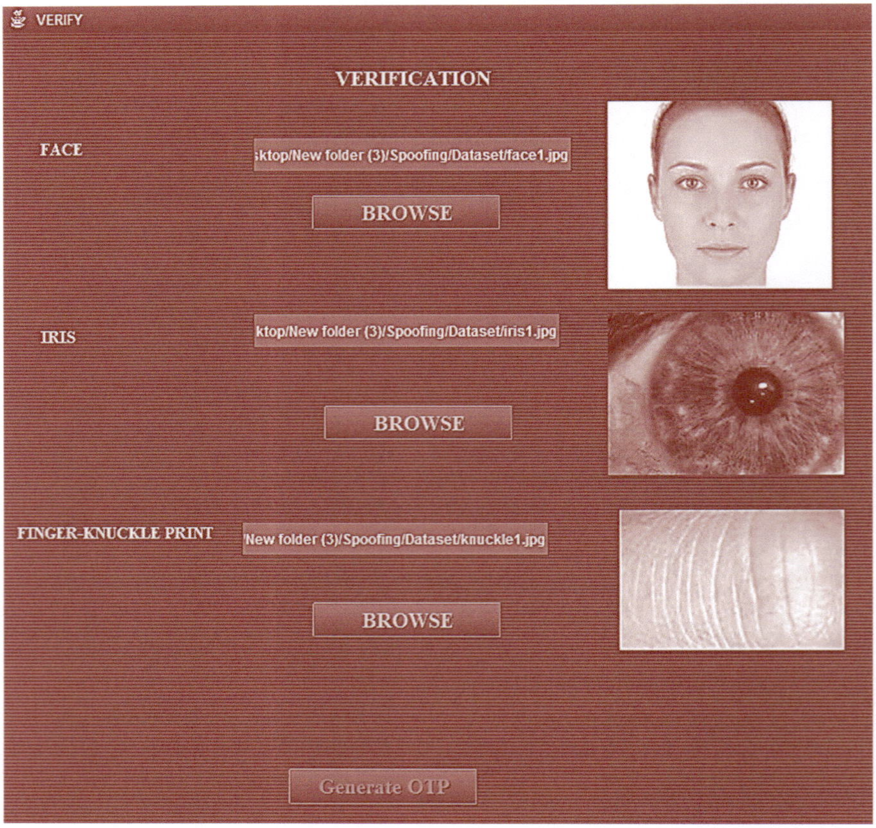

Fig. 3. Verification of multiple biometric traits

Once the verification of OTP is done, the individual gets an access to the system and only then, the users will be able to communicate in an encrypted form, and decryption is done by applying the unique hash values at the destination side, in order to detect and prevent the attacks on the system (Figs. 4, 5 and 6).

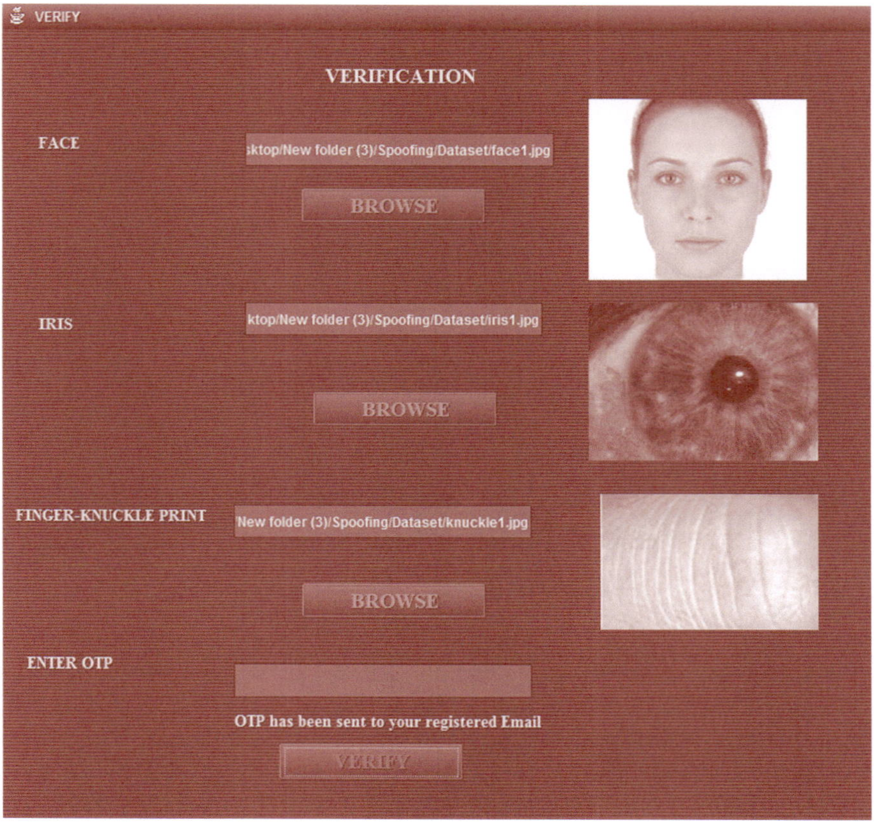

Fig. 4. Generating and verifying the random unique OTP

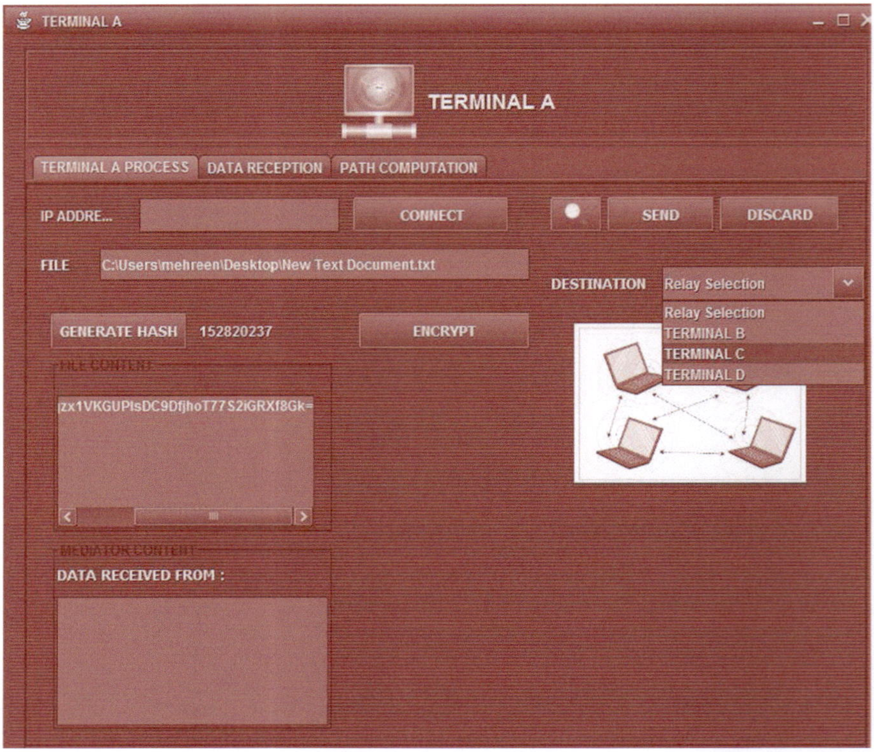

Fig. 5. Encypting and transferring the data at terminals

The data travels from the source node to the destination through an intermediary node, which is usually the shortest path to the destination. The path computation can be done at any stage of the transfer to get the information about who will be acting as an intermediary node (Fig. 7).

If the intermediary node performs an attack on the system, the data packet that was sent is dropped and, an acknowledgment is sent to both the sender and receiver, about data being compromised.

If there is no attack detection, then the corresponding data is successfully transferred to the destination, which at the terminal side applies the hash and decrypts the corresponding received data (Fig. 8).

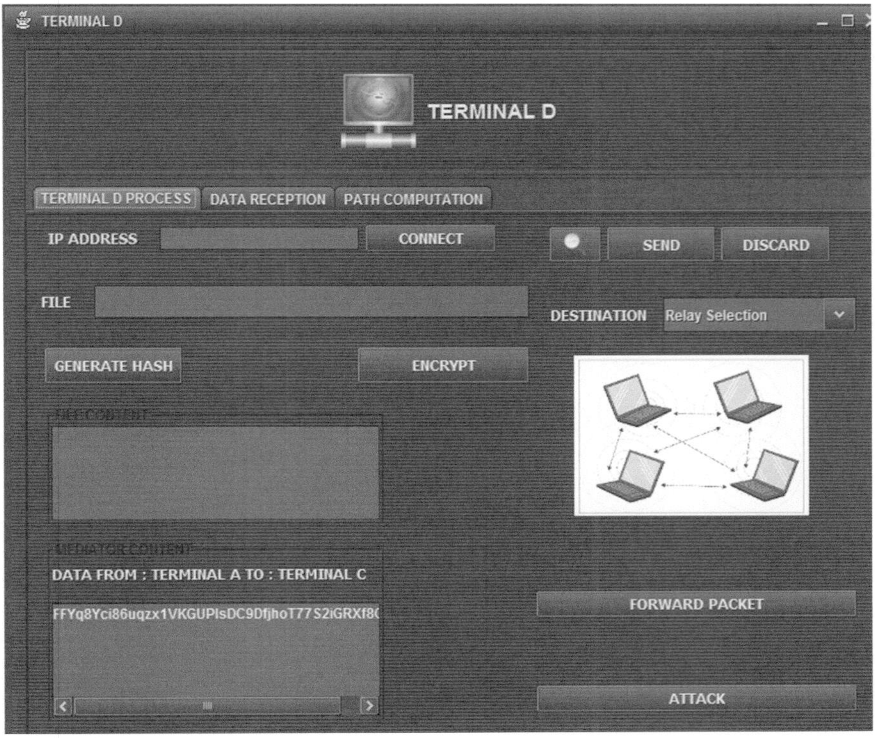

Fig. 6. Intermediary node

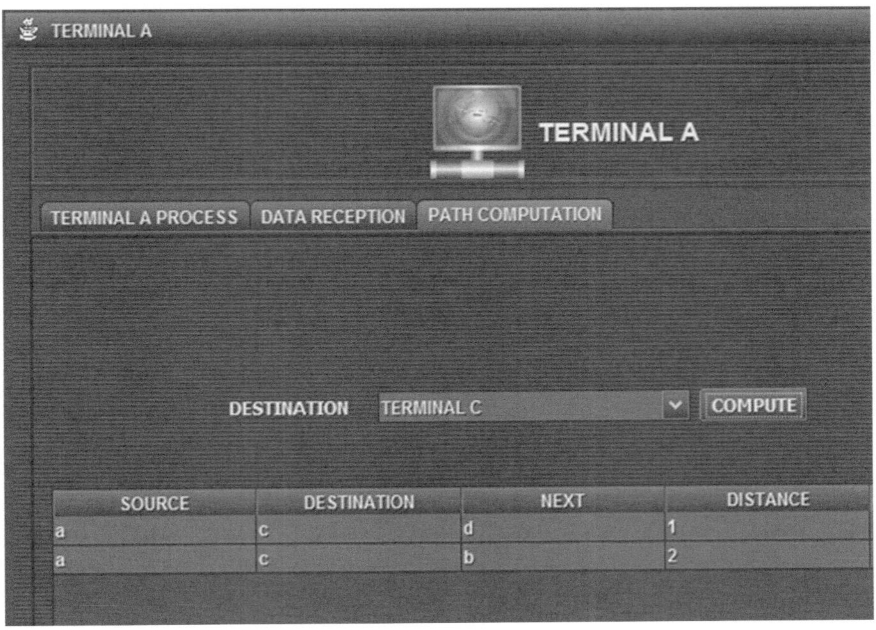

Fig. 7. Shortest path computation at the terminal end

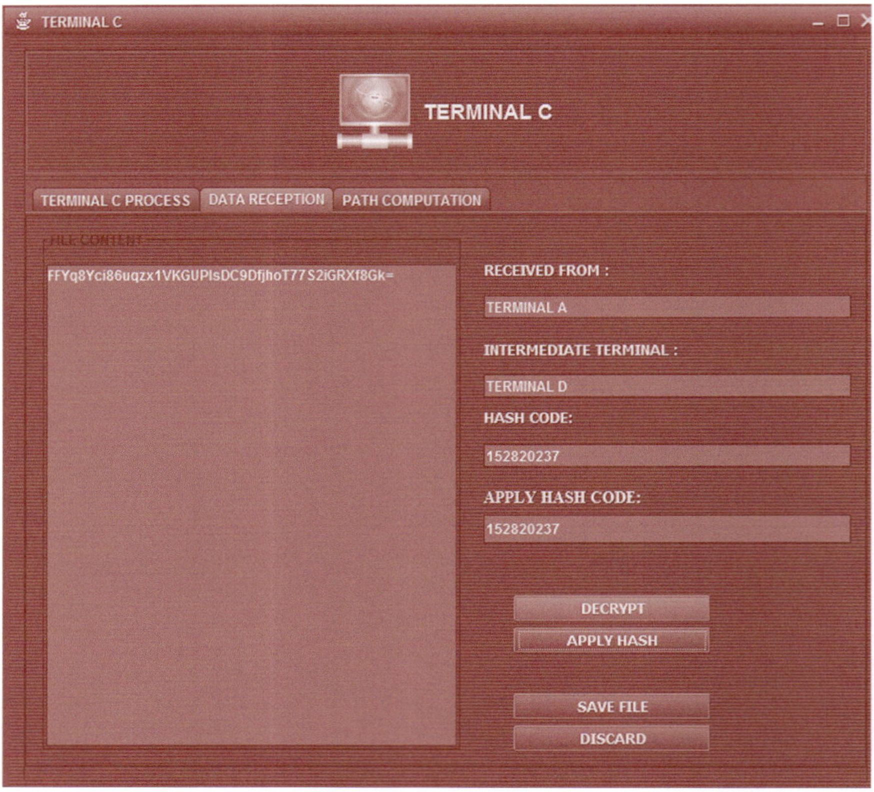

Fig. 8. Destination node

5 Future Scope and Conclusion

This research paper helps to provide an approach that works on giving more importance to security of the system rather than extracting the information from the resources. Extra user interaction, apart from biometric acquisition expects more accurate intrusion detection and prevention. For the detection and prevention of the spoofing attacks, simple pattern linking as well as matching mechanisms are taken into account, therefore, making a system more reliable, trustworthy, and secured.

References

1. Wild P, Radu P, Chen L, Ferryman J (2015) Robust multimodal face and fingerprint fusion in the presence of spoofing attacks. Pattern Recogn 50:17–25
2. Kaustubh D, Vishnu RD, Raut VM, Thakare (2016) Effective methodologies for preventing face spoofing attacks. IJARSE 6(06), June 2016

3. Bagga M, Singh M (2016) Spoofing detection in face recognition: a review. In: 3rd international conference on computing for sustainable global development (INDIACom), pp 2037–2042
4. Kak N, Gupta R, Mahajan S (2010) Iris recognition system. (IJACSA) Int J Adv Comput Sci Appl 1(1)
5. Galbally J, Marcel S, Fierrez J (2014) Biometric antispoofing methods: a survey in face recognition. IEEE Access 2:1530–1552
6. Kiltter J (2018) Multibiometric for identity authentication: issues, benefits and challenges. In: 2nd IEEE international conference
7. Radhika P, Ramya G, Sadhana K, Salini R. Defending man in the middle attacks. Int Res J Eng Technol (IRJET) 4(3). e-ISSN: 2395 -0056, Mar-2017, p-ISSN: 2395-0072
8. Gurumurthy S (2015) Design and implementation of biometrics in networks. JTASR 1(3) eISSN-2454-1788, pISSN-2395-5600. https://doi.org/10.14260//2015/27july-sept 2015
9. Akbulut Y, Sengur A, Budak U, Ekici S (2017) Deep learning based face liveness detection in videos. In: Conference on international artificial intelligence and data processing symposium, September 2017

Design of Dual Notch Band Antenna with Rejected Filter with Defected Slot for WLAN and Satellite Applications

Sanjay Kumar Sahu[1](✉), Bhukya Arun Kumar[1], and Kaibalya Kumar Sethi[2]

[1] School of Electronics and Electrical Engineering, Lovely Professional University, Phagwara, India
sanjay.23393@lpu.co.in
[2] Department of Electronics and Communication Engineering, GITA, Bhubaneswar, India

1 Introduction

Filters at microwaves are extensively used in wireless communication as it adheres to certain properties such as lightweight, low design cost, flexibility, compatibility with integrated circuits and more importantly ease of fabrication. Day by day large numbers of wireless networking techniques are emerging out and those need to be accommodated within the limited spectrum. This stringent demand for modern communication systems can only be filled if it can replace conventional methods by certain advanced and versatile techniques with less investment. For example, the bands like 5.15–5.35 GHz, as well as 5.725–5.775 GHz are unlicensed bands which are meant for indoor use only. But to use it for outdoor purpose, it is required to facilitate the creation of new innovative ideas which may come under wider wireless communication networks [1]. In this context, we may highlight two powerful methods in terms of defected microstrip structure (DMS) and defected ground structure (DGS). Both techniques assume a similar approach of creating some kind of defect either in the signal line or on the ground plane to get something new to the conventional design [2–6]. The beauty in this technique is that it shows resonance properties due to abrupt change in the current path. Though both bring some kind of defect, DMS provides better results than DGS. The advantageous part in DMS [1–6] over DGS is that in the former case the amount of cross talk and ground noise is significantly reduced as per the filter design is concerned. In this design approach, DMS is implemented which refers to band rejection process which is extremely beneficial to reduce noise in local oscillator, satellite, and mobile phone. Due to the presence of defects along signal path, electric, and magnetic field [7–13] are redistributed and this leads to bringing a change in capacitance and inductance. As a consequence that effective permittivity of the substrate is likely to change.

M. N. Mohanty and S. Das (eds.), *Advances in Intelligent Computing and Communication*, Lecture Notes in Networks and Systems 109,
https://doi.org/10.1007/978-981-15-2774-6_8

2 Geometry of the Proposed Structure

The bandstop filter is constructed on RT-Duroid substrate having 0.002 as loss tangent value, depicted in Fig. 1. There are two U slot of different length are removed which is shown clearly in Fig. 2. The dimension is available in table number 1. The resonance frequency can be varied by changing the length (L) and thickness (LX in this case) of U slot. Figure 3 is the equivalent diagram for U-slot in terms of R-L-C. The values of R, L and C can be calculated from the equation given below.

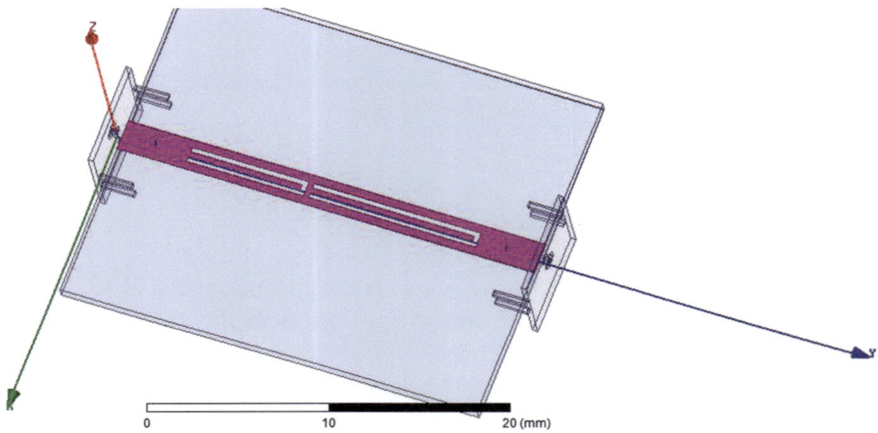

Fig. 1. Structure of proposed filter with defected mechanism

$$C = \frac{f_c}{200\pi}\left(f_0^2 - f_c^2\right) \tag{1}$$

$$L = \frac{1}{\left(4C\pi^2 f_0^2\right)} \tag{2}$$

$$R = \frac{2Z_0}{\sqrt{\left(2\&1/|S11|^2 - [2Z_0(wc - 1/wL)]^2 - 1\right)}} \tag{3}$$

Resistance is normally a value which is estimated from the conductor and dielectric losses in the defect.

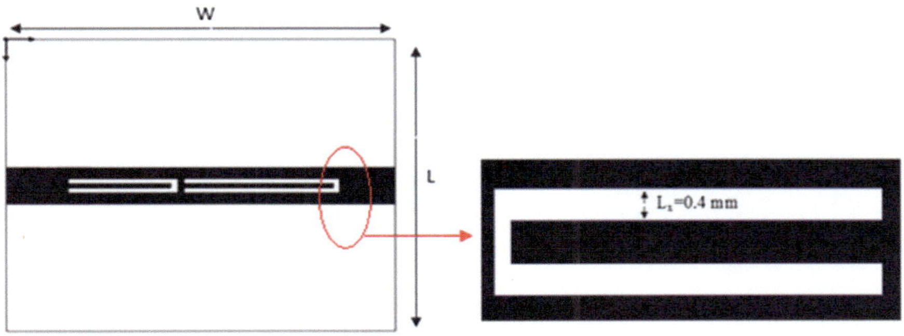

Fig. 2. Sectional view of U-slot

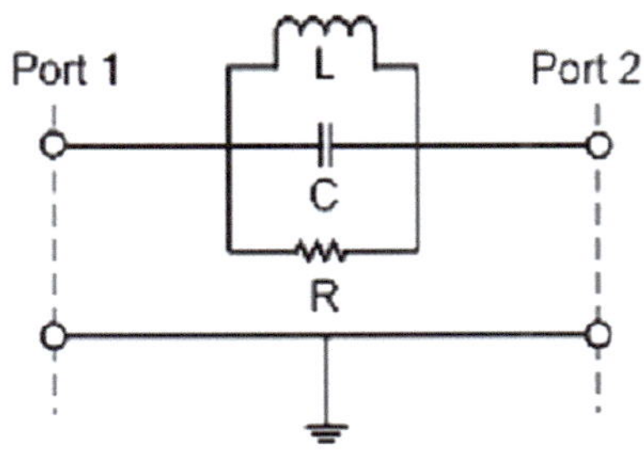

Fig. 3. Equivalent diagram of the defect in terms of R-L-C

Parameters	Quantity	Dimensions (mm)
W	Width of the filter	25
L	Length of the filter	25
h	Substrate height	0.787
Wfeed	Width of signal line	3.2
LX	Width of the U slot	0.4

3 Result Analysis

The proposed structure is simulated using HFSS-13.0 to investigate the frequency response characteristics. The results obtained are presented in terms of S21, Q-factor and VSWR plot [14–16] about two resonant frequencies (5.8 GHz and 8.2 GHz). As per the S21 plot is concern, it is found to be more than −20 dB shown in Fig. 4. For same frequency, we have also calculated Q-factor by using the mathematical formula as = $2\Delta^{0}$

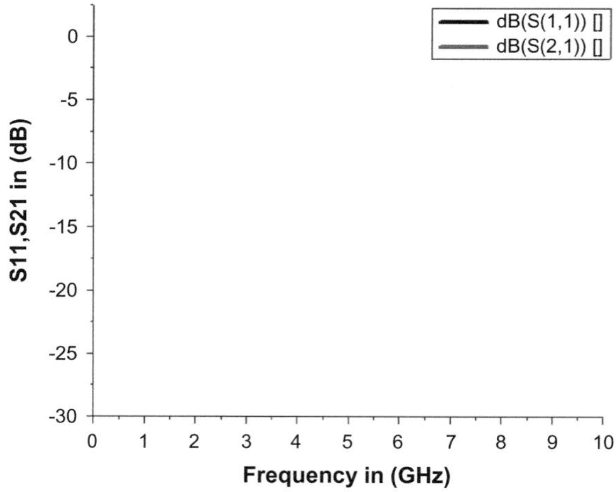

Fig. 4. Frequency response plot (S11 and S21) of the filter

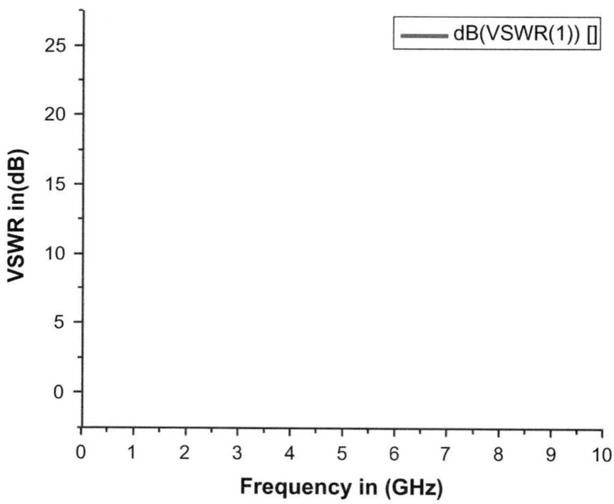

Fig. 5. Required VSWR plot in dB

Where ${}_0$ represents resonance frequency, whereas Δ defines the 3 dB bandwidth for the corresponding frequencies. The value for Q is evaluated which is approximated to 15(14.87). The VSWR result depicted in Fig. 5 explains how good the filter for the above said frequencies are. The graph of VSWR ($\cong$ 25 dB) is a clear indication of representing a bandstop filter.

From the result we obtained, it is observed that DMS technique shows very narrow bandwidth characteristics for a band-reject response of approximately ≥ -20 dB. The presence of U-slot fetches better performance in terms of selectivity with a smaller size. This is one of the important aspects to remove the interference signal for a particular or specific frequency application.

4 Conclusion

A compact dual notch bandstop filter is reported just by employing defect along signal line. The defect being the key factor to realize the above-said filter. An equivalent L-C circuit model is also included in this design. The frequency response plots (S11 and S21) are examined through HFSS-13.0 software. The results are quite good for upper WLAN and X-band satellite communication applications. By varying different dimensions of slots more number of stop band can be achieved. High quality factor, better insertion loss makes this filter efficient for practical use.

References

1. Pozar DM (2010) Microwave engineering, 4th edn. Wiley
2. Salari MA, Abbasiniazare S, Manoochehri O (2012) The effect of electromagnetic waves on multilayer orthogonal microstrip lines with and without defected microstrip structure. IEEE Antennas Wireless Propag Lett 1:1206–1209
3. Liu HW, Zhang ZC, Wang S, Zhu L, Guan XH, Lim JS, Ahn D (2010) Compact dual-band bandpass filter using defected microstrip structure for GPS and WLAN applications. Electron Lett 21:1444–1445
4. Ding C, Guo YJ, Qin P-Y, Yang Y (2015) A compact microstrip phase shifter employing Reconfigurable Defected Microstrip Structure (RDMS) for phased array antennas. IEEE Trans Antennas Propag 63:1985–1996
5. Xiao JK, Zhu YF, Li Y, Ma JG, Li XW (2014) Controllable miniature tri-band bandpass filter using defected microstrip structure. Electron Lett 50:1534–1536
6. Wang J, Zhao J, Li J-L (2014) Compact UWB bandpass filter with triple notched bands using parallel U-shaped defected microstrip structure. Electron Lett 50:89–91
7. Dash RK, Sahu SK, Mishra CS, Sethi K, Palai G, Sahu S (2016) Realization of 'non-linear invisibility cloak' using meta-material. Optik - Int J Light Electron Opt 127:9635–9639
8. Sahu SK, Sahu S, Mishra CS, Palai G (2016) Analysis of reflected frequency band of metamaterial grating at THz frequency: a future application of filter. Optik Int J Light Electron Opt 127:4547–4550
9. Lin C-C, Jin P, Ziolkowski RW (2012) Tri-band-notched Ultrawideband (UWB) antenna using Split Semicircular Resonator (SSR) and Capacitively Loaded Loops (CLL). IEEE Trans Antennas Propag 60:102–109
10. Si L-M, Zhang Q-L, Hu W-D, Yu W-H, Wu Y-M, Lv X, Zhu W (2015) A uniplanar triple-band dipole antenna using complementary capacitively loaded loop. IEEE Antennas Wireless Propag Lett 14:743–746
11. Siddiqui JY, Saha C, Antar YMM (2014) Compact SRR loaded UWB circular monopole antenna with frequency notch charactristics. IEEE Trans Propag 62:4015–4020
12. Panda AK, Satyabrat Sahu K, Mishra RK (2012) A compact triangular SRR loaded CPW line and its use in highly selective wide band Bandpass filter for WiMax communication system. In: International conference on computer and devices for communication
13. Sahu SK, Sahu S, Palai G (2017) Finite difference time domain approach to metamaterial waveguide for IR application as cavity resonator. Optik - Int J Light Electron Opt 130:872–876

14. Liu Q, Yu Y, He S (2013) Capacitively loaded, inductively coupled fed loop antenna with an omnidirectional radiation pattern for UHF RFID tags. IEEE Antennas Wireless Propag Lett 12:1161–1164
15. Sarkar P, Ghatak R, Pal M, Padar DR (2012) Compact UWB bandpass filter with dual notch band using open circuited stubs. IEEE Microw Wireless Compon Lett 9:453–455
16. Hong JS, Lancaster MJ (2001) Microstrip filters for RF/microwave applications. Wiley, New York

Weighted Filtered-s LMS Algorithm for Nonlinear Active Noise Control

Tapasmini Sahoo and Kunal Kumar Das(✉)

Electronics and Communication Engineering, Institute of Technical Education and Research, Siksha 'O' Anusandhan Deemed University, Bhubaneswar 751030, India
kunaldas@soa.ac.in

1 Introduction

Active noise control has been an attractive field of research for many years. But research in this area accelerated only during the last two decades. This is possible with advancement in Digital Signal Processing and availability of fast processors with huge memory. ANC is based on acoustically superimposing a noise with its antinoise. This superposition is carried out with an objective to cancel the noise in real time. For optimum cancellation of noise, antinoise should be equal in magnitude but opposite in phase to that of noise. To generate antinoise an adaptive filter is used which adapt to changes in acoustic paths and noise statistic. ANC systems can be feed forward, feedback or hybrid in nature. Depending on number of error microphones employed to record error signals they are further classified as single channel and multichannel ANCs.

A single channel feedforwad ANC is shown in Fig. 1, where the path between actual noise source and error microphone is called the primary path, the path between the secondary loudspeaker and the error microphone is defined as the secondary path [1]. Antinoise is produced by the controller, which consists of the adaptive filter and update mechanism. The ANC system uses the reference microphone to pick up the reference noise, processes it through the controller to cancel the noise. ANC also uses an error microphone to measure the error which is subsequently used to update the weights of the adaptive filter. Filtered-X LMS (FXLMS) algorithm, which is a modified version of LMS algorithm, is used to adapt the adaptive filter. FXLMS algorithm is the most popular algorithm for ANC, because of its simplicity and low computational requirement. FXLMS algorithm is modified to develop filtered-error LMS (FELMS) algorithm to reduce the computational complexity of multichannel ANCs [2].

When reference noise travels through certain structures the acoustic paths become nonlinear. If the amplitude of the antinoise increases beyond the secondary loudspeaker specifications, nonlinearity results. The presence of nonlinearity in the primary and secondary paths reduces the noise cancellation capability of the FXLMS algorithm drastically. In order to deal with nonlinearity some nonlinear adaptive controller have

M. N. Mohanty and S. Das (eds.), *Advances in Intelligent Computing and Communication*, Lecture Notes in Networks and Systems 109,
https://doi.org/10.1007/978-981-15-2774-6_9

been proposed. Multilayer artificial neural network, fuzzy logic, polynomial filter, Volterra filter, genetic algorithm and particle swarm algorithms [3, 4], etc. have been employed in nonlinear ANC. Recently, functional link artificial neural network (FLANN) has become popular for nonlinear ANC [5]. This interest in FLANN is due to its simple structure and easy to implement filtered-S LMS training algorithm. In order to further improve noise cancellation capability of FLANN based ANC systems, several modifications have also been suggested [6–9].

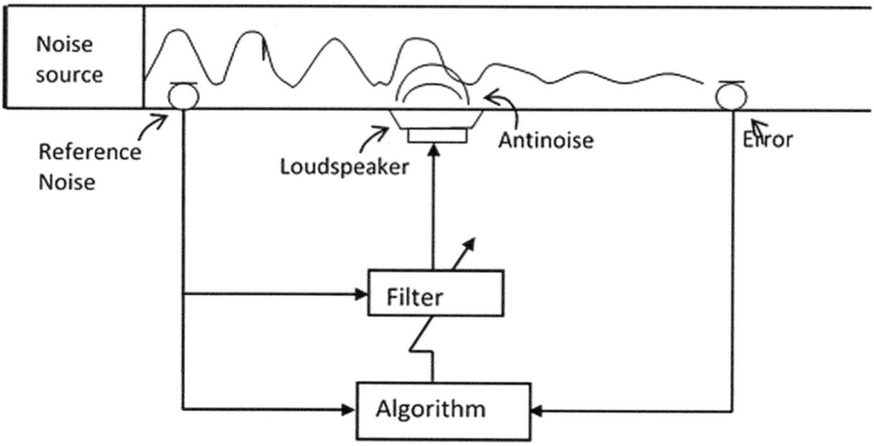

Fig. 1. Schematic diagram of an active noise control system

2 Filtered-s LMS Algorithm

A class of artificial neural networks called functional link artificial neural network (FLANN) has appeared as a strong contender for identifying nonlinear systems. FLANN, was proposed by Y. Pao and has a simple structure compared to other multilayer neural networks [10]. In terms of computational complexity, it has a huge advantage as it requires much less computation in comparison to multilayer neural network using back propagation algorithm. In FLANN, the input is functionally expanded using appropriate basis functions. The basis functions generally used for functional expansion are trigonometric [5], Chebyshev polynomial [11] and Legendre polynomial [12]. In [5] Das and Panda have employed FLANN in a nonlinear ANC and developed a novel filtered-s LMS (FSLMS) algorithm for training the network. Figure 2 represents block diagram of a nonlinear ANC using FLANN. The reference input, $x(n)$, is nonlinearly expanded using trigonometric functions as defined below.

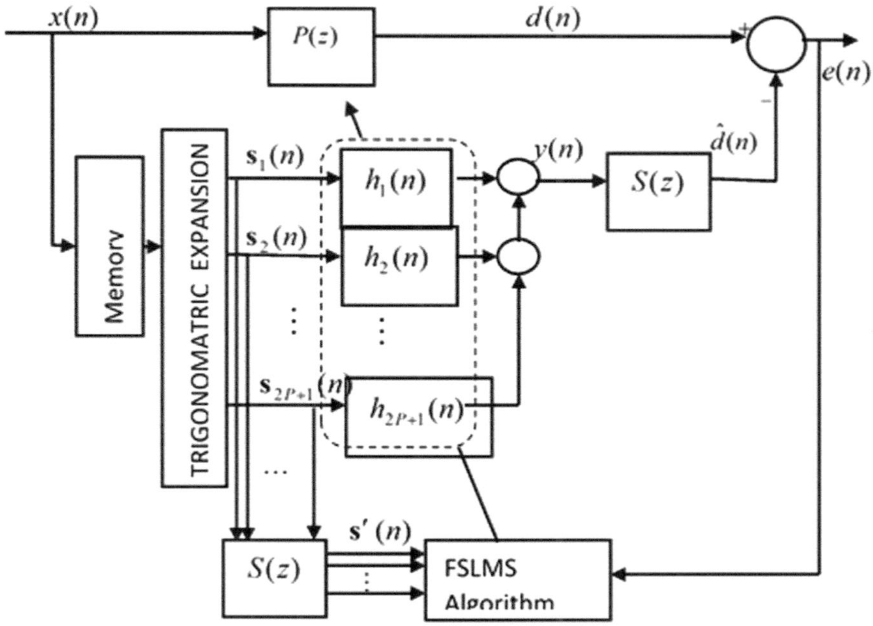

Fig. 2. Block diagram of nonlinear ANC using FSLMS algorithm

$$\begin{aligned}s(n) = [&x(n), \sin(\pi x(n)), \cos(\pi x(n)), \ldots, \sin(\pi P x(n)), \cos(\pi P x(n)),\\ &x(n-1), \sin(\pi x(n-1)), \cos(\pi x(n-1)), \ldots, \sin(\pi P x(n-1)), \cos(\pi P x(n-1)), \ldots,\\ &x(n-N), \sin(\pi x(n-N)), \cos(\pi x(n-N)), \ldots, \sin(\pi P x(n-N)), \cos(\pi P x(n-N))].\end{aligned} \tag{1}$$

where P is the order of functional expansion and N is the memory length. The trigonometrically expanded input signal is rearranged to obtain

$$\begin{aligned} s(n) &= [s_1, s_2, s_3, \ldots, s_{2P+1}]\\ s_1(n) &= [x(n), x(n-1), \ldots, x(n-N)]\\ s_2(n) &= [\sin(\pi x(n)), \sin(\pi x(n-1)), \ldots, \sin(\pi x(n-N))]\\ s_3(n) &= [\cos(\pi x(n)), \cos(\pi x(n-1)), \ldots, \cos(\pi x(n-N))], \ldots,\\ s_{2P} &= [\sin(\pi P x(n)), \sin(\pi P x(n-1)), \ldots, \sin(\pi P x(n-N))];\\ s_{2P+1} &= [\cos(\pi P x(n)), \cos(P\pi x(n-1)), \ldots, \cos(\pi P x(n-N))]\end{aligned} \tag{2}$$

This rearrangement facilitates us to restructure the network as a set of time-delay-line filters, operating in parallel. The output of the time-delay-line filters are defined by $y_i = h_i(n) * s_i(n), i = 1, 2, \ldots, 2P+1$ and $h_i(n)$ are impulse responses of the $2P+1$ adaptive filters. The controller output is obtained by

$$y(n) = \sum_{i=1}^{2P+1} y_i(n). \tag{3}$$

The desired antinoise, $\hat{d}(n)$, is produced by filtering the controller output through the secondary path impulse response. The error recorded by the error microphone is obtained $e(n) = d(n) - \hat{d}(n)$. FSLMS algorithm is employed for updating all the filter weights. The weight update equations can be compactly written as $h_i(n+1) = h_i(n) + \mu_f e(n) s_i'(n)$. The developed FLANN structure has a relatively simpler structure and has lower computational load compared to multilayer artificial neural networks for nonlinear ANCs. But this FLANN structure with FSLMS algorithm inherently slow in convergence. This is because it is based on the LMS algorithm which has a slow convergence capability. Many improvements in the algorithm have been proposed by researchers [13–15]. But research has been mainly focused on reducing the steady-state mean square error performance by taking care of the cross terms. In order to circumvent this drawback of slow convergence, a new structure is proposed in this paper.

3 Weighted Filtered-s LMS Algorithm

In the proposed system a set of new adaptive weights, called combining weights are introduced. The output of the adaptive filters is multiplied with combining weights. The controller output is obtained as follows:

$$y(n) = w_1(n)y_1(n) + w_2(n)y_2(n) + \cdots + w_{2P+1}(n)y_{2P+1}(n) \tag{4}$$

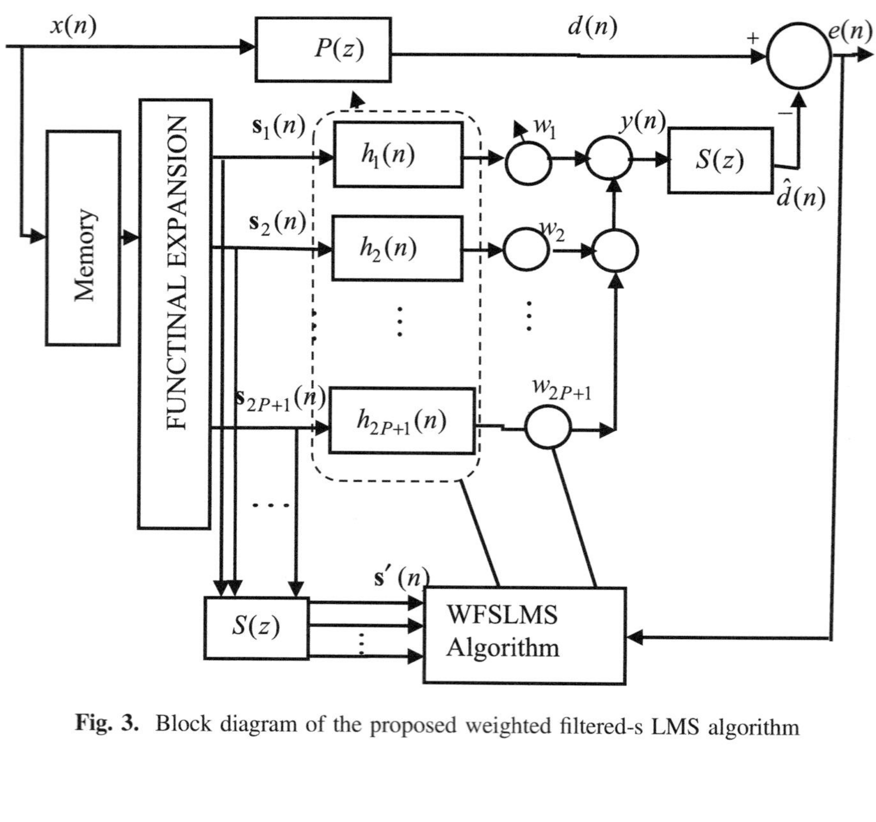

Fig. 3. Block diagram of the proposed weighted filtered-s LMS algorithm

The weights of the adaptive filters are adapted by the FSLMS algorithm and the newly introduced combining weights are updated by the following equation, where μ_c is the step size for newly introduced combining weights.

$$w_i(n+1) = w_i(n) + \mu_c e(n) s_i'(n) \quad (5)$$

4 Simulation Study

The performance of the proposed algorithm for nonlinear ANC is evaluated by conducting a battery of simulation experiments. The noise cancellation capability of the proposed algorithm is compared with that of the standard FSLMS algorithm. Mean Square Error (MSE), defined by $\mathrm{MSE} = 10\log_{10}(E[e(n)^2])$ is considered as the performance measure. In each experiment average, MSE of 100 independent trials is plotted. The reference noise considered in the simulation experiments is a uniformly distributed random noise within the −0.5 to 0.5. It is assumed that the secondary path is accurately identified by system identification techniques (Fig. 3).

4.1 Experiment I

The primary and secondary paths are modelled as FIR filters of length 256 and 128, respectively. The coefficients of the acoustic paths are obtained from the data attached to [1]. The order of the FLANN filter is 1 and memory length is 256. $\mu_f = 0.00005$ and $\mu_c = 1$ where μ_f and is the step size for FLANN filter and μ_c is the step size for combining weights. The Average MSE is shown in Fig. 4.

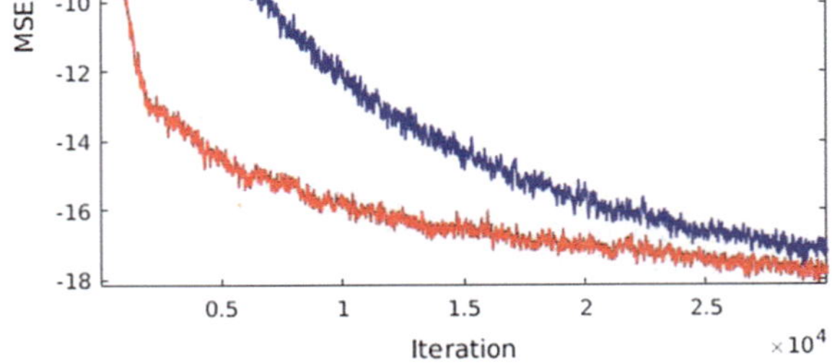

Fig. 4. Average MSE for experiment I

4.2 Experiment II

Primary path and secondary path transfer functions considered in this experiment are $P(z) = z^{-5} - 0.8z^{-6} + 0.7z^{-7}$ and $S(z) = z^{-2} + 1.5z^{-3} - z^{-4}$ [5]. The order of the FLANN expansion is 1 and memory length is 15. The step sizes are $\mu_f = 0.00006$ and $\mu_c = 0.4$. The Average MSE is shown in Fig. 5.

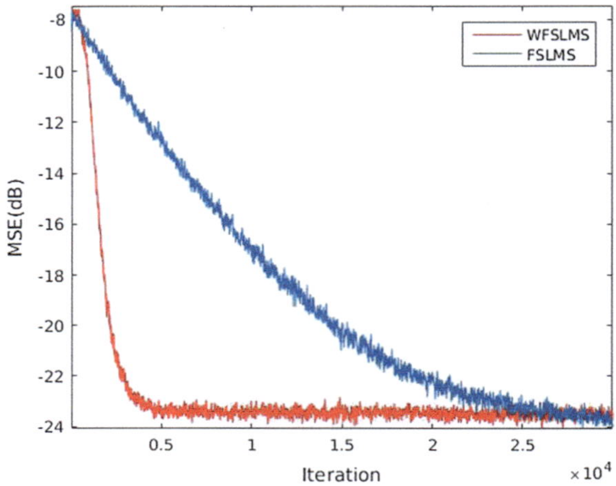

Fig. 5. Average MSE for experiment II

4.3 Experiment III

In this simulation experiment, the reference noise at the error microphone is obtained by first filtering the noise through a filter $R(z) = z^{-3} - 0.3z^{-4} + 0.2z^{-5}$ and then computing $d(n)$, where $u(n) = x(n) * r(n)$ and $r(n)$ is the impulse response of the filter $R(z)$ [16].

The secondary path is having a transfer function $S(z) = z^{-2} + 1.5z^{-3} - z^{-4}$. The order of FLANN expansion is 1 and memory length is 15. The step sizes are $\mu_f = 0.0005$ and $\mu_c = 0.4$. The Average MSE is shown in Fig. 6.

4.4 Experiment IV

In this experiment, both the primary and secondary paths are considered as nonlinear [17]. Here the desired noise signal $d(n)$ and reference noise $x(n)$ are related by

$$\begin{aligned} d(n) = {} & x(n) + 0.8x(n-1) + 0.3x(n-2) + 0.4x(n-3) \\ & - 0.8x(n)x(n-1) + 0.9x(n)x(n-2) \\ & + 0.7x(n)x(n-3) \end{aligned}$$

The secondary path is defined by

$$\hat{d}(n) = y(n) + 0.35y(n-1) + 0.09y(n-2) - 0.5y(n)y(n-1) + 0.4y(n)y(n-2)$$

Where $y(n)$ is the output from the controller. The order of the FLANN expansion is 1 and memory length is 15. The step sizes are $\mu_f = 0.0003$ and $\mu_c = 1$. The Average MSE is shown in Fig. 7.

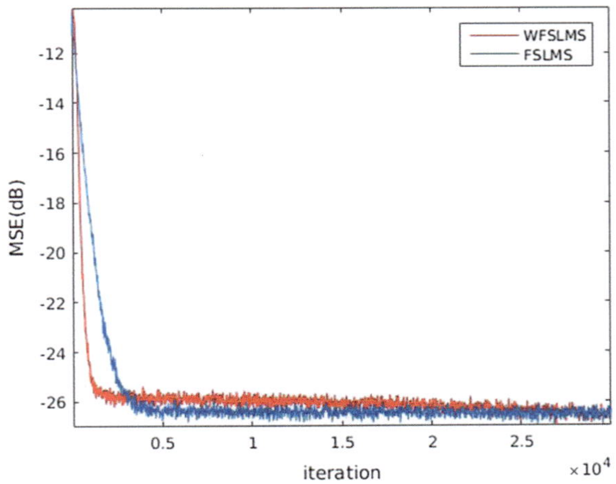

Fig. 6. Average MSE for experiment III

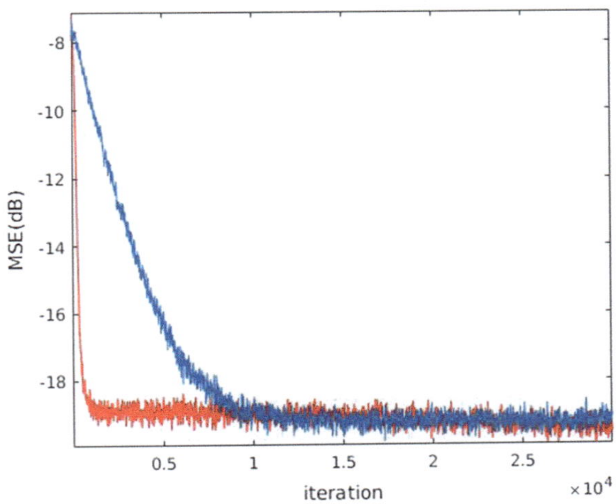

Fig. 7. Average MSE for experiment IV

5 Conclusion

In this paper, we have modified the structure of FLANN by introducing a set of adaptive weights. These weights are used to adaptively combine the output of subfilters of NANC system. A suitable algorithm is also developed for updating these combining weights. These modifications not only alleviate the drawback of slow convergence but also achieve it without sabotaging the structural simplicity of FLANN. The developed algorithm has emerged as a strong competitor of standard FSLMS algorithm as it achieves faster convergence without sacrificing MSE and retain the low computational load of FSLMS algorithm. This structural modification can also be incorporated into a generalized FSLMS algorithm.

References

1. Kuo SM, Morgan DR (1996) Active noise control systems: algorithms and DSP implementations. Wiley, New York
2. Wan EA (1996) Adjoint LMS: an efficient alternative to the Filtered-x LMS and multiple error LMS algorithm. In: Proceedings of IEEE ICASSP, pp 1842–1845
3. Chang CY, Louh FB (2007) Enhancement of active noise control using neural based Filtered-x algorithm. J Sound Vib 305:348–356
4. Tan L, Jing J (2001) Adaptive Volterra filters for active control of nonlinear noise processes. IEEE Trans Signal Process 49(8):1667–1676
5. Das DP, Panda G (2004) Active mitigation of nonlinear noise process using novel Filtered-s LMS algorithm. IEEE Trans Speech Audio Process 12(3):313–322
6. Sicuranzaan GL, Carini A (2012) A generalized FLANN filter for nonlinear active noise control. IEEE Trans Speech Audio Process 19(8):2412–2417
7. Lee DC, Zhang J, Pang Y (2018) A generalized exponential functional link artificial neural network with channel reduced diagonal structure for nonlinear active noise control system. Appl Acoust 139:174–181
8. Lee DC, Zhang J, Pang Y (2018) A bilinear functional link artificial neural network filter for nonlinear active noise control and its stability condition. Appl Acoust 132:19–25
9. Patra JC, Bal RN, Chatterji BN, Panda G (1999) Identification of nonlinear dynamic systems using functional link artificial neural network. IEEE Trans Syst Man Cybern Part B 29(2): 254–262
10. Pao Y (1989) Adaptive pattern recognition and neural network. Addition-Wesley, MA
11. Patra JC, Kot AC (2002) Nonlinear dynamic system identification using Chebyshev functional link artificial neural network. IEEE Trans Syst Man Cybern Part B 32(4):505–511
12. George NV, Panda G (2013) Active control of nonlinear noise processes using cascaded adaptive nonlinear filters. Appl Acoust 74:117–222
13. George NV, Gonzalez A (2014) Convex combination of nonlinear adaptive filters for active noise control. Appl Acoust 76:156–161
14. Xhao H, Zeng X, He Z, Li T (2012) Adaptive extended pipelined second order Volterra filterfor nonlinear active noise controller. IEEE Trans Audio Speech Lang Process 20(4): 1394–1399

15. Zhao H, Zeng X, He Z, Yu S, Chen B (2016) Improved functional link artificial neural network via convex combination for nonlinear active noise control. Appl Acoust 42: 351–359
16. Patel V, George NV (2015) Nonlinear active noise control using spline adaptive filter. Appl Acoust 93:38–43
17. Zhou D, DeBrunner V (2007) Efficient adaptive nonlinear filter for nonlinear active noise control. IEEE Trans Circuits Systems-I 54(3):669–681

Comparison of Classifiers for Speech Emotion Recognition (SER) with Discriminative Spectral Features

Hemanta Kumar Palo[1](✉), Debasis Behera[2], and Bikash Chandra Rout[1]

[1] Siksha O Anusandhan (Deemed to Be University), Bhubaneswar, Odisha, India
hemantapalo@soa.ac.in
[2] C. V. Raman College of Engineering, Bhubaneswar, Odisha, India

1 Introduction

The ambiguity of human behavior in society increases day by day that belongs to the study of human affective states. Among different modalities of expressive emotions such as facial, gestures, and spoken conversation, recognition of emotions from the speech is a complex domain of research as spoken words contribute only 7% in an expressive behavior as compared to 55% and 38% due to facial expression and voice intonation and semantic information, respectively [1]. Acoustic features representing these speech emotions are most informative in the spectral domain. However, spectral features including the MFCCs suffer from embedded pitch filtering and textural dependency that result in poor SER accuracy. The absence of the desired sentence-level information for speech emotion identification motivates researchers to explore the delta (or differential) and double delta (or acceleration) coefficients as additional inputs to the baseline features [2–5]. Nevertheless, the use of a fixed window in these techniques has resulted in an equal time resolution for both high and low frequencies. For better analysis at these frequencies, wavelet analysis with an adaptive window remains an effective time–frequency tool in this field [6]. The spectral features extracted in the wavelet domain have shown to outperform the baseline spectral features in the field of SER.

There are several conventional classifiers applied in the field of SER. Nevertheless, the suitability of a classification algorithm largely depends on the discriminative features and the feature dimension. Three of the classifiers such as the GMM, PNN, and RBFN have been emphasized in this work due to their following attributes: First, it is possible to model any pattern using a sufficient number of RBF neurons in a single hidden layer, unlike MLP, which requires more than one hidden layer [7, 8].

M. N. Mohanty and S. Das (eds.), *Advances in Intelligent Computing and Communication*, Lecture Notes in Networks and Systems 109,
https://doi.org/10.1007/978-981-15-2774-6_10

Like MLP, the RBFN does not require back-propagating the error through multiple hidden layers and hence is faster. Similar to RBFN, the PNN is a neural network-based and is parallel structured. The network is faster by 200,000 to 1 than MLP and two to three orders faster than RBFN [9]. It is simple, requires a single parameter adjustment during the simulation, and can be implemented easily [10]. On the contrary, GMM is a stochastic classifier that uses statistical parameters to model the desired emotion adequately. Ease of learning, bootstrapping from flat data, and faster computation make the classifier compatible for frame-level features [11].

The rest of the sections are organized as follows. The feature extraction techniques proposed in this work are provided in Sect. 2. The simulated results and detailed discussions on the outcomes of the proposed work using the chosen classifiers have been elaborated in Sect. 3. The conclusion and further possible future directions have been briefed in Sect. 4.

2 Proposed Feature Sets

The proposed method of feature extraction has been provided in this section. The MFCC and wavelets have been used as the baseline features to extract the proposed discriminant features with different dimensions. In the first phase, MFCC and wavelets are used to simulate and compare the GMM, RBFN, and the PNN classifiers. Subsequently, different variants of MFCC such as differential MFCC (delta MFCC or DMFCC), wavelet MFCC (WMFCC), and wavelet differential MFCC (WDMFCC) features are simulated and tested with these classifiers for the performance appraisal. Finally, hybrid feature sets such as MFCC + DMFCC and WMFCC + WDMFCC have been formed as a possible means to enhance the classification accuracy of the GMM classifier.

The signal $s_i(m)$ is framed into 30 ms with 10 ms overlapping between frames. The Hamming window has been used to obtain the windowed signal for spectral estimation using FFT [12]. From each frame of an utterance, 16 numbers of MFCC coefficients are extracted. For N frames per utterance and fifty utterances per emotion, the feature matrix has a size $50N \times 16$.

In wavelet analysis, a 3-level decomposition using Daubechies-4(dB-4) wavelets has been used. For 16 bit/sample used in this work, and taking the detailed coefficients D-6 into account, the feature dimension remains approximately 470×1 per utterance. For 50 numbers of utterances per emotion used in this work, the feature size becomes $(50 \times 470) \times 1 = 23500 \times 1$. The proposed derived feature extraction scheme is shown in Fig. 1.

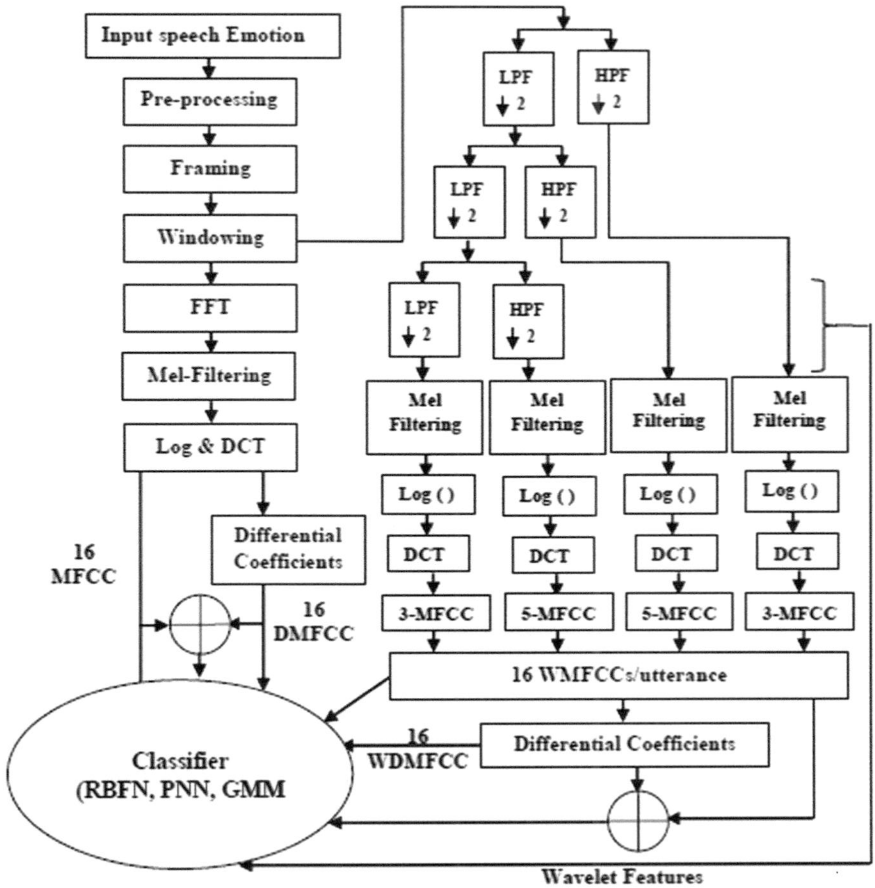

Fig. 1. Proposed method of derive feature extraction

2.1 Wavelet-Based MFCC (WMFCC)

The wavelet-based MFCC (WMFCC) technique is herby explained below.

The signal at the ith level is decomposed into detailed d_i and approximation a_i coefficients, respectively. A three-level decomposition using Daubechies-4 or dB-4 has been chosen. The Surrey Audio-Visual Expressive Emotions (SAVEE) database used in this work has a sampling frequency of 44.1 kHz which is downsampled to 16 kHz with a bandwidth of 8 kHz [13]. Thus, the sub-bands are segmented into (1) 0–1 kHz, (2) 1–2 kHz, (3) 2–4 kHz, and (4) 4–8 kHz range using the filtering. The MFCC coefficients are fed to these bands individually as shown in Fig. 1. From each sub-band 1–2 kHz and 2–4 kHz, five WMFCC coefficients are extracted. For each sub-band 0–1 kHz, 4–8 kHz, three WMFCC coefficients are fetched. The number of WMFCC coefficients in 1–4 kHz frequency bands (middle bands) is more than other bands. The aim is to extract suitable perceptual information from these middle bands as the voice components are more prominent in this band. Sixteen WMFCCs per utterance are collected in this way. For fifty utterances of each emotion, the size of the WMFCC matrix becomes 50 × 16.

2.2 Differential MFCC (DMFCC)

The delta or differential MFCC features signify the dynamism involved or the trajectories of the cepstral-based static MFCC feature concerning time [5]. Thus, including the DMFCC coefficients with the MFCC coefficients will likely to enhance the recognition performance of speech emotion. These velocity or time derivative features of MFCC at time t can be extracted using the regression formula as given by

$$\mathrm{DMFCC}(t) = \frac{\sum_{k=1}^{K} k(\mathrm{MFCC}(i+k) - \mathrm{mfcc}(i-k))}{2\sum_{k=1}^{K} k^2} \tag{1}$$

where MFCC $(t+k)$ and MFCC $(t-k)$ are the static coefficients computed from frame i. The value of K maybe 2 or 3 in practice. However, for delta MFCC $K = 2$ has been chosen here. For 16 MFCC features per frame extracted in this work, there are 16 DMFCC features per frame. The resultant feature size for fifty utterances per emotion becomes 50 N × 16.

2.3 Wavelet-Based Differential MFCC (WDMFCC)

While the wavelet analysis gives information on both the time and frequency content of a signal, the conventional MFCCs provide information on the signal's amplitude or energy [14]. To accommodate the dynamism in the static wavelet-based MFCC features, the differential features of WMFCC have been proposed in this work. The procedure to extract WDMFCC features is enumerated below.

i. Extract the WMFCC features as per the procedure explained earlier.
ii. Apply the derivative algorithm to the WMFCC features extracted as in step (i) as per Eq. (1).

Similar to the 16WMFCC feature extracted per utterance of an emotion, there are 16 numbers of WDMFCC features obtained per utterance. For fifty utterances per emotion, the size of the WDMFCC feature matrix thus becomes 50 × 16.

2.4 Proposed Feature Combination

An attempt is made to combine the static and differential MFCC features for possible enhancement of accuracy with the GMM classifier. For an N number of frames per utterance, and 50 utterances per emotion, the feature matrix size thus becomes 50 N (16MFCC + 16DMFCC) = 50 N × 32. Similarly, the wavelet-based static MFCC and differential MFCC are combined to form another hybrid feature set as indicated by WMFCC + WDMFCC. In this case, the feature dimension using WMFCC and WDMFCC feature becomes 50 × 32. The involvement of both static and dynamic components of the baseline features has proved more beneficial for the recognition task as evidenced in the result section.

3 Results and Discussion

The classification accuracy and the computation time concerning the feature dimensions have been used for comparison. The utterances of emotion are divided into training and testing sets. The utterances used for the training the classifiers are not used for testing purposes. The classification accuracy of individual emotions with the extracted feature sets with the RBFN classifier has been shown in Fig. 2, and that of PNN is shown in Fig. 3.

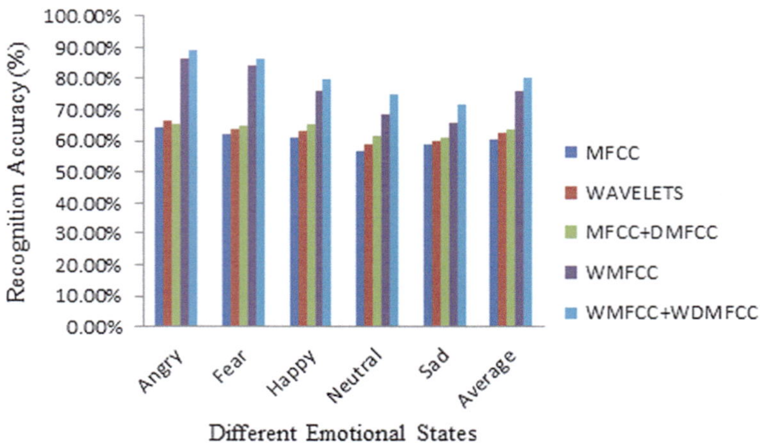

Fig. 2. Recognition accuracy of RBFN with different variants of MFCC features

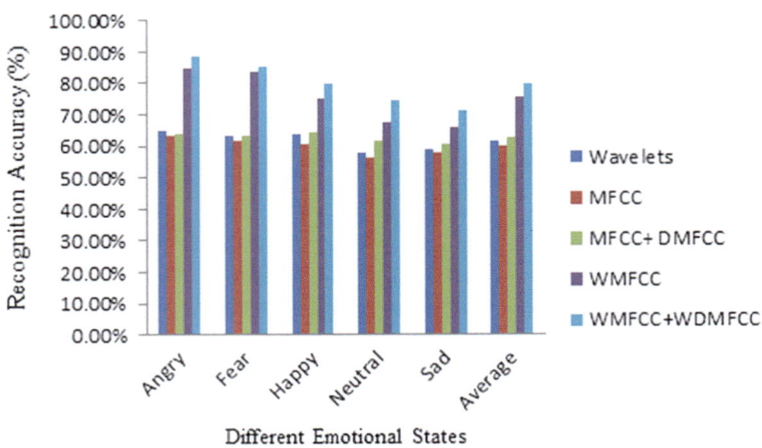

Fig. 3. Recognition accuracy of PNN for different variants of MFCC features

The classification performance of the individual speech emotion with the GMM classifier is shown in Fig. 4. A GMM model is made for each emotion using 16 mean, 16 variances, and one mixture component.

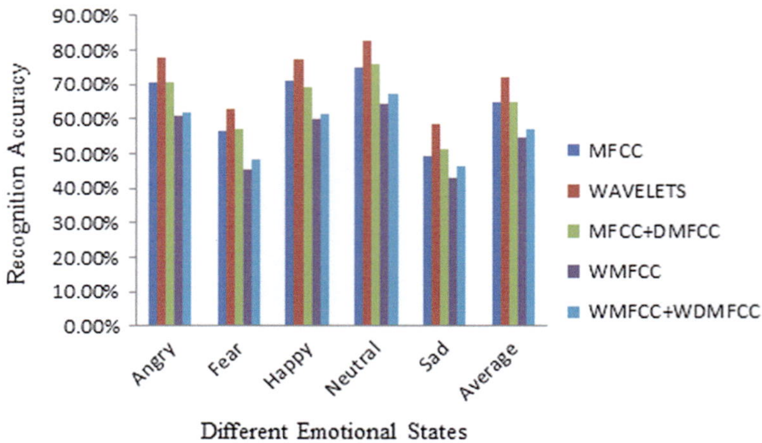

Fig. 4. Recognition accuracy of GMM with different variants of MFCC features

The comparisons of classifiers in terms of lowest classification time, highest average accuracy, and feature reliability have been provided in Table 1.

Table 1. Comparison of classifiers based on recognition accuracy, feature dimension, and computation time with the proposed feature set

Classifiers	Feature	Average accuracy (%)	Classification time (s)	Feature dimension
RBFN	WMFCC + WDMFCC	80.51	28.41	Low
PNN	WMFCC + WDMFCC	80.79	10.56	Low
GMM	WAVELETS	71.82	07.52	High

The following important conclusions can be drawn from Figs. 2, 3, and 4. The classification accuracy of high arousal emotions such as angry, fear, and happy is found to be better due to their well-defined higher energy levels The sad state has the lowest arousal level and found to provide the least classification accuracy. The PNN proves better than RBFN as observed from Figs. 2 and 3. Unlike the RBF network where it is needed to adjust the weights, it is possible to set all the weights of PNN to one. This has made the PNN more reliable to low-dimensional feature sets without any over-fitting as compared to the RBFN when Fig. 3 is compared with that of Fig. 2. However, the difference in accuracy is not significant. The WMFCC and WDMFCC techniques have provided the lowest feature size as compared to other feature extraction techniques as explained in Sects. 2.1 and 2.3. Analysis of Figs. 2 and 3 shows enhanced classification accuracy with these feature extraction methods. The use of the multiresolution capability of wavelet analysis makes these features more reliable. Further, these features provide both time and frequency localized information due to their extraction in the wavelet domain besides the energy information provided by MFCC. These factors have increased the available information of the signal under consideration that has resulted in enhanced SER. The NN-based classifiers outperform the GMM classifier in terms of classification accuracy for small feature sets such as WMFCC and WDMFCC. On the contrary, better emotional modeling has been observed using a GMM classifier than NNs for large feature dimensions. This has been observed when Figs. 2, 3, and 4 have been compared with Table 1. For large training features, the GMM classifier can capture meager differences among distributions. As a result, the classifier is well suited for frame-level spectral features extracted particularly from emotional speech signals. The results in Table 1 also indicate the reliability of the chosen feature sets based on the classification time and highest average recognition accuracy. An attempt is made further to enhance the recognition accuracy of the emotions using feature hybridization in Table 2.

Table 2. Effect of feature combination on the computation time, and classification accuracy concerning the feature dimension using GMM

Features	Feature dimension per emotions	Feature extraction time per utterance (s)	Classification time (s)	Average accuracy (%)
MFCC + DMFCC	50 $N \times 32$	2.38	19.32	80.1
WMFCC + WDMFCC	50×32	2.65	09.42	60.2

The combination approaches have been tested to obtain better GMM modeling as the classifier has proved better for large feature dimensions in our work. The use of multiresolution capability in wavelet analysis has resulted in enhanced accuracy for the WMFCC + WDMFCC as compared to the MFCC + DMFCC combination.

4 Conclusions

The use of both static MFCC and its dynamic (DMFCC) information in the extraction of the features has improved the classification accuracy because of the availability of temporal emotional information. The MFCC features extracted using the multiresolution wavelet analysis have resulted in a more reliable feature than either the baseline MFCC or the wavelets. Further, the involvement of differential algorithm with wavelet-based MFCC features is a novel effort in this direction. The combination of MFCC variants having a larger dimension has resulted in enhanced accuracy in the case of GMM. However, the NNs tend to be better performing with a small set of features. It indicates the feature dimension plays an important role in deciding the classification accuracy in the proposed techniques. Although, it has increased the computational complexity of feature extraction, at the same time the classification time has been reduced. Hence, a trade-off has to be made depending upon the task domain. Similarly, the exploration of other reduction, modification, and combination mechanisms may provide new research directions in this area. Further, the effectiveness of other conventional classifiers to different feature size is an area needs further investigation.

References

1. Mehrabian A (1971) Silent messages, vol 8. Wadsworth, Belmont, CA
2. Deng J, Frühholz S, Zhang Z, Schuller B (2017) Recognizing emotions from whispered speech based on acoustic feature transfer learning. IEEE Access 5:5235–5246
3. Joe CV, Sugi SSS (2016) Optimal feature for emotion recognition from speech. Afr J Basic Appl Sc 8(3):136–144
4. Sato N, Obuchi Y (2007) Emotion recognition using Mel-frequency Cepstral coefficients. Inf Media Technol 2(3):835–848
5. Palo HK, Chandra M, Mohanty MN (2018) Recognition of human speech emotion using variants of Mel-Frequency Cepstral coefficients. In: Advances in systems, control and automation. Springer, Singapore, pp 491–498
6. Palo HK, Mohanty MN (2018) Wavelet-based feature combination for recognition of emotions. Ain Shams Engg J 9:1799–1806
7. Bishop CM (1995) Neural networks for pattern recognition. Oxford University Press
8. Javidi MM, Roshan EF (2013) Speech emotion recognition by using combinations of C5. 0, neural network (NN), and support vector machines (SVM) classification methods. J Math Comput Sc 6:191–200
9. Specht DF, Missiles L, Inc S, Alto P (1990) Probabilistic neural networks and the polynomial Adaline as complementary techniques for classification. IEEE Tran Neural Net 1 (1):111–121
10. Palo HK, Sagar S (2018) Comparison of neural network models for speech emotion recognition. In: 2nd international conference of data science and business analytics, IEEE, pp 127–131
11. Ayadi E, Kamal MS, Karray F (2011) Survey on speech emotion recognition: features, classification schemes, and databases. Pattern Recogn 44(3):572–587 (Elsevier)
12. Palo HK, Behera D, Analysis of speaker's age using clustering approaches with emotionally dependent speech features. In: Critical approaches to information retrieval research, IGI Global, 2020, pp 172–197
13. Haq S, Jackson PJB (2010) Multimodal emotion recognition. In: Wang W (ed) Machine audition: principles, algorithms, and systems. IGI Global Press, chapter 17, pp 398–423
14. Palo HK, Chandra M, Mohanty MN (2017) Emotion recognition using MLP and GMM for Oriya language. Int J Comput Vision Robot 7(4):426–442

Handwritten Odia Numerals Recognition: A Supervised Learning Perspective

Suchismita Behera(✉) and Niva Das

Department of ECE, Siksha ,'O' Anusandhan (Deemed to be University), Bhubaneswar, Odisha, India
{suchismitabehera, nivadas}@soa.ac.in

1 Introduction

Handwritten Recognition (HR) systems measure the ability of a computer or device to efficiently recognize the handwritten documents. HR systems may be developed to operate in offline mode or in online mode. Many devices like smart phones, PDAs and touch screens can take handwriting as input. Online handwriting verification may be employed for password protected systems as it takes less time to write down numbers and text to the screen [1, 2]. Recognition of handwritten documents is considered as one of the challenging areas of research before the image processing community. Some of the popular application areas of the numeral recognition problem include postal automation, bank cheque processing, automatic data entry, etc. Though research has attained a saturation in printed character recognition, it is limited in case of handwritten numerals and characters basically due to large variation observed in shape, scale and format arising out of individual writing styles. Already many pieces of work are available in the area of HR systems for non-Indian language scripts such as Roman, Japanese, Chinese and have been implemented for automatic sorting in postal systems. Though India is a multilingual country and multiple scripts are used, not much work has been done on character recognition. With respect to the Indian scripts, few pieces of work are available in the area of numeral recognition. The major challenges in recognizing the handwritten numerals are the variations in size, style and the shape. These problems can be best resolved by selecting good feature extraction techniques along with an efficient classifier. Many feature extraction and classifier techniques have been reported in the literature for this purpose.

In this paper, we have made an attempt to recognize the handwritten Odia numerals with the help of the Support Vector Machine (SVM) classifier and single decision tree classifier for pixel intensity-based features and Histogram of Oriented Gradient features (HOG) separately. Experimental results on a newly developed Odia handwritten digit database collected from IIT, Bhubaneswar are obtained through a simulation study and presented. The paper has been organized as follows: Sect. 2 provides the literature review of the work done on handwritten Odia numerals, Sect. 3 gives an overview of handwritten Odia numerals. Feature extraction techniques used are highlighted in Sect. 4, experimental results are analyzed in Sect. 5, followed by a Conclusion in Sect. 6.

M. N. Mohanty and S. Das (eds.), *Advances in Intelligent Computing and Communication*, Lecture Notes in Networks and Systems 109,
https://doi.org/10.1007/978-981-15-2774-6_11

2 Literature

Various approaches reported in the literature for handwritten numeral recognition are based on structural [3] and statistical [1] features. Some popular approaches used for this purpose are Back-Propagation Neural Network (BPNN), support vector machines, k-Nearest Neighbor(k-NN), etc. Pal et al. [4] used curvature features to achieve a recognition rate of 94% in Odia numerals. A. Desai [5] used Neural Network for recognition of Gujarati numerals. Bhattacharya et al. [6] used Adaptive Self Organizing Neural Network and MLP network for feature extraction and classification, respectively, for efficient recognition of Bangla handwritten numerals. Dash et al. [7] have reported very high recognition efficiency for Odia handwritten numerals with Stockwell and Slantlet transformation-based features. Roy et al. [8] have used a direction chain code histogram of contour points of the numerals for efficient recognition of Odia digits. In [9], Mishra et al. have reported a combination of Discrete Cosine Transform (DCT) and Discrete Wavelet Transform (DWT)-based feature extraction techniques for Odia handwritten digit recognition. In [10], Pujari et al. used curvature and gradient features to get a recognition rate of 90% with SVM classifier. In [11], Impedevo et al. have reported a classification accuracy of 98% with a k-NN classifier with MNIST numeric dataset.

3 Odia Handwritten Numerals

Odia is the regional language of the Indian state Odisha, spoken by more than 40 million people, belonging to Odisha and the neighboring states like West Bengal, Jharkhand, Chhattisgarh, and Andhra Pradesh. Like other language scripts, Odia script consists of 10 numerals from 0 to 9. One set of handwritten Odia numerals is shown in Fig. 1. Numerals such as one, two, six, seven are seen in two shapes. Nine is also written in a variety of ways by different people. One numeral may also look similar to another. These similarities of shape and structure in different numerals pose challenges for the recognition task.

The handwritten Odia numeral database of IIT, Bhubaneswar is considered for implementing this paper. The database contains more than 250 images for each numeral collected from different sources. As the images are in.tif and.jpeg format, they are converted to a uniform image format for further processing.

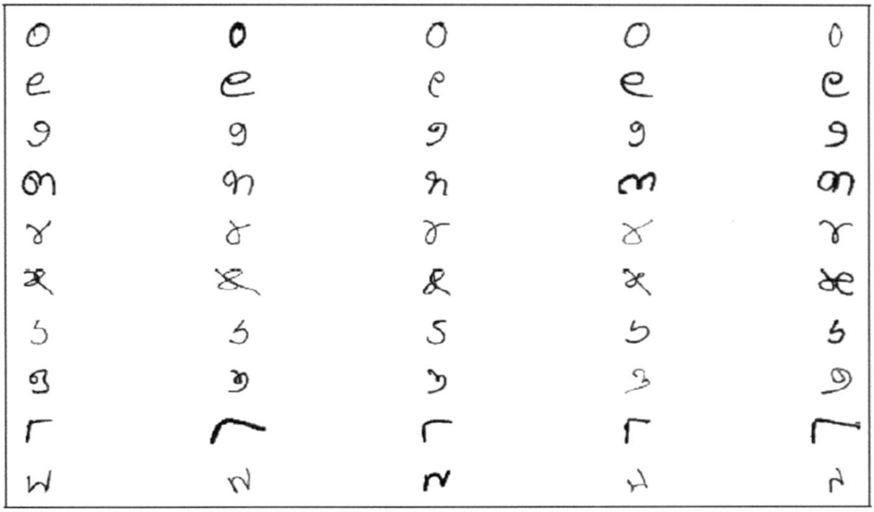

Fig. 1. Sample handwritten Odia numerals from IIT Bhubaneswar database

4 Feature Extraction and Classification

4.1 Feature Extraction

Due to variations in size and presence of noise in the scanned images, some preprocessing is required. The scanned images are first converted to grayscale format and then filtered to remove noise. Then images are normalized in size so as to have a uniform size of (24 × 24) for both training and testing phase. We have used two sets of features for numeral recognition.

The first set is based on intensity information of each pixel. Here, each numeral image is vectorized to a form of a column vector of size 576 × 1. We have considered 400 images of each numeral for training and 100 images for testing. Hence for 10 numerals, the training set is of size 4000 × 1 and testing set is of size 1000 × 1.

The other set is formed by extracting the HOG features of each image. To calculate the HOG features of each image the following steps are adopted:

1. Gradient vector of each pixel is calculated along with magnitude and direction.
2. Each image is divided into cells of size 4 × 4.
3. In each cell, the magnitude and direction of the 16 pixels are represented by a 9-bin histogram as in [12] which is known as HOG descriptors.
4. A block of size 8 × 8 is formed consisting of 4 cells which is represented by a vector of (36 × 1) HOG descriptors. The block is slided horizontally and vertically to cover the whole image which ultimately gives rise to five horizontal and five vertical block positions. Hence for the entire image, the final HOG vector is obtained by concatenation of HOG vectors of all blocks, i.e., 25 × 36 × 1.

4.2 Classification

We have used SVM classifier and single decision tree classifier for classification of the handwritten numerals. In one experiment the pixel intensity information forms the features whereas the HOG descriptors act as features for another experiment for both training and testing set.

5 Results and Discussion

For implementation, we apply the intensity values and HOG descriptor-based feature extraction techniques on handwritten Odia numeral database collected from IIT Bhubaneswar. Two experiments have been carried out on the database. For one experiment pixel intensities are used as the feature set whereas HOG descriptors are used as features for the second experiment. We randomly selected 400 images for training and 100 images for testing for each numeral class. SVM and single decision tree classifiers are applied separately to the features extracted by two different methods in both the experiments. The results are presented in the form of Confusion Matrix which specifies the number of correctly identified numerals along the diagonal. Tables 1, 2, 3 and 4, respectively, show the Confusion Matrix. Table 5 shows the recognition accuracy for each numeral class for four different feature extraction cum classifier combinations. Table 6 shows a comparison of the computational time for different feature extraction cum classifier techniques. It is evident from the results that the decision tree classifier requires less computational time and hence may be recommended for online applications and large database.

Table 1. Confusion matrix of pixel intensity-based single decision tree classifier

0	1	2	3	4	5	6	7	8	9
40	2	0	1	1	1	1	2	0	2
0	30	1	1	6	3	1	2	0	6
0	0	39	0	2	1	2	4	0	2
0	0	1	36	0	6	1	4	1	1
0	0	0	1	45	0	0	3	1	0
0	0	1	3	7	35	0	3	0	1
0	0	1	0	2	2	37	6	1	1
0	0	3	0	2	1	1	43	0	0
0	0	0	1	1	0	1	1	44	0
0	1	0	1	4	2	0	4	1	37

Table 2. Confusion matrix of pixel intensity-based SVM classifier

0	1	2	3	4	5	6	7	8	9
47	1	0	0	0	0	1	0	0	1
1	45	1	0	1	1	0	1	0	0
0	0	50	0	0	0	0	0	0	0
0	0	0	45	0	2	0	1	0	0
0	0	0	0	49	0	0	0	0	1
0	0	0	0	2	48	0	0	0	0
0	0	0	0	0	0	48	2	0	0
1	0	2	0	1	0	0	46	0	0
0	0	0	0	0	0	0	0	50	0
0	1	0	0	0	0	0	0	0	49

Table 3. Confusion HOG feature-based single decision tree classifier

0	1	2	3	4	5	6	7	8	9
48	0	0	0	0	0	2	0	0	0
0	49	0	0	0	1	0	0	0	0
0	1	48	0	0	1	0	0	0	0
0	1	0	45	2	0	0	2	0	0
0	0	0	2	47	0	1	0	0	0
0	1	0	0	1	47	0	1	0	0
0	0	0	0	1	3	46	0	0	0
0	0	1	0	0	1	2	46	0	0
0	0	0	0	3	0	1	0	46	0
0	2	0	1	0	1	1	1	0	44

Table 4. Confusion matrix of HOG feature-based SVM classifier

0	1	2	3	4	5	6	7	8	9
50	0	0	0	0	0	0	0	0	0
0	50	0	0	0	0	0	0	0	0
0	0	50	0	0	0	0	0	0	0
0	0	0	50	0	0	0	0	0	0
0	0	0	0	50	0	0	0	0	0
0	0	0	0	0	50	0	0	0	0
0	0	0	0	0	0	50	0	0	0
0	0	0	0	0	0	0	50	0	0
0	0	0	0	0	0	0	0	50	0
0	0	0	0	0	0	0	0	0	50

Table 5. Recognition accuracy matrix of individual numerals in %

Numeral class	Features + classifier			
	Pixel intensity + single decision tree	Pixel intensity + SVM	HOG + single decision tree	HOG + SVM
0	80	94	96	100
1	60	90	98	100
2	78	100	96	100
3	72	94	90	100
4	90	98	94	100
5	70	96	94	100
6	74	96	92	100
7	86	92	92	100
8	88	100	92	100
9	74	98	88	100
Overall accuracy	77.2	96	93.2	100

Table 6. Computational time for different recognition techniques

Feature extraction + classifiers	Computational time (s)
Pixel intensity-based single tree classifier	7.5001
Pixel intensity-based SVM	134.2926
HOG-based Single tree classifier	4.1538
HOG-based SVM	4.4532

6 Conclusion

This paper presents a comparison of performances of four different feature extraction and classifier schemes for off-line handwritten Odia numeral recognition task. The schemes are tested on a database that contains more than 400 numerals for each class collected from various sources by IIT, Bhubaneswar. The overall recognition accuracy obtained from SVM classifier is more compared to single decision tree classifier. A further single decision tree requires less computational time compared to SVM. Work on this may be extended towards designing a more robust classifier that can perform well in the presence of irregularity and missing parts.

References

1. Mahadevan U, Srihari SN (1999) Parsing and recognition of city, state, and ZIP codes in handwritten addresses. In: Proceedings of the fifth international conference on document analysis and recognition, pp 325–328
2. Kimura F, Sridhar M (1991) Handwritten numeral recognition based on multiple algorithms. Pattern Recogn 24:969–983
3. Plamondon R (2000) Srihari SN (2000) Online and off-line handwriting recognition: a comprehensive survey. IEEE Trans. Pattern Anal. Mach. Intell. 22(1):63–84
4. Pal U, Wakabayashi T, Kimura F (2007) A system for off-line oriya handwritten character recognition using curvature feature. In: 10th international conference on information technology, pp 227–229
5. Desai A (2010) Gujrati handwritten numeral optical character recognition through neural network. Pattern Recogn 33:2582–2589
6. Bhattacharya U, Das TK, Datta A (2002) A hybrid scheme for handprinted numeral recognition based on a self-organizing network and MLP classifiers. Int J Pattern Recogn Artif Intell 16:845–864
7. Dash KS, Puhan NB, Panda G (2015) Handwritten numeral recognitionusing non-redundant stockwell transform and bio-inspired optimal zoning. IET Image Proc 9(10):874–882
8. Roy K, Pal T, Pal U, Kimura F (2005) Oriya handwritten numeral recognition system. In: Proceedings of eighth international conference on document analysis and recognition, vol 2, pp 770–774
9. Mishra TK, Majhi B, Panda S (2013) A comparative analysis of image transformation for handwritten Odia numeral recognition. In: International Conference on Advances in Computing, Communications and Informatics (ICACCI)
10. Pujari P, Majhi B (2015) Recognition of offline handwritten Odia numerals using support vector machine. In: International conference on computational intelligence and networks

11. Impedevo S, Mangini FM, Barbuzzi D (2014) A novel prototype generation technique for handwriting digit recognition. Pattern Recogn 4(3):1002–1010
12. Dalal N, Triggs B (2005) Histogram of oriented gradients for human detection. In: Proceedings of CVPR, vol 1, pp 886–893

An Implementation of Neural Network Approach for Recognition of Handwritten Odia Text

Sachikanta Dash[1(✉)] and Rajendra Kumar Das[2]

[1] Department of CSE, DRIEMS Autonomous Engineering College, Tangi, Cuttack, Odisha 754022, India
sachikanta_dash@rediffmail.com

[2] Department of ENTC, DRIEMS Autonomous Engineering College, Tangi, Cuttack, Odisha 754022, India

1 Introduction

In a practical point of view, all the ongoing innovations based on Computerization, for example, word processors, fax machines and email are having their effect on handwriting. These in-varieties have prompted the adjusting and reinterpreting of the job of handwriting and handwritten messages. HCR is the skill of a machine to get and translate manually written contribution from numerous sources like paper records, photos, and touch screen gadgets, and so on. HCR method is the process of recognizing the writer from a set of handwritings obtained from a lot of writers. Identification and verification are forms that resolve the unique idea of the work of a particular writer, while handwriting interpretation and translation are forms whose goals are to sift through the varieties of processes to determine the message. Recognition of manually written and machine characters is a rising territory of research and finds broad applications in banks, workplaces, and businesses. In the HCR process, an image containing text must be provided adequately and pre-processed. Next, the message text should either undergo segmentation or feature extraction. Small piece of writing will be the outcome, and these must go through various recognition models. Lastly, content-based information should be applied to the recognized data images to verify the result. Machines that process handwriting styles will be dealing with many styles and languages of arbitrary users. Numerous types of analysis, recognition, and translation can link with the handwriting recognition process. The Table 1 encapsulates a few research works proposed by different authors on handwritten characters.

Presently, there is no single methodology that involves it both effectively and efficiently in all contexts. Artificial neural network systems, associated with the handwriting recognition process, don't require profound background knowledge to be able to solve the handwritten recognition issues. The primary goal of this work is to plan a master framework for, HOCR utilizing Neural Network that can viably perceive Odia character kind of arrangement using the Artificial Neural Network approach.

M. N. Mohanty and S. Das (eds.), *Advances in Intelligent Computing and Communication*, Lecture Notes in Networks and Systems 109,
https://doi.org/10.1007/978-981-15-2774-6_12

Table 1. Comparision study on HCR

Author name	Year	Proposal for type of character	Methodology
Vats et al. [1]	2014	Offline handwritten digits	Correlation
Singh et al. [2]	2012	Handwritten Hindi character	Back propagation neural network
Sayyad et al. [3]	2013	Devnagiri script	Back propagation and feed-forward neural network
Mehfuz et al. [4]	2012	Review on offline HCR	Neural network approach
Borde et al. [5]	2012	Handwritten OCR	Fuzzy based approach
Kala et al. [6]	2001	Review on offline HCR	Advanced ANN
Dongare et al. [7]	2014	Devnagiri HCR	Neural network approach
Patil et al. [8]	2012	Devnagiri HCR	Neural network approach
Kaur et al. [9]	2012	Handwritten Gurumukhi character	Back propagation approach
Nohaj et al. [10]	2011	Printed English text	Feed-forward network
Sharma et al. [11]	2013	English letter	Neural network approach
Pradeep et al. [12]	2011	Handwritten alphabets	Multilayer feedforward network
Bhattacharya et al. [13]	2012	Handwritten Bangla character	MLP, MQDF
Pal et al. [14]	2015	Handwritten Bangala character	Convolutional neural network

2 A Brief Review on Odia Script

India is a multi-lingual and multi-script nation, including diverse official languages and has more than a hundred-local script.

Odia is, the sixth most prominent language in India derived from Indo-Aryans language. Odia is one of the most generally spoken languages in the world. More than 35 million individuals in the eastern part of Indian talk in this language. The Odia language derived from the ancient Brahmi script through different alterations. The writing style of Odia language is unique in comparison with some other regional languages in India. The modern Odia script comprises of 13 vowels, 36 consonants and five additional primary characters and ten digits.

Following are some properties of Odia Script:

- The Writing style of Odia is from left to right.
- The idea of upper and lower case (as in English) is missing here.
- Each word in Odia script made out of a few characters isolated from one another.
- In a compound Odia character, the ‘Maatra’ is available at the top, base, left, right or combination of any of these four places of a primary Odia character.

3 Artificial Neural Network

Pattern Recognition is amazingly hard to automate. Animals are familiar with different items and make sense of the enormous measure of visual data with minimal effort. An Artificial Neural Network as the backend is utilized for performing characterization and Recognition assignments. In offline CR frameworks, the Neural Network (NN) has risen as a fast and useful tool for classification towards accomplishing high recognition. NN is broadly categories into two types: feed-forward(FF) and feedback(FB) network. In an FFNN, hubs organized as the input layer, one or more Hidden Layer and an output layer stacked on one above the other. Back Propagation is a learning rule for the preparation of multilayer FFNN. Back Propagation(BP) gets its name from the procedure of propagating the errors in the network in backward from the output layer. For BPNN, it must be accessible to a collection of training data set and the appropriate responses or correct explanation of the set.

In Pattern recognition frameworks, typically, there are uses two passes. The first pass is the feature extractor that includes the extraction of specific features in a character set (for example, discovering bars of pixels inside a picture for CR). The subsequent pass is the classifier that is more general purpose and can be trained using a NN process with sample data sets. Thus, without focusing on building complicated feature extraction and detection algorithms, this paper focuses on implementing a standard BPNN. It also encapsulates and deals with various pre-processing stages for effective recognition.

Following are some benefits of NN over CR:

- The application of NN in HCR brings the idea of reading multiple combinations of handwritings
- HCR will be an effective method as evidence collection in forensic applications.
- Noise can be reduced to a large extent from the original image with the help of NN.
- Here the main aim is to recognize character with more accuracy irrespective of font and size.

4 Design and Implementation of Model

Initially, the Algorithm was designed for the extraction of Character. Here MATLAB is used as a tool for implementing the Algorithm. Then the NN model is developed that would give the optimum results. There was no specific way of finding the accurate Model of NN. It created upon trial and error basis. We have focused on two main phases; Pre-processing and CR. In the initial phase, the pre-processing is done for a scanned document for the extraction of individual Characters from it and normalizing each Character. Initially, we specify an input image file, which opened for reading and pre-processing. The image in jpeg file would be in RGB format (usually), so we convert it into a binary format that is in grayscale format. Then use the threshold to save and convert this grayscale image to binary, i.e., to denote all the pixels below a certain threshold as 0 and above it as 1.

In the implementation process of recognizing handwritten Odia character, firstly the extracted characters are fit into a window without white spaces on all the four sides then feature vector created for each character. A database designed for the characters after normalizing into 32 × 32 pixels. Window to viewport transformation used in the normalization process. Pixel-wise mapping from the original image to the pixel of corresponding normalized image done for every character. After the normalization of the extracted character of the input test image, the 2-D Normalized correlation coefficients approach used to identify similar patterns between a test image and the standard database images.

Following are the steps for the proposed method:

- Input or read the scanned handwritten image document (RGB).
- Convert it to a grey scaled image.
- Pre-processing of the grey-scaled image.
- Character-wise extraction process.
- Mapping of isolated character for recognition using NN.
- Conversion of character images to text after recognition.

5 Experimental Result

For the proposed method various experiments have been performed and tested. The experiments performed on various text images using the software Matlab (R2015a) for Odia Characters recognition [11, 12].

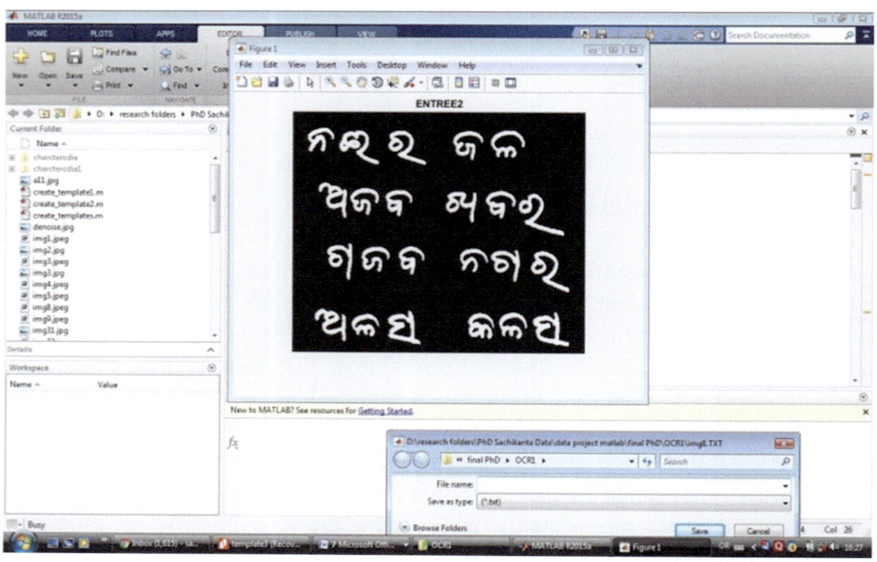

Fig. 1. Scanned image for recognition

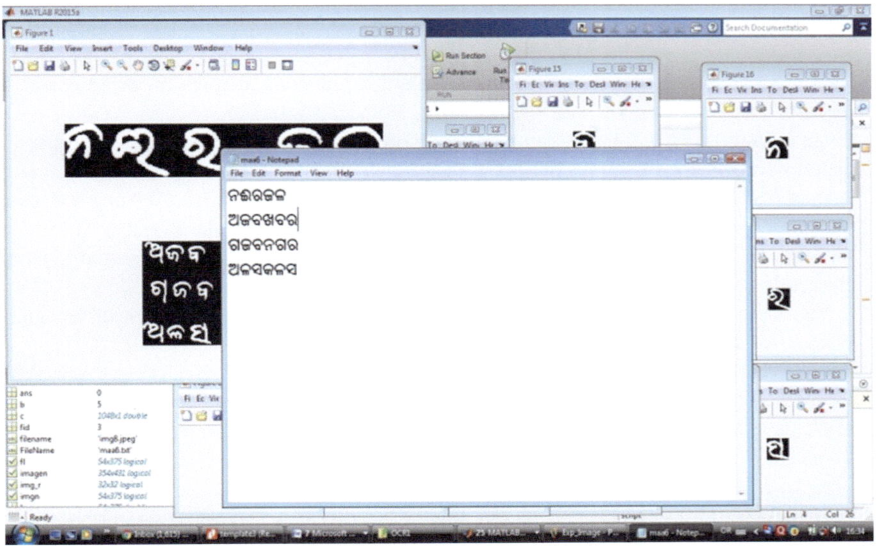

Fig. 2. Recognized characters

The result obtained in generated text file (Fig. 1 and 2) is correct owing to the line and character-wise segmentation method, which starts to read the input from the first line of the texts to the bottom line and then from the left side to the right side. The advantages of using this method are owing to its simplicity but can apply for recognizing [13] various sets of handwriting provided through templates.

6 Conclusion and Future Scope

The recognition process of Handwritten Odia Character using NN has completed effectively and successfully. The significance of the pre-processing stage in this work is enormous and incorporates steps like Skew correction, binarization, noise elimination. The procedure of the NN approach consists of classification along with preparing a set of the database of character for recognizing the Odia character from the handwritten input text. Through this strategy, we can get our results with 99% accuracy. The percentage of accuracy of this work can improvise by including a wide variety of handwritings that enables us to encompass a broad set of variations with the inclusion of more massive data sets and handwriting variation.

References

1. Vats I, Singh S (2014) Offline handwritten english numerals recognition using correlation method. Int J Eng Res Technol (IJERT) 3(6). ISSN: 2278-0181
2. Singh G, Lehri S (2012) Recognition of handwritten hindi characters using back propagation neural network. Int J Comput Sci Inf Technol 3(4):4892–4895. ISSN 0975-9646
3. Sayyad SS, Jadhav A, Jadhav M, Miraje S, Bele P, Pandhare A (2013) Devnagiri character recognition using neural networks. Int J Eng Innovative Technol (IJEIT) 3(1)
4. Mehfuz S, Katiyar G (2012) Intelligent systems for off-line handwritten character recognition: a review. Int J Emerg Technol Adv Eng 2(4)
5. Borde S, Shah E, Rawat P, Patil V (2012) Fuzzy based handwritten character recognition system. Int J Eng Res Appl (IJERA) VNCET 30 Mar'12. ISSN: 2248-9622
6. Kala R, Vazirani H, Shukla A, Tiwari R (2001) An overview of character recognition focused on off-line handwriting. IEEE Trans Syst Man Cybern Part Appl Rev 31(2)
7. MSA Dongare, Dhananjay B. Kshirsagar, Ms. Snehal V. Waghchaure (2014) Handwritten devanagari character recognition using neural network. IOSR J Comput Eng (IOSR-JCE) Ume 16(2):74–79. E-Issn: 2278-0661, P-Issn: 2278-8727 (Ver. X)
8. Patil MB, Narawade V (2012) Recognition of handwritten Devnagari characters through segmentation and artificial neural networks. Int J Eng Res Technol (IJERT) 1(6). ISSN: 2278-0181
9. Kaur M, Kumar S (2012) A recognition system for handwritten gurmukhi characters. Int J Eng Res Technol (IJERT) 1(6). ISSN: 2278-0181
10. Nohaj M, Jaka R (2011) Image preprocessing for optical character recognition using neural networks. Prediction Multidimension Econ Data Civilization Comput Game 30
11. Sharma N et al (2013) Recognition for handwritten english letters: a review. Int J Eng Innovative Technol (IJEIT) 2(7)
12. Pradeep J et al (2011) Diagonal based feature extraction for handwritten alphabets recognition system using neural network. Int J Comput Sci Inf Technol (IJCSIT) 3(1)
13. Bhattacharya U, Chaudhuri BB (2009) handwritten numeral databases of indian scripts and multistage recognition of mixed numerals. IEEE Trans Pattern Anal Mach Intell 31(3):444–457
14. Pal U, Sharma N, Wakabayashi T, Kimura F (2007) Handwritten numeral recognition of six popular Indian scripts. In: IEEE Ninth international conference on document analysis and recognition, vol 2, pp 749–753

Recognition and Extraction of Rain Drops in an Image Using Hough Transform

A. Susmitha[1(✉)], Lipsa Dash[1], and Sunanda Alamuru[2]

[1] New Horizon College of Engineering, Bengaluru, India
[2] SreeNidhi Institute of Science and Technology, Hyderabad, India

1 Introduction

In a dynamic weather condition such as rain, it produces sharp intensity changes in images that can considerably destroy the visibility of scene. The raindrops can refract and reflect the scene radiance and atmospheric illumination produced a complex visual of raindrops. Due to the refraction and reflection, the raindrops are brighter than its background. Raindrops are randomly distributed and moving at high velocity. The movement of raindrops with very high velocity known as rain streaks will produce spatial and temporal intensity variation in image and video. Therefore, detecting and removing raindrops will improve the quality of the image. To detect and remove raindrops and then restore the image back ground K-means clustering, Hough Transform and Gaussian filter are used. The k-means clustering gives maximum correct clustering rate and Hough transform is used to extract and remove raindrops from the image.

2 Related Work

In past few years, many methods have been proposed for the removal of the rain. The different papers and the researchers who contributed to it are discussed below.

- Henderson [1] proposed that weather is defined as the current atmospheric condition such as temperature, rainfall, wind, and humidity. Bad weather conditions can be classified into two which are the static or steady weather conditions and dynamic weather conditions. The static weather condition including fog, mist and haze whereas the dynamic weather conditions are rain, hail and snow. Figure 1 shows the classification of bad weather conditions.
- Narasimhan and Nayar [2] developed a model to detect bad weather conditions in images. The detection algorithm can be classified into four; no explicit detection, per-pixel detection, patch-based detection and frequency-based detection. Figure 2 summarizes the detection algorithm that has been used by researchers.

M. N. Mohanty and S. Das (eds.), *Advances in Intelligent Computing and Communication*, Lecture Notes in Networks and Systems 109,
https://doi.org/10.1007/978-981-15-2774-6_13

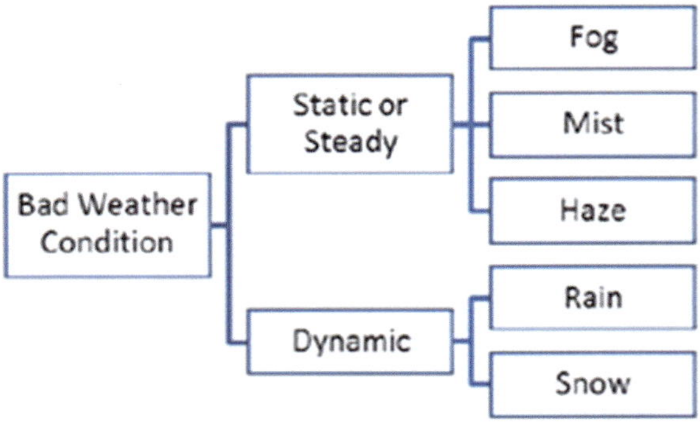

Fig. 1. Bad weather classification

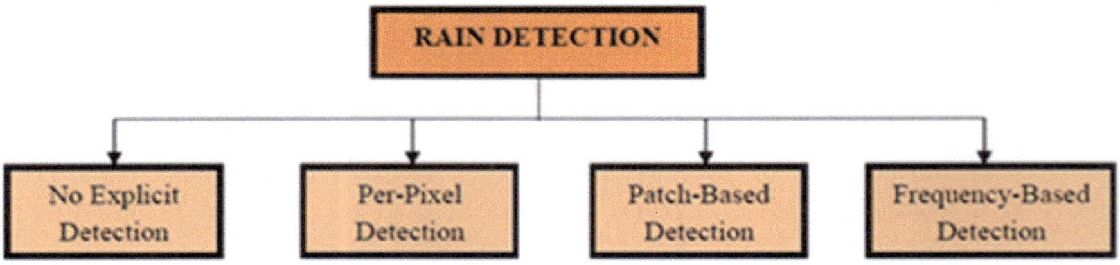

Fig. 2. Taxonomy of rain detection in image [3]

(a) No Explicit Detection

Some researchers did not perform the detection algorithm to remove rain from images Narasimhan and Nayar [2]. This detection dealt with single image of mild rain. Since the faraway image of rain was taken, the image looked like in a foggy condition. They used the same algorithm to restore the contrast of the rainy image. Some researchers simply applied filtering to remove the rain from the images [4, 5]. Coincidentally, this technique will perform poorly when it is applied to the moving objects scene.

(b) Per-Pixel Detection

The pixels of an image can be divided into rain and non-rain parts. Zhang et al. implement k-means to the grayscale intensity of each pixel to divide the pixels into the rain part and non-rain part. Since the rains have temporal properties, some pixels are misclassified.

(c) Patch-Based Detection.

This detection detects the region of rain in an image. By applying the constraints of photometric model in each frame in the video, Garg and Nayar [6] manage to detect the candidate rain pixels. For a three frames static background scene, the intensities of previous frame (In − 1) must be equal to the intensities of the next frame (In + 1) and the change of intensities in the current frame (nth) must satisfy the constraint, c (threshold: minimum change in intensity produced by a raindrop that detected in the presence of noise). When this constraint of photometric model applied to a scene with object motions, this algorithm will detect some false positives due to the movement of the object 6. Barnum et al. [3] found that this algorithm fails to detect individual streaks in an image [3]. Figure 4 shows the detection of rain by Garg and Nayar [6].

(d) Frequency-Based Detection

Barnum et al. [7] used spatio-temporal frequency method to detect rain in an image. The inverse transform of the estimated proportion of energy of rain is used to detect the rain streak in a single frame.

- Tripathi and Mukhopadhyay [8] proposed rain removal techniques can be divided into two main approaches which are time domain-based approach and frequency domain-based approach. In the time domain-based approach basically considered the properties of rain itself including the chromatic, temporal or both. Figure 3 shows the taxonomy of rain removal techniques.

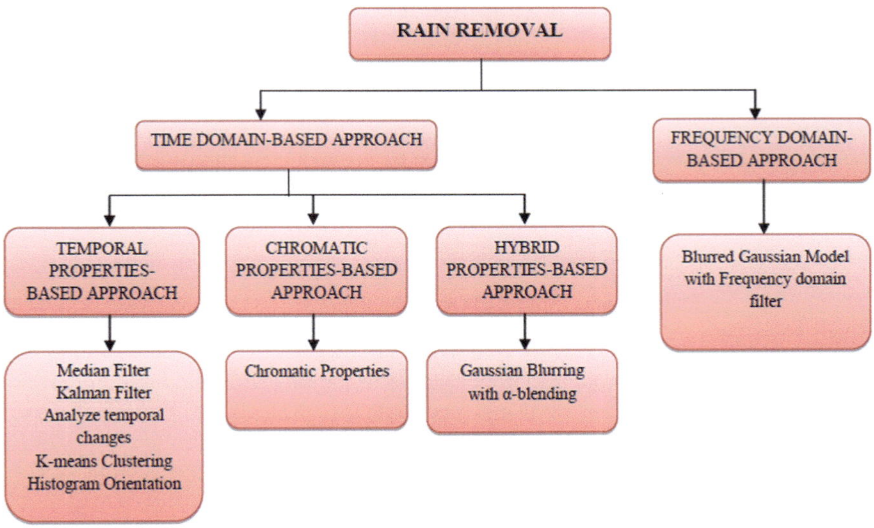

Fig. 3. Taxonomy of rain removal techniques [8]

(a) Temporal Properties-Based Approach

Due to the temporal properties, rains are not remaining fixed at one time. For a static scene, Starik and Werman [4] applied temporal median filter for each pixel to reduced rain effects from images. Even though this technique is one of the easiest techniques to remove rain from images, but this technique will produce blurring when applied in a dynamic scene. By taking the advantage of temporal properties of rain, Park and Lee [9] proposed a new algorithm which is the intensity estimation by using Kalman filter in a real-time situation. Since this algorithm estimated that the video captured by using a static camera with static background, therefore it fails to estimate the pixels' intensity in a dynamic background.

(b) Chromatic Properties-Based Approach

The chromatic properties-based approach proposed by Liu et al. (2008) deals with colored pixel images. By using the chromatic properties, two relations between the same pixels in two consecutive frames can be developed (Liu et al. 2008). As the raindrop falling to the ground, it will affect the pixels' intensity. Therefore, the first relation is between the background pixel and the rain-affected pixel where the rain-affected pixels are brighter than the background and there are two related rain-affected pixels in the consecutive frames.

To restore the rain-affected pixels, the intensity of neighborhood is used to estimate the background value. The disadvantage of this technique as stated by Tripathi and Mukhopadhyay [8], this technique fails to differentiate between the rain pixels and the moving object pixels.

(c) Hybrid Properties-Based Approach

Basically, a hybrid properties-based approach is the composite of both temporal and Chromatic based approach. Due to the temporal properties of rain, the raindrops are not fixed at a particular pixel in all frames and the chromatic properties of raindrops stated by Zhang et al. [7] show that the intensity changes of R, G and B channels due to the raindrops are approximately the same. The raindrop pixels are detected by applying the K-means clustering to the intensity histogram of the video. In the process of removing the raindrops, Zhang et al. (2006) applied dilation with Gaussian blurring to the detected pixels and then used them as the channel. The raindrops are removed from the videos by using-blending technique.

(d) Frequency Domain-Based Approach

This approach basically analyze the frequency information of each frame. Proposed by Barnum et al. [3] can be applied in either single frame or multiple consecutive frames. After the rain streaks detected by using rain detection, the removing of rain pixels can be done by replacing them with their temporal neighbors. This method rarely removes the rain streak completely but it will modify the brightness of rain streak by reducing the brightness.

- Garg and Nayar [10] proposed a method by adjusting the camera parameters. Here exposure time is increased or the depth of field is reduced. However, this method fails to handle heavy rain and fast-moving objects which are close to the camera.
- Zhang et al. [11] proposed a method based on the chromatic and temporal properties. This method uses k-means clustering to estimate the non-rain-affected pixel value to inpaint the rain-affected pixels. This clustering method is effective only in removing rain from static background when there is no moving object. This method uses all the frames available in a sequence for the removal of the rain.
- Barnum et al. [7], Barnum and Narasimhan [3] proposed a method for the detection and removal of rain streaks by using frequency information of each frame. Here a blurred Gaussian model is used to approximate the blurring produced by the raindrops. This model is suitable when the rain streaks are prominent, but this blurred Gaussian model fails to detect the rain streak when it is not sharp enough.

- Liu et al. [12] proposed a method for the removal of rain by using chromatic based properties in rain-affected videos. It fails to detect all possible rain streaks. The reason could be that chromatic property is not satisfied in practice as described in the previous discussion. This method requires at least three consecutive frames for the removal of rain.

3 Proposed Method

The proposed hybrid method is shown in Fig. 4.

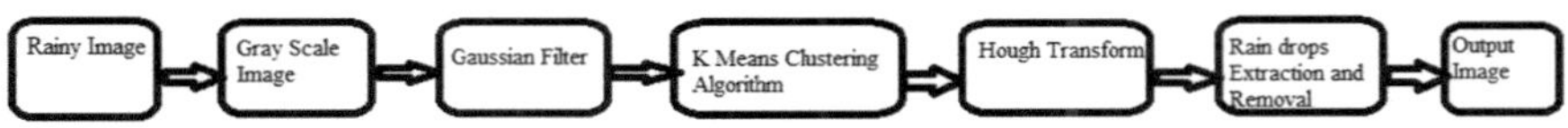

Fig. 4. Block diagram of rain drop recognition and extraction

3.1 Color Image to Grayscale Image Conversion

The input image is a rainy image which is in color. Grayscale image is convenient for processing and to make clusters. Hence the color image must be converted into grayscale image.

3.2 Gaussian Filter

After converting the color image into grayscale image, it needs to be processed that is the noise present in the image has to be removed. For noise removal, Gaussian filter is used. Filtering is an operation of removing frequency components from an image selectively. The Gaussian Filter is smoothing filter that is used to 'blur' images and remove detail and noise. It is similar to the mean filter, but it uses a different kernel that represents the shape of a Gaussian ('bell-shaped') hump.

3.3 K-Means Clustering Algorithm

Clustering is performed after noise removal. cluster is a collection of objects which are "similar" between them and are "dissimilar" to the objects belonging to other clusters. To achieve this, the k-means algorithm is used. It is the most important unsupervised learning algorithm. It determines the best separation of observations, based on the minimizing function from each input parameter to the cluster centroid. In our proposed method the image is divided into clusters of 2, 3…10 to find the correct cluster number K. In this method, the first step of k-means clustering is to randomly choose 2 (In this case where $k = 2$) arbitrary means. Second, all observations are assigned to 1 of the two clusters, based on their distance to each mean. Third, the mean of each cluster is updated based on associated observations. Steps 2 and 3 are iteratively repeated until a stopping criteria is hit (i.e., no observations change membership, the sum of the distances is minimized or the maximum number of iterations is reached).

Algorithm

1. First, initialize k points, called means, randomly.
2. Categorize each item to its closest mean and update the mean's coordinates, which are the average of.
3. the items categorized in that mean so far.
4. Repeat the process for a given number of iterations until clusters are formed.

3.4 Hough Transform

The Hough transform is a technique that is used to isolate features of a particular shape within an image. Because it requires that the desired features be specified in some parametric form, the classical Hough transform is most commonly used for the detection of regular curves such as lines, circles, ellipses, etc. A generalized Hough transform can be employed in applications where a simple analytic description of a feature(s) is not possible. Due to the computational complexity of the generalized Hough algorithm, we restrict the main focus of this discussion to the classical Hough transform. Despite its domain restrictions, the classical Hough transform retains many applications, as most manufactured parts contain feature boundaries which can be described by regular curves. The main advantage of the Hough transform technique is that it is tolerant of gaps in feature boundary descriptions and it is relatively unaffected by image noise. The Hough technique is particularly useful for computing a global description of a feature(s), given local measurements. We can use Hough transform to detect other features with analytical descriptions. For instance, in the case of circles, the parametric equation is $(x - a)^2 + (y - b)^2 = r^2$ where a and b are the coordinates of the center of the circle and r is the radius. In this case, the computational complication of the algorithm initiates to increase. We research with the Hough circle detector on raindrop image. One way of reducing the computation required to perform the Hough transform is to make use of gradient.

4 Experimental Result

The input image is a color image. To perform clustering, input image should be in grayscale form. So the YCBCR color space is used to convert image into grayscale form. YCBCR gives an accurate gray component. It is also useful to reconstruct the original RGB color image once raindrop detection and removal process are over. Once the conversion is performed we will move towards noise removal. After noise removal, we perform clustering using k-means clustering algorithm. Then after we use Hough transform to detect and remove the raindrops from the single image. Once we detect and remove the raindrop from the image, we will get the output image which is devoid of the rain drops. The output image which we obtain through various techniques such as RGB to YCbCr conversion, K-means clustering, Gaussian filter, Hough Transform is in the form of a GUI (Graphical User Interface) which is as shown Figs. 5, 6 and 7.

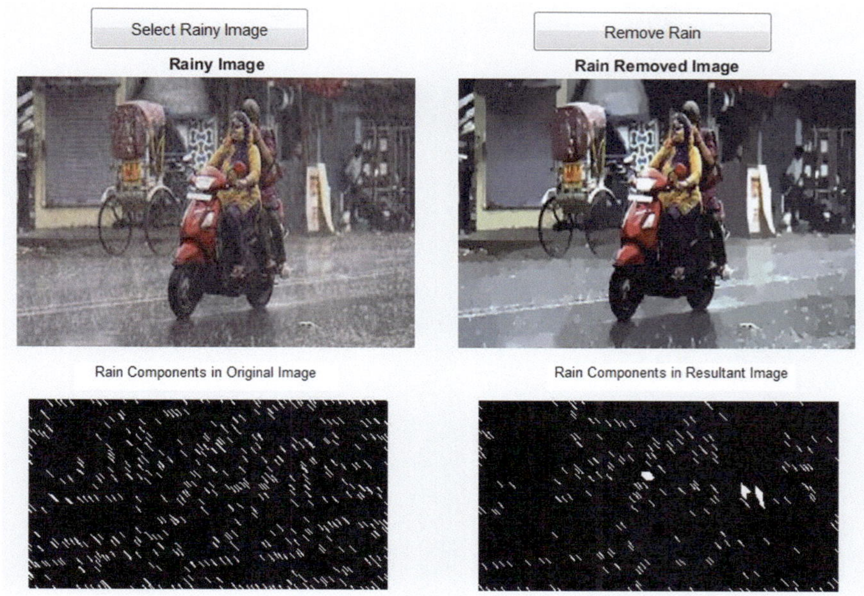

Fig. 5. OUTPUT1

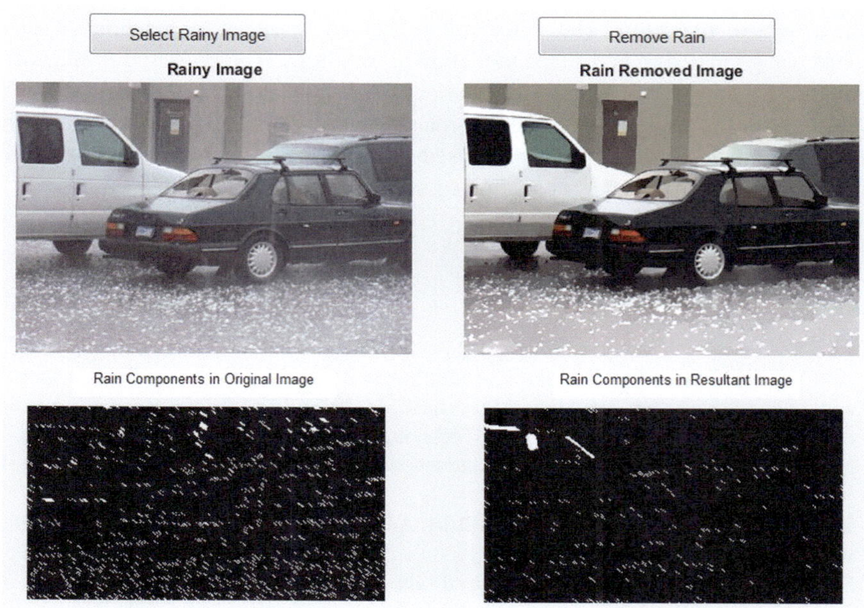

Fig. 6. OUTPUT2

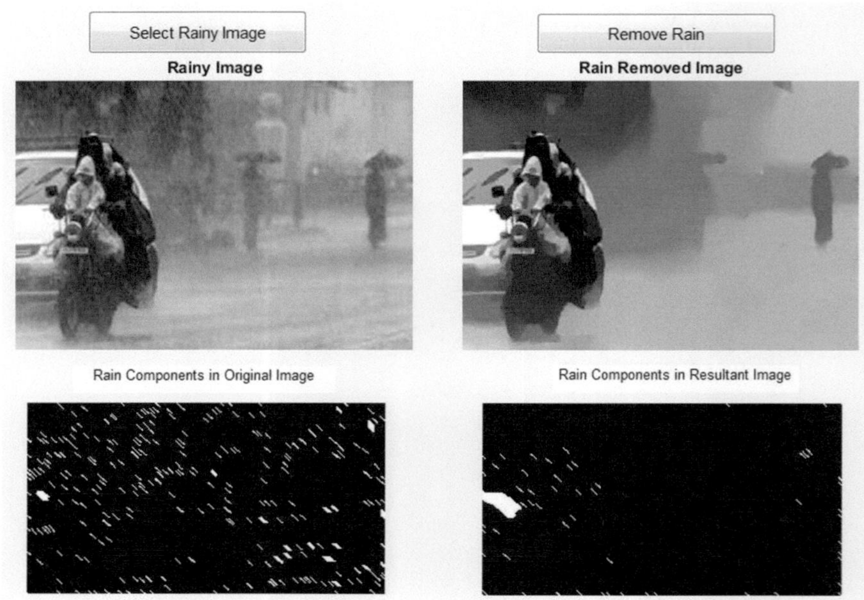

Fig. 7. OUTPUT3

5 Conclusion

In this paper, we have offered a new approach of recognition and extraction of raindrop using a hybrid algorithm. The algorithm is framed in order to recognize the rain droplets using clustering and shape modeling of raindrops. The proposed framework is based on K-means clustering and Gaussian filter for the efficient retrieval of rain droplets from the rainy image. The k-means clustering results in highest correct clustering rate and Hough transform is used to detect and remove the raindrops from single image. The proposed system showed better results than available approaches for raindrop recognition and deletion.

References

1. Abd Wahab MH, Su CH, Zakaria N, Abdul Salam R (2013) Review on raindrop detection and removal in weather degraded images. In: CSIT 2013: The 5th International Conference on Computer Science and Information Technology, 27–28 Mar 2013. Amman, Jordan
2. Narasimhan and Nayar (2003) IEEE International Conference on Computer Vision, vol 2, pp 1067–1074
3. Barnum P, Narasimhan SG, Kanade T (2010) Analysis of rain and snow in frequency space. Int J Comput Vis 86(2–3):256–274
4. Starik S, Werman M (2003) Simulation of rain in videos. In: Proceedings of Texture: The 3rd International Workshop on Texture Analysis and Synthesis, 17 Oct 2003. Nice, France, pp 95–100

5. Shariah S, Marhaban MM, Ramli AR, Mohd Noor SBB (2011) Rain detection and removal using a new algorithm. In: Proceedings of the National Geoscience Conference, 11–12 Jun 2011. The Puteri Pacific Johor Bahru, Johor, Malaysia
6. Garg K, Nayar SK (2004) Detection and removal of rain from videos. IEEE Comput Soc Conf Comput Vis Pattern Recognit 1:528–535
7. Barnum P, Kanade T, Narasimhan SG (2007) Spatio temporal frequency analysis for removing rain and snow from videos. In: Workshop on Photometric Analysis For Computer Vision (PACV), in conjunction with ICCV
8. Tripathi AK, Mukhopadhyay S (2014) Removal of rain from videos: a review. SIViP 8:1421–1430
9. Park WJ, Lee KH (2008) Rain removal using Kalman filter in video. In: International Conference on Smart Manufacturing Application, pp 494–497
10. Garg K, Nayar SK (2005) IEEE International Conference on Computer Vision, vol 2, pp 1067–1074
11. Zhang X, Li H, Qi Y, Leow WK, Ng TK (2006) Rain removal in video by combining temporal and chromatic properties. In: IEEE International Conference on Multimedia and Expo
12. Liu P, Xu J, Liu J, Tang X (2009) Pixel based temporal analysis using chromatic property for removing rain from videos. Comput Inf Sci 2(1):53–60
13. Fattal R (2008) Single image dehazing. In: SIGGRAPH
14. Halimeh JC, Roser M (2009) Raindrop detection on car windshields using geometric-photometric environment construction and intensity—based correlation. In: IEEE Intelligent Vehicles Symposium, 2009, pp 610–615. 978-1-4244-3504-3/09/$25.00 ©2009
15 Hassim R, Bade A (2015) Taxonomy of rain detection and rain removal techniques. Transact Sci Technol, 28–35
16. He K, Sun J, Tang X (2008) Single image haze removal using dark channel prior. In: IEEE Computer Society Conference on Computer Vision and Pattern recognition (CVPR)
18. Ramesh Kanthan M, Naganandini Sujatha S (2013) Automatic grayscale classification using histogram clustering for geometric active contour models. Int J Curr Eng Technol 3(2). ISSN 2277-4106
19. Ramesh Kanthan M, Naganandini Sujatha S (2015) Rain drop detection and removal using K-means clustering. In: 2015 IEEE International Conference on Computational Intelligence and Computing Research. 978-1-4799-7849-6/15
20. Roser M, Geiger A (2009) Video-based raindrop detection for improved image registration. In: 2009 IEEE 12th International Conference on Computer Vision Workshops, ICCV Workshops
21. Tan RT (2008) Visibility in bad weather from a single image. In: IEEE Conference on Computer Vision and Pattern Recognition (CVPR 08)
22. You S, Tan RT, Kawakami R, Ikeuchi K (2013) Adherent raindrop detection and removal in video. In: Proceedings of the IEEE Conference on Computer Vision and Pattern Recognition (CVPR)
23. You S, Tan RT, Kawakami R, Mukaigawa Y, Ikeuchi K (2014) Raindrop detection and removal from long range trajectories. In: IEEE Conference on Computer Vision and Pattern Recognition (CVPR)
24. You S, Tan RT, Kawakami R, Mukaigawa Y, Ikeuchi K (2016) Adherent raindrop modelling, detection and removal in video. IEEE Trans Pattern Anal Mach Intell 38(9)
25. Webster DD (2014) Automatic rain drop detection for improved sensing in automotive computer vision applications. M.Sc. Thesis, Cranfield University

A Novel Technique for Target Recognition Using Multiresolution Technique

Tamanna Sahoo and Bibhuprasad Mohanty(✉)

Department of Electronics and Communication Engineering, ITER, Siksha 'O' Anusandhan University (Deemed to be University), Bhubaneswar, India
{tamannasahoo, bibhumohanty}@soa.ac.in

1 Introduction

Long ago, computer vision area has seen a rapid development in automation which has enabled researchers to develop models or algorithms [1] using human visual system and its basic components as important part of the models, providing better detection or tracking results. The main objective is the identification of specified targets in a scene under different categories (stationary or moving) and environment (atmospheric turbulence) [2]. Identification of specific targets is one of the difficult tasks with these restriction which led to many researches and approaches such as pattern recognition, traditional thresholding methods, wavelets [3], texture-based approaches, descriptor-based methods, and so on [4]. This paper mainly concentrates on approaches based on wavelet transform and its wavelet coefficient features(WCFs).

One of the major advantages of multiresolution technique (wavelet transform) is that it can be implemented both in pyramidal or tree structure. It also provides similar subband decomposition in consecutive scales. Further properties like rapid processing, adaptation to changing local image statistics, exact position information and adaptation to high background noise, are certain advantages of Discrete Wavelet Transform.

In this paper, the main objective is to identify the target object in a scene, which is to be achieved by calculating features using gray level co-occurrence matrix(GLCM), of detail coefficients that is subband of transformed sub-block images(non-overlapping and adjacent blocks) of same sizes of the input image. The approach for computing GLCM of detail subband is given by:

(i) In the first approach, GLCM of each single detail subband image is computed and summed to get the GLCM of whole sub-block image [1].
(ii) Second, the dynamics [5] of the sub-block image are first created by applying inverse DWT to detail coefficient and then calculating GLCM from the dynamic image of sub-block image.

The calculation of GLCM has led to computation of combination of wavelet coefficient features (WCFs) (cluster prominence, cluster shade, and contrast), which are used for this paper. The highest value obtained by combining the feature values is recognized as seed block. Further seed point is extracted from seed block which is used in region growing process to obtain the specified target object.

M. N. Mohanty and S. Das (eds.), *Advances in Intelligent Computing and Communication*, Lecture Notes in Networks and Systems 109,
https://doi.org/10.1007/978-981-15-2774-6_14

The remainder of the paper is prepared accordingly. In Sect. 2, the scope of approach has been discussed. Further in Sect. 3, the proposed system via Discrete Wavelet Transform [6] and wavelet coefficient features are described. The experimental results are shown and the details are discussed in Sect. 4, followed by a conclusion and forthcoming work on the proposed work has been explained in Sect. 5.

2 Scope of Approach

The core approach of this work is to develop a constructive system which will be able to recognize the target object under diverse types of atmospheric turbulence and various kind of object. And taking the advantage of multiresolution analysis of Discrete Wavelet Transform, it has been used for this paper.

3 Proposed Method

The steps of the proposed system are shown in figure above and it is described below.

- Here, for the division of input image of size N × M into sub-blocks of 32 × 32 or 16 × 16, the input image is first resized into 512 × 512. Further, the input image is converted to gray-scale image(if it is a color one) for better processing (Fig. 1).

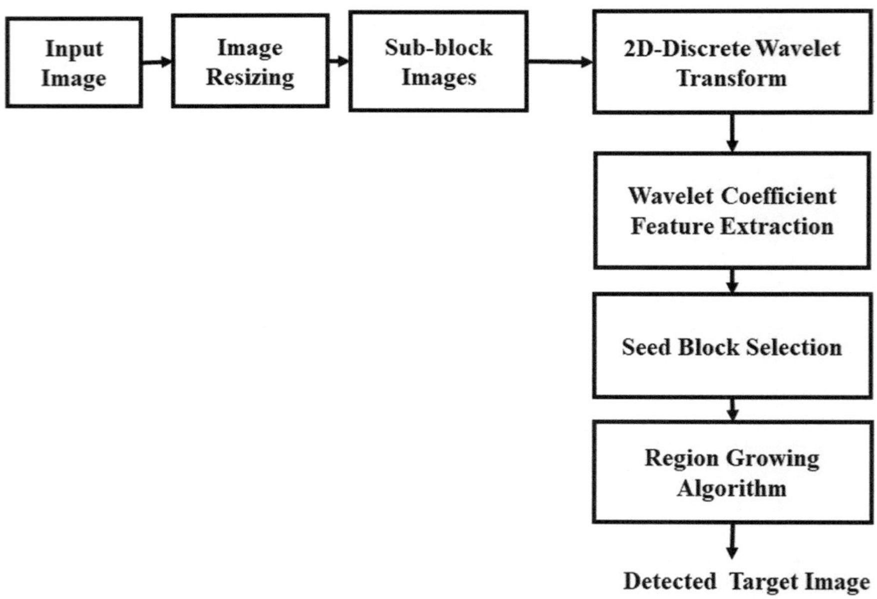

Fig. 1. Basic building block of the proposed detected target system

- The division of sub-block image is carried out in such a way that it is non-overlapping and adjacent blocks. The distinct sub-block images, starting from the top-left corner, are decomposed into two-level two dimensional discrete wavelet transform [7].

$$F_{IA}^{-(j+1)}(x,y) = \left(\bar{h}^{*}(-x) * \left(\bar{h}^{*}(-y)F_{IA}^{-(j)}(x,y)\right) \downarrow 2y\right) \downarrow 2x \tag{1}$$

$$F_{IH}^{-(j+1)}(x,y) = \left(\bar{h}^{*}(-x) * \left(\bar{g}^{*}(-y) * F_{IH}^{-(j)}(x,y)\right) \downarrow 2y\right)2x \tag{2}$$

$$F_{ID}^{-(j+1)}(x,y) = \left(\bar{g}^{*}(-x) * \left(\bar{h}^{*}(-y) * F_{ID}^{-(j)}(x,y)\right) \downarrow 2y\right)2x \tag{3}$$

$$F_{IV}^{-(j+1)}(x,y) = \left(\bar{g}^{*}(-x) * \left(\bar{g}^{*}(-y) * F_{IV}^{-(j)}(x,y)\right) \downarrow 2y\right)2x \tag{4}$$

where F_{IA}, F_{ID}, $F_{IV,}$ and F_{IH} are the wavelet decomposition subbands (approximation, diagonal, vertical and horizontal coefficient).

- From the transformed coefficient, gray level co-occurrence matrix (GLCM) is calculated. GLCM [8] is the gray level occurrence in the configuration which is represented as matrix, consisting of relative frequencies Co (p, q, d, Θ). The p, q represents the two gray level pixels and how frequently these two pixels appear in the window in the direction Θ and separation between the pixels is given by a displacement vector d. For instance, if the displacement vector is (1, 0) then in the direction of 90 it is represented as one pixel below and zero pixel to the left [1].

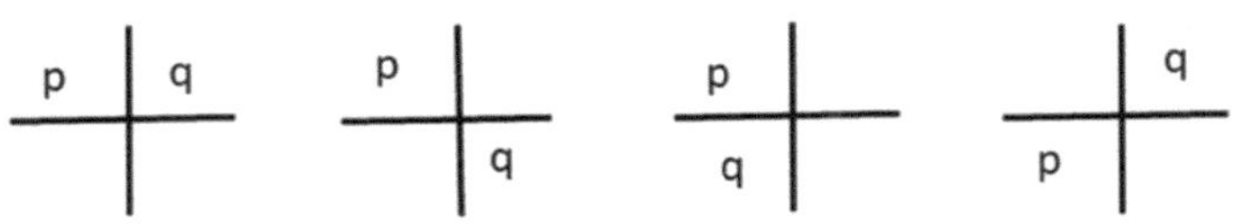

Fig. 2. Co-occurrence matrix orientation. **a** $\Theta = 0$, $d = (0, 1)$, **b** $\Theta = 45$, $d = (1, 1)$, **c** $\Theta = 90$, $d = (1, 0)$, **d** $\Theta = 135$, $d = (1, -1)$

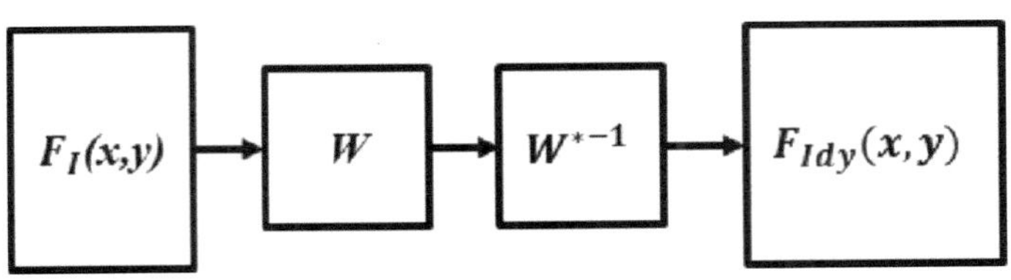

Fig. 3. Dynamics in wavelet domain

The figure below demonstrates the co-occurrence matrix in different orientations. In the present work, it is derived from $\Theta = 0$ and $d = (0, 1)$ (i.e., zero pixel below and one pixel to the right) for detail sub-bands (Figs. 2 and 3).

- The calculation of GLCM [9] is one of the major tasks of the proposed system and is performed only on the detail sub-bands of both the levels of the sub-block image in two ways:

  (a) **Method 1**: It is done by taking computing GLCM of each detail coefficient image and then combining them to provide the GLCM of sub-block image.
  (b) **Method 2**: By creating the dynamics which is given by the formula below [5] for each sub-block image using the formulae given below and then computing the GLCM.

$$W^{*-1} = [W] - F_{IA}^{-j}(x, y) \tag{5}$$

- From the obtained GLCM, the evaluation of wavelet coefficient features (cluster prominence, cluster shade, and contrast) using the formula presented below are calculated and this is performed for each sub-block.

$$\textbf{Contrast} = \sum_{p,q=1}^{N} (p-q)^2 \mathrm{Co}\,(p, q) \tag{6}$$

$$\textbf{Cluster shade} = \sum_{p,q=1}^{N} (p - M_x + q - M_y)^3 \mathrm{Co}\,(p, q) \tag{7}$$

$$\textbf{Cluster prominence} = \sum_{p,q=1}^{N} (p - M_x + q - M_y)^4 \mathrm{Co}\,(p, q) \tag{8}$$

where,

$$M_x = \sum_{p,q=1}^{N} p\mathrm{Co}(p, q), M_y = \sum_{p,q=1}^{N} q\mathrm{Co}\,(p, q) \tag{9}$$

Contrast feature values are acquired by normalizing linearly whereas cluster prominence and cluster shade are acquired by normalizing logarithmically, depending upon the dynamic ranges of the features.

- The next step is the selection of seed block which is the sub-block image having maximum of combined normalized feature value (cluster prominence, cluster shade, and contrast). The concept behind the selection of seed block is that the high value of wavelet coefficient feature represents that it is a part of the target.

The seed block is used along with region growing algorithm for the identification of target object. Region growing is one of the segmentation process (based on region), usually depends on some predefined criteria (adjacency and threshold) sub-regions are grown into larger regions. In the present work, seed point is chosen as the criteria where the center point of the seed block [2]. Further, in the process, a comparison of all the unallocated neighboring pixels is created by differencing the pixel's intensity value with the region's mean(seed point) to grow the region. This can be used as a measure of similarity. The smallest pixel difference measured during the process is assigned to the respective region. The method's continuity depends on intensity comparison between region mean and new pixel and thresholding.

3.1 Algorithm of Proposed System

Input: Input image(size: $M \times N$)
Output: Detected target image

(a) Read the input image and resize it to 512 × 512.
(b) Obtain 32 × 32 sub-block image, initiating from the top-left corner.
(c) Decompose the sub-block images using 2D-DWT.
(d) Derive the Gray level co-occurrence matrix (GLCM) for detail subbands of transformed sub-block images using both the methods.
(e) Calculate and combine features (contrast, cluster prominence, and cluster shade) in wavelet domain, from GLCM.
(f) Reiterate the steps (c) to (e) for each sub-block images.
(g) Select a sub-block image as seed block (having maximum of combined wavelet coefficient feature value).
(h) Apply seed point (center of seed block) in region growing algorithm to obtain the target object.

4 Result and Discussion

In the previous section, the proposed algorithm has been discussed and applied on natural single-target image and the image chosen is having a clear background. Further, the DWT decomposition level used for the purpose is two-level and filter used is haar filter. In the region growing technique, the mean distance method using seed point is used which is obtained from the seed block. The detailed information of single-target image (a xylophone image) has been presented below.

Original image size = 360 × 240 × 3
Resized image size = 512 × 512
Method 1
Seed block = 105th sub-image block,
Seed points = (208,272),
Sum of WCF = 7.4292e + 03.

Method 2
Seed block = 89th sub-image block,
Seed points = (176,272),
Sum of WCF = 2.2956e + 03.

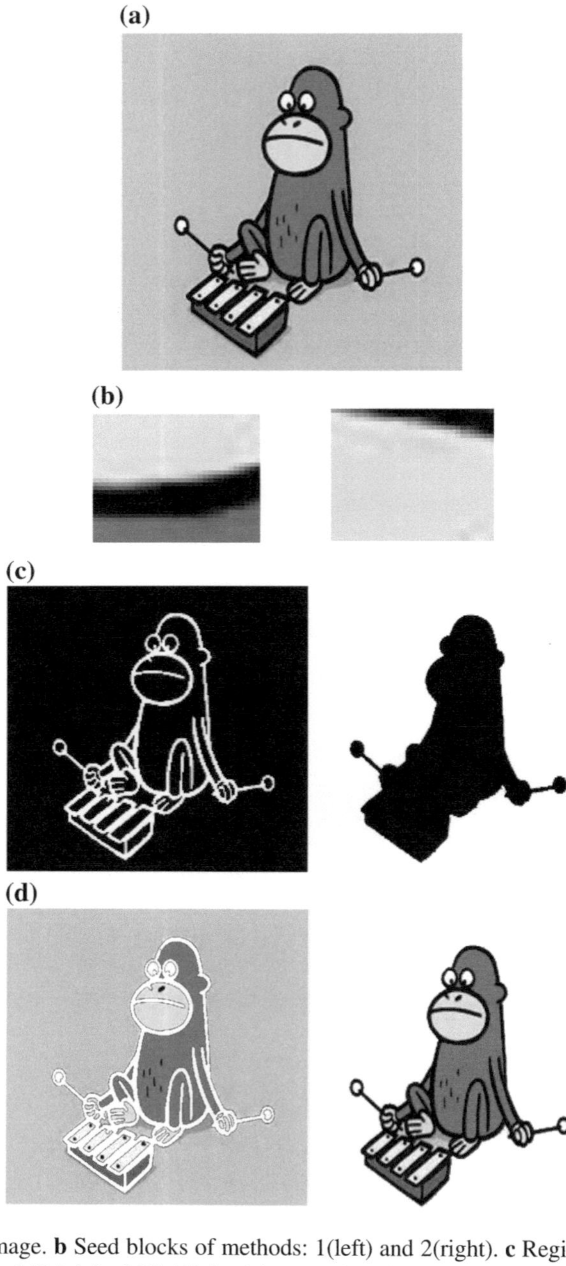

Fig. 4. **a** Input image. **b** Seed blocks of methods: 1(left) and 2(right). **c** Region grown image of methods: 1(left) and 2(right). **d** Highlighted image of methods: 1(left) and 2(right)

The results of proposed methods are shown in Fig. 4 where (a) shows the original resized image of 512 × 512 while (b), (c) shows the seed blocks obtained from two methods of wavelet coefficient feature values, (d) shows the image after region growing process using the seed points value and (e) represents the highlighted image.

5 Conclusion

Wavelet-based approach using dynamics for finding the coefficient features and seed block in the proposed algorithm is found to be convincing with respect to visual inspection in a single target image [10]. The proposed method can be extended for multiple targets which can be either stationary or moving and with cluttered backgrounds. In future, this proposed technique in wavelet domain can be extended at different steps of visual attention based moving object detection which is presently our research area [11].

References

1. Arivazhagan S, Ganesan L (2004) Automatic target detection using wavelet transform. EURASIP J Appl Sig Process 17:2663–2674
2. Pulla Rao C, Guruva Reddy A., Rama Rao CB (2016) Target detection using multi resolution analysis for camouflaged images. Int J Cybern Inf 5(4)
3. Espinal F, Huntsberger TL, Jawerth BD, Kubota T (1998) Wavelet-based fractal signature analysis for automatic target recognition. Opt Eng 37(1):166–174
4. Mahmoodabadi SZ, Ahmadian A, Abolhasani MD, Eslami M, Bidgoli JH (2005) Feature Extraction based on multiresolution wavelet transform. In: IEEE Proceedings, pp 3902–3905
5. Sahoo T, Mohanty B (2018, September) Moving object detection using background subtraction in wavelet domain. In: 2nd International Conference on Data Science and Business Analytics (ICDSBA)
6. Gomes JPP, Brancalion JFB, Fernandes D (2008) Automatic target recognition in synthetic aperture radar image using multiresolution analysis and classifiers combination. IEEE
7. Arivazhagan S, Ganesan L (2003) Texture classification using wavelet transform. Pattern Recogn Lett 24(9–10):1513–1521
8. Hazra D (2011) Texture recognition with combined GLCM, wavelet and rotated wavelet features. Int J Comput Electr Eng 3(1):17938163
9. Shan Z, Aviyente S (2005) Image denoising based on the wavelet co-occurrence matrix. ICASSP2005, IEEE, pp 645–648
10. Pan Y, Chen Y, Fu Q, Zhang P, Xu X (2011) Study on the Camouflaged target detection method based on 3D convexity. Proc Modern Appl Sci 5:152–157
11. Singh SK, Dhawale CA, Misra S (2013) Survey of object detection methods in camouflaged image. In: Proceedings of IERI, Elsevier, pp 1–6

A Novel Technique of Shadow Detection Using Color Invariant Technique

Leeza Panda and Bibhuprasad Mohanty(✉)

Department of Electronics and Communication Engineering, ITER, Siksha 'O' Anusandhan University (Deemed to be University), Bhubaneswar, India
{leezapanda, bibhumohanty}@soa.ac.in

1 Introduction

When object occludes light from a source of light which appears as surface edges, shadows occur in the image. Shadow detection and removal over the past decades cover many specific applications such as traffic surveillance [1, 2], face recognition [3–5] and image segmentation. Object shadow detection has been an active field of research for several decades. A shadow in an image has become a serious issue as it causes image distortion, object merging, object loss, etc. Based on the intensity of light, shadows are classified into two types: cast shadow and self-shadow. Cast shadows occur when an object is projected and a dark area forms in the direction of light whereas self-shadows are a certain example of cast shadow where the shadow of the body is extrapolated onto itself. Self-shadows are indefinite which do not have a prominent edge whereas cast shadows have sub-region like shading and interreflection (Fig. 1).

Images with shadows can decrease the performance of certain computer vision applications such as object detection, tracking, surveillance, etc. Hence it is necessary to find out the methods for shadow detection and removal, and moreover, it is also important to keep the original information intact while removal.

2 Brief Literature Review

Finlayson et al. [6] came up with a method where the original image with illumination invariant image recognizes the shadow features. Firstly the features are set to zero and then the representation of edges is done to get a shadowless image. During reintegration, error propagation can be reduced by closing the shadow features before the reintegration process takes place. This process is proved by Fredembach and Finlayson [7]. Lalonde et al. [8] proposed a method where geometrical features are being matched which detects shadow with good accuracy. One of the challenging problems here is when the shadow isn't on the ground. It appears in larger variation when the shadow

M. N. Mohanty and S. Das (eds.), *Advances in Intelligent Computing and Communication*, Lecture Notes in Networks and Systems 109,
https://doi.org/10.1007/978-981-15-2774-6_15

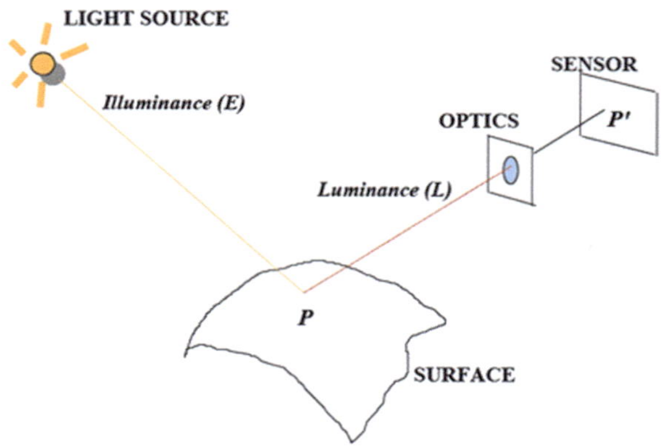

Fig. 1. Shadow occurrence illustration

isn't on the ground. Wu and Tang [9] proposed a method to detect vague shadows in an image using derivatives of the input image. A method to remove the shadows from curved areas retaining the background texture is proposed in [9]. The removal of shadows is achieved by calculating different scale factors for shadow regions and penumbra regions to cancel the effect of shadows.
Shadows detection can be categorized into five algorithms:

1. Invariant based shadow detection: 2D chromaticity projection was used along a line upright in the change in direction of illumination. This was used to obtain a 1D grayscale illumination invariant representation of an image which was the most prominent contribution of Finlayson et al. [7]. Edge detection was performed both on illumination invariant and original images. The shadow edges were the result of a difference in edge-maps.
2. Feature-based shadow detection: An algorithm that started with a hypothesis testing over a shadowed area on which they are formed was modeled by Salvador et al. [10]. Detection of Shadow pixels in HIS, YCbCr color spaces and hue difference of background and foreground regions were confirmed by the algorithm. Edge information was used for the detection using Sobel operator and subsequently Boolean AND was used to combine the detection results.
3. Region-based shadow detection: A region-based technique to detect shadows from natural scenes was modeled by Guo et al. [11]. A single region classifier incorporating SVM with a kernel and a pairwise classifier incorporating SVM with RBF kernel was trained by them. An energy function which amalgamated the single region classifier and pairwise classifier predictions were then minimized by using graph cut. Regions of the same material with the same illumination and different illumination were used as their pair classifiers.
4. Color model-based shadow detection: Salvador et al.'s [12] contribution to this category is one of the earliest ones. The generation of object contours was achieved using an edge map of the c1c2c3 color invariant of the actual image. As the luminance image is affected by shadows, dark regions that are shadow candidates are extracted using the edge map of a luminance image.

5. Interactive-based shadow detection: A shadow extraction technique based on quad map input by the user was proposed by Wu and Tang [9]. The quad map necessitated a user to mark four different regions in an image: shadow, non-shadow, uncertain and excluded regions. The input was subsequently used in a Bayesian framework to extract the shadow.

Shor and Lischinski proposed one of the simplest interactive methods in which a bare mouse click by the user marked the shadow region [13]. The shadow mask was then derived from the clicked area by iterative region growing and an illumination invariant distant measure. The technique worked only for a surface that had both shadowed and non-shadowed areas.

The process of Shadow detection encompasses two significant affirmations—detecting the shadowed region and re-establishing the illumination in the shadowed area. Some degree of image understanding is involved in the detection task which is used to determine whether a pixel is dark due to a shadow or caused due to the reflection at the concerned scene point. Accomplishing this task mandates some assumptions about the shadowed surfaces in the scene and expecting some hints from the user. The illumination restoration task is also appreciably challenging, as it attempts to remove any noticeable differences between the originally lit and the restored parts of the images. To be more particular, it is difficult to avoid the differences in local contrast and the amount of noise between the original and restored areas.

3 Methodology

The concept of invariant image is based on the theory that the shadows appear in lower light condition areas. This illumination invariant technique finds a depiction image where shadows don't exist. But two conditions need to be satisfied; the original image must be noiseless and uncompressed. The edge map which is created from the original image by the edge detection process can also hamper in the shadow detection process. The illumination invariant algorithm for shadow detection which consists of five parts: Image preprocessing which includes conversion of RGB image with the application of median filter followed by the calculation of shadow ratio and the threshold value. After the threshold value is calculated, a shadow mask is created. After the mask is been created the area size threshold is adjusted or tuned followed by dilation of the grayscale scale image which results in shadow boundaries. The RGB to greyscale conversion is given by the formula

$$Y = \log\left(\frac{\max[F_R F_G F_B]}{\min[F_R F_G F_B] + 1}\right) \tag{1}$$

4 Results and Analysis

The above technique has been tested in Matlab R2016b. Figure 2 shows original test images with the presence of shadows. Figure 3 shows the converted binary image. Figure 4 shows us the binary image with shadow detected followed by Fig. 5 RGB images with shadow detected.

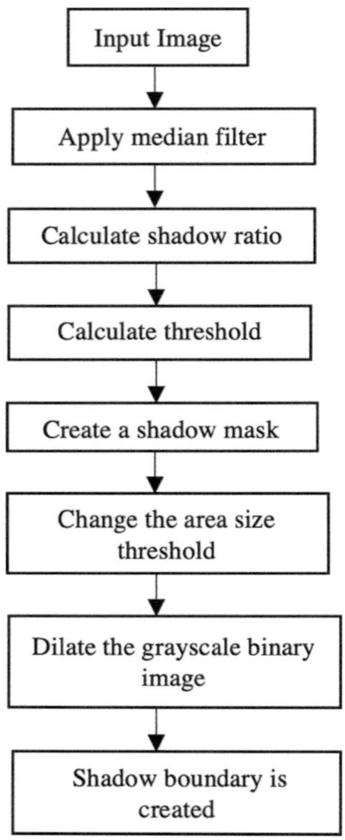

Fig. 2. Flowchart representation for shadow detection

The above method has been implemented on Matlab R2016b and the illumination invariant method has been used to detect the shadows in the images.

Fig. 3. Original images with the presence of shadows

Fig. 4. Binary images with shadows detected

Fig. 5. RGB images with shadows detected

5 Conclusion and Future Work

This paper carries forward the various works done in the field of shadow detection. Various methodologies can be adopted for the determination of the contours of the shadow and differentiating the shadows with the nearby objects. The experiment illustrated the flow algorithms and illustrations for shadow detection. The illumination invariant method was used for shadow detection. The method takes into account the shadow ratios, calculates the shadow masks and finally dilates the grayscale binary image of the shadow which in turn results in shadow boundaries. The illustrations and algorithms can be further enhanced and utilized to work on automatic driver assistance. The functionality of shadow detection being a very significant part of lane detection in automatic driver assistance.

References

1. Wang JM, Chung YC, Chang CL, Chen SW (2004) Shadow detection and removal for traffic images. In: 2004 IEEE international conference on networking, sensing and control, vol 1, 21–23 Mar 2004, pp 649–654
2. Bevilacqua A (2003) Effective shadow detection in traffic monitoring applications. WSCG
3. Chen T, Yin W, Zhou XS, Comaniciu D, Huang TS (2005) Illumination normalization for face recognition and uneven background correction using total variation based image models. CVPR 2(2005):532–539
4. Adini Y, Moses Y, Ullman S (1997) Face recognition: the problem of compensating for changes in illumination direction. IEEE Trans Pattern Anal Mach Intell 19(7):721–732
5. Zhao W, Chellappa R (1999) Robust face recognition using symmetric shape-from-shading. Technical report. Center for Automation Research, University of Maryland
6. Finlayson GD, Hordley SD, Drew MS (2002) Removing shadows from images. In: Proceedings of the 7th European conference on computer vision, pp 823–836
7. Fredembach C, Finlayson, GD (2004) Fast re-integration of shadow free images. In: Proceeding color imaging conference
8. Lalonde J-F, Efros AA, Narasimhan SG (2010) Detecting ground shadows in outdoor consumer photographs. In: Proceeding 11th European conference computer vision
9. Wu TP, Tang CK (2005) A Bayesian approach for shadow extraction from a single image. In: Proceedings of the IEEE international conference on computer vision, Beijing, China, vol 1, 17–21 Oct 2005, pp 480–487
10. Salvador E, Cavallaro A, Ebrahimi T (2004) Cast shadow segmentation using invariant color features. Comput Vis Image Underst 95(2):238–259
11. Guo R, Dai Q, Hoiem D (2013) Paired regions for shadow detection and removal. IEEE Trans Pattern Anal Mach Intell 35(12):2956–2967
12. Salvador E, Cavallaro A, Ebrahimi T (2001) Shadow identification and classification using invariant color models. In: Proceedings of IEEE international conference on acoustics, speech, and signal processing, Salt Lake City, USA, vol 3, 7–11 May 2001, pp 1545–1548
13. Shor Y, Lischinski D (2008) The shadow meets the mask: pyramid-based shadow removal. Comput Graph Forum 27(2):577–586

Comparison of Performance Metrics of Star Topology and Ring Topology in Wireless Sensor Network

Laxmipriya Moharana(✉), Bedanta Kumar Biswal, Raman Raj, and Suraj Naik

Electronics and Communication Engineering, Faculty of Engineering, Siksha 'O'Anusandhan (Deemed to be University), Bhubaneswar, India
{laxmimoharana, surajnaik}@soa.ac.in

1 Introduction

In today's world, wireless sensor networks (WSNs) are an emerging area of wireless computing, gathering information, and networking. A WSN comprises of a collection of nodes organized into a cooperative network. The recent advancement in WSN is observed in the field of railways for safety and security purposes. [1–4]. A sensor node consists of an ADC, Microcontroller, Memory, power device and a Transceiver as in Fig. 1 [5]. Energy conservation to be maintained within a wireless sensor network to improve the lifetime of sensor nodes [6]. To decrease the energy consumption and to improve the lifetime of a sensor network, the sensor deployment in a certain geometrical area to be taken care of by implementing different topology criteria. The well-suited topology is to be used for sensor deployment. In this paper, we have proposed two different topology models i.e. star topology and ring topology and calculated and plotted results of throughput of sensor nodes, average throughput, delay and jitter. The result shows a detailed comparison between two topologies which comprises of six sensor nodes in WSN connected in wireless medium based upon the above-mentioned matrices.

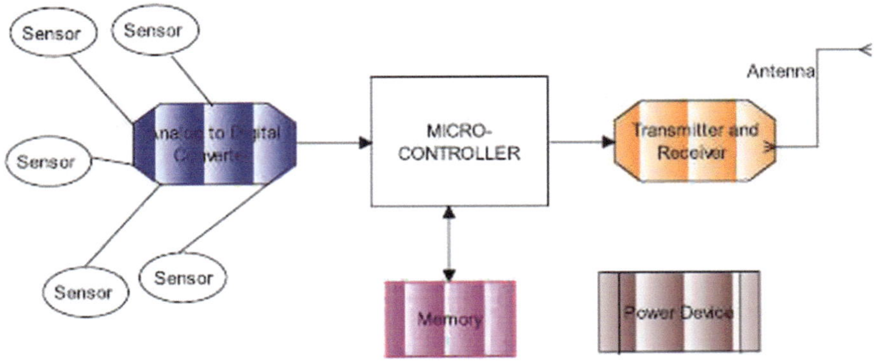

Fig. 1. Basic block diagram of a sensor node

M. N. Mohanty and S. Das (eds.), *Advances in Intelligent Computing and Communication*, Lecture Notes in Networks and Systems 109,
https://doi.org/10.1007/978-981-15-2774-6_16

2 Literature Survey

A model is to be initiated to observe the in-network computation in a sensing field and to find the optimum achievable throughput. Single path routing technique is helpful to achieve such an optimal throughput. Another important aspect of sensor networks is to maximize the lifetime of a sensor node [7]. By using matrices such as overall network throughput and total energy consumption, the performance of topology control algorithm can be frame worked [8]. An efficient topology may be designed such that the energy consumed by each and every sensor nodes of a sensor network is to be economized; a possible way to achieve that is by coverage maintenance and graph connectivity [9]. If we analyze the energy aware routing schemes by adopting different topologies then we can say that Routing Information Protocol (RIP) is more energy efficient compared to Intrazone Routing Protocol (IARP) and STAR [10].when we compare the quality of service of star topology with multi-hop topology outdoor propagation we can find that multi hope topology is superior than star topology in certain constraints [11].

3 Testing, Analysis, and Evaluation

Each node is represented with green color as shown in Fig. 2, which is placed as per the configuration provided in the NS-2 tcl file. The nodes are placed with a wireless connection set up having some specific node parameters and are connected in space with specific co-ordinates.

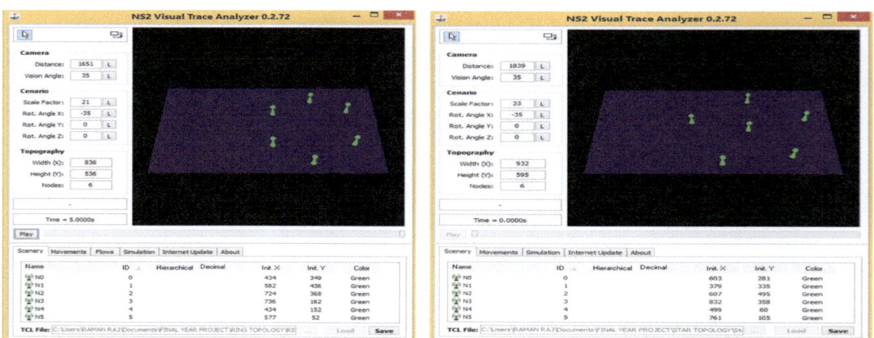

Fig. 2. Scenario of ring and star topology in NS 2 visual trace analyzer

In ring topology data packets are sent around the circular path until the packets will be their final destination. 1st part of Fig. 3 represents a ring network formed by ns-2.35. Here the source is node-2 and the destination is node-4. So, packets are transmitted from node-2 to node-4. In either direction, the packets can travel to its destination. When a node will be out of order during data transmission, data packets will be traveled in another route to reach their destination for that this topology is called self-healing network. In the Star Topology, a central connection point is there called a hub which is a computer hub or sometimes just a switch. 2nd part of Fig. 3 represents a star network. Here the hub is node-0 and from node-2 packets are transmitted to node-4 through node-0. In a Star Network, a failure occurs in the cable one computer be affected but the entire network will be safe. But when any problem arises to hub, the whole system will fail.

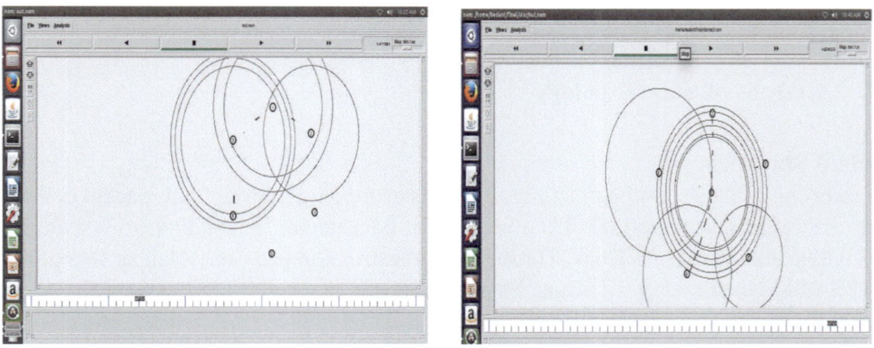

Fig. 3. Scenario of ring and star topology of NAM file in NS 2.35 with simulation time of 1.4 s (approx.)

As data transmission took place from node 2 to node 4, the node 2 is considered as source node and node 4 as destination node, the trace file of ns2 code is analyzed to calculate the throughput, packet delay and packet jitter at node-4. As we know throughput is defined as a measure of how fast we actually send data through a network without any loss. It is the calculation of no. of bits received without any error per sec. Here it is calculated as (no. of bits received + hold rate)/(data rate × sampling time). It can also be calculated as Size of the packets/Transmission time, so unit of throughput is expressed as bits/s. Delay can be termed as the total time data packets need to transfer from the source node to the destination node. The sources of delay in WSN are source and destination processing delay, propagation delay and network delay. The given formula can be used to calculate the average Delay, where Arrival time is the time when packet "*i*" reaches the destination and Start time is the time when packet "*i*" leaves the source. "*n*" is the total number of packets. Average delay = (Arrival Time − Start Time)/*n*. Average jitter for data transmission in a WSN can be defined as the fluctuation in the delay introduced by the different devices along the communication path. It is the variation in the time between packets arriving. Jitter is generally associated with network consistency and stability. Average Jitter = ((Arrival Time + 1) − (Start Time + 1)) − ((Arrival Time) − (Start Time))/*n* − 1 [12].

3.1 Analysis for Star and Ring Topology

The topology is formed to transform data from node 2 to node 4. In ns2 visual trace analyzer, the star topology and ring topology was formed by taking 5 nodes with certain co-ordination that is shown in Fig. 2, also in ns2 the NAM (network animation) file was generated to show the formation of Star and ring topology and the corresponding data transmission shown in Fig. 3. Here we were assuming data transmission is taken place from node 2 to node 4 through node 0, since node 0 is considered as the central hub for star topology and for Ring topology data packets are transmitted from node 2 to node 4 through node 1 and node 0.

3.2 Analysis of Star Topology

Node 0 Statistics

By referring to the second part of Fig. 3 i.e. a star topology throughput, packet delay and jitter are calculated at node 0. Total number of packets are transmitted from node 2 are 177 with a data rate of 93 Kbps. Through the transmission path, the total packets dropped are 23 with a data rate of 12 Kbps. By referring Fig. 4 we got minimum packet delay is 0.055531097 ms and maximum packet delay is 0.40498733 ms. So, the average delay at node 0 is calculated as 0.28157563 ms. From Fig. 5 we got the packet jitter values as follows: minimum jitter is 0.00002 ms, maximum packet jitter is 0.22034536 ms. So, the average packet jitter is 0.11018268 ms. Figure 6 shows the received throughput at node 0 the statistic is as follows: minimum throughput is 6000 bytes/s, maximum throughput is 42,000 bytes/s. So, the average throughput is 24,000 bytes/s.

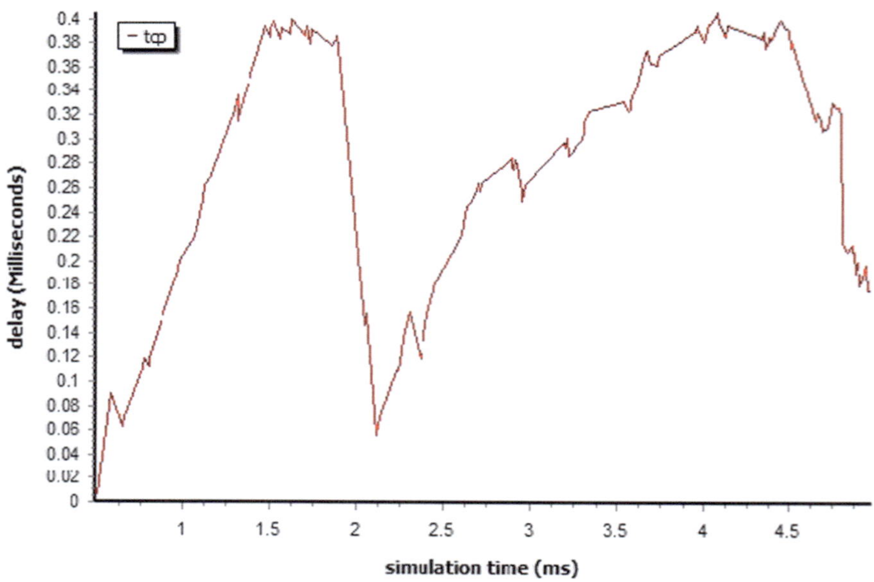

Fig. 4. Delay over time at Node 0-star

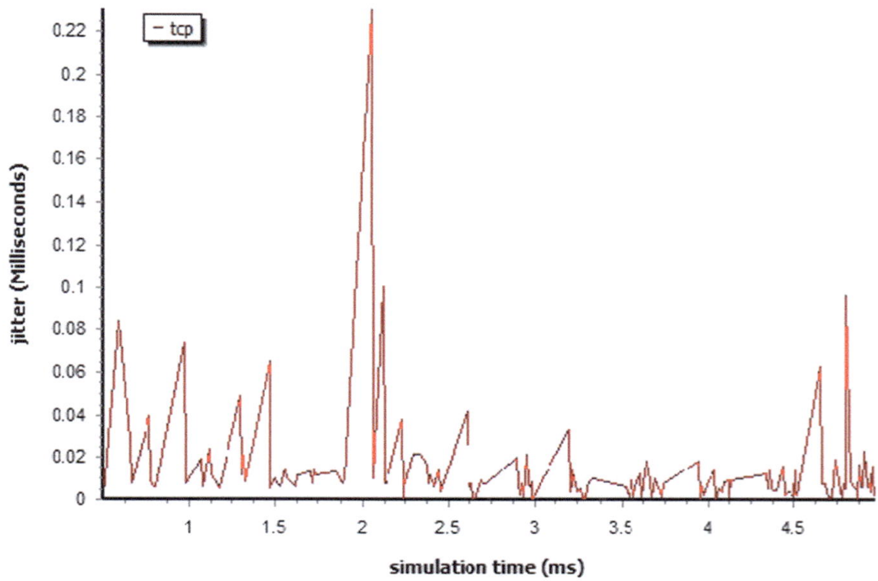

Fig. 5. Jitter over time at Node 0-star

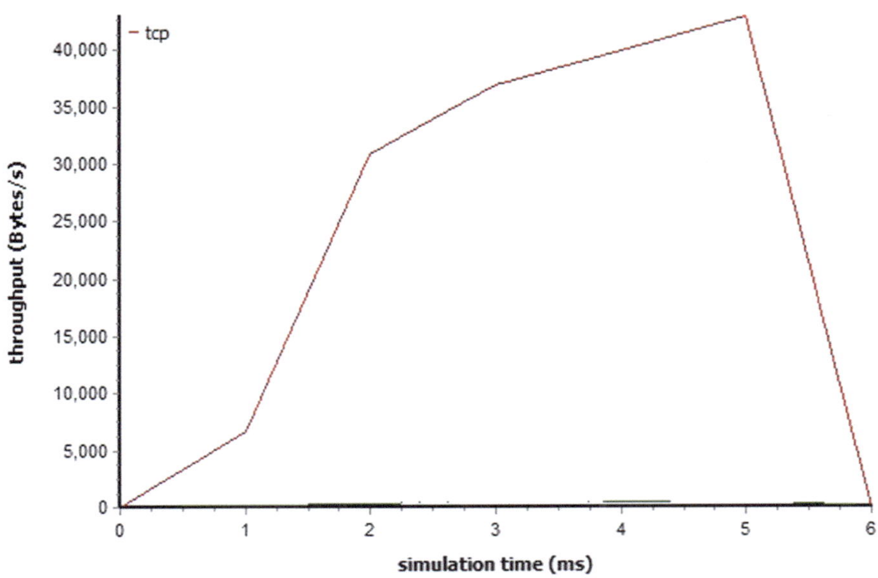

Fig. 6. Throughput received at Node 0-star

Node 4 Statistics

From node 0 the transmitted packets are 101 with 53 Kbps data rate, in between the transmission the total packets dropped are 9 with 5 Kbps data rate. By referring Figs. 7, 8 and 9.

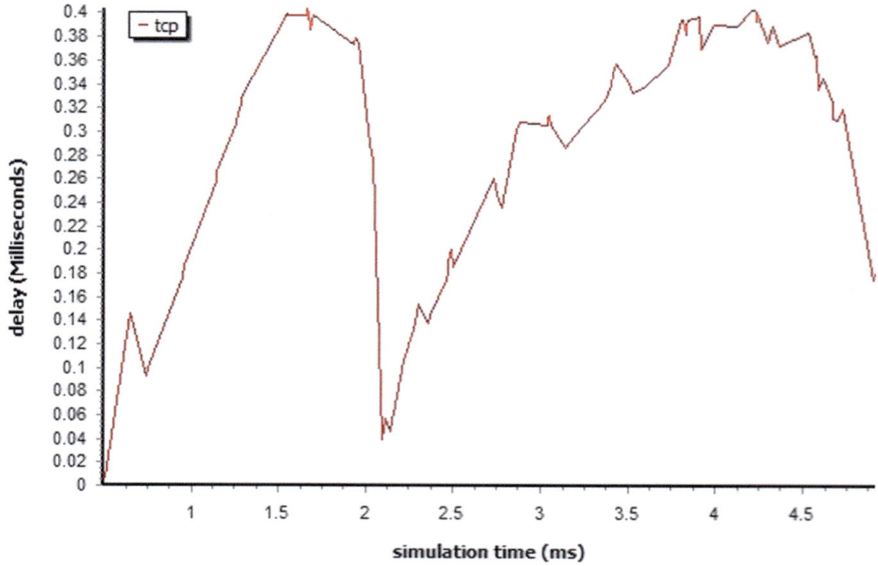

Fig. 7. Delay over time at Node 4-star

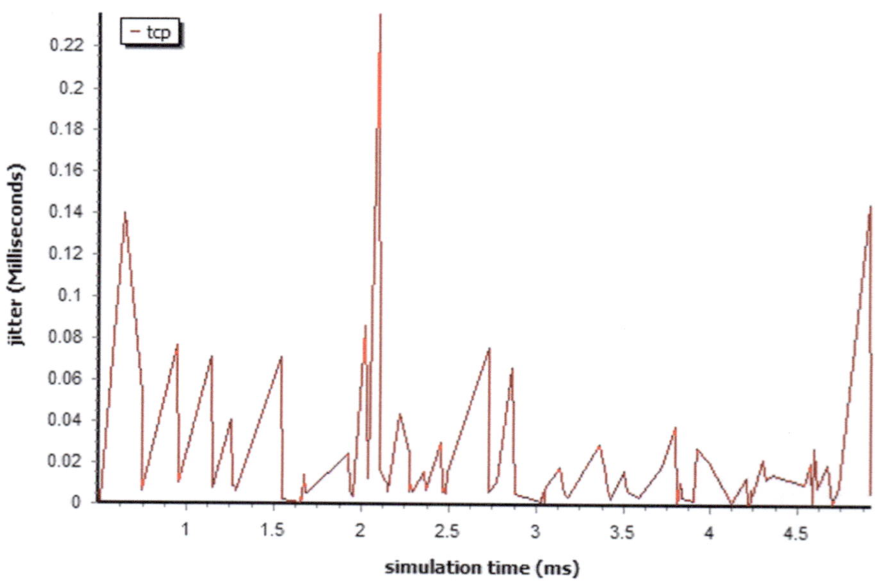

Fig. 8. Jitter over time at Node 4-star

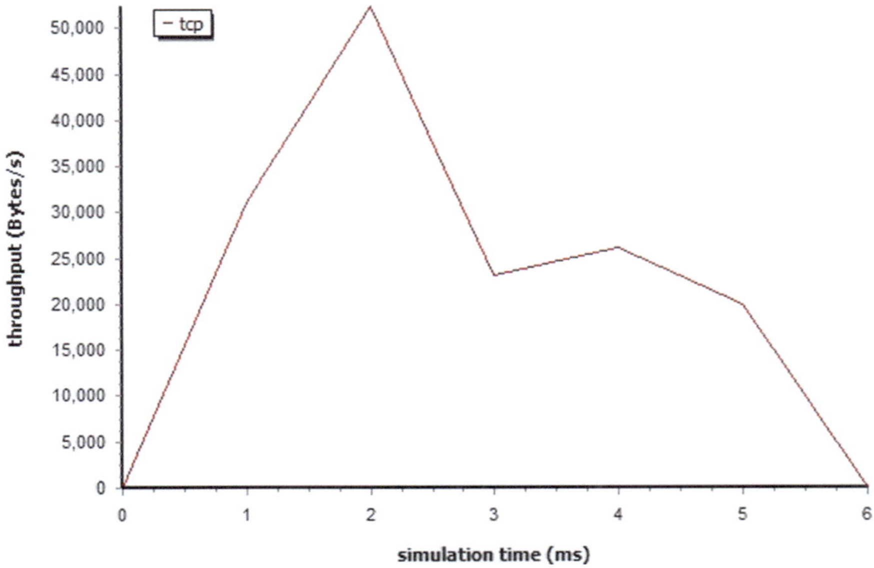

Fig. 9. Throughput received at Node 4-satr

The destination node i.e. the node 4 having the packet delay, packet jitter and throughput statistics as follows: minimum packet delay is 0.039644157 ms, maximum packet delay is 0.4034061 ms and the average packet delay is 0.28857745 ms. Minimum packets jitter is 0.00018 ms, maximum jitter is 0.288336 ms. and the average jitter is 0.144258 ms. Minimum received throughput at node 4 is 6000 bytes/s, maximum throughput 52,000 bytes/s and average throughput is 29,000 bytes/s.

3.3 Analysis of Ring Topology

Node 1 Statistics

By following part 1 of Fig. 3 the ring topology different performance metrics are calculated at nodes 1, 0 and 4. which are given below. From node 2 the packets are transmitted are 58 with a data rate 86 Kbps, then the total dropped packets at node 1 are 7 with data rate 9 Kbps. At node 1 Minimum packet delay is 0.12554 ms, maximum packet delay is 1.1548 ms and the average delay is 0.64017 ms, minimum jitter is 0.00008 ms, maximum jitter is 0.62123 ms and average jitter is 0.3106 ms, minimum throughput is 500 bytes/s, maximum throughput is 8000 bytes/s and average through is 4250 bytes/s as shown in Figs. 10, 11 and 12.

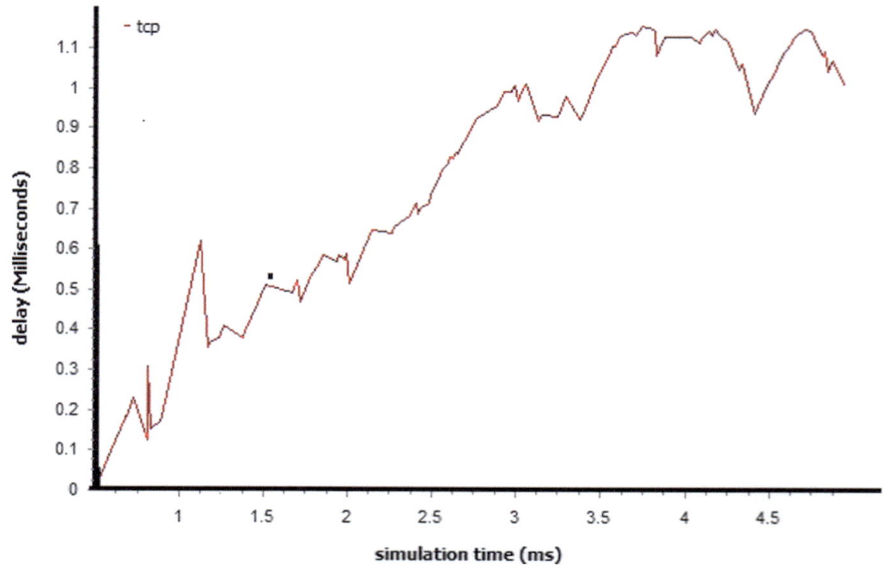

Fig. 10. Delay over time at Node 1-ring

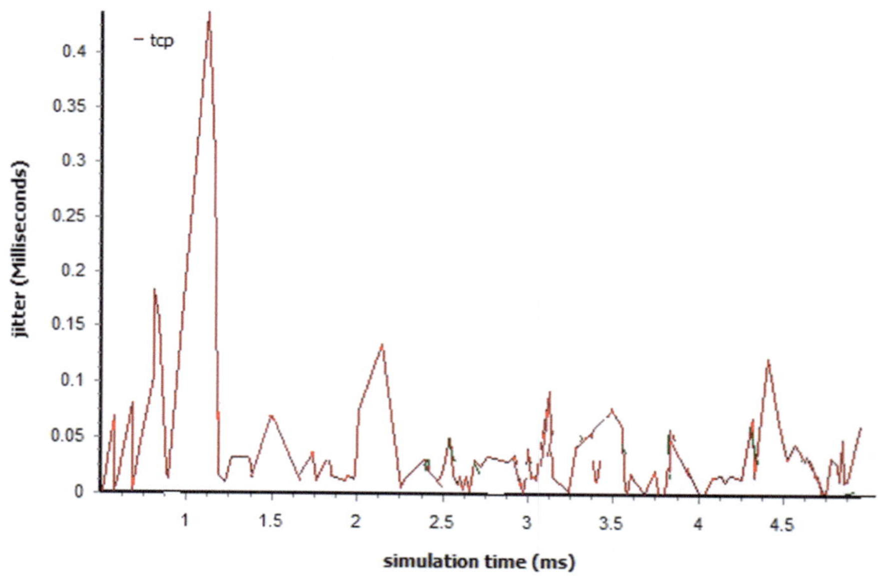

Fig. 11. Jitter over time at Node 1-ring

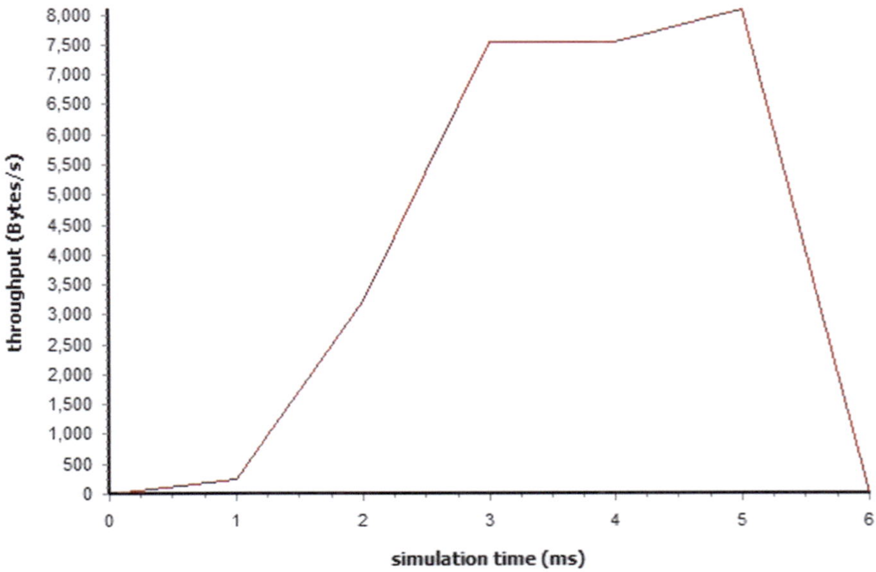

Fig. 12. Throughput received at Node 1-ring

Node 0 Statistics

By analyzing the plots from Figs. 13, 14 and 15. From node 1 the packets are transmitted are 102 with a data rate 152 Kbps. then the total dropped packet at node 0 is 1 with data rate 2 Kbps. At node 0 Minimum packet delay is 0.01388597 ms, maximum

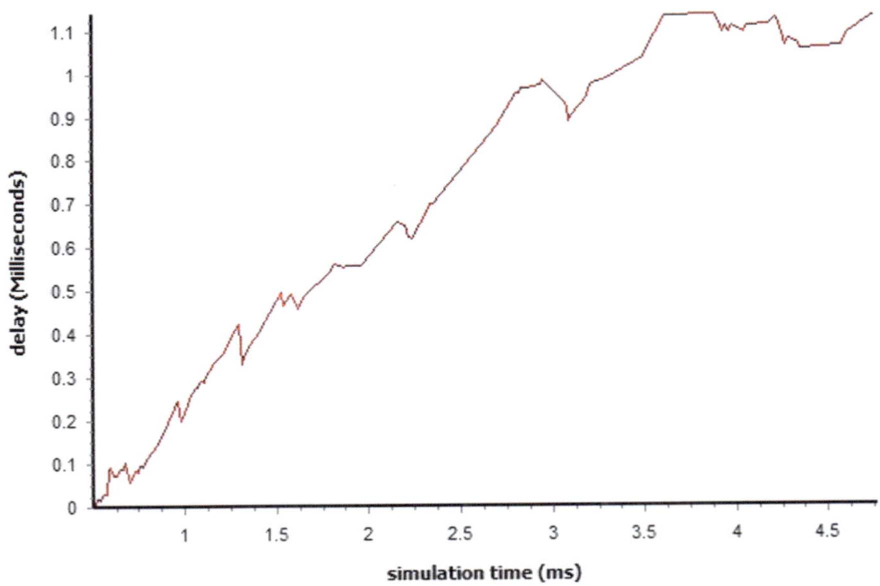

Fig. 13. Delay over time at Node 0-ring

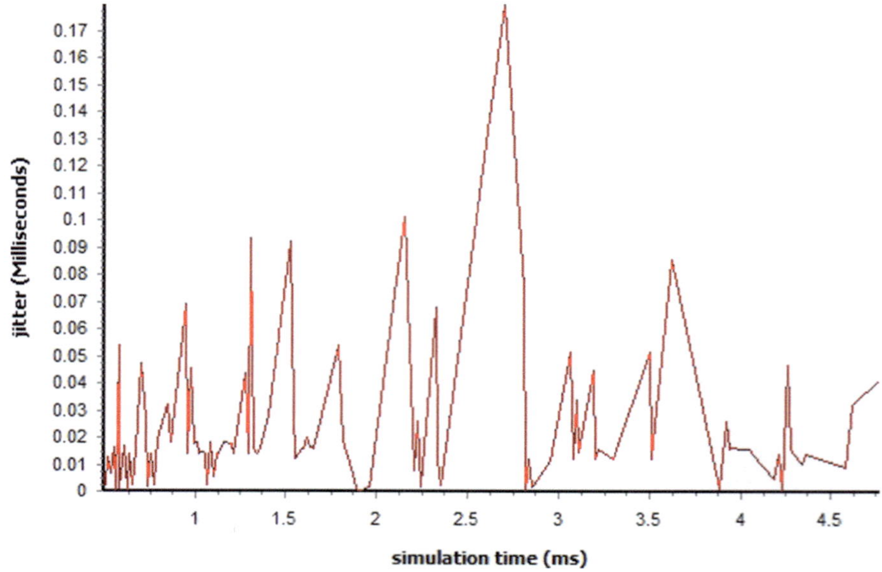

Fig. 14. Jitter over time at Node 0-ring

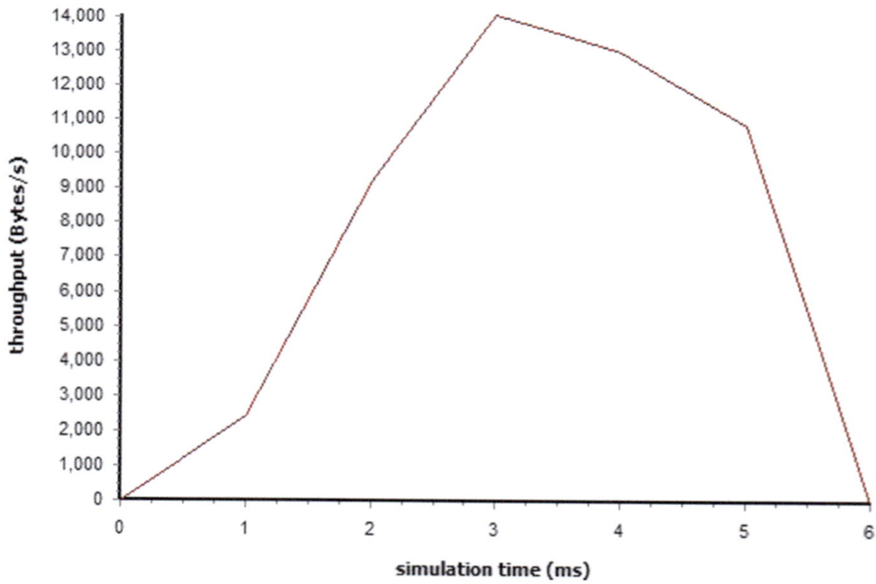

Fig. 15. Throughput received at Node 0-ring

packet delay is 1.1409631 ms and the average delay is 0.577424 ms, minimum jitter is 0.00006 ms, maximum jitter is 0.19961004 ms and average jitter is 0.099835 ms, minimum throughput is 2000 bytes/s, maximum throughput is 15,000 bytes/s and average through is 8500 bytes/s.

Node 4 Statistics

By analyzing the plots from Figs. 16, 17 and 18. From node 0 the packets are transmitted are 11 with a data rate 15 Kbps. then the total dropped packets at node 4 are 2 with data rate 3 Kbps. At node 4 Minimum packet delay is 0.735275 ms, maximum packet delay is 1.544114 ms and the average delay is 1.139694 ms, minimum jitter 0.016051866 ms, maximum jitter is 0.60370114 ms and average jitter is 0.3098765 ms, minimum throughput is 100 bytes/s, maximum throughput is 3200 bytes/s and average through is 1650 bytes/s.

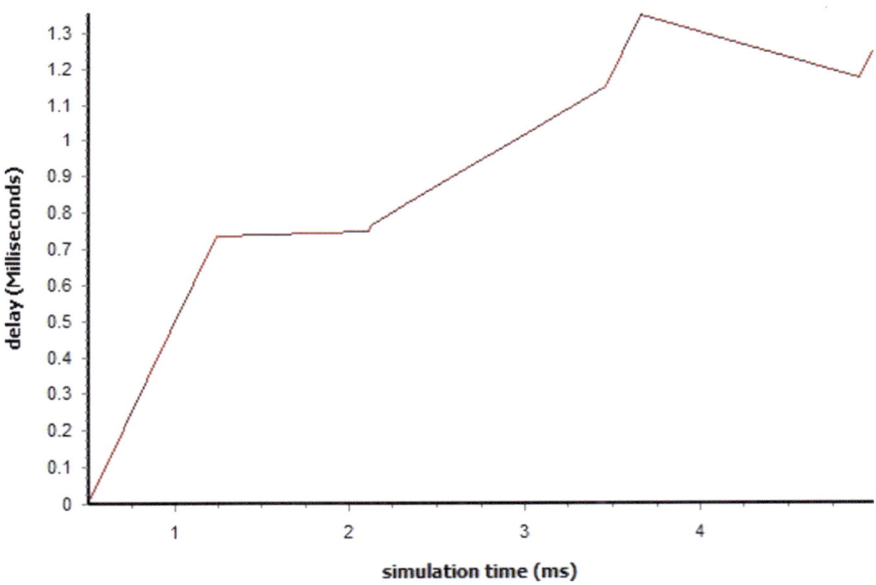

Fig. 16. Delay over time at Node 4-ring

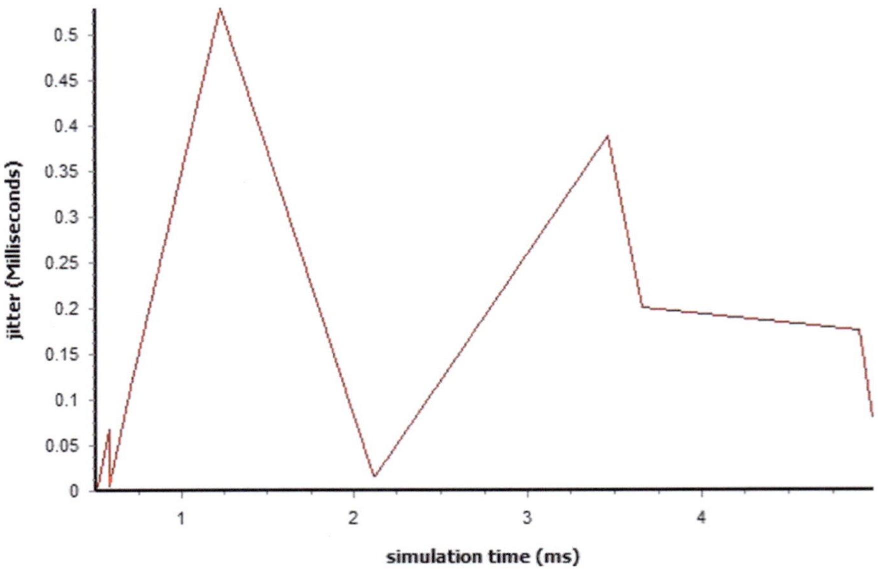

Fig. 17. Jitter over time at Node 4-ring

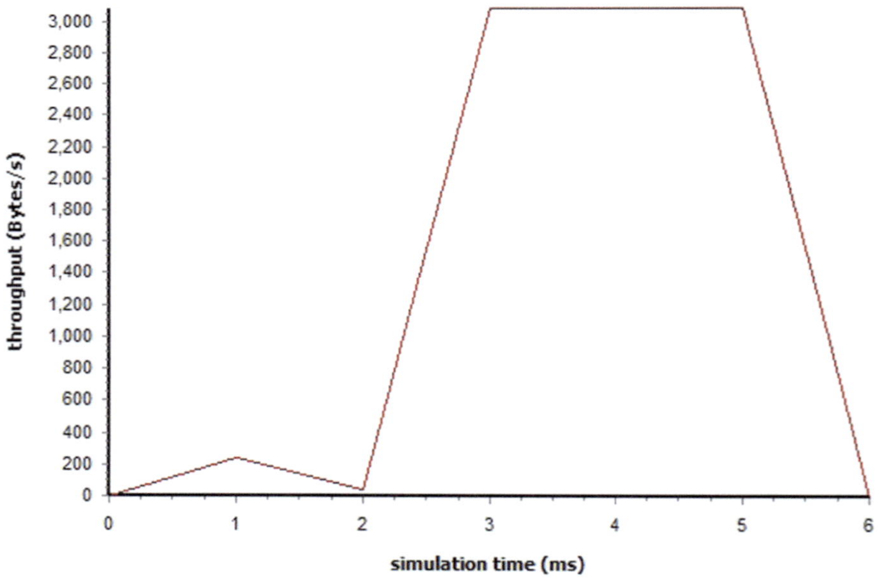

Fig. 18. Throughput received at Node 4-ring

4 Conclusion

Wireless Sensor Networks (WSN) promises a new domain on the way computers and humans interact with our environment and play a vital role in area of new research and development, invention and investigations. From the above topology design and analysis, we concluded that the performance metrics of star topology is much better than ring topology. The packets sent, received, throughput, end to end delay, jitter was much higher in star topology. In real life scenario, we can deploy the sensor nodes by forming ring topology because here each node can be treated as the normal node as well as a hub so it is lowering the cost of implementation.

References

1. Hodge VJ, O'Keefe S, Weeks M, Moulds A (2015) Wireless sensor networks for condition monitoring in the railway industry: a survey. IEEE Trans Intell Transp Syst 16(3):1088–1106
2. Nuffer J, Bein T (2006) Application of piezoelectric materials in transportation industry. In: Proceedings of Global symposium on innovative solutions for the advancement of the transport industry, San Sebastian, Spain, pp 1–11
3. Pourghodrat A (2011) Energy harvesting systems design for railroad safety. M.S. thesis, Department of Mechanical and Materials and Engineering, University of Nebraska–Lincoln, Lincoln, NE, USA
4. Lynch JP, Loh KJ (2006) A summary review of wireless sensors and sensor networks for structural health monitoring. Shock Vibrat Dig 38(2):91–130
5. Tolani M, Singh SRK, Shubham K, Kumar R (2017) Two-layer optimized railway monitoring system using Wi-Fi and zigbee interfaced wireless sensor network. IEEE Sens J 17(7):2241–2248
6. Nguyen T-T, Shieh C-S, Dao T-K, Wu J-S, Hu Robot W-C (2013) Prolonging of the network lifetime of WSN using fuzzy clustering topology. In: Second international conference on vision and signal processing (RVSP)
7. Sapidi R, Girard A (2013) Maximum achievable throughput in a wireless sensor network using in network computation for statistical functions. IEEE/ACM Trans Netw 21(5):1581–1594
8. Sinha P (2003) Throughput and energy efficiency in topology-controlled multi-hop wireless sensor networks. In: 2nd Second ACM international workshop on wireless sensor networks and applications (WSNA), San Diego, California, USA
9. Santha Meena S, Manikandan J (2017) Study and evaluation of different topologies in wireless sensor network. In: International conference on wireless communications, signal processing and networking (WiSPNET), Chennai, India, March 2017
10. Alam S, De D, Ray A (2015) Analysis of energy consumption for IARP, RIP and STAR routing protocols in wireless sensor networks. In: Second international conference on advances in computing and communication engineering, Dehradun, India, May 2015
11. Pramono S, Putri AO, Warsito E, Budi Basuki S (2017) Comparative analysis of star topology and multihop topology outdoor propagation based on quality of service (QoS) of wireless sensor network (WSN). In: IEEE international conference on communication, networks and satellite (Comnetsat), Semarang, Indonesia, Oct 2017
12. Birla J, Sah B (2012) Performance metrics in ad-hoc network. IJLTET 1(1)

A Framework for Real-Time Lane Detection Using Spatial Modelling of Road Surfaces

Pankaj Prusty and Bibhuprasad Mohanty(✉)

Electronics and Communication Engineering, Faculty of Engineering, ITER,
Siksha 'O' Anusandhan (Deemed to be University), Bhubaneswar, India
{pankajprusty, bibhumohanty}@soa.ac.in

1 Introduction

In this paper, a robust and novel method using spatial neighbourhood properties of non-lane pixels and comparing the testing pixels with these pixels for detection of lane followed by morphological operations has been proposed. This method works perfect for various types of lane markings like straight, curved, continuous and discontinuous, etc.

Lane detection basically is a method of identifying the lane markers on a road. For automatic driver assistance systems, lane mark is an important and essential part. The objective of lane mark detection is to give a warning to the driver if the vehicle is approaching towards the lane mark or near the lane mark. Nowadays because of highly increase in traffic many accidents are caused due to lane departure and many of these lane departures are due to drivers fault. So therefore, there is a high need for research to develop robust systems for automated detection of lane marks. Lanes are of various types which include road surface markings, zebra crossings, arrow markings, etc. Our system is robust in capturing all these types of lane markings. The importance of lane detection is it makes the system intelligent and can be a good part of driverless system. But the basic requirement for this is that it should be accurate, robust and capable of all types of lane detection.

Our spatial modelling of background algorithm works well for continuous and discontinuous lane marking in real-time. Here we do not require any prior information about the road surfaces and also no data base is required here. All works are done intensity plane. Less memory requirement and no need for data base make the system robust. Our algorithm is capable of detecting the lane immediately when the frame arrives.

Previous work has been discussed in Sect. 2. Section 3 contains our proposed method followed by experimental results in Sect. 4.

M. N. Mohanty and S. Das (eds.), *Advances in Intelligent Computing and Communication*, Lecture Notes in Networks and Systems 109,
https://doi.org/10.1007/978-981-15-2774-6_17

2 Literature Survey

Basic image features like histogram, edges, and gradients play a measure role in the lane detection process which is based on vision. Supervised learning methods have been used in papers [1–5]. In [3] a system is proposed which calculates the departure degree from lane in correspondence with the boundaries angular relationship. They have used the shadow of the front vehicle to locate the position of the vehicle in front of them. Then they have used the neuro-fuzzy network to estimate the actual distance. The authors of [6] have been proposed a spatiotemporal method with curve fitting properties. Authors of [4] have been proposed a method where they have used a base classifier based on some properties of the environment. The system does not work in dark shadow region, reflector region etc. Authors of [5] have combined the self-clustering algorithm, fuzzy C-mean and fuzzy rules to improve the performance in various light conditions. Author [7], in their paper, used Hough transformation along with Kalman filtering method for 3° Bezier splines.

3 Proposed Work

In this section, we discuss our method for detection of lane mark from road surfaces. A camera is mounted on the bonnet of a vehicle and it is assumed that this camera is capable of sending the captured video frames to a computer which is on board. We apply our spatial road surface modelling algorithm and detection method. First, the road surface background is modelled by considering the non-lane pixels and then the lane is detected by comparing the lane frame with the background.

3.1 Region of Interest Selection

We first select the region of interest in order to reduce the miss and false classification rate. This makes our system faster. We select that part of the image which contains the road surfaces only not the sky, pedestrian, etc., since these parts do not provide any meaningful information to us. The region of selection can be better if the placement of the camera is fixed and known.

3.2 Gray-Scaling of the Image

Then the region of interest is converted to grayscale where each pixel represents the intensity of the original image. Working on gray scale will help us in analyzing the variation in intensity which makes the system robust.

3.3 Spatial Modelling of Road Surfaces

As the statistical correlation between neighbouring pixels of an image is nearly equal, so we have to consider the spatial neighbours of a pixel where no lane mark is there to model the road surfaces. Initially, we have to assume that the first frame does not contain any lane mark to model the road surfaces. Also, we can model the road surface by considering the non-lane areas. The process of modelling the road surfaces is done as per the procedure explained below.

$$P = \{q_1, q_2, q_3 \ldots q_N\} \tag{1}$$

where q_i is the randomly selected N samples from the spatial $n \times n$ neighbourhood of pixel with location (x, y). We will apply this concept in the entire image so that we have a set of pixels Q.

$$S = \{P^1, P^2, P^3, \ldots, P^S\} \tag{2}$$

where S represents the total number of pixels in the image. Choosing the value of n more than five may degrade the performance of the algorithm due to the fact that the more we go away from the pixel the resemblance between pixels decreases. N is taken to be nearly 80% samples from $n \times n$. The model gets updated or new model is created when the detected frame has no lane marks at all and discards all the previously stored data values which make the system robust.

3.4 Lane Mark Identification

As explained in Eq. (2), we have now a spatial model of the road surfaces. We want to classify a new pixel value in respect to its neighbouring value that whether this is the background or lane mark. A new pixel is now compared to the background samples and if it is a non-lane pixel then its value must be close to some of the values of the corresponding spatial model of P^i. The classification of the pixel is done as per the following equations.

$$X = |\text{test}(x, y) - \text{sm}(x, y)| > \text{Th1}, \text{for all } N \tag{3}$$

where 'test' is the frame for testing with pixel location (x, y) and 'sm' is the spatial modelled intensities of location (x, y). N represents the number of intensity values present for the location (x, y) in 'sm' and 'Th1' is empirically decided. Experimentally it has been found that the value of 'Th1' is 150. Once all the value of X for location (x, y) is calculated, then its cardinality is calculated.

The decision is taken as per the following equation. If the following equation is satisfied then the pixel with location (x, y) is classified as non-lane pixel else it is classified as lane pixel.

$$|X| \geq \text{Th2} \tag{4}$$

where $|.|$ represents the cardinality of a set and Th2 is to be empirically calculated. It has been found that 4 or more value of Th2 shows good results with least variation.

4 Experimental Results

Our proposed algorithm has been implemented in MATLAB 2016a on Windows 10 operating system. The implemented algorithm's performance has been evaluated on two videos shoot in a Nikon camera mounted on the hood of a vehicle and one video from YouTube [3]. Our algorithm was run on a laptop with Intel Corei3 3110 M Cpu with 2.4 GHz clock speed and 4 GB RAM. Figure 1 shows the videos used for simulation.

Fig. 1. Videos used for simulation

First, two videos are taken from a camera mounted on the hood of a vehicle. Here the camera is mounted in such a way that it focuses the road surfaces and lane only. Whereas the third video is a YouTube video where the scene contains road surfaces, sky, other vehicles, etc. The region of interest needs to be identified before proceeding towards the algorithm. Figure 2 shows one frame of the YouTube video and its representation after ROI selection.

Fig. 2. (Left) Frame 6 from original video (right) Same frame after ROI selection and grayscale conversion

First row of Fig. 3 is a video taken by a camera mounted on the hood of a vehicle. For simulation purpose, each frame is converted to 256 × 256 pixels. The result obtained is compared with STIC method [6]. The morphological operations are then applied on the detected lane pixels to give a clear boundary. Second row represents frame 6 of a video downloaded from YouTube. In this video, we have no prior information about the road surface and camera mounting condition. We simply downloaded it as it is from a moving video and simulated it. The results obtained are shown in Fig. 3.

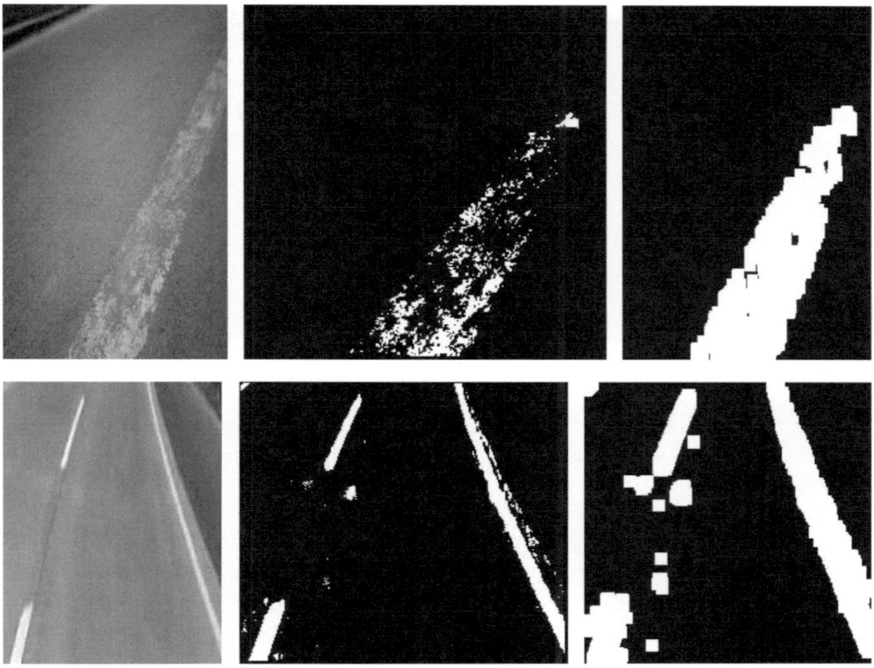

Fig. 3. Results of Lane detection of two videos (first row: video taken by us, second row: video from YouTube), From left to right of each row: grayscale of the frame, result after lane detection, result after morphological operation

To compare the performance of the algorithm, we have taken the help of four parameters: true positive (TP) and false positive (FP). Where true positive is the number of correctly classified pixels and false positive is the number of pixels falsely classified as lane. Table 1 represents the performance of the proposed method with the STIC method.

Table 1. Comparison between the STIC method and proposed method with the following performance metrics

Videos	STIC method		Proposed method	
	TP (%)	FP (%)	TP (%)	FP (%)
Video 1 (taken from a camera)	93	5	97	2.3
Video 2 (taken from a camera)	94	4.23	98.2	0.7
Video 3 (YouTube video)	92	3.04	91.89	4.56

5 Conclusion

We have proposed a real-time lane detection method based on the concept of spatial modelling method. The system is robust because it does not require any preexisting data base and also less memory requirement is there. All the works have been done in intensity plane. The method is capable of detecting various types of lane marks like dotted discontinuous, etc. The results show that the proposed method is robust and efficient in comparison to other methods. The method does not consider the poor light condition, which can be the future scope of the work.

References

1. Zhou S, Lagnemma K (2010) Self-supervised learning method for unstructured road detection using fuzzy support vector machine. In: IEEE international conference on Intelligent robots and systems (IROS)
2. Zhou S, Gong J, Xiong G, Lagnemma L (2010) Road detection using support vector machine based on online learning and evaluation. In: Intelligent vehicle symposium (IV)
3. Wu C, Lin C, Lee C (2012) Applying a functional neuro fuzzy network to realtime lane detection and front vehicle distance measurement. IEEE Trans Syst Man Cybern 42(4)
4. Kuhnl T, Kummert F, Fritsch J (2012) Spatial ray features for real time ego-lane extraction'. IEEE Trans Intell Transp Syst USA, 16–19 Sept 16–19
5. Wang J, Lin C, Chen S (2010) Applying fuzzy method to vision based lane detection and departure warning system. Expert Syst. Appl. 37(2010):113126
6. Gupta A, Choudhary A (2017) Real-time lane detection using spatio-temporal incremental clustering. In: Proceedings of IEEE 20th international conference on intelligent transportation system, pp 1–6
7. Wang Y, Teoh EK, Shen D (2004) Lane detection and tracking using b-snake. Image Vis Comput 22(4):269–280

Mobile Robot Path-Planning Using Oppositional-Based Improved Firefly Algorithm Under Cluttered Environment

Mohit Ranjan Panda[1], Susmita Panda[2(✉)], Rojalina Priyadarshini[1], and Pradipta Das[3]

[1] School of Computer Engineering, KIIT University, Bhubaneswar, Odisha, India
[2] Institute of Technical Education and Research, SOA Deemed to be University, Bhubaneswar, Odisha, India
susmitapanda@soa.ac.in
[3] Veer Surendra Sai University of Technology, Burla, Odisha, India

1 Introduction

Robotics is one of the most important areas of research not only for its industrial applications but also for its rapid involvement in many real-world aspects. And in Robotics path planning is still a challenge for the mobile robots when the environment is uncertain and unpredictable. Path-planning is a NP-complete problem which deals with finding out the optimal and collision-free path from the source to the destination either in a static environment or dynamic environment. Environment can be expressed as the surrounding of the robot. If the environment is unchanged while a robot is deliberating, then it is a static environment or else it is a dynamic environment. Previously traditional algorithms such as visibility graph-based, potential field method, Voronoi-diagram, etc. are being used to find out the optimum path, but due to complex calculations and having no guarantee that we can get the optimum solution, researchers started to use various nature-inspired algorithms which showed better results.

There are many local and global path-planning techniques proposed by the researchers for the navigation of the robots. When the environment is known, global path-planning techniques are used. But the drawback of this is the inability to cope up with the uncertainty of the environment. So to overcome this problem, various local path-planning techniques are used. When the environment is not known or partially known to the robot then it is called local path-planning. Various local path-planning approaches are there like Genetic Algorithm (GA), Ant Colony Optimization (ACO), Particle Swarm Optimization (PSO) and many more.

GA is a class of metaheuristic algorithm which is encouraged by the natural selection procedure that belongs to the group of evolutionary algorithm and is used in optimization problems to generate high-quality solutions. ACO is used to solve computational problem and it uses the probabilistic technique. PSO is a method that is used in optimization problems. It uses an iterative approach to improvise the candidate solution with regard to given quality parameters. Like this, many more optimization

M. N. Mohanty and S. Das (eds.), *Advances in Intelligent Computing and Communication*, Lecture Notes in Networks and Systems 109,
https://doi.org/10.1007/978-981-15-2774-6_18

algorithms are there. In this paper, we are concentrating on the Firefly Algorithm (FA) which is presented by Dr. Xin-She Yang in the year 2007 at Cambridge University. This is again a type of metaheuristic algorithm which is motivated by the flashing nature of the fireflies. FA is having simple and efficient features along with firefly characteristics which makes it more feasible to be used in various optimization problems and it has achieved good results. Because of its high convergence rate and robustness, it is self-adoptable to the changing environment and unfavorable situation in a complex system [1].

Although FA shows better performance, it has been observed from the experiments that sometimes it suffers from premature convergence, i.e., it gets trapped in local optima and this makes the optimization precession low or it may lead to the failure. The loss of population diversity in the early stage is the reason behind this. To overcome this problem, FA is enhanced to get higher convergence speed and greater accuracy. To overcome that, we added a mutation to the FA-based approach, which involves oppositional and dimensional learning.

2 Basic Firefly Algorithm

Firefly algorithm is a simple and effective optimization algorithm that is inspired by the bioluminescence process of the fireflies by which they emit the light and by this light they attract other fireflies towards them. Attraction between the fireflies depends on the intensity of the light they emit. The main advantage of the FA is, it uses the random movement of the individuals, and this movement depends on the mutual influence among the fireflies at the same time. One firefly will be attracted towards the other firefly if and only if the other firefly has greater light intensity. And this concept acts as the basics of the firefly algorithm. There are three constraints that act as a base for the firefly algorithm. They are given as follows:

(1) Fireflies will move towards the brighter firefly irrespective of the gender, i.e., they are not gender-specific.
(2) The power of attraction of the fireflies is dependent upon the intensity of the same, i.e., more the brightness, more eye-catching the fireflies will be. And the firefly having less brightness will be attracted towards the brighter firefly.
(3) As the objective function is dependent on the brightness of the fireflies, it is estimated by calculating the brightness of the firefly.

There are some definitions which we need to understand before getting into the firefly algorithm, such as

Attractiveness function

It is represented by the symbol β. This is the objective function that plays a very vital role in this algorithm. This attractiveness function is directly dependent on the brightness of the fireflies. And with the increase in distance between the fireflies, the brightness decreases. So the attractiveness function of the firefly is represented by an unmodulating decreasing function, i.e.

$$\beta(r) = \beta_0 \times \mathrm{e}^{-\gamma r^m}, (m \geq 1) \tag{1}$$

where $\beta(r)$ is the power of attraction which is at a distance r, β_0 is the power of attraction of the firefly at distance 0, γ is one coefficient called as the light absorption coefficient.

Here in exponential function $\mathrm{e}^{-\gamma r^m}$ r is present in the exponent and it shows when the distance will increase the brightness will decrease.

The distance between the fireflies can be calculated by using the formula,

$$r_{ij} = \left|\left|x_i - x_j\right|\right| = \sqrt{\sum_{k=1}^{d} \left(x_{i,k} - x_{k,j}\right)^2} \tag{2}$$

where r_{ij} is the distance between ith firefly and jth firefly, x_i and x_j are the position of the ith and jth firefly.

Movement of the Firefly

A firefly will be attracted to the other firefly if the other firefly has greater brightness. So this movement can be expressed as

$$x_i = x_i + \beta_0 \times \mathrm{e}^{-\gamma r_{ij}^2} \times \left(x_i - x_j\right) + \alpha \times \left(\mathrm{rand} - \frac{1}{2}\right) \tag{3}$$

There are different parameters that define the Firefly Algorithm and those parameters are Absorption parameter, denoted by the symbol γ and Random parameter, identified by the symbol α.

Working of Firefly Algorithm

Begin
Initialize the Fitness function, f(z) where $Z = (z_1, z_2, z_3, \ldots\ldots, z_n)$
Formulate an initial population of fireflies Z_i $(1,2,3, \ldots\ldots, n)$
Intensity of light L_i at z_i is determined by $f(z_i)$
Initialize γ
While (N < Iteration ^{max})
    For i = 1 to n-eachand every fireflies
        For j = 1 to n-each and every fireflies
            If ($L_j > L_i$)
                firefly i will move towards firefly j in d-dimension
            End if
            with the distance r attractiveness varies via $\beta_0 e^{-\gamma r^2}$
            the new solution update the light intensity
        End for j
    End for i
    best firefly will be found after ranking all fireflies
End while
Best Firefly will be found out
End

Sticking to local minima is a problem which original FA suffers from mainly while handling dimensional rich problems. The reason is that, as it doesn't take into account each dimension from every firefly in isolation with others. By which some moves towards the global optimal solution are missed. The proposed improved FA helps to get rid of the problem of getting stuck in the local best solution, by considering the global best solution in every generation. During the position updation process, the best solutions are taken into account.

The limitation of the basic Firefly Algorithm which sometimes shows low convergence rate and stuck at local optima, i.e., it is not so efficient in case of local search. So to improvise the basic FA a new strategy is being applied, i.e., oppositional-based learning which is described below.

3 Oppositional-Based Learning

Oppositional-based learning is a new way within field of robotics. It is being used to increase the speed of convergence of various optimization algorithms. This technique is developed by Tizhoosh which is applied successfully to various algorithms and which showed better results. The concept behind this technique is to consider the currently generated candidate solution with the corresponding opposite solution to get better optimization results. Here opposite solution is also taken into consideration because it has been proven by the researchers that sometimes opposite solution is a better solution. Let x is a real number and x' is its opposite number and $x \in [a, b]$ then x' will be defined as

$$x' = a + b - x \tag{4}$$

One may extend this description to higher dimensions, i.e., let $P(x_1, x_2, x_3, \ldots, x_n)$ be the n-dimension vector, where $x_i \in [a_i, b_i]$ and $i = 1, 2, 3, \ldots, n$. So the opposite vector of p will be represented as $P'(x'_1, x'_2, x'_3, \ldots, x'_n)$ and $x'_i = a_i + b_i - x_i$.

4 The Hybridized Algorithm

At the starting of the algorithm, the firefly population is set and fireflies are generated randomly, so the diversity is highest at that time. Gradually diversity in the population decreases with the increase in iteration number due to selection pressure or sample generating bias. So when this occurs, the FA's exploration capability will be reduced and this will be called premature convergence. So to increase the searching ability of the improved firefly algorithm, hereby combining the IFA with the oppositional-based search technique, a new hybridized algorithm is being proposed so that the performance will be increased. To overcome this problem, a way is to increase the diversity in the population [13]. For this, we can replace the worst firefly i.e. the firefly having the least brightness will be replaced by a newly constructed firefly. By this, the diversity will be increased and this will eliminate a firefly from the local optima. As a new firefly is added, the previous search routine is broken and this leads to the enhancement of the diversity. This new approach can be formulated as,

$$x_{\text{worst}}(t+1) = \begin{cases} x_{\text{best}}(t) & \text{if rnd()} < p \\ x_{\max} + x_{\min} - x_{\text{worst}}(t) & \text{otherwise} \end{cases} \tag{5}$$

where worst and best is the representation of the worst and the best firefly respectively. $x_{\text{worst}}(t)$ and $x_{\text{best}}(t)$ is the position of the best and the worst firefly respectively at tth iteration. rnd() is a uniform random number which ranges from 0 to 1. p whose value stands between 0 and 1 is the probability that the worst firefly will move towards the brightest firefly. After that, the elimination of the worst firefly from the normal path will be done. And the new firefly will be constructed by taking some elements from brightest firefly.

Proposed Algorithm
After combining the basic firefly algorithm with the oppositional-based learning approach the new algorithm is described below.

1: **Begin**
2: Generate an input space (initial population) of fireflies x_i (1,2,3, ……, n)
3: Calculate the light intensity of each firefly i.e. I_i (1,2,3, ……, n)
4: Define the fitness function i.e. light absorption coefficient γ
5: Initialize Gbestas the best firefly from the initial fireflies fitness evaluation
6: **While** (t < Max Iteration Number)
/* start update G_{best} by considering best in each dimension */
7: For i = 1 to D, D represents the Dimension
8: For j = 1 to N, where N is the number of firefly
9: Y←$G_{bestpos}$
10: Y ← x (j, i)
11: if f(G_{best}< f(Y)
12: G_{best}← f(Y)
13: $G_{bestpos}$← Y
14: End
15: End
16: End
/* end */
17: Update the new G_{best}
18: Find the worst firefly by considering the fitness function
19: Find out the value of the $rnd()$ and probability p
20: **If** $(rnd() < p)$
21: A new firefly will be constructed according to the position of the brightest firefly
22: **Else**
23: Calculate the opposition number of the worst firefly using Eqn (5)
24: The new firefly will be constructed according to the opposition number
25: **For** i = 1 to n for all fireflies
26: Shift firefly i in the direction of G_{best} in d-dimension
27: Compute distance among i^{th} firefly and G_{best} by equation

$$r_{ij} = \sqrt{\sum_{k=1}^{d}(x_{i,k} - x_{k,j})^2}$$

28: Determine attractiveness which varies with distance r via $\beta(r) = \beta_0 * e^{-\gamma r^2}$
29: light intensity will be updated after evaluation
30: **End for** i
31: Update the solution set
32: End While
33: **End**

OIFA is formed by using both the above modifications, i.e. inclusion of dimensional-based generation of Gbest and including oppositional-based optimization. The above Algorithm explains the pseudo-code of OIFA. Lines 1–5 explain the initialization and building of the input space of fireflies and The used fitness function $f(x_i)$

is giving a value which is the level of light intensity of a particular firefly x_i. Lines 7–16 explain the way for the updation of global best firefly by applying the context vector Y. lines 18–24 shows the application of oppositional number for optimization. In lines 25–30, the Euclidean distance is computed among each firefly and best fire-fly. Then each firefly is going to follow the Gbest.

5 Computer Simulation

A simulated environment has been created to test the multi-robot path-planning algorithm. C Programming is used to satisfy the purpose of a Pentium processor with 14 spherical robots. The first round of code is executed in a group of size of 50 for simulating four different algorithms such as OIFA, IFA and FA, each algorithm runs 30 times to get the simulated output. The input space is taken as a graphical area, where each robot's start and goal points are fixed priory. As robots are of a circular shape, their radius is defined by taking 6 pixels. The obstacles are also marked and are taken in varieties of seven unique shapes, where velocities of each robot for the experiment are kept as fixed for a single run of the program, but again were modified for different runs within the same program. In the experimental work, the very first world map configuration is shown in Fig. 1. Figure 2 presents the intermediary stages of the robot movements. Finally, Fig. 3 is shows all the robots are reaching towards their predefined goal position in the world map.

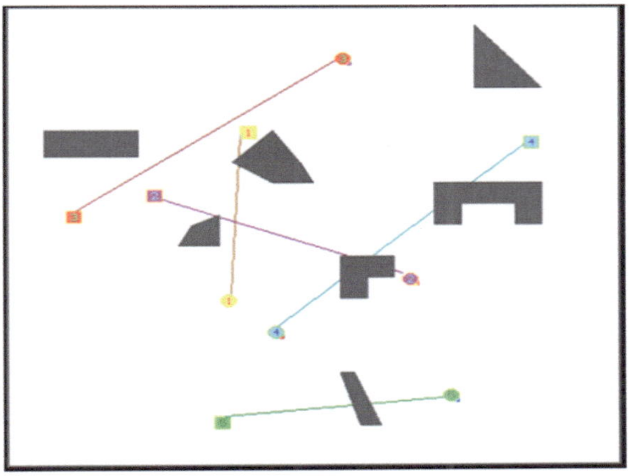

Fig. 1. Initial world map with 7 obstacles and 5 robots

For evaluation and validation of the proposed model's performance, (ATTD) Average total trajectory deviation and (AUTD) Average Untraveled trajectory Distance are considered. From Fig. 4 it is observed that there is a change in AUTD with a change in velocity and the number of steps. Here it could be observed more resources

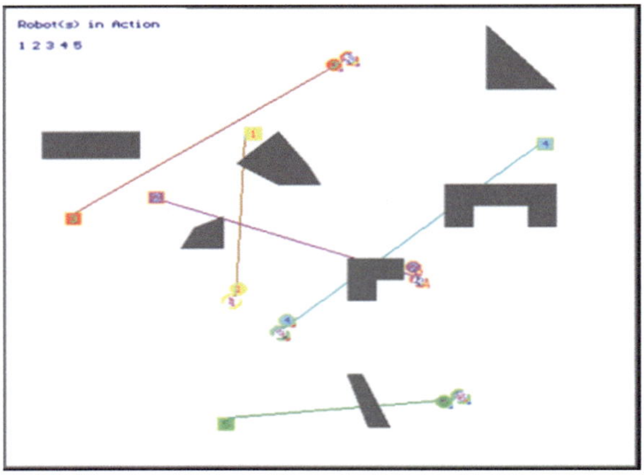

Fig. 2. Intermediate state of the world map using OIFA after 4 steps

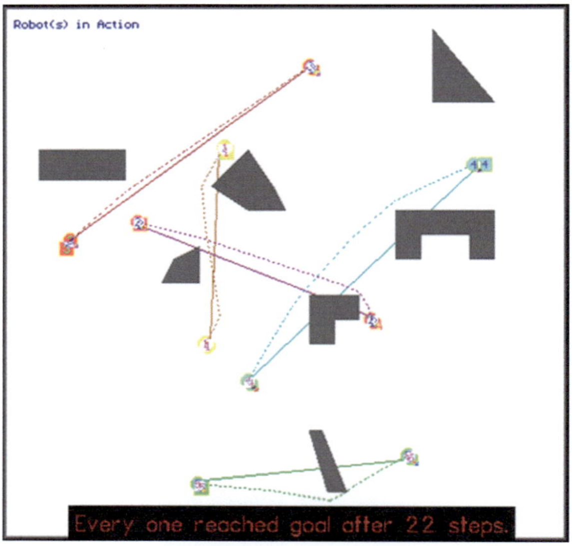

Fig. 3. Five robots reached their respective predefined goal after 22 steps

are necessary to measure AUTD to progress towards the convergence with number of iterations. It can be observed from Fig. 5 that if the number of robots increases, convergence rate decreases. Hence, this leads to a fall in AUTD. The exact path is drawn for all three techniques e.g. OIFA, IFA, and FA. Figure 6 gives an overview of calculated ATTD with varying the number of robots mostly between 1 and 8. From the below graph, it can be concluded that OIFA performs better compared to IFA and FA.

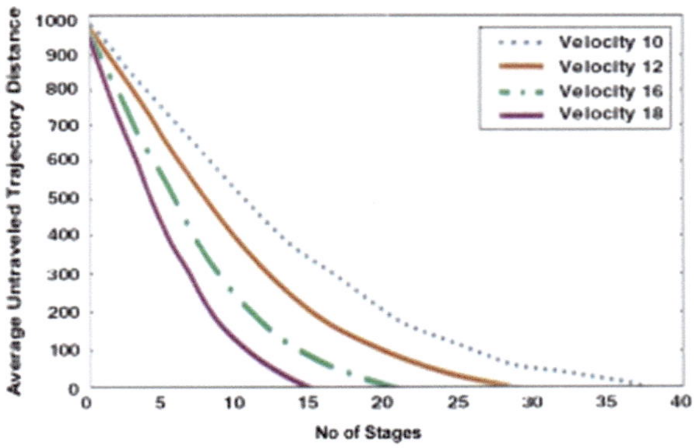

Fig. 4. Average untraveled trajectory distance versus no of stages with variable velocity and fixed number of obstacles

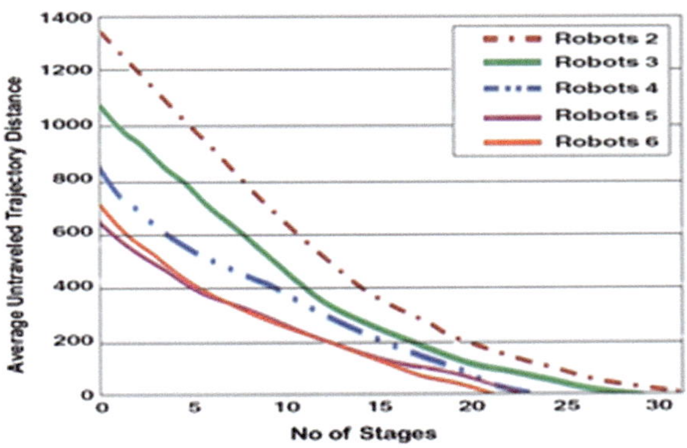

Fig. 5. Average untraveled trajectory distance versus no of stages with fixed number of obstacles and varying the number of robot

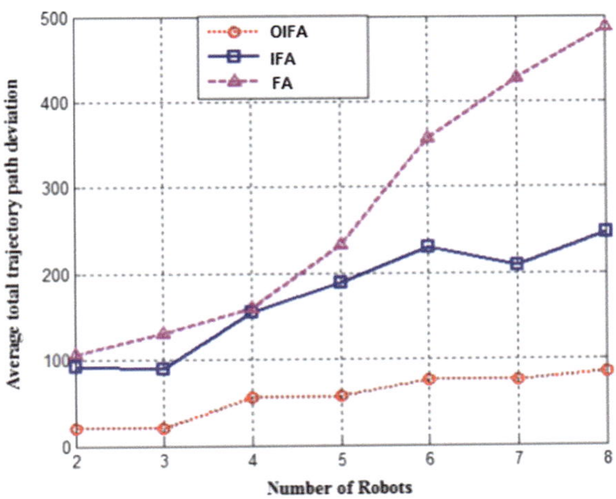

Fig. 6. Average total trajectory path deviation versus no of obstacles with varying the number of obstacles and velocity

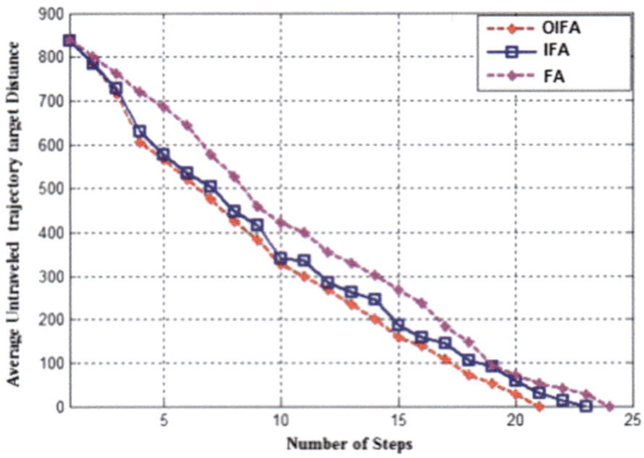

Fig. 7. Average untraveled trajectory target distance versus no of steps required under different algorithms

The minimum count of ATTD is sufficient to identify OIFA as better in comparison to IFA and FA with varying the robot numbers. The behavior of the proposed model is analyzed to establish the relation between the number of hops and AUTD, thus a graph is drawn which is given in Fig. 7. This graph is generated by considering all three of the mechanisms with 7 obstacles and 5 robots.

6 Conclusion and Future Work

In this work, an oppositional-based improved firefly algorithm was proposed for getting an optimal path for mobile robots which will also be collision-free, with a well-defined start and end position for each of the robot present in the environment. The simulation result is showing that the proposed approach provides a better result than the conventional firefly algorithm. In the considered environment, the environment is a static one where the landscape and obstacles, both are static in comparison to the robots; but the robots are dynamic. Research will be carried out in the future using dynamic obstacles other than robots during multi-robot route preparation, such as moving cars, animals and onboard camera.

References

1. Yang XS (2009) Firefly algorithms for multimodal optimization. In: Stochastic algorithms: foundations and applications, lecture notes in computer sciences, vol 5792, pp 169–178
2. Farahani SM, Abshouri AA, Nasiri B, Meybodi M (2012) Some hybrid models to improve firefly algorithm performance. Int J Artif Intell 8(S12):97–117
3. El-Sawy AA, Zaki EM, Rizk-Allah R (2012) A novel hybrid ant colony optimization and firefly algorithm for solving constrained engineering design problems. J Nat Sci Math 6(1)
4. Kasdirin HA, Yahya NM, Tokhi MO (2015) Hybridizing firefly algorithm with invasive weed optimization for engineering design problems. In: 2015 IEEE international conference on evolving and adaptive intelligent systems (EAIS). IEEE, pp 1–6
5. Hidalgo-Paniagua A, Vega-Rodríguez MA, Ferruz J, Pavón N (2017) Solving the multi-objective path planning problem in mobile robotics with a firefly-based approach. Soft Comput 21(4): 949–964
6. Cen Y, Song C, Xie N, Wang L (2008) Path planning method for mobile robot based on ant colony optimization algorithm. In: 3rd IEEE conference on industrial electronics and applications (ICIEA), Singapore, pp 289–301
7. Rahnamayan S, Tizhoosh H, Salama M (2008) Opposition-based differential evolution. IEEE Trans Evol Comput 12(1):64–79
8. Rashid M, Baig A (2010) Improved opposition-based PSO for feed forward neural network training. In: Proceedings of international conference on information science and application, pp 1–6
9. Shaw B, Mukherjee V, Ghoshal SP (2012) A novel opposition-based gravitational search algorithm for combined economic and emission dispatch problems of power systems. Int J Electr Power Energy Syst 35(1):21–33
10. Chatterjee A, Ghoshal SP, Mukherjee V (2012) Solution of combined economic and emission dispatch problems of power systems by an opposition-based harmony search algorithm. Int J Electr Power Energy Syst 39(1):9–20
11. Panda MR, Das P, Dutta S, Pradhan S (2018) Optimal path planning for mobile robots using oppositional invasive weed optimization. Int J Comput Intell 1–29, Wiley
12. Panda MR, Dutta S, Pradhan S (2017) Hybridizing invasive weed optimization with firefly algorithm for multi-robot motion planning. Arab J Sci Eng 1–11, Springer
13. Panda MR, Das P, Pradhan S (2017). Hybridization of IWO and IPSO for mobile robots navigation in a dynamic environment. J. King Saud Univ-Comput Inf Sci Elsevier

Design and Modal Analysis of Few-Mode Fibers for Spatial Division Multiplexing

Bhagyalaxmi Behera[1], S. K. Varshney[2], and Mihir Narayan Mohanty[1(✉)]

[1] ITER, Siksha 'O' Anusandhan (Deemed to be University), Bhubaneswar, India
[2] Department of Electronics and Electrical Communication Engineering, IIT Kharagpur, Kharagpur, India

1 Introduction

Optical communication has been around since the '70s with the fiber links utilizing multi-mode optical fibers with less bit rate. However, with the advent of single-mode fibers, it has been possible to go for a high bit rate (>few Gbps per channel) as well as long distance. With the help of wavelength division multiplexing (WDM) or dense WDM (DWDM). It is possible to achieve a data rate of order of terabits per second. There is a huge demand for high bandwidth and higher bit rate at the metro and access networks where spatial-mode multiplexing or mode-division multiplexing may play a pivotal role.

A theoretical limit on the information-carrying capacity of a fiber link, due to the non-linear nature of light was investigated by approximating a non-linear channel to a linear channel with multiplicative noise. It is also observed that the non-linear effects happen due to intensity-dependent refractive index variation is one of the important cause of the reduction in spectral efficiency of a single-mode fiber [1].

With an aim to overcome the upcoming capacity crisis of current optical fiber links several approaches have been evolved, such as space-division multiplexing (SDM) [2], and mode-division multiplexing (MDM) [3]. Use of few-mode fibers and multi-core fibers (MCFs) are gaining special attention in optical transmission using MDM and SDM, where the transmission capacity increases substantially as compare to SMFs [2, 4]. MDM has attracted the attention of the scientific community to increase the transmission capacity of a fiber link using FMFs. Several MDM transmission systems using few-mode fibers have been proposed [5, 6]. In this regard, A 2×10 Gbps MDM transmission system was realized over a distance of 10 km with the help of a two-mode (LP_{01}, LP_{11}) fiber [5]. Authors in [6] have reported the effective transmission of five modes in a mode-division multiplexed transmission, where they have concluded that the performance of the MDM transmission system is limited by the inter-modal crosstalk between the LP modes. Transmission of 32Wavelength Division Multiplexed (WDM) channels using four-LP modes or 12 spatial/polarization modes (LP_{01}, LP_{02}, LP_{11}, LP_{21}) over a distance of 177 km was demonstrated using few-mode fiber in [7], where authors claimed a spectral efficiency of 32 bit/s/Hz per core for MDM transmission. In their concluding remark, they have reported the strongly coupled transmitted signals are recuperated using a 12×12 multi-input–multi-output (MIMO)

M. N. Mohanty and S. Das (eds.), *Advances in Intelligent Computing and Communication*, Lecture Notes in Networks and Systems 109,
https://doi.org/10.1007/978-981-15-2774-6_19

signal processing. A spectral efficiency-distance product of 31,500 bit/s/Hz km has been obtained in [8] by using 33 WDM channels and three-mode graded-index fiber over 3500-km with low differential group delay (DGD) low mode-dependent loss (MDL)at a cost of 6 × 6 MIMO processing. A data rate of 280 Bits/s was reported in [9], with the help of a mode multiplexed system using 3-mode graded-index few-mode fibers. The transmission capacity of a MDM system increases with the increasing number of LP modes in an FMF at cost of computational complexity and power consumption of MIMO signal processing. The crosstalk due to strong mode coupling between the transmitted modes and the large DGD dependent distortion in FMFs puts a challenge for the researcher. The use of graded-index FMF is an approach to lower down the group delay and/or the Differential Mode Delay (DMD) [10, 11]. From the literature review, one can observe that FMF is the best solution to enhance the capacity limit through MDM as long as the numbers of modes are few with lesser mode coupling and low DGD.

FMF is a special type of multi-mode fiber that supports a few tens of guided modes, where each individual mode is treated as a unique data channel. The FMF was designed in the year 1978, with a graded-index profile to support LP_{01} and LP_{11} mode. Its modal properties were demonstrated with a core diameter of 16.3 μm, and the normalized frequency parameter was set with an approximation of $(V) = 4.45 \pm 0.11$ under the operating wavelength of 1.25 μm [12]. The initial motivation of this design is to introduce a large core to compensate for the splicing loss present in SMFs. The importance was lost through the invention of SMF with a splicing loss of 0.1 dB, and it took more than 30 years to regain the importance of FMF in current research to facilitate the high transmission capacity of optical fiber links.

In this article, the FMF design over-view of the modified-graded-index profile for a few-mode operation is discussed. The fiber parameters are set to support four-LP modes (LP_{01}, LP_{02}, LP_{11}, and LP_{21}) with low DGD, large mode area A_{eff}, and low bending loss for MDM application *C*-band from 1530 to 1565 nm.

2 Design Principle

The few-mode fiber profile structures to be discussed in this article are based on the following principles:

- The first three higher-order LP modes (LP_{02}, LP_{11}, and LP_{21}) are allowed to propagate along with the dominant LP_{01} mode.
- The modes are guided with less mode coupling by maintaining a large difference between effective indices of the guided modes.
- The DGD between the above four guided modes is kept small.
- V_{eff} is selected just below the cutoff *V*-value of LP_{31} mode.

The variation of refractive index with respect to the core radius of a graded-index FMF (GI-FMF) and modified-graded-index FMF (MGI-FMF) is illustrated in Fig. 1a, b, respectively.

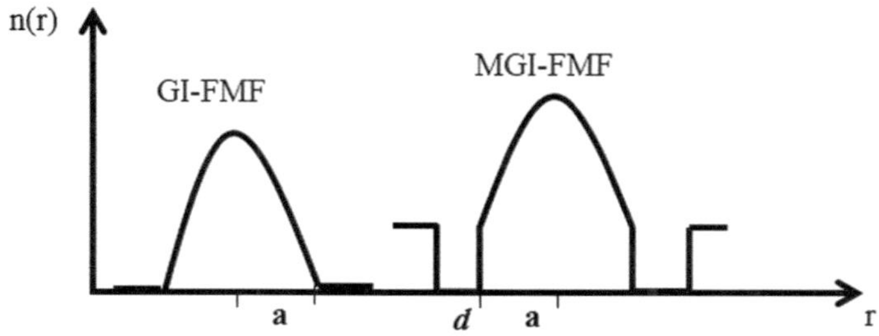

Fig. 1. Refractive index profile of GI-FMF and MGI-FMF

The GI-FMF consists of a graded-index core and a single cladding, whereas the MGI-FMF is defined by a graded-index core with up doping ($\Delta^+ = 0.3\%$), and followed by a depression in the cladding which is down doped ($\Delta^- = 0.7\%$). The graded-index core helps to maintain equalized group delay between the guided modes, plus a large effective index difference between the guided modes helps to reduce the mode coupling. The relative refractive index difference (Δ^+) between the core and cladding is kept small to achieve a large mode area. The depression in the cladding is introduced to reduce the bending loss. The parameters of the proposed fiber are core radius a, width of the depression in the cladding d, the graded-index profile parameters α and ρ, and the refractive index of the core and cladding n_{co} and n_{cl}, respectively. Where (Δ^+) is defined in Eq. (1). The core is assumed to be pure silica. The variation of graded-index profile w.r.t core radius is expressed in Eq. (2) GI-FMF and in Eq. (3) MGI-FMF.

$$\Delta^{+} = \begin{cases} \left(n_{co}^2 - n_{cl}^2\right)/2n_{co}^2 & \text{for GI\_FMF} \\ n_{co} - n_{cl}/n_{co} & \text{for MGI\_FMF} \end{cases} \tag{1}$$

$$n(r)|\mathrm{GI}_{\mathrm{FMF}} = \begin{cases} n_{co}\left[1 - \rho\Delta^{+}(r/a)^{\alpha}\right] & 0 \le r \le a \\ n_{co}\left[1 - 2\Delta^{+}\right]^{1/2} & r > a \end{cases} \tag{2}$$

$$n(r)|\mathrm{MGI\_FMF} = \begin{cases} n_{co}\left[1 - \rho\Delta^{+}(r/a)^{\alpha}\right](0 \le r \le a) \\ n_{co}\left(1 - \Delta^{-}\right)(a \le r \le d) \\ n_{co}\left(1 - 2\Delta^{+}\right)(r > d) \end{cases} \tag{3}$$

The graded-index core of GI-FMF varies from a parabolic profile to a step-index profile with a variation of $\alpha = 2$ to $\alpha = \infty$, respectively. Whereas ρ defines a sharper or slower transition between n_{co} and n_{cl}. For our design simplification, we have assumed $\rho = 1$. The effective mode indices and corresponding modal behavior of the proposed fiber have been calculated using a finite difference method. In order to achieve low bending loss, we have proposed MGI-FMF with a depression in the cladding just after the core of width d. The depression is down doped by $\Delta^- = 0.7\%$. The index profile of MGI-FMF is expressed in Eq. (3).

3 Result and Analysis

It has been observed that the GI-FMF profile supports LP_{01} and LP_{11} modes at 1.55 μm for any value of $\alpha \geq 2$. The effective index variation (n_{eff}) w.r.t α, and the mode field distribution of LP_{01} and LP_{11} modes in GI-FMF is demonstrated in Fig. 2a–c, respectively. With this result, we conclude that this fiber profile exhibits weak mode coupling between LP_{01} and LP_{11} mode by maintaining a large index difference.

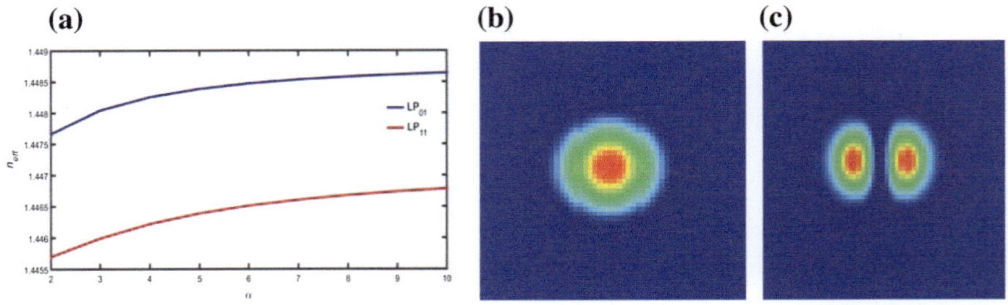

Fig. 2. **a** Variation of n_{eff} of LP_{01} and LP_{11} modes with respect to α, mode field distribution of **b** LP_{01}, **c** LP_{11} modes in GI-FMF with a core radius of 8 μm, $\Delta^+ = 0.3\%$ at $\lambda = 1.55$ μm

The modal characteristic of GI-FMF at 1.55 μm is exhibited in Table 1. Here the simulation results are obtained by the finite difference method. The graded-index profile of GI-FMF, described in Eq. (2) supports only LP_{01} and LP_{11} modes with a core radius of 8 μm, $\Delta^+ = 0.3\%$ for any value of $\alpha \geq 2$, where ρ is assumed to be 1. The results are verified for the entire *C*-band center around 1.55 μm. From Table 1 we conclude that the GI-FMF structure satisfies smaller mode coupling, large effective area, and small DGD between the modes. Hence this fiber structure claims to support mode multiplexing in *C*-band. But this design is suffered by large bending loss as well.

Table 1. Characteristics of GI-FMF profile with $\alpha = 5$ in Fig. 1 at 1.55 μm

Property	Value
Δn_{eff} of LP_{01} ($n_{eff} - n_{cl}$)	3.8×10^{-3}
Δn_{eff} of LP_{11} ($n_{eff} - n_{cl}$)	1.7×10^{-3}
Δn_{eff} LP_{01} − Δn_{eff} LP_{11}	2.1×10^{-3}
The cutoff wavelength of LP_{11}	1843 nm
DGD between LP_{11} and LP_{01}	−0.0007 ps/km
Bending loss of LP_{01} (at bending radius $R = 5$ mm)	9.244×10^{1}db/km
Bending loss of LP_{11} (at bending radius $R = 5$ mm)	2.319×10^{2}db/km
A_{eff} of LP_{01}	135 μm^2
A_{eff} of LP_{11}	189 μm^2
Dispersion in LP_{01}	13.5 ps/km nm
Dispersion in LP_{11}	11.4 ps/km nm

In this phase, we have discussed MGI-FMF. This fiber profile effectively guides four-LP modes (LP_{01}, LP_{11}, LP_{21}, and LP_{02}). The effective indices of the guided modes depend on both a and *d*. Hence, the n_{eff} versus normalized (*d*/2*a*) is plotted in Fig. 3. With this, we conclude that the n_{eff}-variation is significant for three values of *d* (=1.6, 3.2, 4.8 μm). Hence, the rest of the simulations are performed with these *d* values.

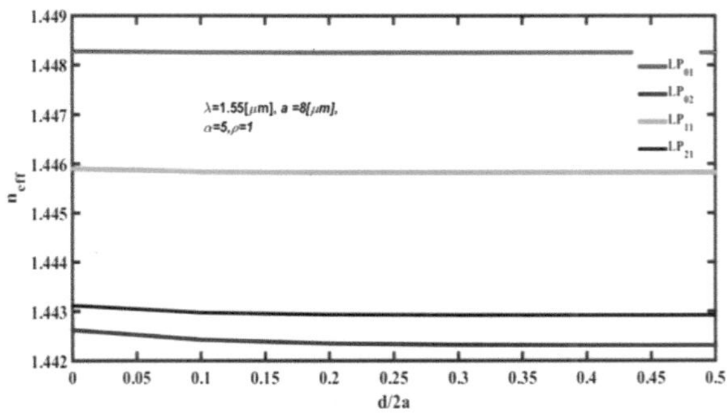

Fig. 3. Variation of n_{eff} of LP_{01}, LP_{11}, LP_{02}, and LP_{21} modes as a function of *d*/2*a*, in MGI-FMF with a core radius of 8 μm, $\alpha = 5$, $\Delta^{+} = 0.3\%$, $\Delta^{-} = 0.7\%$ at $\lambda = 1.55$ μm

The MGI-FMF structure allows us to guide effectively, first three higher-order modes (LP_{11}, LP_{02}, LP_{21}) along with the dominant LP_{01} mode for any value of α between 1 and ∞, and $\rho = 1$ at operating wavelength 1.55 μm. Figure 4 shows the variation of n_{eff} as a function of α, for $d = 1.6$ μm. A large index difference (Δn_{eff}) is aimed to maintain between the modes to achieve less mode coupling for MDM applications in *C*-band communication.

The variation of normalized propagation parameter *b* w.r.t the normalized frequency parameter *V* for different values of *d* (=1.6, 3.2, 4.8 μm) are depicted in Fig. 5a–c, respectively. Where the *V* and *b* are calculated from Eq. (4) and (6), respectively. With this result, we conclude that MGI-FMF allows guiding five LP modes (LP_{01}, LP_{02}, LP_{11}, LP_{21}, and LP_{31}). But we have set V_{eff} just below the cutoff *V*-value of LP_{31} to effectively guide the first four modes. We find that *b* of LP_{21} and LP_{02} moves closer with the increasing value of *V*. But a distinct gap is maintained between the *b* of LP_{21} and LP_{02} by increasing the width *d*.

A large difference in the *b* of the guiding modes ensures less coupling between the modes to avoid the crosstalk in MDM.

$$V = \frac{2\pi a}{\lambda}\,\mathrm{NA} \tag{4}$$

$$\mathrm{NA} = \begin{cases} \mathrm{NA}(0)[1-(r/a)^{\alpha}]^{1/2} \cong \left[n_{co}^{2}(r) - n_{cl}^{2}\right]^{1/2} & r \le a \\ 0 & r > a \end{cases} \tag{5}$$

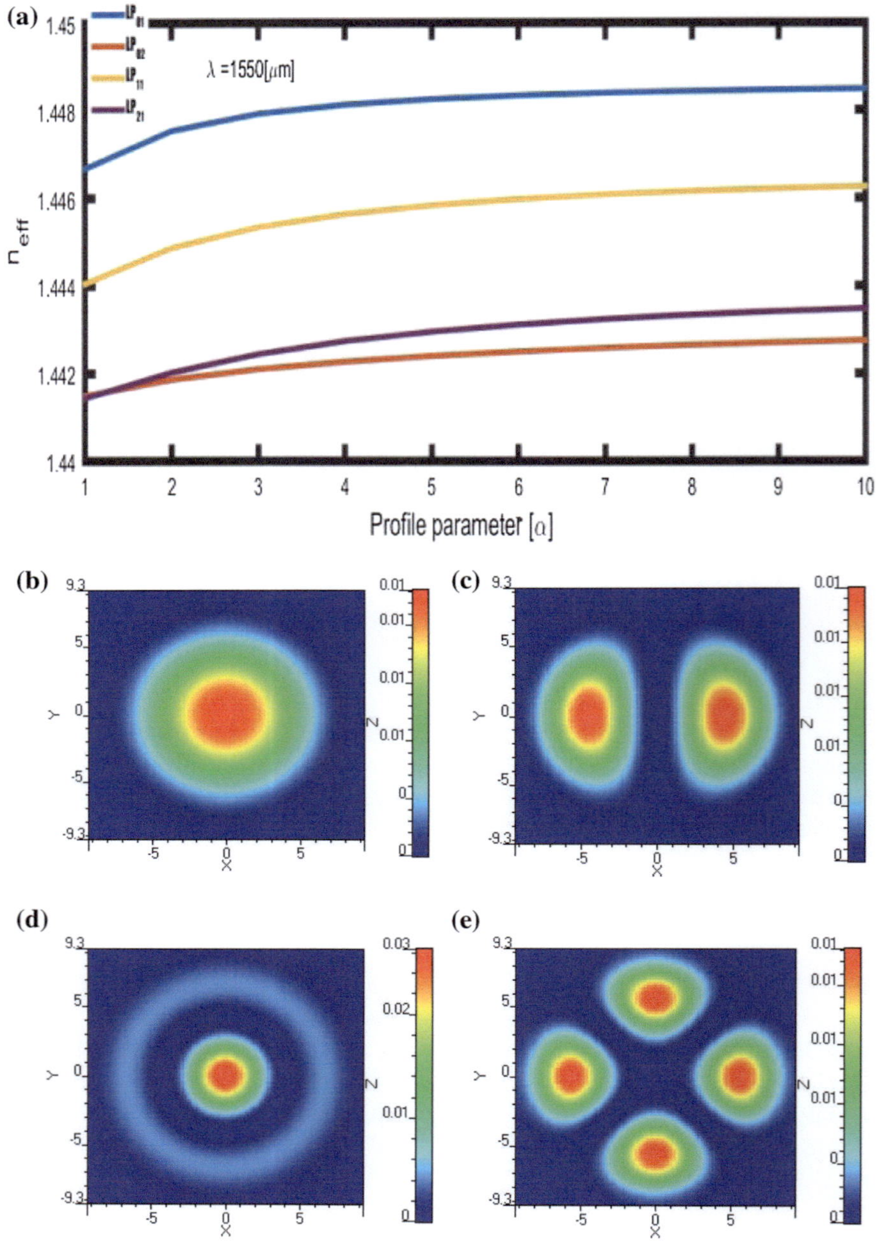

Fig. 4. **a** Variation of n_{eff} of LP_{01}, LP_{11}, LP_{02}, and LP_{21} modes with respect to α, mode field distribution of **b** LP_{01}, **c** LP_{11}, **d** LP_{02}, **e** LP_{21} modes in MGI-FMF with a core radius of 8 μm, $d = 1.6$ μm, $\Delta^{+} = 0.3\%$, $\Delta^{-} = 0.7\%$ at $\lambda = 1.55$ μm

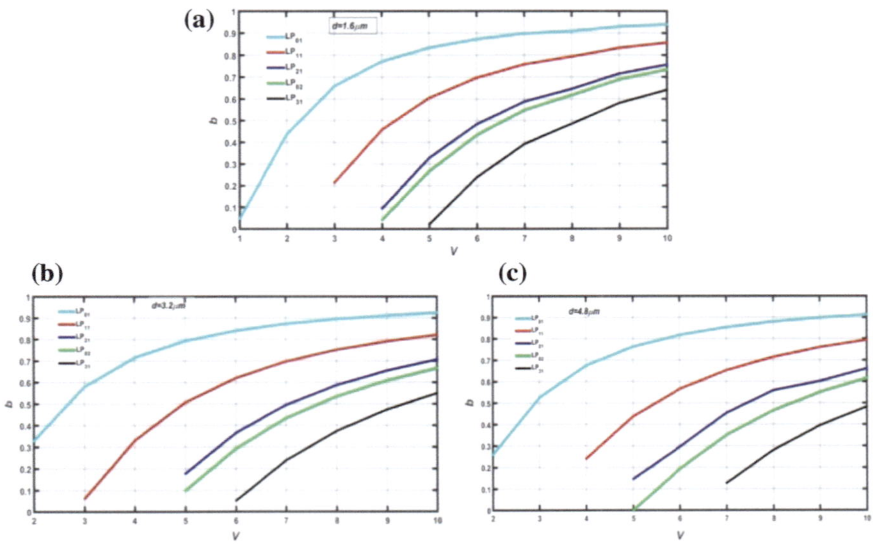

Fig. 5. Variation of b of LP_{01}, LP_{11}, LP_{02}, LP_{21}, and LP_{31}modes as a function of V **a** $d = 1.6$ µm, **b** $d = 3.2$ µm **c** $d = 4.8$ µmin MGI-FMF with a core radius of 8 µm, $\Delta^{+} = 0.3\%$, $\Delta^{-} = 0.7\%$ at $\lambda = 1.55$ µm

$$b = \frac{n_{\text{eff}}^2 - n_{\text{cl}}^2}{n_{\text{co}}^2 - n_{\text{cl}}^2} \tag{6}$$

Table 2 summarizes the modal characteristics of MGI-FMF at 1.55 µm. The results have been calculated by using the finite difference method. The simulation results are obtained with the following fiber parameters, core radius $a = 8$ µm, depression width $d = 4.8$ µm, $\Delta^{+} = 0.3\%$, $\Delta^{-} = 0.7\%$, $\alpha = 5$. In comparison with GI-FMF, MGI-FMF has lower bending loss at the cost of a fractional reduction in the mode area. MGI-FMF guides four-LP modes effectively with low DGD and less mode coupling between the modes.

Further, the bending loss of GI-FMF and MGI-FMF is compared in Fig. 6. The bending loss is calculated at a bending radius of $R = 5$ mm and $\lambda = 1.55$ µm. From Fig. 6 we conclude, that the low index contrast of depression in the cladding of MGI-FMF decreases the penetration of optical power into the cladding and thus, reduces the bending loss excessively.

4 Conclusion

Though the graded-index fibers have been designed earlier, still there was a conflict about the few-mode operation. In this work the modified-graded-index structure which guides four modes have been simulated successfully. The performance shows the

Table 2. Characteristics of MGI-FMF profile with $\alpha = 5$, $d = 4.8$ μm in Fig. 1 at 1.55 μm

Property	Value
Δn_{eff} of LP_{01} ($n_{eff} - n_{cl}$)	8.3×10^{-3}
Δn_{eff} of LP_{11} ($n_{eff} - n_{cl}$)	5.9×10^{-3}
Δn_{eff} of LP_{21} ($n_{eff} - n_{cl}$)	3.0×10^{-3}
Δn_{eff} of LP_{02} ($n_{eff} - n_{cl}$)	1.4×10^{-3}
Δn_{eff} LP_{01} − Δn_{eff} LP_{11}	2.4×10^{-3}
Δn_{eff} LP_{11} − Δn_{eff} LP_{21}	2.9×10^{-3}
Δn_{eff} LP_{21} − Δn_{eff} LP_{02}	1.6×10^{-3}
The cutoff wavelength of LP_{11}	1943 nm
The cutoff wavelength of LP_{21}	1813 nm
The cutoff wavelength of LP_{02}	1734 nm
DGD between LP_{11} and LP_{01}	−0.003 ps/km
DGD between LP_{21} and LP_{11}	−0.002 ps/km
DGD between LP_{02} and LP_{21}	−0.001 ps/km
Bending loss of LP_{01} (at bending radius $R = 5$ mm)	1.743×10^{-1} dB/km
Bending loss of LP_{11} (at bending radius $R = 5$ mm)	5.687×10^{1} dB/km
Bending loss of LP_{21} (at bending radius $R = 5$ mm)	4.088×10^{4}dB/km
Bending loss of LP_{02} (at bending radius $R = 5$ mm)	2.318×10^{6}dB/km
A_{eff} of LP_{01}	117 μm^2
A_{eff} of LP_{11}	121 μm^2
A_{eff} of LP_{21}	149 μm^2
A_{eff} of LP_{02}	170 μm^2
Dispersion in LP_{01}	24.1 ps/km nm
Dispersion in LP_{11}	23.3 ps/km nm
Dispersion in LP_{21}	19.4 ps/km nm
Dispersion in LP_{02}	19.1 ps/km nm

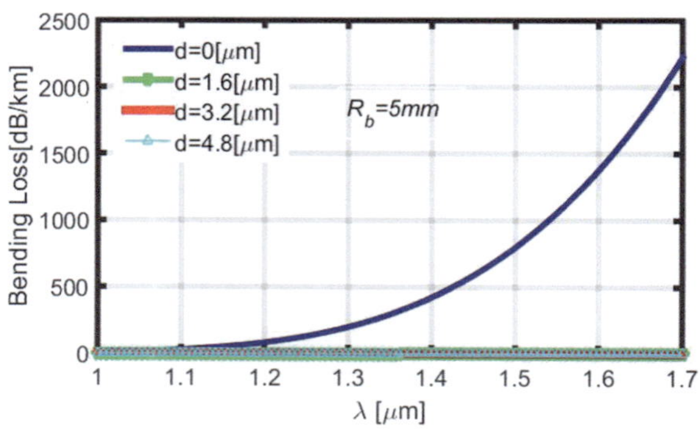

Fig. 6 Comparison of bending loss in GI-FMF and MGI-FMF as a function of wavelength at a bending radius $R = 5$ mm

efficiency of our proposed method. It is well observed that the suitability condition for mode-division multiplexing in *C*-band is possible. Whereas, the fiber parameters are under study to support more than four modes with a *V*-value at which the DGD of the guided modes are equalized. Simultaneously, the width of the depression in the cladding, and its distance from the core must be taken care of for better accuracy; the work is kept for future work.

References

1. Mitra PP, Stark JB (2001) Nonlinear limits to the information capacity of optical fibre communications. Nature 411(6841):1027–1029
2. Richardson DJ, Fini JM, Nelson LE (2013) Space-division multiplexing in optical fibres. Nat Photon 7(5):354
3. Berdagué S, Facq P (1982) Mode division multiplexing in optical fibers. Appl Opt 21 (11):1950–1955
4. Kitayama KI, Diamantopoulos NP (2017) Few-mode optical fibers: original motivation and recent progress. IEEE Commun Mag 55(8):163–169
5. Hanzawa N, Saitoh K, Sakamoto T, Matsui T, Tomita S, Koshiba M (2011). Demonstration of mode-division multiplexing transmission over 10 km two-mode fiber with mode coupler. In: Optical fiber communication conference. Optical Society of America, p. OWA4
6. Salsi M, Koebele C, Charlet G, Bigo S (2012). Mode division multiplexed transmission with a weakly-coupled few-mode fiber. In: OFC/NFOEC, IEEE, pp 1–3
7. Ryf R, Randel S, Fontaine NK, Montoliu M, Burrows E, Chandrasekhar S, Gnauck AH, Xie C, Essiambre RJ, Winzer P, Delbue R (2013) 32-bit/s/Hz spectral efficiency WDM transmission over 177-km few-mode fiber. In: Optical fiber communication conference. Optical Society of America, pp PDP5A-1
8. Rademacher G, Ryf R, Fontaine NK, Chen H, Essiambre RJ, Puttnam BJ, Luis RS, Awaji Y, Wada N, Gross S, Riesen N (2017). 3500-km mode-multiplexed transmission through a three-mode graded-index few-mode fiber link. In: 2017 European conference on optical communication (ECOC). IEEE, pp 1–3
9. Rademacher G, Luís RS, Puttnam BJ, Ryf R, Furukawa H, Maruyama R, Aikawa K, Maruta A, Awaji Y, Wada N (2018). 93.34 Tbit/s/mode (280 Tbit/s) transmission in a 3-mode graded-index few-mode fiber. In Optical fiber communication conference. Optical Society of America, pp W4C-3
10. Ferreira F, Fonseca D, Silva H (2013) Design of few-mode fibers with arbitrary and flattened differential mode delay. IEEE Photon Technol Lett 25(5):438–441
11. Gruner-Nielsen L, Sun Y, Nicholson JW, Jakobsen D, Jespersen KG, Lingle R, Palsdottir B (2012) Few mode transmission fiber with low DGD, low mode coupling, and low loss. J Lightwave Technol 30(23):3693–3698
12. Sakai J, Kitayama K, Ikeda M, Kato Y, Tatsuya K (1978) Design considerations of broadband dual-mode optical fibers. IEEE Trans Microw Theory Tech 26(9):658–665

Speech Enhancement Using Wiener Filter Based on Voiced Speech Probability

Rashmirekha Ram, Abhisek Das, and Saumendra Kumar Mohapatra(✉)

ITER, Siksha 'O' Anusandhan (Deemed to be University), Bhubaneswar, India

1 Introduction

Communication through speech is the fundamental and basic needs of human beings. Due to clarity of communication, speech emerges as the most effective and advantageous source. At the point, when the speaker and audience are near to one other, communication is simple and with less noise. Nevertheless, in case of distance communication, the listener faces difficulty to listen due to involvement of various noises that degrades the original signal. The quality of speech needs to be prominent for accuracy of information exchange. Background interfering noise and channel noise are some of those which drastically degrade the performance of the system. However, because of the intricacies of the speech signal, increasing the quality creates challenge in speech processing research and communication system applications [1].

The characteristics of speech signal depend on time. Various kinds of noise present in the background degrade the clarity and quality of the speech signal. To improve the quality and to reduce the noise level, researchers have proposed numerous methods, for example, adaptive techniques, filtering algorithms, statistical and subspace methods [2–5]. Upadhyay et al. designed Wiener filter with recursive noise estimation for enhancement. For performance measures, additive noise and pink noise were evaluated. Nevertheless, speech signal was distorted [6]. To overcome this, a conventional filtering was proposed based on voiced and unvoiced frames of the speech. The filter was incorporated with empirical mode decomposition (EMD) and adaptive center weighted average (ACWA) filter. The additive Gaussian noise was tested for the proposed method. However, residual noise was present in the enhanced signal because of EMD. The filtering can be improved by using proper classification methods to distinguish voiced frames [7]. Authors used gamma tone filter bank for signal sub-band decomposition. This phase-dependent low-delay filter bank analyzed the signal frame by frame, and the fundamental period was estimated. For estimation, periodicity degree (PD) was proposed to detect the period and to estimate the SNR in each frame. The comb filter ratio and the normalized autocorrelation were multiplied to calculate the PD in each sub-band frame signal unit. The noise energy level and the Wiener gain applied in each frame formed the enhanced signal. The combination of the filter bank and the Wiener filter provided better enhanced signal compared to other baseline statistical-based methods [8].

M. N. Mohanty and S. Das (eds.), *Advances in Intelligent Computing and Communication*, Lecture Notes in Networks and Systems 109,
https://doi.org/10.1007/978-981-15-2774-6_20

Wavelet methods have occupied a special attention in speech enhancement area because of their noise reduction characteristics. Depending on this, an enhancement method was designed by utilizing voiced speech probability with Wavelet decomposition. The speech probability was decided on the voiced and unvoiced frames depending on VAD. The probability was computed between two Gaussian mixture models (GMM) and the likelihood ratio test. The mean square error was minimized by the gain estimator and the wavelet decomposition method. TIMIT database was used to test the results with five different noisy conditions. The speech presence probability (SPP) plays an important role in frequency domain speech enhancement techniques. Most SPP methods reduced the noise variance by wavelet denoising methods. However, this method could not optimize the multitaper spectrum and denoise the speech signals. To overcome this disadvantage, wavelet transform was utilized to the periodogram, and then, an oracle was used to indicate the noise spectrum. The generalized likelihood ratio in SPP estimated the noise spectrum and provided better improvement in quality and intelligibility [9, 10]. Nevertheless, in stationary environment, single-channel speech enhancement is quite difficult, and for this, Wiener filter is more preferable. The filter transfer function used in frequency domain and the local mean and variance were calculated sample to sample. In this method, noise was estimated by using present and previous power values of spectrum. The smoothing parameter utilized in the Wiener filter enhanced the signal preferably better. Both male and female speakers from the NOIZEUS database were taken to evaluate the results, and it provided significantly better results than other conventional methods [6, 11].

The paper is organized as follows. Speech enhancement and some related literatures of enhancement techniques are discussed in Sect. 1. The proposed method is enlightened in Sect. 2. Different noisy signals from NOIZEUS database are tested, and the results are depicted in Sect. 3. Lastly, Sect. 4 concludes the work with some future scope.

2 Methodology

Voice activity detection is used in this work to find the voiced speech probability. For this, decision directed approach is chosen, and the decision rule is achieved from likelihood ratio test [12]. The noise is estimated from unvoiced region of the speech by minimum mean square error estimation (MMSE) technique. For filtering the signal, Wiener filter is used. As the novelty, the output of the Wiener filter is obtained from the *a priori* SNR of the signal which is estimated from decision directed approach. It provides the smoother estimates of the signal. As a result, better enhanced signal is obtained. This section is subdivided into two subsections: (A) voiced speech probability (VSP) using voice activity detection, (B) Wiener filtering using VSP for speech enhancement.

2.1 Voiced Speech Probability (VSP) Using VAD

Consider the clean speech signal $s(t)$ is degraded by uncorrelated additive noise $n(t)$. For kth frequency bin, the discrete Fourier transform (DFT) coefficient vectors of noisy speech $y(t)$ can be represented as Y_k. Similarly, S_k and N_k are the DFT coefficients of clean speech and noise. For VAD, two hypotheses may arise in each frame are:

$$\begin{aligned} H_0 &: \text{speech absent} : Y_k = N_k \\ H_1 &: \text{speech absent} : Y_k = N_k + S_k \end{aligned} \tag{1}$$

The probability density functions (*pdfs*) for DFT coefficients are asymptotically and independent Gaussian random variables. The *pdfs* on H_0 and H_1 can be expressed as [12]

$$p(Y|H_0) = \prod_{k=0}^{M-1} \frac{1}{\pi\alpha_N(k)} \exp\left\{ -\frac{|Y_k|^2}{\alpha_N(k)} \right\} \tag{2}$$

$$p(S|H_1) = \prod_{k=0}^{M-1} \frac{1}{\pi[\alpha_N(k) + \alpha_S(k)]} \cdot \exp\left\{ -\frac{|S_k|^2}{\alpha_N(k) + \alpha_S(k)} \right\} \tag{3}$$

where $\alpha_S(k)$ and $\alpha_N(k)$ represent the variances of S_k and N_k, respectively. M denotes the signal dimension of DFT coefficient vectors. The decision rule is derived from the mean of likelihood ratio test for each frequency band. The likelihood ratio for the kth frequency band is

$$L_k \triangleq \frac{p(Y_k|H_1)}{p(Y_k|H_0)} = \frac{1}{1+\beta_k} \exp\left\{ \frac{\lambda_k \beta_k}{1+\beta_k} \right\} \tag{4}$$

where β_k is the a priori SNR and λ_k is the a posteriori SNR. These two parameters can be expressed as

$$\beta_k \triangleq P_S(k) / P_N(k) \tag{5}$$

and

$$\lambda_k \triangleq |Y_k|^2 / \alpha_N(k) \tag{6}$$

where $P_S(k)$ is the power spectra of clean speech and $P_N(k)$ is the power spectra of noise The block diagram of VAD is shown in Fig. 1.

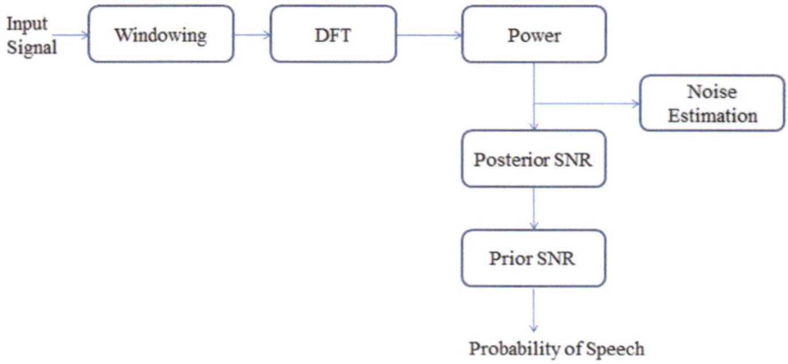

Fig. 1. Block diagram of voice activity detection

We assume that $\alpha_N(k)$ coefficients are known by noise estimation, and the a priori SNR β_k needs to be estimated. To estimate β_k, decision directed (DD) method is chosen and is expressed as [13]

$$\hat{\beta}_k(m)^{(DD)} = \mu \frac{\hat{A}_k(m-1)}{\alpha_N(k, m-1)} + (1-\mu)P[\lambda_k(m) - 1] \tag{7}$$

where m is the frame index and $\hat{A}_k(m-1)$ = signal amplitude of previous frame that is estimated by MMSE. Equation (7) provides the DD estimates of a priori SNR. It reduces the fluctuations arise in the estimated likelihood ratio of noise periods. An example of probability of speech presence by VAD is shown in Fig. 2.

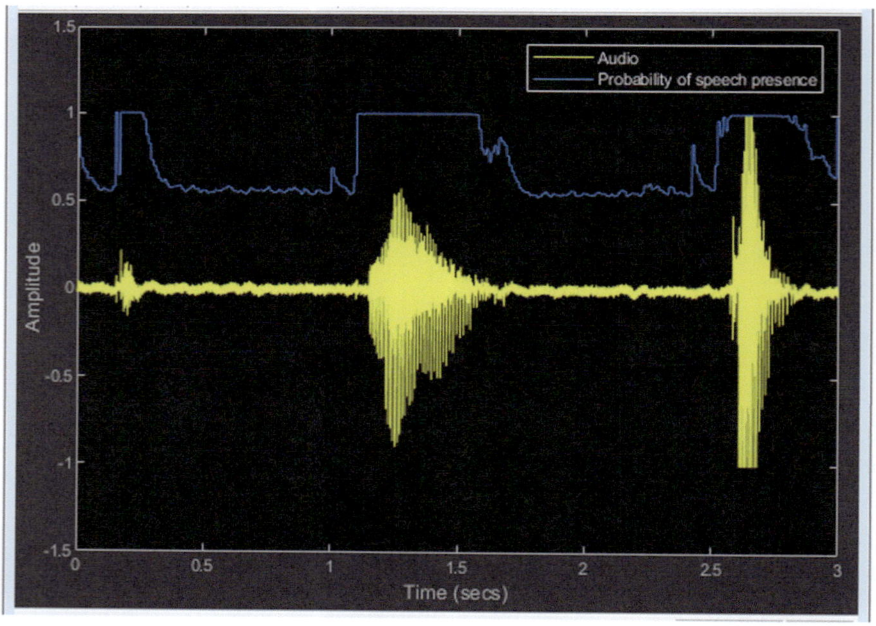

Fig. 2. Speech presence probability for a noisy input signal

2.2 Wiener Filtering Using VSP for Speech Enhancement

After finding the speech from the noisy speech signals, the noise spectrum is updated. The MSE estimator is used to estimate the signal as it is based on conditional probability. To estimate the SNR, the gain function is used, and it depends on a priori and a posteriori estimates. Again, both these SNR values depend on the estimated conditional speech and noise spectral variance. To reduce the MSE, the spectral component of noisy signals and the spectral gain function is multiplied. The speech variance and the noise spectral variance are estimated from the voiced frames and the unvoiced frames, respectively. The spectral gain function used in MMSE estimation is considered from [14, 15]. To estimate the a priori SNR, the gain function predicts the spectral variance of clean signal from the noisy signals. The block diagram of the proposed speech enhancement technique is shown in Fig. 3.

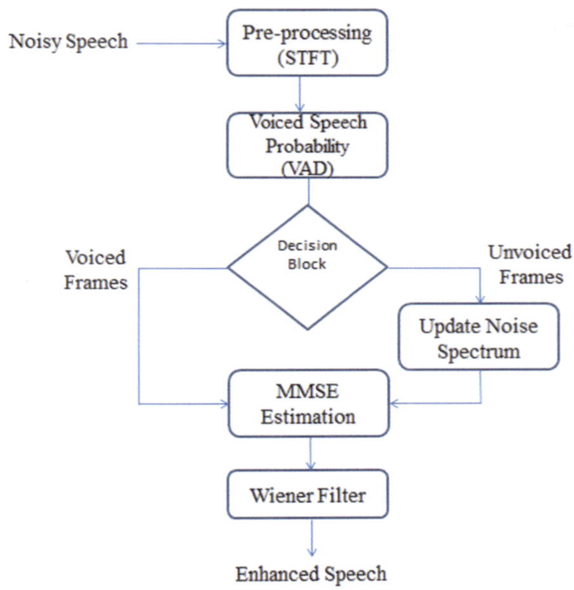

Fig. 3. Block diagram of proposed speech enhancement technique

Further to minimize the estimation error optimally, Wiener filter is used. The objective is to generate an estimate of the clean signal spectrum S_k from the desired signal D_k. The Wiener filter output in frequency domain can be written as,

$$H(k) = \frac{P_S(k)}{P_S(k) + P_N(k)} \tag{8}$$

The filter coefficients are real, even and non-negative because both the power spectra are greater than zero and are even symmetry [2]. This signifies that the impulse response $h(k)$ is even and indicates that $h(k)$ is not causal. Consequently, Eq. (8) can be expressed as

$$H(k) = \frac{\beta_k}{\beta_k + 1} \tag{9}$$

The values of β_k are estimated using Eq. (7) and can be expressed as

$$H(k) = \frac{\hat{\beta}_k(m)^{DD}}{\hat{\beta}_k(m)^{DD} + 1} \tag{10}$$

The above equation provides the enhanced signal by Wiener filter based on voiced speech probability. This proposed technique emphasizes the portion of the speech spectrum wherever the SNR value is high and attenuates the portion of the signal spectrum wherever the SNR value is low. Different signals are tested to prove the efficacy of the proposed technique, and the results are shown in the next section.

3 Results and Discussion

For experimentation, clean speech and noisy speech signals are taken from NOIZEUS database [2]. It has been generally used for benchmark experimentation. This database consists of 30 IEEE phonetically balanced sentences. These sentences are produced by both male and female speaker. Eight different noises with four different SNR levels are there. In our experimentation, Babble noise of SNR 0–15 dB is considered with an increment of 5 dB. For simulation, MATLAB is used. The sentence '*He knew the skill of the great young actress*' is uttered by a male speaker and is taken as the clean signal. The spectrogram of the clean speech signal is shown in Fig. 4. Its corresponding noisy signal corrupted by Babble noise of SNR 5 dB is shown in Fig. 5.

To prove the effectiveness, we compare the enhanced signal of proposed algorithm and the output of Wiener filter. For pre-processing, 16 ms of frame size with 10 ms skip rate is used. Hamming window of length 256 is considered. The windowed frame is then analyzed using DFT. For speech probability, VAD is used with decision directed approach. The proper separation of voiced and unvoiced frames provides better estimation of the gain function. Unlike the MMSE gain function, Wiener gain function does not depend on λ_k. Therefore, estimation of β_k is very much important to estimate the noise from the unvoiced region of the speech. The output of the voiced speech probability is shown in Fig. 2.

The Wiener filter output and the proposed technique are shown in Figs. 6 and 7, respectively. It is observed from the figures that the frequency distribution is less in the proposed technique as compared to Fig. 6. Because when $H(k) = 0$, $\beta_k \to 0$. This signifies very low SNR regions. Similarly, $H(k) = 1$ when $\beta_k \to \infty$ and signifies very high SNR regions.

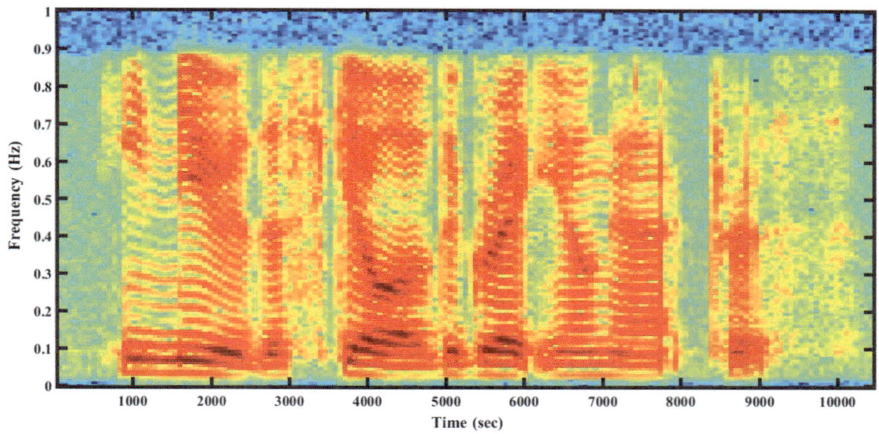

Fig. 4. Spectrogram of clean speech signal (‘*He knew the skill of the great young actress*’)

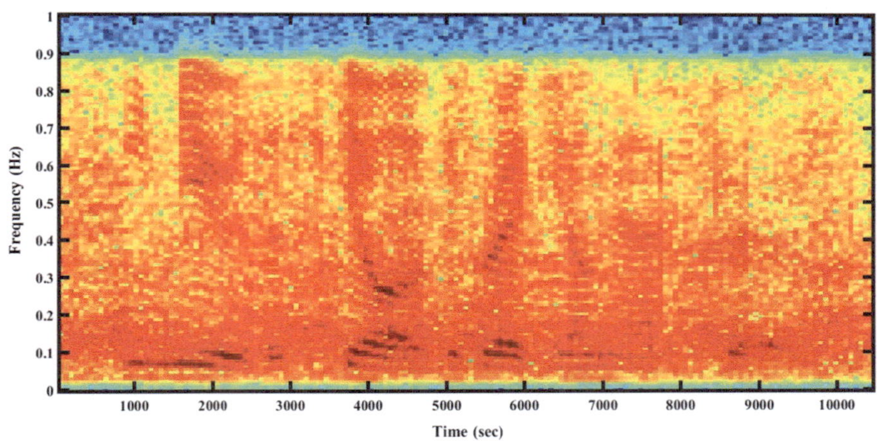

Fig. 5. Spectrogram of noisy speech (Babble noise of SNR 5 dB)

To examine the performance of the enhancement techniques, SNR and PESQ are used [16]. The maximum SNR improvement is achieved 2.54 dB (17.54–15 dB). Similarly, the maximum PESQ score is obtained as 2.78. In case of Wiener filter, the maximum PESQ score is 2.25 (Table 1).

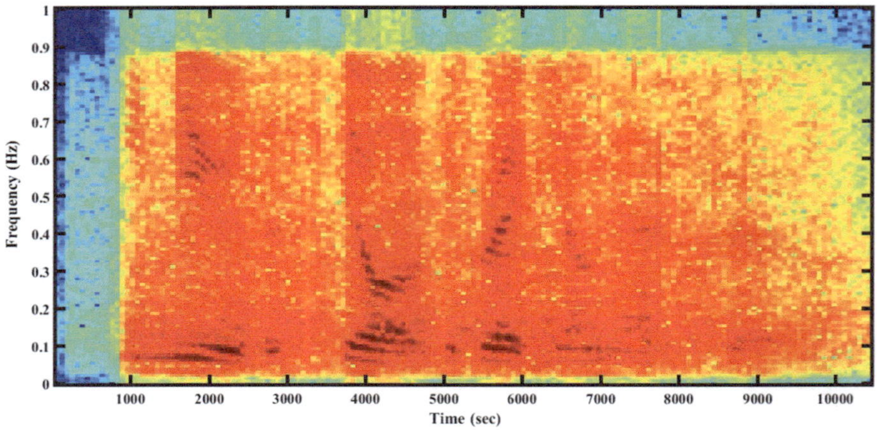

Fig. 6. Spectrogram of enhanced signal using Wiener filter

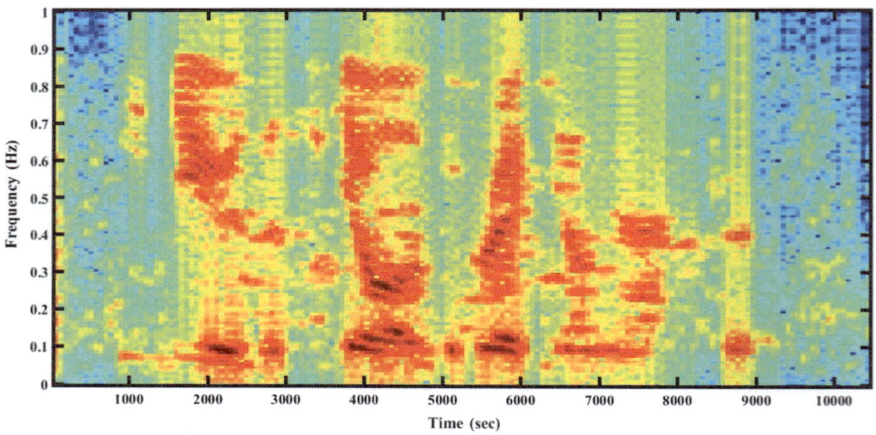

Fig. 7. Enhanced signal using proposed method

Table 1. Comparison between Wiener filter and proposed method in terms of SNR and PESQ score

Babble noise		Wiener filter	Proposed method
SNR			
0 dB	SNR (dB)	1.23	2.32
	PESQ	1.65	1.92
5 dB	SNR (dB)	6.58	7.23
	PESQ	1.99	2.38
10 dB	SNR (dB)	11.67	12.29
	PESQ	2.23	2.45
15 dB	SNR (dB)	16.88	17.54
	PESQ	2.25	2.78

4 Conclusion

In this work, Wiener filter with VSP is used for enhancement. The decision directed approach with likelihood ratio determines the voiced and unvoiced region of the speech. The noise is estimated from the unvoiced region and for this MMSE is used. Both a priori and a posteriori SNR are estimated. However, in both MMSE and Wiener estimator does not depend on a posteriori estimation. Therefore, the a priori SNR estimated from DD approach used in Wiener filter provides better enhanced signal. The proposed method is compared with Wiener filter. The proposed method provides a maximum SNR of 17.54 dB and a PESQ score of 2.78. Results show that the proposed Wiener filter outperforms better than Wiener filter. In the future, different VAD method can be applied. Other filtering methods may be tested for quality improvement.

References

1. Quatieri TF (2002) Discrete-time speech signal processing: principle and practice. Prentice Hall, New York
2. Loizou PC (2013) Speech enhancement: theory and practice. CRC Press
3. Ram R, Mohanty MN (2016) Performance analysis of adaptive algorithms for speech enhancement applications. Indian J Sci Technol 9(44):1–9
4. Ram R, Mohanty MN (2017) Design of fractional fourier transform based filter for speech enhancement. Int J Control Theory Appl 10(7):235–243
5. Ke Y, Hu Y, Li J, Zheng C, Li X (2019) A generalized subspace approach for multichannel speech enhancement using machine learning-based speech presence probability estimation, vol 146. Audio Engineering Society Convention, Audio Engineering Society
6. Upadhyay N, Jaiswal RK (2016) Single channel speech enhancement: using wiener filtering with recursive noise estimation. Procedia Comput Sci 84:22–30
7. Khaldi K, Boudraa AO, Turki M (2016) Voiced/unvoiced speech classification-based adaptive filtering of decomposed empirical modes for speech enhancement. IET Signal Process 10:169–180
8. Chen Z, Hohmann V (2015) Online monaural speech enhancement based on periodicity analysis and *a Priori* SNR estimation. EEE/ACM Trans Audio Speech Lang Process (TASLP) 23(11):1904–1916
9. Bhowmick A, Chandra M (2017) Speech enhancement using voiced speech probability based wavelet decomposition. Comput and Electr Eng 62:706–718
10. Lun DPK, Shen TW, Hsung TC, Ho DK (2016) Wavelet based speech presence probability estimator for speech enhancement. Digit Signal Process 22(6):1161–1173
11. El-Fattah MAA, Dessouky MI, Abbas AM, Diab SM, El-Rabaie ESM, Al-Nuaimy W, … El-Samie FEA (2014) Speech Enhancement with an adaptive wiener filter. Int J Speech Technol 17(1):53–64
12. Sohn J, Kim NS, Sung W (1999) A statistical model-based voice activity detection. IEEE Signal Process Lett 6(1):1–3
13. Ephraim Y, Malah D (1984) Speech enhancement using a minimum-mean square error short-time spectral amplitude estimator. IEEE Trans Acoust, Speech, Signal Process 32 (6):1109–1121
14. Mowlaee P, Scheran D, Stahl J, Wood SU, Kleijn WB (2019) Maximum a posteriori speech enhancement based on double spectrum. Proc. Interspeech 2019:2738–2742

15. Ram R, Mohanty MN (2019) Use of radial basis function network with discrete wavelet transform for speech enhancement. Int J Comput Vis Robot 9(2):207–223
16. Ram R, Mohanty MN (2018) Performance analysis of adaptive variational mode decomposition approach for speech enhancement. Int J Speech Technol 21(23):69–381

Analysis of Different Parameter of FSO Communication System by Using Robust Model Design

Shubham Mahajan and Amit Kant Pandit(✉)

School of Electronics & Communication, Shri Mata Vaishno Devi University, Katra, Reasi, J&K, India

1 Introduction

"FSO is an observable pathway innovation which uses lasers and photo-identifiers to give optical associations between two points—without the fiber. FSO can transmit information, voice, or video at velocities equipped for arriving at 2.5 Gbps. Items equipped for paces up to 10 Gbps are relied upon to hit the business sectors inside one year. FSO units comprise an optical handset with a laser (transmitter) and a photo-identifier (beneficiary) to give full duplex (bidirectional) capacity [1]. FSO systems utilize imperceptible infrared laser light wavelengths in the 750–1550 nm territory. FSO gives point-to-point transmission of communication data through the climate utilizing the optical flag as the transporter frequencies. It has attracted consideration media transmission industry, because of its cost adequacy—simple establishment, snappy foundation of communication interface particularly in the calamity the executive's situation, high data transfer capacity provisioning, and wide scope of uses. The scope of frequencies where it works makes FSO communication free from authorizing [2]. With FSO communication, most extreme information moves rates up to 2.5 Gbps is conceivable, not normal for the greatest information move paces of 622 Mbps offered by RF communication systems. FSO includes the optical transmission of voice, video, and information utilizing air as the mode of transmission (Fig. 1).

Transmission utilizing FSO innovation is generally straightforward [3, 4]. It includes two systems each comprising of an optical handset which comprises a laser transmitter and a collector to give full duplex (bidirectional) capacity. Each FSO system utilizes a high-control optical source (e.g., laser) in addition to a telescope that transmits light through the climate to another telescope that gets the data. By then, the accepting telescope interfaces with a high affectability recipient through an optical fiber.

2 Literature Review

"Throughout the most recent two decades free-space optical communication (FSO) has turned out to be increasingly intriguing as an aide or choice to radio recurrence correspondence. This chapter gives an outline of the difficulties a system fashioner needs

M. N. Mohanty and S. Das (eds.), *Advances in Intelligent Computing and Communication*, Lecture Notes in Networks and Systems 109,
https://doi.org/10.1007/978-981-15-2774-6_21

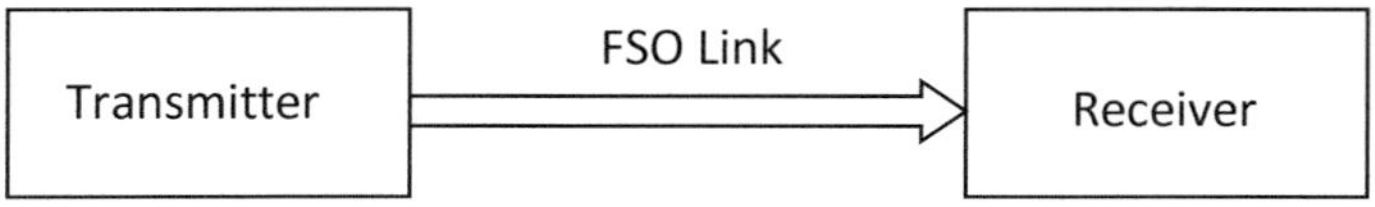

Fig. 1. Basic FSO communication system link

to consider while executing an FSO system. Commonplace increases and misfortunes along the way from the transmitter through the medium to the collector are presented in this chapter [5]. Nitty-gritty discourses of these points can be found in this uncommon issue of the Radio building Journal. Its high transfer speed abilities and low lessening attributes make it perfect for gigabit transmission. Fiber-optic communication has altered the broadcast communications industry. It has additionally made its essence broadly felt inside the information organizing network too. Utilizing fiber-optic link, optical communication shave empowered broadcast communications connects to be made over a lot of more noteworthy separations and with much lower levels of losing the transmission medium and conceivably most imperative to fall, fiber-optical interchanges have empowered a lot of higher information rates to be obliged. Fiber-optic systems are significant media transmission foundation for overall broadband systems. Wide data transmission signal transmission with low postponement is a key prerequisite in the present-day applications. Optical filaments furnish gigantic and fantastic transmission transfer speed with unimportant inactivity and are presently the transmission mode of decision for long separation and high information rate transmission in media transmission systems. Free-space optics is a medium with high transfer speed having most extreme information rates and security issues supporting its advancement for the present period. Fierce air influences the presentation of the connection. Stickiness, water vapor, signals ingestion, bar shine, spreading, and meandering are a portion of the elements which cause laser pillar debasement [6]. Keeping up a free-space optical connection between two intersections is an extreme test and needs improvement in its highlights.

3 Methodology

FSO Link: The transmitter balances information on to the prompt intensity of an optical pillar. We consider power modulated direct recognition channels utilizing on/off keying (OOK) balance, which is broadly utilized in functional systems [7, 8]. The got photograph current sign is identified with the occurrence optical power by the finder responsivity r.

$$\text{Gotten Signal}\, Y = \text{Hr}\, X + N$$

X Transmitted intensity
H Channel state
Y Resulting electrical signal
N Signal-independent AWGN.

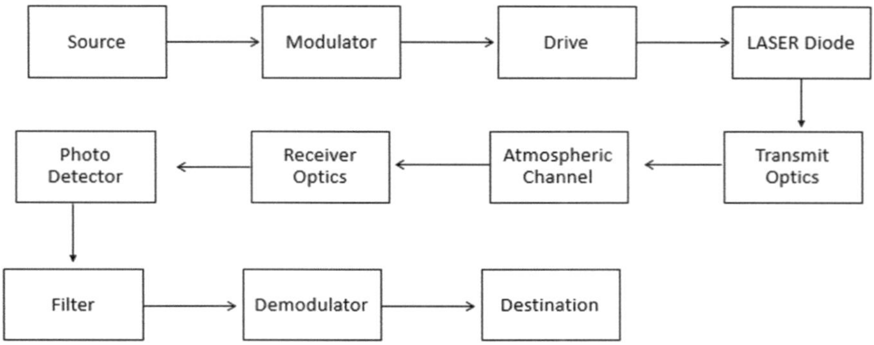

Fig. 2. FSO communication system

Parameters: We assessed the bit error rate (BER) execution of the proposed system changing with various system parameters, for example, distance transmitter power path loss factor, atmospheric turbulence received signal power. BER is profoundly corrupted on extreme environmental turbulence condition ever for a short separation of free-space channel. The impact of way misfortune factor because of thick haze is additionally extreme on the BER despite the fact that the disturbance impact and free-space separation are short. The ideal transmitted info power and collector reception apparatus span at which the BER is least is emphatically subject to free-space separation and climatic parameters. In FSO, there are different parameters which we need to dissect, and for it, we need to utilize different squares from reenactment library with the goal that goal can be accomplished. During this the need to overcome the errors, dispersion, scattering and other atmospheric conditions (Figs. 2 and 3).

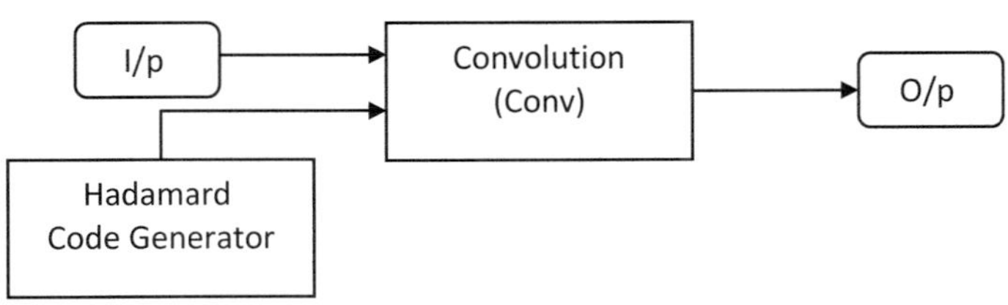

Fig. 3. Block diagram of Hadamard code generator

Free-space optics (FSO) innovation-based remote systems are not without difficulties. The key restriction of free-space optical interchanges emerges from the earth through which it engenders. Albeit moderately unaffected by downpour and day off, space optical communication systems can be seriously influenced by haze and environmental disturbance. The primary plan difficulties in free-space optical interchanges are delineated in figure.

Atmospheric Turbulence: For weak turbulences, the probability of changing likelihood density function is demonstrated as a log-normal dispersion the log-amplitude of the optical intensity has a Gaussian PDF with log-amplitude difference given by

$$\sigma_x^2 \approx \frac{\sigma_r^2}{4}$$

Haze and free-space optics: The primary test is mist. Mist is vapor made out of water droplets, which are just a couple of hundred microns in distance across however can adjust light attributes or totally impede the section of light through a blend of retention, dissipating, and reflection. This can prompt a decline in the power thickness of the transmitted pillar, diminishing the successful separation of a free-space optical connection [9, 10].

Scintillation and free-space optics: Scintillation is the spatial variety in light force brought about by environmental turbulences. Such turbulences are brought about by wind and temperature slopes that make pockets of air with quickly fluctuating densities and, accordingly, quick changing records of optical reflection. These air pockets act like focal points with time-changing properties and can prompt sharp increments in the bit-mistake paces of free-space optical communication systems, especially within the sight of direct daylight (Fig. 4).

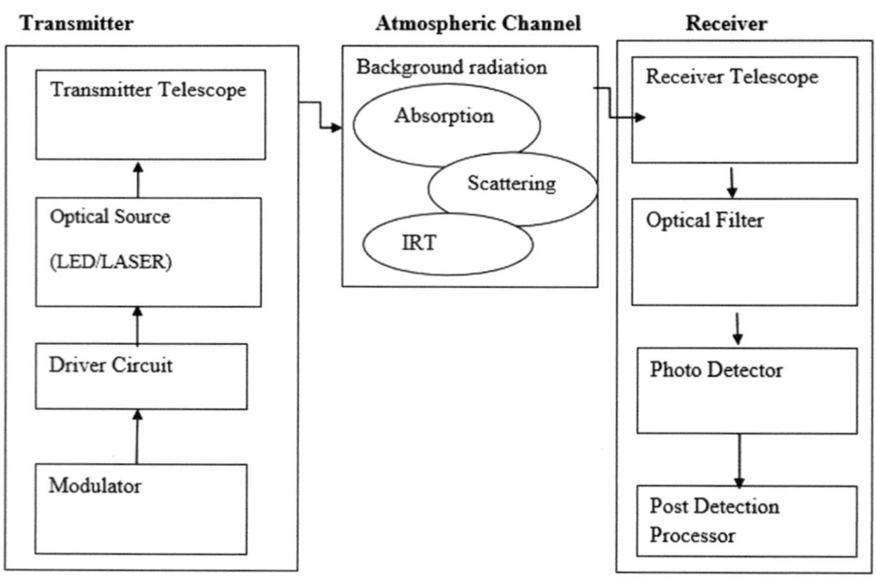

Fig. 4. Block diagram of atmospheric effects on FSO communication system

Beam wander and free-space optics: Beam meander emerges when tempestuous breeze current (whirlpools) bigger than the distance across of the transmitted optical bar cause a moderate, however critical, relocation of the transmitted bar. Shaft meander may likewise be the aftereffect of seismic movement that causes a relative relocation between the situation of the transmitting laser and the getting photograph finder (Fig. 5).

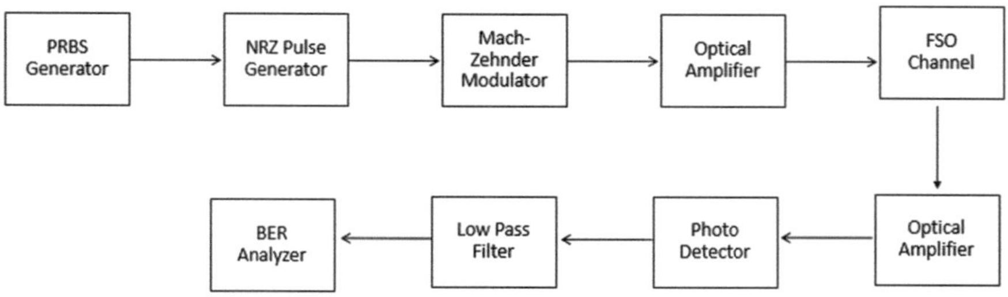

Fig. 5. Block diagram of transmitter and receiver FSO communication system

FSO CHALLENGES

A. Absorption and scattering loss

In process of propagation through earth's atmosphere, the laser beam has to interact with particles and gas molecules present in atmosphere. Absorption and scattering loss are major reasons of loss in atmosphere channel. Attenuation in atmosphere is represented as below:

$$\alpha = \mathrm{e}^{-\sigma d}$$

distance of measurement is denoted by d; coefficient attenuation particular to length per unit is denoted by σ.

$$\sigma \cong \frac{3.912}{V}\left(\frac{\lambda}{550}\right)^{-q}$$

Visibility is denoted by V and is measured in km; wavelength is denoted by λ and measured in nm; diffusing particle distribution is of size q attained by Kim and Kruse.

B. Fog

It results in atmospheric attenuation leading to absorption and scattering losses. It will result in high attenuation 350 dB/km in dense fog conditions.

$$\alpha_{\text{Fog}} = \frac{10 \log V\%}{V(\text{km})}\left(\frac{\lambda}{\lambda_o}\right)^{-q}$$

Visibility is denoted by V, measured in km; $V\%$ is percentage of air drop; wavelength is denoted by λ in nm; range of visibility is denoted by λ_o (550 nm); and diffusing particle distribution is of size q. Now, by Kruse model parameter of q is defined as

$$q = \begin{cases} 1.6 \quad \text{if} \quad V > 50\ \text{km}; \\ 1.3 \quad \text{if}\ 6\ \text{km} < V < 50\ \text{km}; \\ 0.585V \quad \text{if} \quad V < 6\ \text{km}. \end{cases}$$

Kim [9] has defined q parameter as below:

$$q = \begin{cases} 1.6 & V > 50\ \text{kms} \\ 1.3 & 6\ \text{kms} < V < 50\ \text{kms} \\ 0.16\,V + 0.34 & 1\ \text{kms} < V < 6\ \text{kms} \\ V - 0.5 & 0.5\ \text{kms} < V < 1\ \text{kms} \\ 0 & V < 0.5\ \text{kms} \end{cases}$$

C. Rain

Its brunt is not more than fog. Wavelength of FSO is smaller than drops of rain. It has been suggested that 2.5 mm/hr. corresponds to light rain and 25 mm/hr corresponds to heavy rain [10], [15]. According to Carbonneau.

$$\text{Attenuation of rain} = 1.076 * R_3^2$$

D. Snow

Attenuation because of snow is more than rain, and not as much as fog as size of snow particles is in the middle of fog and rain. Attenuation because of snow is 30–350 dB/km.

Software Used MATLAB 2015: It is amazing programming that gives a domain to numerical calculation just as graphical showcase of yields. In MATLAB, the information is in the ASCII design just as paired configuration. It is elite language for specialized registering coordinate calculation, representation, and programming in a straightforward manner where issues and arrangements are communicated in commonplace scientific documentation.

- Acquisition, Data Exploration, Analyzing, and Visualization
- Engineering complex drawing and logical illustrations
- Analyzing of algorithmic structuring
- Mathematical and Computational capacities
- Modeling and mimicking issues prototyping
- GUI (graphical UI) building condition.

Utilizing MATLAB, you can take care of specialized figuring issues all around effectively and efficient when contrasted with customary programming dialects, for example, C, C++, and FORTRAN. The name MATLAB represents network research center.

Features of MATLAB:-

- MATLAB is an abnormal state language utilized for numerical calculation, perception, and application advancement.
- It makes all around well-disposed condition for iterative investigation, plan, and critical thinking.
- Mathematical capacities for understanding conventional differential conditions, Fourier examination, direct variable-based math, insights, separating, enhancement, numerical mix.
- Development instruments for upgrading code quality and augmenting execution.
- Tools for structure applications with custom graphical interfaces (GUI).
- Functions for incorporating MATLAB-based calculations with outside applications and we can ready to create code in hex document, c, implanted and so on.

4 Conclusion

In the wake of examining different research papers, we will dissect the effect of different parameters like air disturbance, way misfortune factor, and pointing blunder on the exhibition of free-space optical communication system. To acquire our objective, we will recreate a model of free-space communication system in MATLAB utilizing Simulink. From transmitter to recipient, we will utilize different square which can give us better outcome intended to state when we transmit information from transmitter; then, it is proliferated viva remote channel and certainly different parameters like haze, lessening component, downpour, and environmental turbulences influence our information so that at collector side we got twisted information it implies BER will be there and to decrease it is our fundamental target. Other than this another significant factor is control that must be least. At that point, the BER is assessed by shifting the few parameters, for example, separation, transmitter control, way misfortune factor, got signal power, and climatic disturbance. Way misfortune factor is extreme in nearness of thick haze. The impact of climatic turbulences was likewise seen at most extreme disturbance conditions; the BER is additionally roughly 12–28. In research, our fundamental center will be to dissect the effect of different parameters on information transmission."

Acknowledgements. This work in itself is an acknowledgment to the inspiration, drive, and the technical assistance contributed to it by many people. It would have never been possible without the help and guidance that it received from my teachers. Many thanks go to my colleagues around me for providing a friendly and entertaining working environment. Special thanks to my friends for her personal support.

On a personal level, I am forever indebted to my parents for the love they provided me through my entire life.

References

1. Fidler F, Knapek M, Horwath J, Leeb WR (2010) Optical communications for high-altitude platforms. IEEE J Sel Top Quantum Electron 16(5)
2. Al-Habash MA, Andrews LC, Philips RL (2001) Mathematical model for the irradiance probability density function of a laser propagating through turbulent media. Opt Eng 40(8): 1554–1562
3. Noshada M, Rostami A (2012) FWM minimization in WDM optical communication systems using the asymmetrical dispersion managed fibers. Int J Light Electron Opt 123(9):758–760
4. Wang X, Kitayama K (2004) Analysis of beat noise in coherent and incoherent time-spreading OCDMA. IEEE/OSA J Light Technol 22(10):2226–2235
5. Zhu X, Kahn JM (2003) Performance bounds for coded free-space optical communications through atmospheric turbulence. IEEE Trans Commun 51(8):1233–1239
6. Ahmed A, Hranilovic S (2007) Outage capacity optimization for free-space optical links with pointing errors. IEEE J Light Wave Technol 25(7)
7. Shake TH (2005) Confident performance of encoded optical CDMA. IEEE/OSA J Light Technol 23:1652–1663
8. Sharma Prachi et al (2013) A review of the development in the field of fiber optic communication systems. Int J Emerg Technol Adv Eng 3(5):113–119
9. Takaaki O (2010) Recent status and trends in optical submarine cable systems. NEC Tech J 5(1):4–7
10. Kaushal H, Kaddoum G, Optical communication in space: challenges and mitigation techniques. IEEE Commun Surveys Tutorials. https://doi.org/10.1109/comst.2016.2603518
11. Zhu X, Kahn JM (2002) Free space optical communication through atmospheric turbulence channels. IEEE Trans Commun 50(8):1293–1300
12. Naboulsi MA, Sizun H, de Fornel F (2004) Fog attenuation prediction for optical and infrared waves. Opt Eng 43(2):319–329
13. Milev A, Minchev C (2007) A simulation model of an optical communication CDMA system. Int Conf Comput Syst Technol Compo Sys Tech
14. Theodore S, Rappapo R (2007) Wireless communications: principles and practice, 2nd edn. Prentice Hall

This grasshopper optimization algorithm (GOA) is claimed to provide a balance exploration and exploitation in the search space of optimization problem. However, GOA quickly moves towards the current optimal point in the search space and thereby increases the risk of finding the local optimum point. The current work deals with the same problem of GOA, and an attempt has been made by providing two unique modifications in the approach of conventional GOA. The improved GOA (IGOA) proposed in this work has been tested for 19 benchmark test functions. Various performance measures have been considered to verify the superiority of IGOA over GOA and have been discussed in detail in this paper.

Further, this paper is divided in five sections. A brief introduction and the need for the improvement of GOA are presented in Sect. 1. Implementation and brief details of GOA are presented in Sect. 2. Section 3 presents the improvements made in GOA and implementation aspects of IGOA in detail. Section 4 presents and compiles the experimental results for considered benchmark test functions. Finally, Sect. 5 concludes the paper with inference drawn from the work carried out in the paper.

2 Grasshopper Optimization Algorithm (GOA)

Saremi et al. [23] related the movements of different grasshoppers to solve optimization problem by giving equivalence to these movements to exploitation and exploration in a search space.

A mathematical model as proposed in [23] is presented here

$$X_i = r_1 S_i + r_2 G_i + r_3 A_i \tag{1}$$

where X_i is the position of the ith grasshopper, S_i is the social interaction, G_i is gravity force on the ith grasshopper, and A_i shows the wind advection. r_1, r_2, and r_3 are random numbers in [0, 1] to provide random weight to different factors in (1). Equation 2 defines the social interaction [23], obtained by social forces as given by $s(r)$ function given by Eq. (3). d_{ij} is the distance between ith and jth grasshopper.

$$S_i = \sum_{\substack{j=1 \\ j \neq i}}^{N} s\left(d_{ij}\right) \widehat{d_{ij}} \tag{2}$$

$$s(r) = f e^{-\frac{r}{l}} - e^{-r} \tag{3}$$

where f is intensity of attraction and l is attractive length scale. The set of equations from Eqs. (1) to (3) are then further used to mathematically model some aspects of grasshopper's movement in order to solve optimization problems. It may be noted that the second and third terms in Eq. 1 are less significant in the developed mathematical model in [23], and is represented as,

$$X_i^d = c\left[\sum_{\substack{j=1 \\ j \neq i}}^{N} c\frac{(ub_d - lb_d)}{2} s\left(x_j^d - x_i^d\right)\frac{x_j - x_i}{d_{ij}}\right] + \widehat{T_d} \tag{4}$$

where ub_d and lb_d are upper bound and lower bound in dth dimension. c is a coefficient to shrink the comfort, repulsion, and attraction zone. A detailed representation of these zones is explained in [23]. The factor c plays a very important role in mathematical modelling of the two important aspects of the optimization procedure, i.e., exploration and exploitation. The c factor is multiplied two times in update Eq. (4), which provides a balance between the above said two procedures. It may be noted that the c factor is not a constant value and is varied as the optimization procedure progresses further. The original version of GOA decreases the c factor linearly from unity to a minimum value of 10^{-5} with iterations. However, GOA has a strong tendency to move very quickly towards the current optimal point and the choice of factor like r and l becomes very crucial in order to determine an optimal balance between random walks and targeted walks of search agents. Though provisions and care have been taken while proposing GOA, many times, it gets stuck in local optimal points. In order to avoid this issue, a modified and improvised version of GOA has been proposed in this paper, which is discussed in subsequent section.

3 Improved Grasshopper Optimization Algorithm (IGOA)

GOA is prone to quickly converge and confines the search space to move towards the current optimal solution. This behaviour of GOA increases the risk of entrapment in local optimal solution which should be avoided. In order to address this issue, two modifications are proposed in this work. These modifications enable the IGOA to converge faster and avoid local optimal solutions in contrast to conventional GOA. These modifications are in the form of,

(a) update equation for c factor
(b) random walk application in grasshopper swarm.

The improved version of GOA, i.e., IGOA makes use of an interesting update equation for c factor to be used in update of search agents. Since c factor is supposed to provide a balance between exploration and exploitation, it is essential that c factor remains close to unity for initial iterations and then decreases and then maintains at a minimum level to support the exploitation procedure. This paper adopts the above-mentioned methodology using an interesting c factor update equation as given in Eq. 6. It is worth mentioning here that the value of c factor will always start from unity and may reach to very low values as the iteration values progress. The value of σ plays an important role in variation of c factor as it tells the rate at which the c factor will decrease. A lower value of σ will decrease the decrement rate and the value of c factor will remain high for a high number of iterations. Similarly, a very high value of σ (greater than 6) will provide a steeper decrement in c factor and will largely support the exploitation procedure. A typical value of three to four can be taken for σ. This paper adopts the value of σ as 4 for all the optimization procedures. A typical variation of c factor for 200 iterations is shown in Fig. 1 for GOA (linear c factor variation) and for IGOA (exponential variation). It may be seen that for IGOA, the values of c for iterations lesser than 50 support more exploration than GOA and later support a balanced approach and then emphasize more on exploitation. This increases the chances of IGOA to explore the search space in a more judicial manner. The simulation results support this claim and are presented in Sect. 4.

$$c = \begin{cases} \exp\left(-0.5\left\{\frac{l}{L/\sigma}\right\}^2\right) & \text{if} \quad c > c_{\min} \\ c_{\min} & \text{otherwise} \end{cases} \tag{5}$$

Further, another step is also added to IGOA to avoid premature convergence of optimization procedure. A biased random walk nature, inspired by the cuckoo search algorithm [12], is added to conventional GOA to increase the exploration possibility. A randomly selected grasshopper in the swarm is moved towards another randomly selected grasshopper with a random step size. This random walk of grasshoppers in a swarm will increase the exploration possibility and will avoid entrapment of procedure in local minima. It is worth mentioning here that this random walk applied to a grasshopper is associated with a probability p lying in range of [0, 1]. A random number is generated in each iteration which is compared with p, and if the value of random number is lesser than p, then the random walk is initiated.

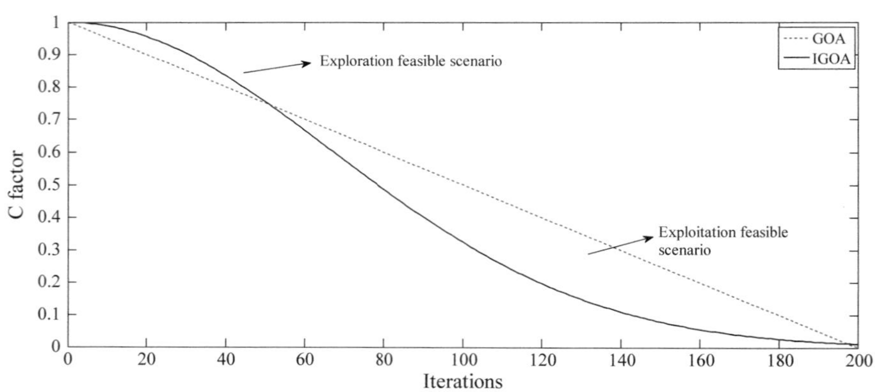

Fig. 1. Variation of c factor for GOA and IGOA

4 Simulation Results

In order to test the efficacy of the proposed IGOA over GOA, nineteen benchmark test functions were optimized using both algorithms. The results were compared on the basis of achieved best values, mean, and standard deviation from 20 independent trial runs. These benchmark test functions range from unimodal to multimodal function with higher dimensions as well. Since it was essential to maintain the same platform for GOA and IGOA, both algorithms are implemented in LabVIEW environment. Benchmark test functions [27] listed in Table 1 are optimized by GOA and IGOA for 200 iterations with a swarm size of 20 for all the functions except F11 and F18, whose dimensions are considered as 10 in contrast of 2, which is for all other function. For these function (F11 and F18), the swarm size is taken as 30 and maximum number of iterations as 500. Further, for the optimization procedures the values of all other parameters of GOA is taken same as in [23], and also for IGOA, wherever relevant. Figure 2 shows the convergence graph for F1 obtained by GOA and IGOA. It may be clearly seen from this graph that the IGOA performance is far better than GOA, as it converges very fast in comparison with GOA and also obtains the minimum value better than GOA. It should be noted here that the convergence graphs plotted for different functions are for the best result obtained from the 20 independent trial runs. The faster convergence of IGOA in comparison with GOA is due to the modified variation in c factor in optimization procedure. It enabled the IGOA to search more randomly in initial phase of optimization, and then for later parts, it searched with a smaller value of c factor to facilitate the exploitation procedure. This resulted in faster and more steep convergence of IGOA in contrast to GOA which only had a steeper convergence at very late stage as shown in Fig. 2. Further, due to space constraint, convergence curves are not shown for other functions; however, quantitative comparison is shown between IGOA and GOA for considered functions.

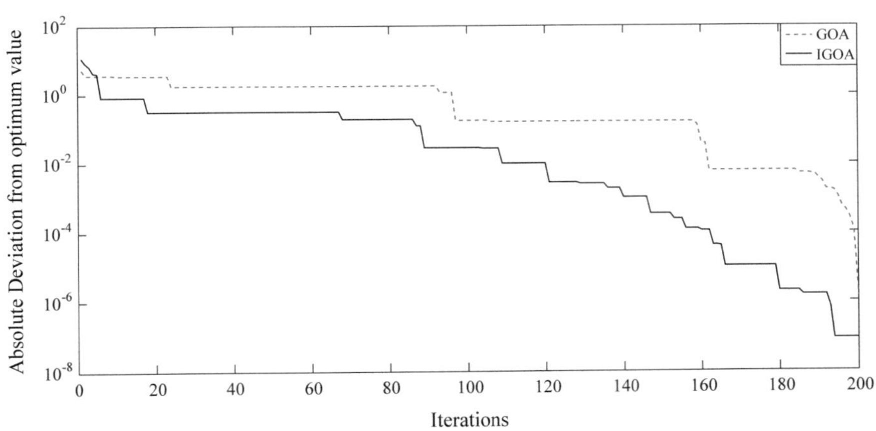

Fig. 2. Convergence graphs for absolute deviation from optimum values for F1 function

To evaluate and support the supremacy of IGOA over GOA, a detailed comparative analysis was also done using 20 independent trial runs for each test function. Different factors such as mean value, standard deviation, and percentage improvement in mean values (PIMV). The expressions to calculate PIMV, which is used to rank the algorithms, are given as follows:

$$\mathrm{PIMV}(\%) = \frac{|\mathrm{GMV} - \mathrm{Mean}_{\mathrm{GOA}}| - |\mathrm{GMV} - \mathrm{Mean}_{\mathrm{IGOA}}|}{|\mathrm{GMV} - \mathrm{Mean}_{\mathrm{GOA}}|} \times 100 \quad (6)$$

where GMV is the global minimum value for the specified test function, and Mean is the average value obtained by GOA or IGOA, for the specified test function. This performance measure is an interesting criterion to look for. This performance criterion looks for two important things: First it incorporates the IGOA performance to locate the global minimum value, and second it also incorporates the relative improvement of the IGOA over GOA. Similarly, performance criterion for logarithmic improvement in best value obtained by GOA and IGOA has also been taken and is calculated as,

$$\mathrm{LIBV} = \log_{10}\left(\frac{|\mathrm{GMV} - \mathrm{Best}_{\mathrm{GOA}}|}{|\mathrm{GMV} - \mathrm{Best}_{\mathrm{IGOA}}|}\right) \quad (7)$$

These logarithmic performance criteria are helpful to determine the IGOA's improvement over GOA in terms of mean value and best value. These values must be positive which indicate that the IGOA is superior in terms of deviation of mean or best value from the global minimum of a specified test function. Details of best values obtained by GOA and IGOA for different considered test functions are shown in Table 1. It may be noted that the best values obtained from 20 independent trial runs by IGOA is always superior or equivalent to conventional GOA. It may be noted from the data provided in Table 1 that none of the values obtained by IGOA is worse than GOA and all the LIBV performance measures are positive. To further investigate the performance of considered GOA and IGOA, the statistical analysis using the 20 independent runs test data has to be taken into consideration. The PIMV seems to be a better performance criterion, since it incorporates the mean values information from a set of independent test runs. It may also be possible that the optimized values are dispersed in a large range and the mean value gets close to the global minimum values. To make this into consideration, the standard deviation of the optimized values has also been taken into consideration, to find whether the algorithm has the necessary repeatability feature in it or not. To showcase the effectiveness of IGOA over GOA in this manner, the values of mean and standard deviation are listed in Table 2 for different test functions. It may be noted from data provided in Table 2 that the standard deviation values for IGOA are very close to zero, which tells that IGOA performance is repeatable in nature. While GOA has a tendency to stick in local minima for some of the test functions and thereby has a very high value of standard deviations for some test functions. The high value of standard deviations for GOA can be observed particularly for F2, F8-10, F13, F15-16 and F19.

Table 1. Best values as obtained by GOA and IGOA for considered test functions

S. No.	Function name	Best optimized values from 20 trial runs		Logarithmic improvement (LIBV)
		GOA	IGOA	
F1	Ackley	1.3805E−06	8.6150E−08	1.204788
F2	Beale	4.92E−13	0	NA
F3	Bohachevsky	6.4208E−10	1.00E−14	4.807594
F4	Bohachevsky 1	5.9789E−10	4.30E−14	4.143159
F5	Bohachevsky 2	1.5328E−10	1.30E−14	4.071542
F6	Booth	1.047E−12	0	NA
F7	Branin	0.397887357	0.397887358	0.431408
F8	Easom	−0.99999999994437	−0.99999999999999	4.444563
F9	Eggholder	−959.640662720651	−959.64066272085	2.32E−06
F10	Goldstein Price	3.00000000000375	2.99999999999991	1.619052
F11	Griewank	0.109267741	0.02546913	0.632478
F12	Matyas	5.00E−15	0	NA
F13	Rosenbrock	1.10129E−07	0	NA
F14	Schaffer	2.00E−15	0	NA
F15	Schwefel	2.545556E−05	2.545513E−05	7.40E−06
F16	Shubert	−186.730908828	−186.730908831	NA
F17	Six-Hump camelback	−1.031628453	−1.031628453	NA
F18	Sumsquares	4.13675E−07	6.5711E−11	3.798625
F19	Zakharov	1.75422E−04	3.69E−13	8.677059

This data shows that the repeatability of the IGOA is higher than GOA. Further, PIMV can indicate more about the efficacy of IGOA while optimizing different test functions. It may be noted that the maximum value for PIMV can be 100 and the values which are closer to the maximum PIMV support the excellent performance shown by IGOA while optimizing different test functions. Positive PIMV values indicate that the IGOA performance is better than the GOA; however, a value close to 100 indicates that the IGOA performance is not only better than GOA but also the mean value obtained is either very close or equal to the global optimum value for a particular benchmark test function. The PIMV presented in Table 2 also indicates the supremacy of IGOA over GOA for all considered benchmark test functions.

Table 2. Mean and standard deviation values for different functions

S. No.	Mean values		Standard deviations		PIMV (%)
	GOA	IGOA	GOA	IGOA	
F1	2.25E−05	3.37E−07	1.17E−05	1.27E−07	98.50
F2	0.330795669	7.95E−15	0.417640307	3.41E−14	100
F3	6.28E−09	6.92E−13	6.48E−09	6.13E−13	99.99
F4	2.36E−08	2.69E−12	2.60E−08	2.52E−12	99.99
F5	3.99E−08	1.01E−12	7.43E−08	7.37E−13	99.99
F6	4.59E−11	4.00E−15	8.32E−11	3.73E−15	99.99
F7	0.397887358	0.397887358	3.48E−11	1.70E−15	99.97
F8	−0.45	−1	0.510417785	2.79E−13	100
F9	−753.5810622	−954.674431	151.5400571	5.69E−02	97.58
F10	8.4	3	18.7880141	3.17E−14	100
F11	0.2065592	0.11519567	0.088591911	6.42E−02	44.23
F12	3.03E−12	0	6.94E−12	0	100
F13	0.240502042	2.01E−07	0.737820273	7.83E−07	99.99
F14	4.82E−13	0	3.81E−13	0	100
F15	169.76381516	2.55E−05	141.1030576	8.16E−12	99.99
F16	−172.9084996	−186.730908	42.54447783	7.01E−12	100
F17	−0.786779065	−1.03162845	0.383729877	7.73E−15	100
F18	4.04E−04	8.37E−06	8.57E−04	2.62E−05	97.92
F19	1.0128319	1.68E−07	3.965683405	6.09E−07	99.99

5 Conclusion

This work proposed an improved version of grasshopper optimization algorithm (GOA). The improved GOA (IGOA) proposed in this work eliminates a few shortcomings of GOA. The variation of c factor throughout the optimization procedure in a linear fashion is an important functionality of GOA which is claimed to provide a balanced exploration and exploitation of the search space. This paper improves this balance by proposing a new method for the variation of c factor which helps in determining the optimum point more quickly and accurately. To further increase the effectiveness of IGOA, another step which provides the random walks to grasshopper in its swarm has also been proposed. This random walk further reduces the risk of IGOA to get trapped in local minimum points. These two modifications have significantly improved the efficacy of IGOA over GOA, and the same has been verified by performing optimization procedure on 19 benchmark test functions of different nature. To avoid any biasing towards a particular algorithm, 20 independent trial runs were made and parameters like best, mean, and standard deviation were calculated. To showcase the effectiveness of IGOA over GOA, different performance measures were also considered, namely percentage improvement in mean value, logarithmic improvement in best values. Based on rigorous experimentation and obtained performance measures, it was observed that the IGOA was able to provide a significant improvement over GOA and the mean values for all functions were very close to global optimum values. Further, repeatability of IGOA was also substantially better than GOA for all the considered test functions. Based on the obtained results, it may be concluded that IGOA provides more accurate and repeatable results than GOA and can be applied to real-world optimization problems.

References

1. Spall JC (2005) Introduction to stochastic search and optimization: estimation, simulation, and control, vol. 65. John Wiley & Sons
2. Harik GR, Lobo FG, Goldberg DE (1999) The compact genetic algorithm. IEEE Trans Evol Comput 3(4):287–297
3. Holland JH (1992) Genetic algorithms. Sci Am 267(1):66–73
4. Kennedy J (2011) Particle swarm optimization. In: Encyclopedia of machine learning. Springer USA, pp 760–766
5. Poli R, Kennedy J, Blackwell T (2007) Particle swarm optimization. Swarm Intell 1(1):33–57
6. Yang XS (2010) A new metaheuristic bat-inspired algorithm. In: Nature inspired cooperative strategies for optimization (NICSO 2010). Springer, Berlin, Heidelberg, pp 65–74
7. Yang XS, Hossein Gandomi A (2012) Bat algorithm: a novel approach for global engineering optimization. Eng Comput 29(5):464–483
8. Yang XS (2012) Flower pollination algorithm for global optimization. In: International conference on unconventional computing and natural computation. Springer, Berlin, Heidelberg, pp 240–249
9. Yang XS, Karamanoglu M, He X (2014) Flower pollination algorithm: a novel approach for multiobjective optimization. Eng Optim 46(9):1222–1237
10. Mirjalili S, Mirjalili SM, Lewis A (2014) Grey wolf optimizer. Adv Eng Softw 69:46–61
11. Mirjalili S, Saremi S, Mirjalili SM, Coelho LDS (2016) Multi-objective grey wolf optimizer: a novel algorithm for multi-criterion optimization. Expert Syst Appl 47:106–119
12. Yang XS, Deb S (2009) Cuckoo search via Lévy flights. In: Nature and biologically inspired computing, 2009. NaBIC 2009. World Congress on. IEEE, pp 210–214
13. Yang XS, Deb S (2010) Engineering optimisation by cuckoo search. Int J Math Model Numer Optim 1(4):330–343
14. Mirjalili S (2015) Moth-flame optimization algorithm: a novel nature-inspired heuristic paradigm. Knowl-Based Syst 89:228–249
15. Mirjalili S (2015) The ant lion optimizer. Adv Eng Softw 83:80–98
16. Mirjalili S, Jangir P, Saremi S (2017) Multi-objective ant lion optimizer: a multi-objective optimization algorithm for solving engineering problems. Appl Intell 46(1):79–95
17. Gandomi AH, Alavi AH (2012) Krill herd: a new bio-inspired optimization algorithm. Commun Nonlinear Sci Numer Simul 17(12):4831–4845
18. Wang GG, Guo L, Gandomi AH, Hao GS, Wang H (2014) Chaotic krill herd algorithm. Inf Sci 274:17–34
19. Dorigo M, Birattari M (2011) Ant colony optimization. In: Encyclopedia of machine learning. Springer, Boston, MA, pp 36–39
20. Afshar A, Haddad OB, Marino MA, Adams BJ (2007) Honey-bee mating optimization (HBMO) algorithm for optimal reservoir operation. J Franklin Inst 344(5):452–462

21. Fu Y, Ding M, Zhou C (2012) Phase angle-encoded and quantum-behaved particle swarm optimization applied to three-dimensional route planning for UAV. IEEE Trans Syst, Man, Cybern-Part A: Syst Humans 42(2):511–526
22. Mishra P, Kumar V, Rana KPS (2015) A fractional order fuzzy PID controller for binary distillation column control. Expert Syst Appl 42(22):8533–8549
23. Saremi S, Mirjalili S, Lewis A (2017) Grasshopper optimisation algorithm: theory and application. Adv Eng Softw 105:30–47
24. Chattaraj N, Ganguli R (2017) Multi-objective optimization of a triple layer piezoelectric bender with a flexible extension using genetic algorithm. Mech Adv Mater Struct 1–9
25. Roberge V, Tarbouchi M, Labonté G (2013) Comparison of parallel genetic algorithm and particle swarm optimization for real-time UAV path planning. IEEE Trans Industr Inf 9 (1):132–141
26. Mishra P, Kumar V, Rana KPS (2018) An efficient method for parameter estimation of a nonlinear system using backtracking search algorithm. Eng Sci Technol Int J 21(3):338–350
27. Surjanovic S, Bingham D (2013) Virtual library of simulation experiments. https://www.sfu.ca/~ssurjano/optimization.html. Last Accessed on 29 May 2018

Design of Array Antenna for Body Area Network

Sarmistha Satrusallya, Shaktijeet Mahapatra, and Mihir Narayan Mohanty(✉)

Department of ECE, ITER, Siksha 'O' Anusandhan (Deemed to be University), Bhubaneswar, India
{sarmisthasatrusalya, shaktijeetmahapatra, mihirmohanty}@soa.ac.in

1 Introduction

Antennas with smaller profile and lightweight are developed nowadays to be used in various applications. One of the applications is the development of wearable antenna. These antennas are widely used in the field of medicine, military, and different scientific community. The wearable antenna in the UWB range (3–10.6 GHz) provides better performance in terms of bandwidth and gain.

Researchers have developed different antennas in the desirable band such as coplanar waveguide fed disc antenna and microstrip fed annular slot antenna [1]. Authors also presented conical dielectric resonator antenna for use in body area network operating in the range of (3.4–5 GHz). The antenna provides a good performance in frequency and time domains [2]. A cavity slot antenna was proposed at 2.45 GHz resulted in high efficiency of more than 50% in the vicinity of the human body [3]. Similarly, the radiation of three types of UWB antenna was experimentally characterized. Among omnidirectional, monopole antenna and printed diversity antenna, the latter exhibits improved quality of radio frequency link for on-body communication [4]. A CPW-fed ultra-wideband antenna and button-shaped antenna based on a microstrip Yagi array were also presented for the same purpose. The antennas were suitable for operation in the range of 2.45 GHz and had application in the fields of telemedicine [5, 6]. Kang et al. designed a folded antenna with reduced backward radiation and proximity effect in the human bodies. A switchable beam textile antenna capable of generating beam steering in four directions was also proposed. The antenna is attached in safety jackets and rain coats to be used for search and rescue operation [7, 8]. Different array antennas were also proposed for on-body communication. Millimetre wave spectrum has application in health monitoring and live streaming. A frequency scanning array antenna was designed to provide reconfigurability in the direction of main beam pattern [9–11].

In this paper, we present an array antenna of five elements with circular slots cut on the centre of the patch. The antenna is small and is easy to fabricate. A defective ground plane has been used for better performance.

M. N. Mohanty and S. Das (eds.), *Advances in Intelligent Computing and Communication*, Lecture Notes in Networks and Systems 109,
https://doi.org/10.1007/978-981-15-2774-6_23

The paper is organized into the following sections: Sect. 2 presents the design of the antenna; simulation results are described in Sect. 3, and in the last section, conclusions are drawn.

2 Design Considerations

The proposed antenna is an array having five elements. The substrate is chosen to be FR4-epoxy having a relative permittivity of 4.4 and dielectric loss tangent of 0.02. The rectangular patches are placed on a substrate having dimension 30 mm × 180 mm 1.6 mm. Each rectangular patch has a circular slot of radius 2 mm. The ground plane dimensions are 15 mm × 180 mm. The array is fed by a microstrip fed with 50 Ω impedance. Figure 1 shows the proposed antenna array (Table 1).

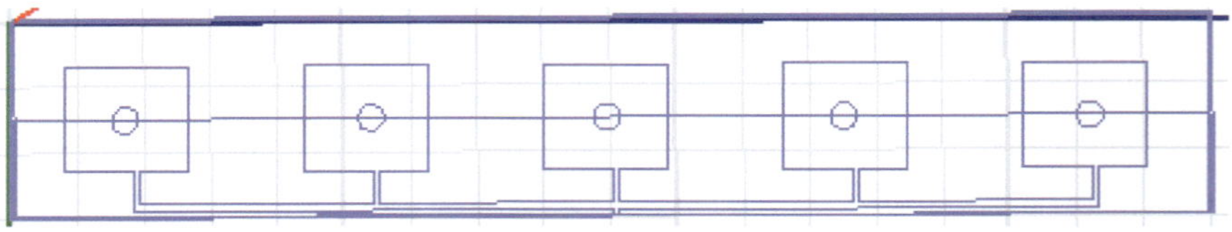

Fig. 1. Design of the proposed antenna array

Table 1. Design parameters

Parameters	Values (in mm)
Length of substrate	180
Width of substrate	30
Height of substrate	1.6
Length of patch	16
Width of patch	18.7
Radius of the slot	2

3 Simulation Results and Discussions

Figure 2 shows the impedance bandwidth of the proposed design. The antenna resonates at two frequencies, 3.1 and 3.6 GHz. The impedance bandwidths are 0.2, 0.8 GHz with a return loss of −13 dB and −31 dB, respectively.

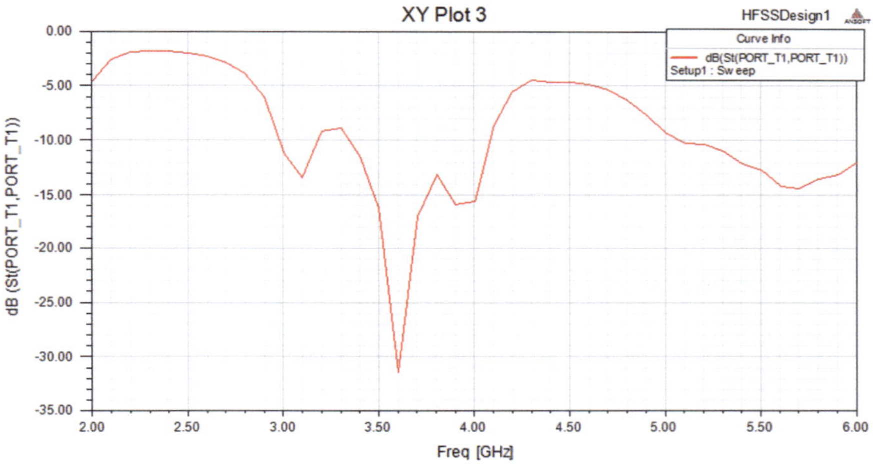

Fig. 2. Impedance bandwidth of the proposed antenna

Figure 3 shows the VSWR of the design. It can be clearly seen that the VSWR at the resonating frequencies lies between 1 and 2.

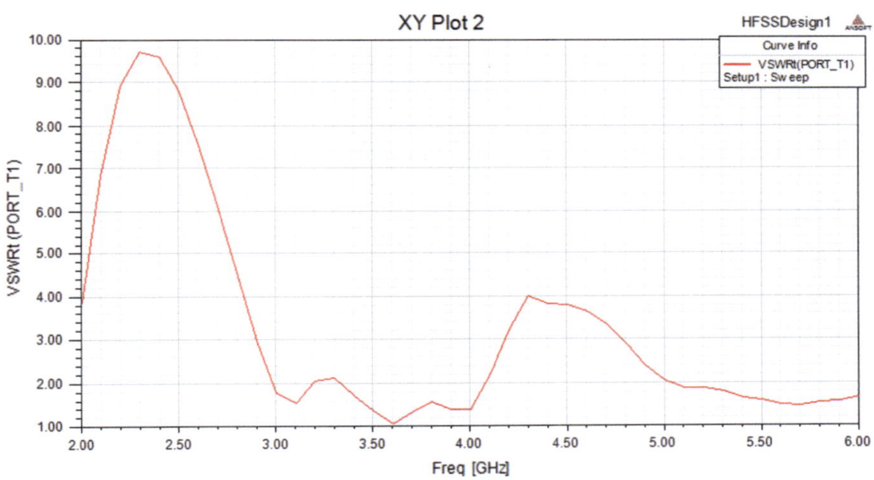

Fig. 3. VSWR of the proposed antenna

Figure 4 represents the gain of the antenna. The gain is found to be 3.85 db which is suitable for on-body communication.

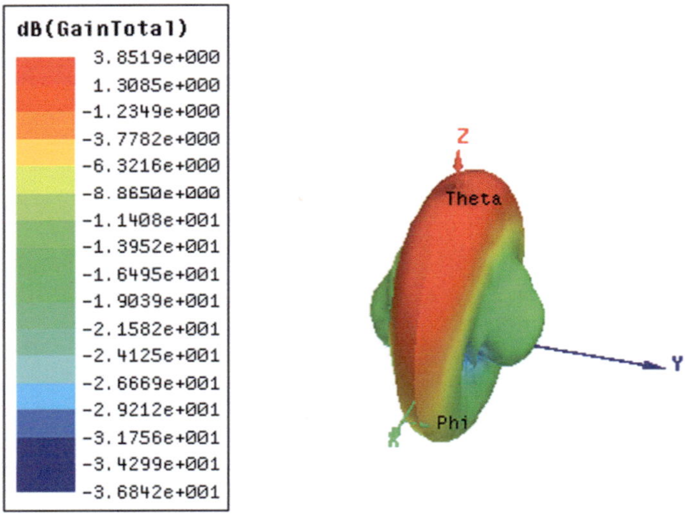

Fig. 4. 3D polar plot of the proposed design

4 Conclusions

An array antenna has been designed for wearable application in the range of (3.1–5 GHz). The antenna resonates at 3.1 and 3.6 GHz. The defective ground structure provides a better performance of the proposed design. The antenna can be used for on-body and off-body communication.

References

1. Klemm M, Troester G (2006) Textile UWB antennas for wireless body area networks. IEEE Trans Antennas Propag 54(11):3192–3197
2. Almpanis G, Fumeaux C, Frohlich J, Vahldieck R (2008) A truncated conical dielectric resonator antenna for body-area network applications. IEEE Antennas Wirel Propag Lett 8:279–282
3. Haga N, Saito K, Takahashi M, Ito K (2009) Characteristics of cavity slot antenna for body-area networks. IEEE Trans Antennas Propag 57(4):837–843
4. See TS, Chen ZN (2009) Experimental characterization of UWB antennas for on-body communications. IEEE Trans Antennas Propag 57(4):866–874
5. Shin H, Kim J, Choi J (2009) A stair-shaped CPW-fed printed UWB antenna for wireless body area network. In: 2009 Asia pacific microwave conference, pp 1965–1968
6. Khaleel HR, Al-Rizzo HM, Rucker, DG, Elwi TA (2010) Wearable Yagi microstrip antenna for telemedicine applications.In: 2010 IEEE Radio and Wireless Symposium (RWS), pp 280–283
7. Kang CH, Wu SJ, Tarng JH (2011) A novel folded UWB antenna for wireless body area network. IEEE Trans Antennas Propag 60(2):1139–1142

8. Jais MI, Jamlos MFB, Jusoh M, Sabapathy T, Kamarudin MR, Ahmad RB, Ishak NLK (2013) A novel 2.45 GHz switchable beam textile antenna (SBTA) for outdoor wireless body area network (WBAN) applications. Prog Electromagn Res 138:613–627
9. Park J, Tak J, Choi J (2014) A low-profile dipole array antenna for on-body communications. In: 2014 IEEE international workshop on electromagnetics (iWEM), pp 4–5
10. ElSalamouny MY, Shubair RM (2015) Novel design of compact low-profile multi-band microstrip antennas for medical applications. In: 2015 loughborough antennas & propagation conference (LAPC), pp 1–4
11. Chauhan SS, Abegaonkar MP, Basu A (2017) Millimeter wave frequency scanning antenna for body centric communication. In: 2017 international symposium on antennas and propagation (ISAP), pp 1–2

Optimal Load Condition for a Resonant Wireless Charging System

Jagadish C. Padhi, Durga P. Kar, Praveen P. Nayk(✉), Biswaranjan Swain, Dipti Patnaik, and Satyanarayan Bhuyan

Department of Electronics and Communication Engineering, SOA Deemed to be University, Bhubaneswar, Odisha, India
{praveennayak, satyanarayanbhuyan}@soa.ac.in

1 Introduction

Recent years have witnessed the vast improvement in the wireless charging system for low-power electrical/electronic appliances of few mW to high-power application such as electric vehicle (EV) of kW level. This is due to the convenience, safety, and utmost suitable in all weather conditions [1–8]. Charging can be carried out through conventional inductive coupling but not so far achieved its height due to the short-range energy transfer efficiency and limitation in coil size. Providentially, the alternative in the form resonant inductive coupling-based wireless charging system came into the picture to enhance the power transfer efficiency as well as the transfer range with adequate dimension that can be embedded into the vehicle [9–13]. But to make the wireless charging system more practical, it is important to the issues dealt with the operating frequency, load, and the power transfer range; there is a need of experimental investigation. So, in order to mitigate the issues related to transfer load power and transfer efficiency, the optimal load condition is very much essential. Hence, load dependencies of power transfer efficiency have been experimentally investigated in this paper. The solution will provide a noteworthy advancement in the field of wireless charging under non-ideal scenario.

2 Experimental Setup

The experimental setup consists of a plastic rod graded with a scale, mounted on a base along which circular acrylic plates which can slide up or down the plastic rod. On these plates, required coils of various dimensions can be mounted. This setup makes it convenient to swap coils and take experimental results with accuracy. The material used in making this apparatus is non-magnetic and does not affect the experiment in any way. Copper conductor of gauge SWG12 (1.3208 mm radius) is wound into circular coils, and capacitors are connected to the ends of the transmitting and receiving

M. N. Mohanty and S. Das (eds.), *Advances in Intelligent Computing and Communication*, Lecture Notes in Networks and Systems 109,
https://doi.org/10.1007/978-981-15-2774-6_24

coils. A high-frequency function generator is used as a power source to drive the source coil. The dimensions of the coil are chosen such that they are easy to handle and capable of demonstrating wireless power, fixed as two cylindrical coils of 7.5 cm radius made of copper wire SWG12 (conductor radius 0.13208 cm), with 5 turns each (Fig. 1).

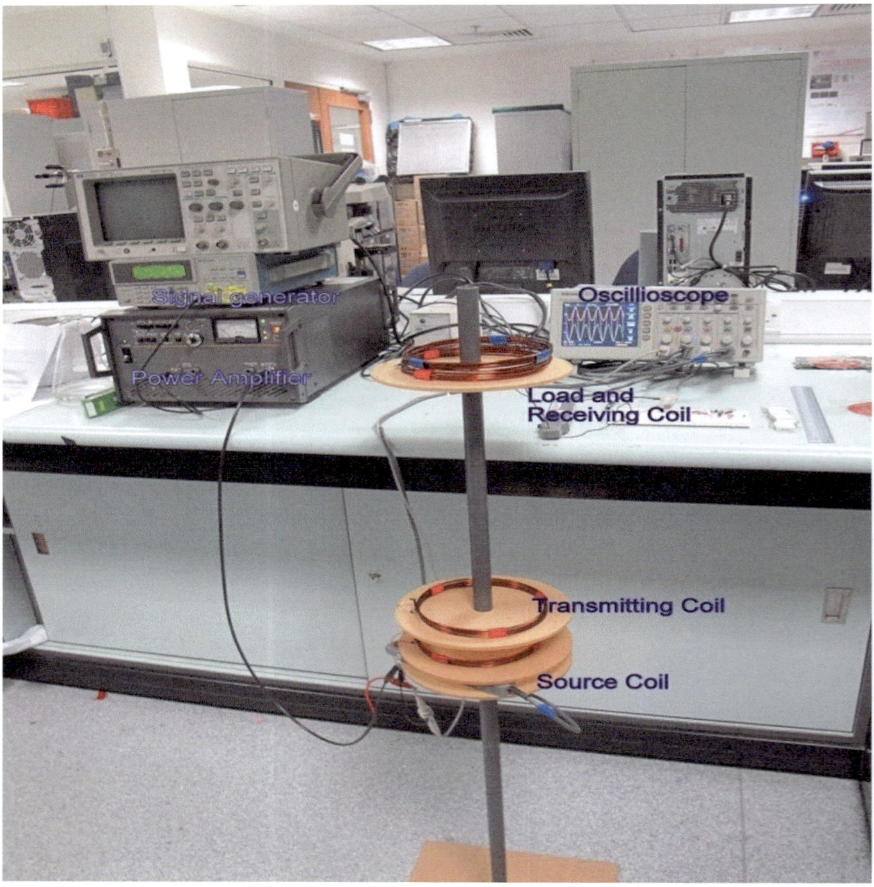

Fig. 1. Experiment setup of wireless power transfer system

The value of inductance is 6.8μH which gives a resonant frequency of 1.93 MHz with a capacitance of 1000pF, but experimentally, it is later shown that resonant frequency is 2.05 MHz. This small discrepancy is due to the structures not being perfect cylinders. With this setup experiments are conducted to determine the effect of size, shape, distance, and frequency on efficiency of power transfer.

3 Results and Discussion

Experiments are conducted to observe the variation of output power with regard to three parameters, namely frequency, load, and distance. The frequency dependence on the output power is shown in Fig. 2.

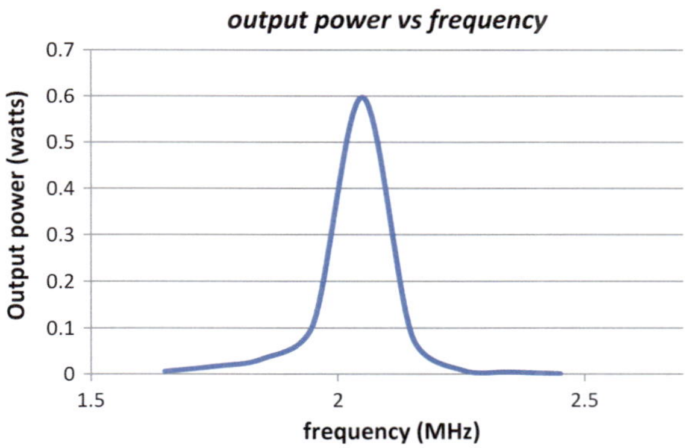

Fig. 2. Frequency characteristics of output power at a fixed load

In this case, the frequency is varied with the load set at 220 Ω and distance at 10 cm. It is expected that at resonance frequency, output power is maximum since evanescent coupling would take place between the resonators. The graph proves that the resonant frequency for this system is 2.05 MHz.

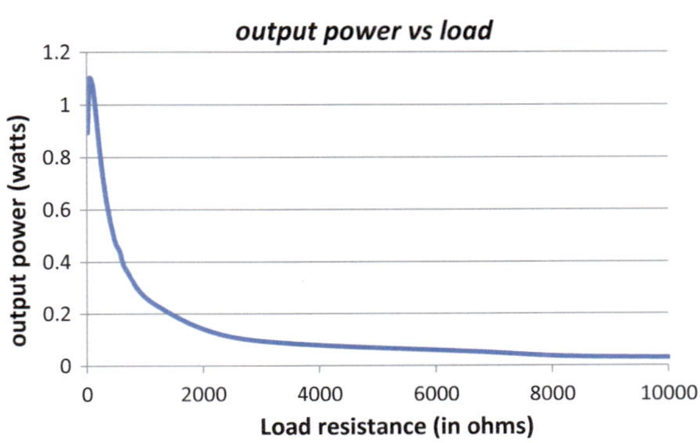

Fig. 3. Electric load characteristics of output power

The electric load characteristic is shown in Fig. 3. The load is varied from 10 to 10,000 Ω with the frequency set at 2.05 MHz, and distance between the transmitter and receiver is set at 10 cms. It is observed that maximum output power is achieved at 220 Ω (Fig. 4).

An Improved Grasshopper Optimization Algorithm for Solving Numerical Optimization Problems

Puneet Mishra[1(✉)], Vishal Goyal[2], and Aasheesh Shukla[2]

[1] Department of Electrical and Electronics Engineering, Birla Institute of Technology and Science, Pilani, Pilani Campus, Jhunjhunu, Rajasthan 333031, India

[2] Department of Electronics and Communication Engineering, GLA University, Mathura, India

aasheesh.shukla@gla.ac.in

1 Introduction

To avoid the entrapment of optimization techniques in local optimal points, a class of optimization techniques known as stochastic optimization has arisen [1]. The stochastic optimization problems rely on random generation of candidate solution and are free from gradient-based techniques to solve the optimization problem. A subcategory of stochastic optimization algorithms is bio-inspired optimization algorithm. The bio-inspired optimization algorithms have proven their efficacy to solve real-world challenging optimization problems in recent past. Some of the bio-inspired optimization algorithms are genetic algorithm (GA) [2, 3], particle swarm optimization (PSO) algorithm [4, 5], bat algorithm (BA) [6, 7], flower pollination algorithm (FPA) [8, 9], grey wolf optimizer (GWO) [10, 11], cuckoo search algorithm (CSA) [12, 13], ant-lion optimizer (ALO) [14–16], Krill Herd optimization algorithm (KHOA) [17, 18], moth flame optimization algorithm (MFOA) [14, 15], ant colony optimization (ACO) [19], honey bee mating optimization [20], and many more.

It may be noted that the search algorithms relying on natural physical laws are adequately efficient in searching the global optimum points; however, bio-inspired algorithm development and their application have become more popular. The application areas of these algorithms range in a variety of fields such as route planning [21], controller parameter optimization [22], structure design problems [23], micro-electronic mechanical system design [24], UAV path planning [25], and system identification [26]. It may be noted that, though optimization algorithms based on the salient features of different species have evolved in literature, the quest for a more efficient algorithm is always continued, as it is very hard to say, that which algorithm will be applicable and suitable to a wide variety of optimization problems. In the same trend, a bio-inspired optimization algorithm which mimics the swarm and food search behaviour of grasshoppers has been recently proposed in the literature [23].

M. N. Mohanty and S. Das (eds.), *Advances in Intelligent Computing and Communication*, Lecture Notes in Networks and Systems 109,
https://doi.org/10.1007/978-981-15-2774-6_22

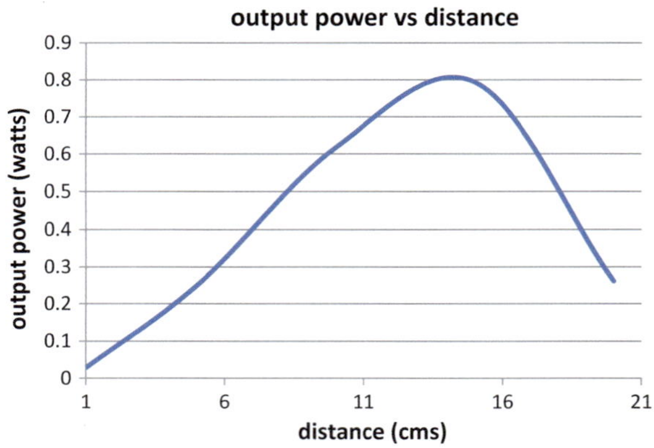

Fig. 4. Dependency of output power with distance

The distance between the transmitter and receiver is varied from 1 to 20 cm with the frequency of power supply set at 2.05 MHz and the load resistance set at 220 Ω. From the result, it is clearly observed that output power initially increases and attains its maximum at a separation distance of 14 cm and then gradually decreases afterward. This confirms the counter-EMF effect in the experimental setup. This also shows a key difference from simple electromagnetic induction where maximum power output would be achieved when distance is minimum. From the graph, it is noted that at 14 cm distance the output power is maximum at 0.8 W.

The power transfer efficiency corresponding to frequency, separation distance, and load is discussed and given below. The frequency characteristic of power transfer efficiency is illustrated in Fig. 5. The high-frequency signal from the signal generator is varied with the load set at 220 Ω and distance at 10 cm. It is found that at resonance frequency, efficiency attains its maximum since evanescent coupling would take place between the resonators and which confirms that the resonant frequency for this system is 2.05 MHz.

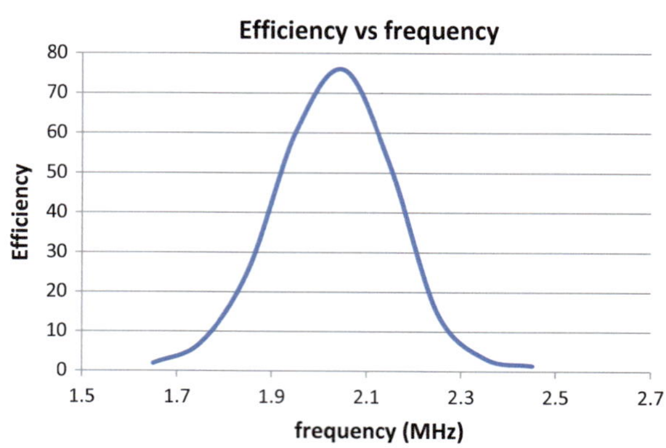

Fig. 5. Frequency characteristics of wireless power transfer efficiency

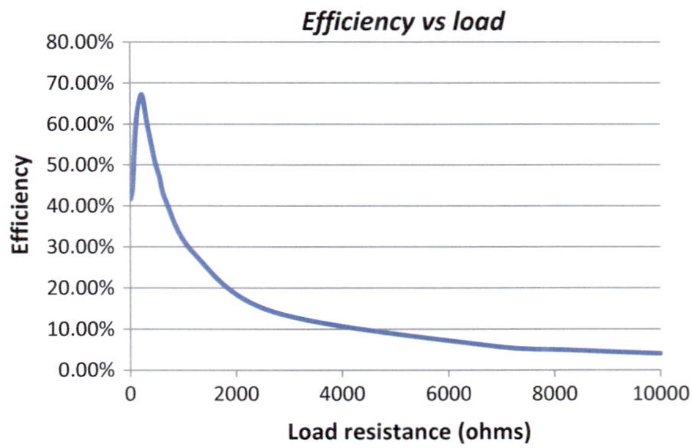

Fig. 6. Dependency of efficiency on the load resistance

The load variation corresponding to the power transfer efficiency is shown in Fig. 6. The load is varied from 10 to 10,000 Ω with the frequency set at 2.05 MHz and distance between the transmitter and receiver set at 10 cms. It is observed from the result that the maximum efficiency is achieved at 220 Ω. This gives the indication that power levels do not change the resonance peak. This value indicates the maximum output power point.

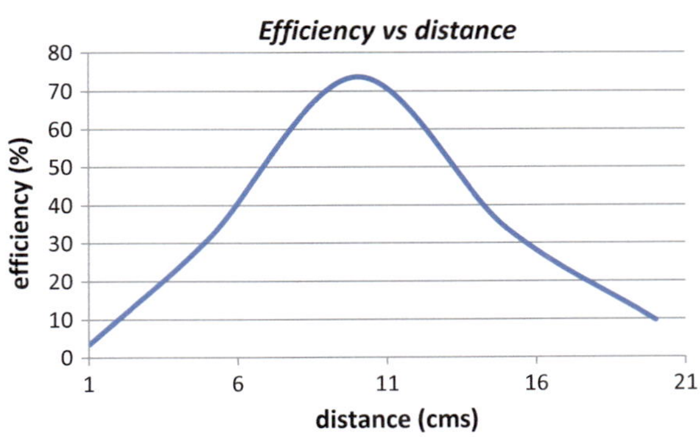

Fig. 7. Effect of separation distance on efficiency

The effect of separation distance on the power transfer efficiency is shown in Fig. 7. The distance between the transmitter and receiver is varied from 1 to 20 cm with the frequency of power supply set at 2.05 MHz and the load resistance set at 220 Ω. It is observed that efficiency initially increases and starts decreasing then. This confirms the counter-EMF effect in the experimental setup. In this case efficiency of 81% is achieved at about 10 cm distance.

From the above three experiments, it is concluded that a maximum efficiency of 73% is achieved at the resonance frequency of 2.05 MHz with load resistance 220 Ω and 10 cm separation.

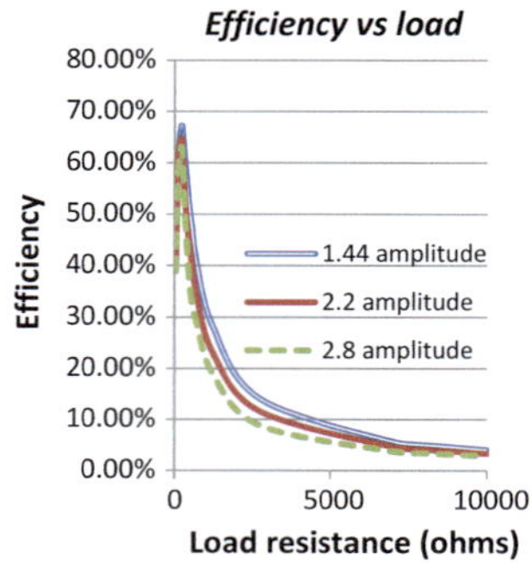

Fig. 8. Effect of load on efficiency for different signal amplitude

The effects of load on efficiency for different signal amplitude are shown in Fig. 8. Comparing to the previous result, the difference is the input power is increased by increasing the amplitude in the signal generator which is then amplified 10× times by the power amplifier. It was observed that increasing power levels has minimal effect on efficiency and the small reduction in efficiency is due to the heating of the ceramic capacitor and wires which increases the resistance of the circuit.

4 Conclusion

The power transfer to load and the power transfer efficiency corresponding to frequency, load, and separation air gap have been experimentally investigated. The power transfer efficiency is measured for different load values. It is found that both the power delivered to the load and the power transfer efficiency are very much sensitive to electrical load connected to the receiver coil. Moreover, the resonance frequency is seen to be independent of the power level. From the result, it is also confirmed that there exists an optimal load for which maximum power transfer is achieved. From the experimental result, the maximum efficiency 73% is achieved for an optimal load 220 Ω at the resonant frequency 2.05 MHz for a separation distance of 10 cm.

References

1. Pedder DAG, Brown AD, Skinner JA (1999) A contactless electrical energy transmission system. IEEE Trans Ind Electron 46:23–30
2. Chang-Gyun K, Dong-Hyun S, Jung-Sik Y, Jong-Hu P, Cho BH (2001) Design of a contactless battery charger for cellular phone. IEEE Trans Industr Electron 48:1238–1247
3. Covic GA, Boys JT (2013) Modern trends in inductive power transfer for transport applications. IEEE J Emerg Sel Top Power Electron 1(1):28–41
4. Li S, Mi CC (2015) Wireless power transfer for electric vehicle applications. IEEE J Emerg Sel Top Power Electron 3(1):4–17

5. Kurs A, Karalis A, Moffatt R, Joannopoulos JD, Fisher P, Soljacic M (2007) Wireless power transfer via strongly coupled magnetic resonances. Science 317:83–86
6. Ho SL, Wang J, Fu W, Sun M (2011) A comparative study between novel witricity and traditional inductive magnetic coupling in wireless charging. IEEE Trans Magn 47:1522–1525
7. Kuipers J, Bruning H, Bakker S, Rijna H (2012) Near field resonant inductive coupling to power electronic devices dispersed in water. Sens Actuators, A 178:217–222
8. Sample AP, Meyer DA, Smith JR (2011) Analysis, experimental results, and range adaptation of magnetically coupled resonators for wireless power transfer. IEEE Trans Industr Electron 58(2):544–554
9. Hui SYR, Ho WWC (2005) A new generation of universal contactless Battery Charging platform for portable Consumer Electronic equipment. IEEE Trans Power Electron 20:620–627
10. Mon G, Arcuti P, Tarricone L (2015) Resonant inductive link for remote powering of pacemakers. IEEE Trans Microw Theory Tech 63(11):3814–3822
11. Orekan T, Zhang P, Shih C (2018) Analysis, design, and maximum power-efficiency tracking for undersea wireless power transfer. IEEE J Emerg Sel Top Power Electron 6 (2):843–854
12. RamRakhyani AK, Mirabbasi S, Chiao M (2011) Design and optimization of resonance-based efficient wireless power delivery systems for biomedical implants. IEEE Trans Biomed Circuits Syst 5:48–63
13. Kar DP, Nayak PP, Bhuyan S (2015) Bi-directional magnetic resonance based wireless power transfer for electronic devices. Appl Phys Lett 107(13):3901

Design of Microstrip Antenna with Corner Truncation for UWB Application and Its Analysis

Swarnaprava Sahoo and Laxmi Prasad Mishra(✉)

Department of Electronics and Communication Engineering, Institute of Technical Education and Research, Siksha 'O' Anusandhan (Deemed to be University), Bhubaneswar, Odisha, India
laxmimishra@soa.ac.in

1 Introduction

Nowadays, the microstrip patch antenna is very much important in wide range of multifunctional wireless communication systems like UWB, WLAN, spacecraft, radar, and applications of telemetry in biology. These systems have motivated the invention of miniaturizing UWB radiators with high gain, good radiation efficiency, and patterns. The possibilities of abilities of these applications are because of popular advantages like low profile, lightweight, compact in size, planar configuration, easy fabrication, inexpensive manufacturing, and integration with other microwave circuitries. The main disadvantage of patch antenna is its narrow bandwidth. So, many researchers have developed different familiar methods for improvement of microstrip antenna impedance bandwidth by sustaining the patch radiator size as compact as feasible, by applying thicker substrate and low permittivity value [1], inserting slots in radiating patch [2], using different feeding methods and impedance matching [3], by ground plane modification [4]. The impedance bandwidth and radiating patch radiator size are generally complementary contrary to each other. So, the characteristics development of one normally results in deterioration of the other. Under these conditions, the research workers developed many different methods to enhance such limitations, basically improving the impedance bandwidth. So, different antenna designs are proposed. They are inserting slot and analysis of U-slot loading patch [5], by loading stacked patch [6]. The impedance bandwidth of above antennas is up to 40%.

The integration of all the important characteristics like high gain, wide bandwidth, and good radiation pattern into a single antenna design is very complicated task. Some limitations are overcome by this innovative work which is based on ultra-wideband U-slot insertion and incorporation of two identical corner truncations in metallic patch. In this research article, the proposed antenna is analyzed for improvement of impedance bandwidth. The antenna characteristics are analyzed on the basis of corresponding circuit approximation, cavity model analysis, and circuit theory concept. The coaxial feed is placed nearby center of metallic patch for better excitation of the intended radiator over wide bandwidth. The parameters like gain and radiation patterns of the intended radiator are also deliberated.

M. N. Mohanty and S. Das (eds.), *Advances in Intelligent Computing and Communication*, Lecture Notes in Networks and Systems 109,
https://doi.org/10.1007/978-981-15-2774-6_25

2 Antenna Geometry

Figure 1 presents the geometrical consideration of the intended radiator with optimized dimensions. The intended radiator consists of an inserted U-slot and a pair of truncated corners rectangular microstrip patch. The dimension of the rectangular patch is $L \times W = 35.5 \times 26\,\text{mm}^2$ with center frequency 7.2 GHz. The truncated length is $\delta l =$ 9.5 mm. The rectangular patch is designed on Rogers R04350 (tm) dielectric substrate (size: $71 \times 52\,\text{mm}^2$) with dielectric constant (ε_r) 3.6 of height (h) 6 mm, loss tangent $(\tan\delta)$ 0.004, and full ground plane. The U-slot of dimensions $(l_{vs} \times w_{hs} \times t = 17.3 \times 12 \times 2.1\,\text{mm}^3)$ is inserted in the center of the patch. The three arms of the U-slot are symmetrically placed w.r.t the feed point y_f = 1.5 mm.

3 Theoretical Consideration

Based on cavity expansion model [7], the equivalent circuit of the radiating element is $R_{\text{patch}} \| L_{\text{patch}} \| C_{\text{patch}}$ as shown in Fig. 2.

$$
\begin{aligned}
&R_{\text{patch}} = Q_{\text{patch}} / \left(\omega_{\text{patch}} C_{\text{patch}}\right), L_{\text{patch}} = 1 / \left(\omega^2_{\text{patch}} C_{\text{patch}}\right), \\
&C_{\text{patch}} = [\varepsilon_0 \varepsilon_{\text{eff}} LW / 2h] \cos^{-2}(\pi y_f / L)
\end{aligned}
\tag{1}
$$

where ω_{patch} = angular design frequency, $Q_{\text{patch}} = c\sqrt{\varepsilon_{\text{eff}}}/4fh$, and f = design frequency.

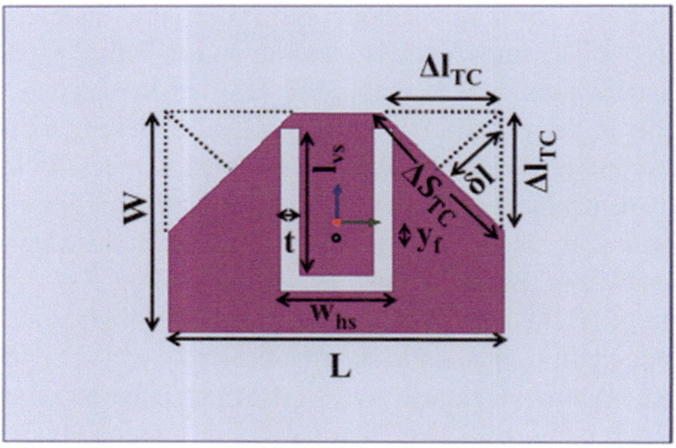

Fig. 1. Proposed antenna geometry for UWB application

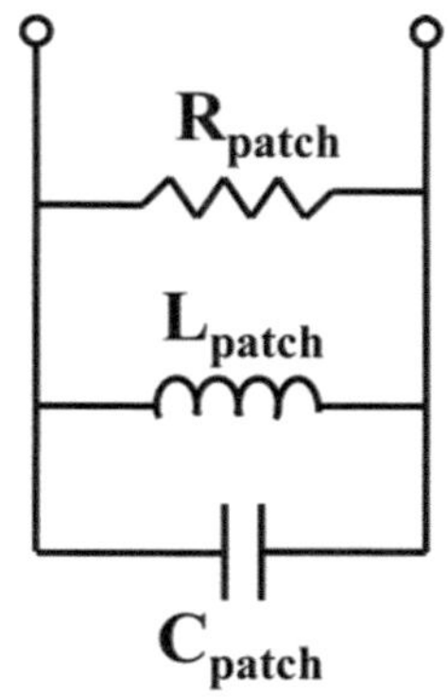

Fig. 2. Circuit of rectangular microstrip antenna

The impedance of the rectangular microstrip radiator is:

$$Z_{\text{patch}} = 1/\left[\left(1/R_{\text{patch}}\right) + \left(1/\omega L_{\text{patch}}\right) + j\omega C_{\text{patch}}\right] \tag{2}$$

In rectangular patch shown in Fig. 1,

$$\Delta S_{\text{TC}} = \Delta l_{\text{TC}}^{2} \tag{3}$$

The effect of capacitance due to truncating corner is:

$$C_{\text{TC}} = \varepsilon_0 \varepsilon_{\text{eff}} \Delta S_{\text{TC}}/h \tag{4}$$

The impedance of the truncated rectangular microstrip radiator is:

$$Z_{\text{patch}} = R_{\text{patch}} + jX_{\text{patch}} = 1/\left[\left(1/R_{\text{pach}}\right) + \left(1/j\omega L_{\text{patch}}\right) + j\omega C_{\text{T}}\right] \tag{5}$$

where $C_{\text{T}} = C_{\text{patch}} + C_{\text{TC}}$.

The corresponding circuit of the inserted U-slot presented in Fig. 3 is based on duality relationship of slot and dipole. The resonant frequency is changed due to the insertion of U-slot in the radiating element that changes the radiator electrical dimension. The path flow length increases due to the current flow throughout the inserted slot and on the boundaries. This consequence can be modeled as series connection of radiation resistance (*R*) and reactive component (*X*) of the two perpendicular and one parallel arms of the inserted U-slot. Then, the individual series combination will be the parallel connection of radiation resistance and reactive components $(R_{\text{vs}}, X_{\text{vs}}\ \text{and}\ R_{\text{hs}}, X_{\text{hs}})$. So, the corresponding circuit of the inserted U-slot can be evaluated [8] which is presented in Fig. 3.

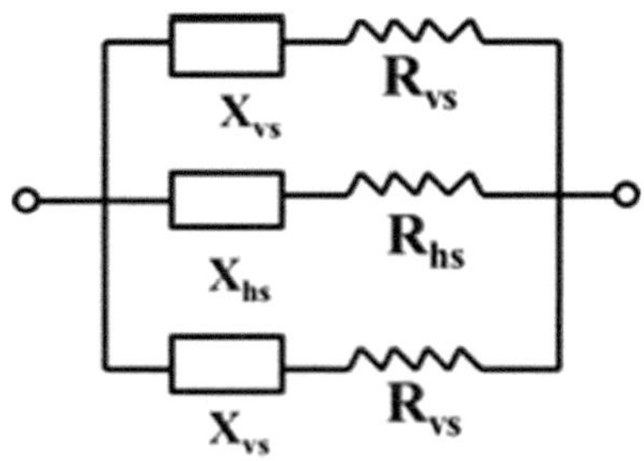

Fig. 3. Corresponding circuit of inserted U-slot

In this research work, for the analysis of inserted U-slot rectangular microstrip patch radiator, the capacitive reactance is taken parallel to the reactance of the radiating element. So, the impedance of inserted U-slot can be evaluated [7] as:

$$Z_{\mathrm{Uslot}} = z_{\mathrm{vs}} z_{\mathrm{hs}} / (z_{\mathrm{vs}} + 2z_{\mathrm{hs}}) \tag{6}$$

where z_{vs} and z_{hs} are impedances of the two vertical arm slots and one horizontal arm slot, respectively.

The corresponding circuit of the intended radiator can be represented as the parallel connection of truncated radiating element and inserted U-slot as given in Fig. 4. So, the input impedance of the intended radiator is given as:

$$Z_{\mathrm{patch}} \| Z_{\mathrm{Uslot}} = Z_{\mathrm{patch}} Z_{\mathrm{Uslot}} / \left(Z_{\mathrm{patch}} + Z_{\mathrm{Uslot}} \right) \tag{7}$$

The radiation characteristics of the intended radiator can be determined [9] from Eqs. (8) and (9). The gain of the proposed antenna can be obtained [7, 10] from Eq. (10).

$$\begin{aligned} E_\theta = & -\frac{jK_0 W V_0 e^{-jk_0 r}}{\pi r} \cos(kh\cos\theta) \frac{\sin((k_0 W/2)\sin\theta\sin\varphi)}{(k_0 W/2)\sin\theta\sin\varphi} \\ & \cos\left(\frac{k_0 L}{2}\sin\theta\sin\varphi\right)\cos\varphi \quad \text{for } 0 \le \theta \le \pi/2 \end{aligned} \tag{8}$$

$$\begin{aligned} E_\varphi = & -\frac{jk_0 w v_0 e^{-jk_0 r}}{\pi r} \cos(kh\cos\theta) \frac{\sin((k_0 W/2)\sin\theta\sin\varphi)}{(k_0 W/2)\sin\theta\sin\varphi} \\ & \cos\left(\frac{k_0 L}{2}\sin\theta\sin\varphi\right)\cos\varphi\sin\varphi \quad \text{for } 0 \le \theta \le \pi/2 \end{aligned} \tag{9}$$

where v_0 = radiating edge voltage, r = distance of an arbitrary point, $k_0 = K\sqrt{\varepsilon_r}$, $K = 2\pi/\lambda$,

$$G = \eta \cdot D \tag{10}$$

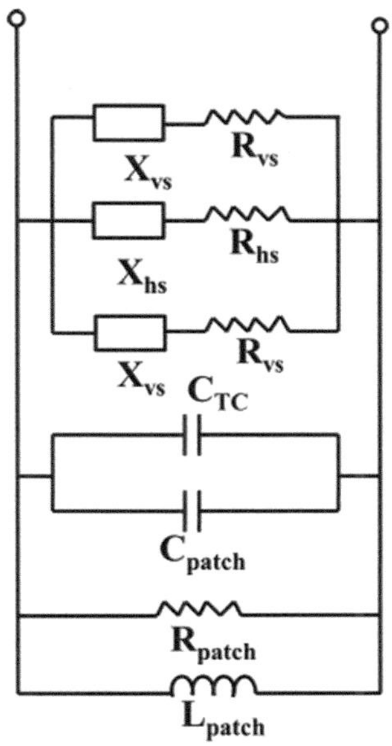

Fig. 4. Equivalent circuit of the intended antenna

where,
η = efficiency of antenna = Radiated power, P_r/Input power, P_t,
$P_r = 1/(2\omega_{patch}C_{patch})$ and C_{patch} is defined in Eq. (1),
$P_t = 1/(\omega_{patch}C_{patch})$,
D = directivity of the proposed antenna = $4W^{22}\Pi/I_1\lambda_0^2$,
$I_1 = \int_0^{\pi} \sin^2((k_0 W \cos\theta)/2)\tan^2\theta \sin\theta d\theta$.

4 Results and Discussion

The return loss (S_{11}) value less than −10 dB of microstrip patch radiator without U-slot varies from 2.5 to 2.73 GHz (simulated B.W.% = 8%) and 2.54 to 2.98 GHz (theoretical B.W.% = 15.94%) with center frequencies at 2.6 and 2.71 GHz which is $(f_l + f_h)/2$, where f_l and f_h are lower and upper limit frequencies, respectively, presented in Fig. 5a, whereas the corresponding impedance bandwidths of the proposed antenna determined from -10 dB are 10 GHz: 120.48% from 3.3 to 13.3 GHz (theoretical), 7.6 GHz: 105.5% from 3.4 to 11 GHz (simulated), respectively, presented in Fig. 5b. The resonant frequencies are spaced nearly to each other which provide UWB characteristic. This bandwidth is also applicable for S-, C-, X-, and Ku-band modern wireless communication systems. Also, the impedance bandwidth of proposed antenna has been improved from 15.94% to 120.48%. Some discriminates occur in between theoretical and simulated results. This is due to as the suggested assumption is suitable for thin material but in this research work large material thickness is used which influences the radiation resistance and feed inductance. The larger impedance bandwidth is due to the insertion of U-slot in the radiating element.

Figure 6 shows that gain of the proposed antenna varies from 1.5 to 5.5 dBi (theoretical), 1.6 to 5.9 dBi (simulated) over the whole impedance bandwidth.

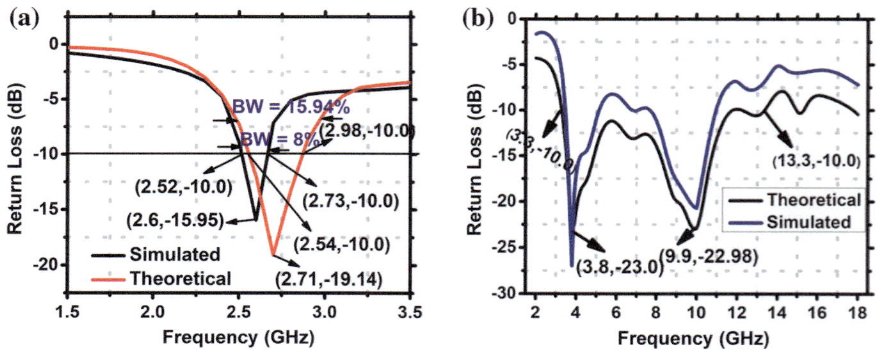

Fig. 5. Simulated and theoretical return loss plot against frequency **a** Rectangular microstrip patch antenna, **b** intended antenna

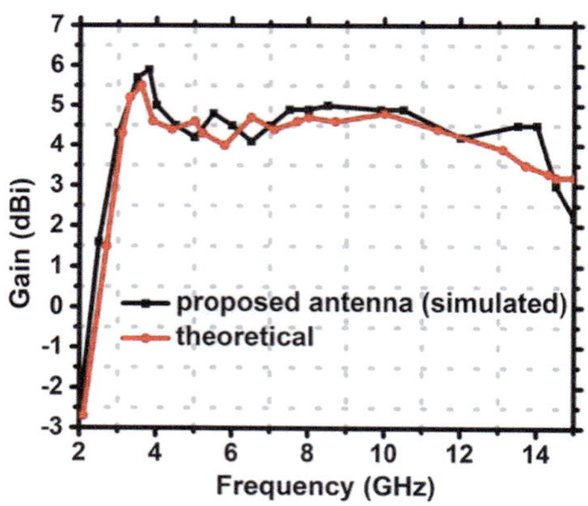

Fig. 6. Gain plot against frequency of intended antenna

Figure 7 presents the radiation patterns of the intended radiator which are quite stable throughout the operating band. The E-plane cross-polarization level is higher than that of H-plane due to feed location on the *Y*-axis. In H-plane, the slight increase of cross-polarization level is observed at 3.8 GHz. The corner truncation in rectangular radiating element may create the small variance among the suppositional and simulation radiation characteristics. Again, the small broadside nature of the radiation patterns is also changed due to the assumptions proposed in the theory.

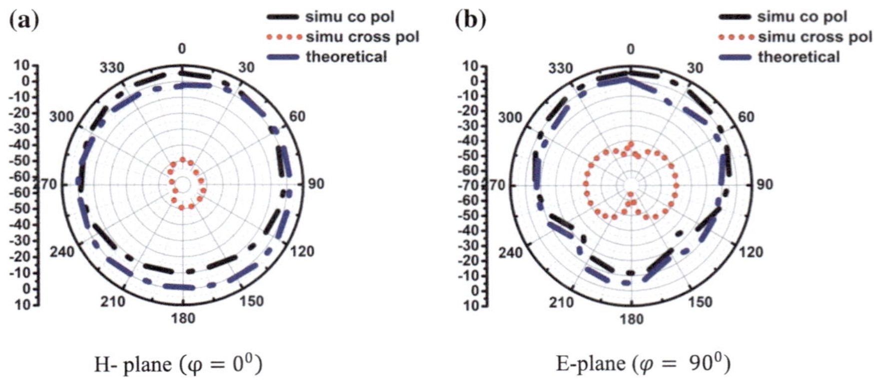

Fig. 7. Simulated and measured radiation characteristics of the intended radiator at resonant frequencies 9.9 GHz

5 Conclusion

This research article demonstrates bandwidth enhancement analysis of microstrip antenna with equivalent circuit approximate using cavity expansion model for ultra-wideband application. The intended radiator is successfully designed and simulated with bandwidth 105.5% (simulated: from 3.4 to 11 GHz), 120.48% (theoretical: from 3.3 to 13.3 GHz). The proposed antenna with large bandwidth is suitable for multi-functional wireless S-band (from 2 to 4 GHz), C-band (from 4 to 8 GHz), X-band (from 8 to 12 GHz), and Ku-band (from 12 to 18 GHz) applications.

References

1. Schaubert DH, Pozar DM, Adrian A (1989) Effect of microstrip antenna substrate thickness and permittivity: comparison of theories and experiment. IEEE Trans Antennas Wirel Propag Lett 37:677–682
2. Sahoo S, Mohanty MN, Mishra LP (2018) Bandwidth improvement of compact planar antenna for uwb application with dual notch band performance using parasitic resonant structure. Prog Electromagnet Res M 66:29–39
3. Shakib MN, Moghavvemi M, Mahadi WNL (2015) A low profile patch antenna for ultra wideband application. IEEE Antennas Wirel Propag Lett 14:1790–1793
4. Sahoo S, Mishra LP, Mohanty MN, Mishra RK (2018) Design of compact uwb monopole planar antenna with modified partial ground plane. Microwave Opt Technol Lett 60:575–578

5. Khidre A, Lee KF, Elsherbeni AZ, Yang F (2013) Wide band dual beam U-slot microstrip antenna. IEEE Trans Antennas Propag 61(3):1415–1418
6. Yang W, Zhou J, Yu Z, Li L (2014) Single fed low profile broadband circularly polarized stacked patch antenna. IEEE Trans Antennas Propag 62(10):5406–5410
7. Wolf EA (1988) Antenna analysis. Artech House, Norwood (USA)
8. Narayan S, Sharma S, Vishvakarma BR (2005) Analysis of slot loaded rectangular microstrip patch antenna. Indian J Radio Space Phys 34:424–430
9. Derneryd AG, Lind AG (1979) Extended analysis of rectangular microstrip resonator antennas. IEEE Trans Antennas Propag 27(6):846–849
10. Kumar G, Ray KP (2003) Broadband microstrip antennas. Artech House, Norwood, MA

Analysis of Vector-Controlled Onshore DFIG with STATCOM for Power Compensation

Sovit Kumar Pradhan[1(✉)], Radhakrishna Das[2], Debswarup Rath[3], and S. S. P. B. K. Prasad[3]

[1] Department of Electrical Engineering, NIT, Silchar, India
[2] Department of Electrical Engineering, GIET, Bhubaneswar, India
[3] Department of Electrical & Electronics Engineering, GIET, Bhubaneswar, India

1 Introduction

With ever-increasing desire for variable speed operation, the doubly fed induction generator (DFIG) is used along with wind turbines. This system is in great demand. Voltage control and reactive power management are the concerns for power system, while grid-connected DFIG is also affected due to continuous voltage fluctuations.

All equipment terminal voltages in the system are within suitable and operatable boundary conditions. Both the equipment from utilities and customer are designed to function at a predefined standard operatable voltage level. Overuse or operation of the equipment at voltages outside the permissible or boundary limit array should unfavorably upset their performance and cause them harm [1–3].

System steadiness is improved to make the most use of transmission system arrangement. control have an important role For system stability analysis, voltage and reactive power are major concern for researcher [4].

The reactive power flow is reduced to decrease secondary resistance and secondary reactance to applied lowest. This safeguards the transmission system operations proficiently for active power transfer.

Because of nonlinear quality of transmission line due to loading, it consumes or absorbs reactive power [5]. This creates problematical situations for transmission lines.

2 Background

2.1 Doubly Fed Induction Generator (DFIG)

A doubly fed induction generator (DFIG) exhibits more advantages in comparison with other generators that applied in wind turbines [6]. In DFIG, as its name suggests rotor and stator both are connected with grid for supply of power irrespective of synchronous or subsynchronous operations [7]. The rotor currents create the magnetic field in the rotor, and the rotor magnetic field interacts with the stator magnetic field to develop torque in the machine. The torque is produced due to above interaction of magnetic fields and can be considered as vector multiplication of the stator field and rotor field.

M. N. Mohanty and S. Das (eds.), *Advances in Intelligent Computing and Communication*, Lecture Notes in Networks and Systems 109,
https://doi.org/10.1007/978-981-15-2774-6_26

2.2 DFIG Operation

The AC–DC–AC converters are the voltage-source converters used on the connection from rotor to grid named as rotor-side converter (RSC) and grid-side converter (GSC). This is also known as back-to-back converter. For the reduction of ripple voltage between converters, a DC-link capacitor is placed [7]. VSC with IGBT application are used for this purpose [8]. Simulations have been performed with considering converter on both sides (Fig. 1).

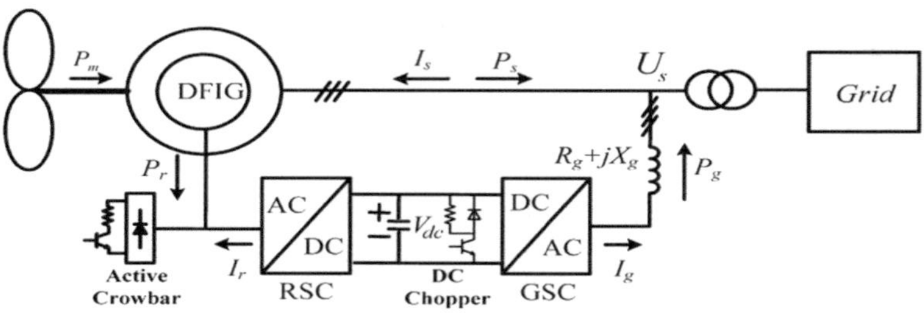

Fig. 1. Wind turbine and doubly fed induction generator system [7]

2.3 Static Synchronous Compensator

The static synchronous compensator (STATCOM) is a shunt-connected reactive compensation equipment preferred in DFIG as it provides better result than shunt capacitors. In STATCOM, power electronic switching devices are used to provide controllability in magnitude and phase angle of the voltages. It also gives support to bus voltages with modulation results improved transient characteristics such as better transient stability limit and reduced oscillations in response.

Usually, a STATCOM is implemented in the networks that have reduced power factor and less voltage stability. The active power capability of the network can be increased with placement of energy storage device across the DC capacitor. The STATCOM also provides better reactive power support in comparison with SVC in case of low voltage.

2.4 Reasons for Selection of STATCOM

Shunt capacitors in wind firm can provide required reactive power but not responsive to voltage change. At higher than the voltage limit of 1.05 pu, the required reactive power compensation cannot be provided through fixed capacitor [9].

In case of imbalances in reactive power, which critically affects the entire power systems along with consumers should be normalized. The STATCOM shows better result regarding compensation to grid [10]. In this paper, PWM technique applied to VSC STATCOM is planned to soothe grid-connected DFIG-based wind turbines.

2.5 Voltage-Source Converter in DFIG

For improvement, most DFIG converters use a two-level six-switch converter as shown in Fig. 2. A two-level converter output voltage has two levels of voltage. The insulated gate bipolar transistors (IGBTs) are considered as most appropriate switch for this operation. The output voltage of six-switch converter should have magnitude and frequency within the determined value that the max value of voltage is less than the DC-link voltage. The switching frequency of PWM converter is considered in such a way that control of output voltage is possible.

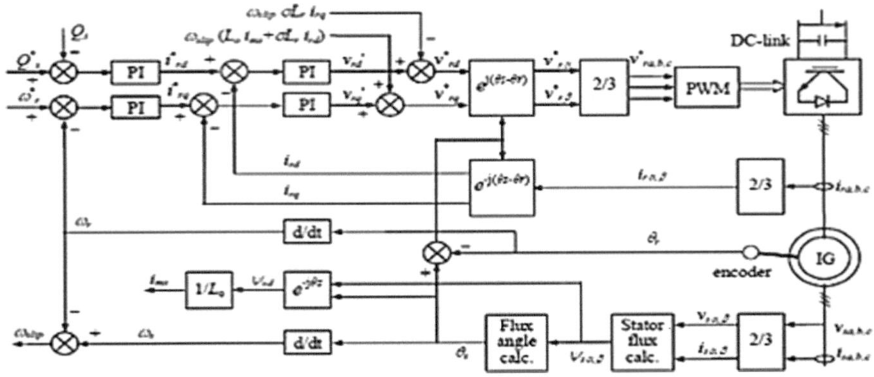

Fig. 2. Vector structure of rotor-side converter [7]

2.6 Vector Control of Both Side Converter Circuit

The Rotor-Side Converter (RSC): The role of the rotor-side converter is to administer the rotor currents effectively to produce necessary torque due to interaction between fluxes produced at stator and rotor. The torque is normalized the wind turbine output power. In order to achieve maximum power or track the maximum power point of DFIG, the controller of power is required. A proportional–Integral (PI) regulator or controller is implemented to minimize the error to zero at current loop. The controller output is considered as reference for rotor current that added in the winding current of rotor through rotor-side converter [7]. This quadrature part controls the electro- magnetic torque. With comparison between irq and irqref, the error can be found out and the fault is reduced to zero.

The Grid-Side Converter (GSC): The grid-side converter targets to achieve the proper control of the voltage. Furthermore, it is allowable to take up reactive power for voltage holdup requirements with two control closed paths. Outer closed path consists of path through the voltage regulator. The icdref is the reference current for regulator [7]. The inner closed path is through the current regulator, and current regulation can be performed with reference to both direct and quadrature axis dc current (Fig. 3).

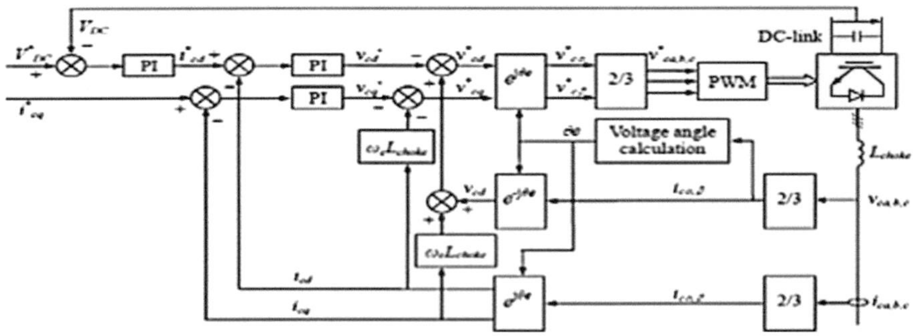

Fig. 3. Vector structure of grid-side converter [7]

3 Simulation and Results

3.1 Simulation Model

Vector control of DFIG has been performed for analysis purpose of output voltage. The value of parameters used in simulation analysis is considered with real values of 99 MW capacity of Damanjodi windmill, India (Fig. 4).

Proportional–integral controller is used in DFIG controller for better result as PI controller improves dynamic response. Here, vector control is implemented so that active power and reactive power can be controlled by current components. Phase and the magnitude of the three phase are calculated and fed to the PI controller. Harmonics can be compensated by using this model. This stator flux field-oriented control improvises the output by diminishing the fluctuations. Simulations have been performed using MATLAB software.

3.2 Result Analysis

Figure 5a shows simulated result of compensated graph of STATCOM applied to DFIG, and it indicates the maximum value of 400 occurs at time 0. 04 s. The minimum value is around 264 at 0.15 s (approx). The setting time is 0.02 s. In Fig. 5b, rotor angle characteristics can be analyzed and steady-state condition is achieved at 0.04 s. The DC output of grid-side converter is shown in the next figure, and the difference in between maximum voltage and minimum voltage decreases as the time increases. From that, it can be assumed that the total harmonic distortion is also decreasing. The average voltage value is near about 1150 V.

From the above discussion, stator flux-oriented control of PI controller is working better in DFIG with STATCOM for power compensations. The DC output voltage fluctuation is approximately 22 V after 1 s with an average output voltage of 1150 V. The frequency of rotor angle is remaining stable in this vector-controlled model.

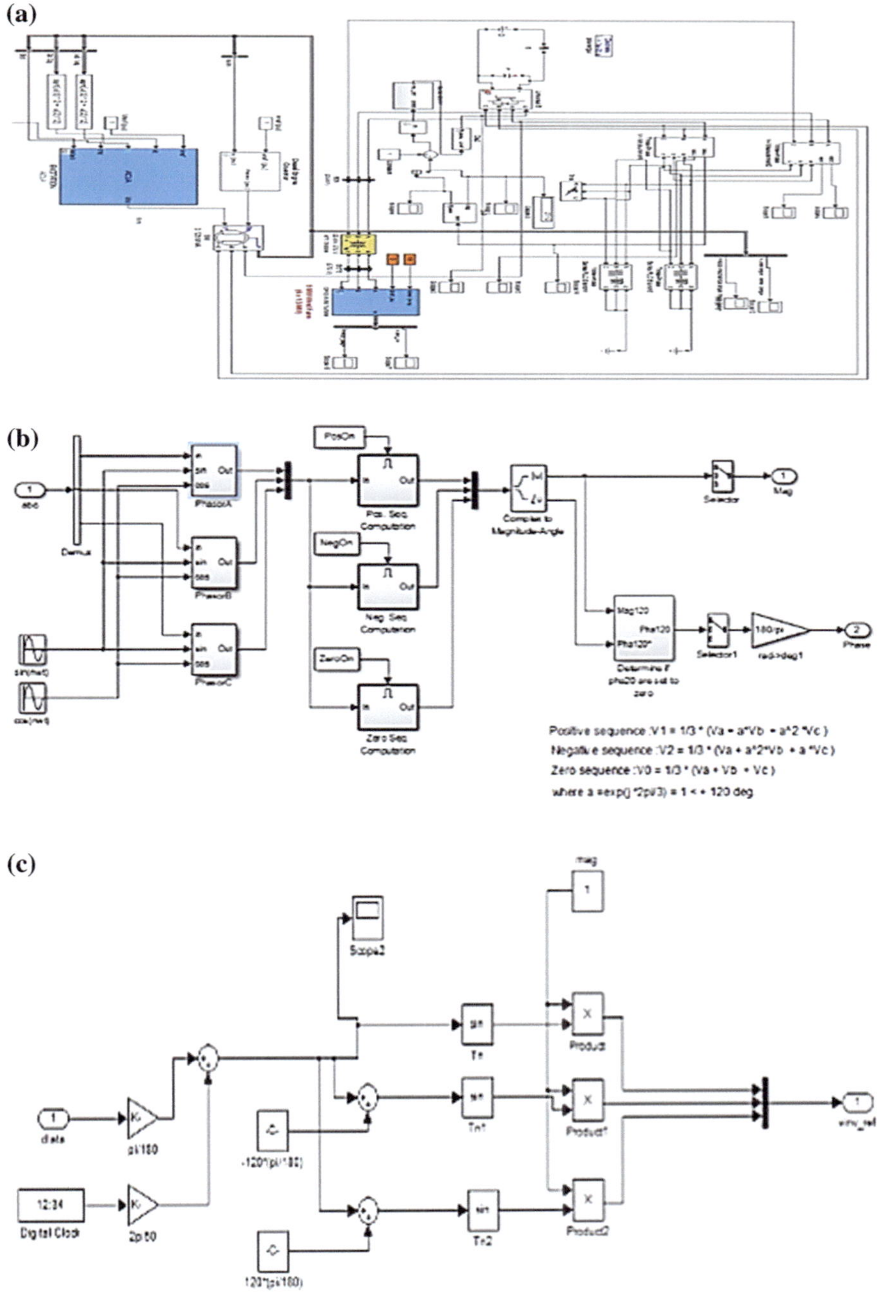

Fig. 4. **a** Simulation model of DFIG with STATCOM, **b** proportional–integral controller used in DFIG, **c** subsystem model of STATCOM

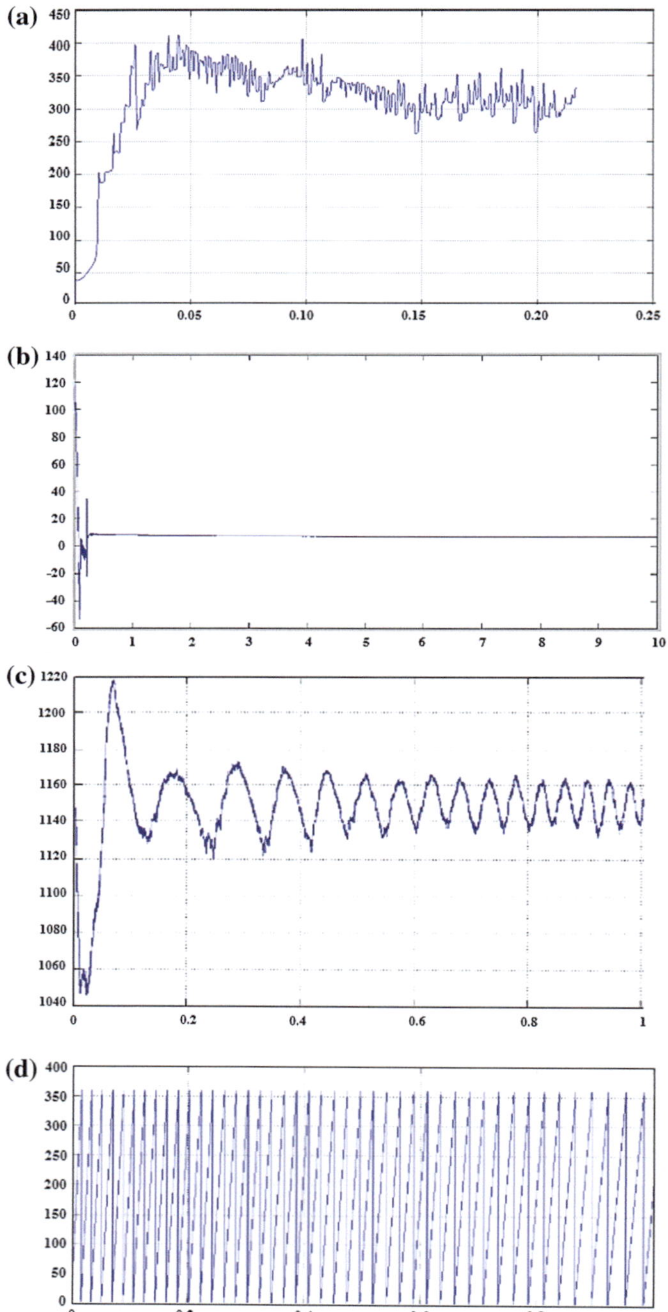

Fig. 5. **a** Compensate graph of the STATCOM, **b** load angle characteristics with respect to time, **c** DC output voltage of converter, **d** rotor angle characteristics

4 Conclusion

In this paper, vector control strategy to PI controller with reactive and active power compensation is performed. The control of DC bus voltage and reactive power is achieved through controlling the rotor current with help of back-to-back converter. The model includes the variation in wind speed for the simulation of characteristics of the wind turbine. In relation with the results, this model is capable of delivering stable output voltage with maintaining the power quality and this is achieved by eliminating harmonics of the utility grid by vector control. For future work, various experiments will be performed to verify the other uses of PI regulator with vector control schemes.

References

1. Yasa Y, Sozer Y, Mese E (2013) Harmonic analysis of doubly fed induction generator based utility interactive wind turbine systems during fault conditions. In: IEEE energy conversion congress and exposition, pp 2270–2276
2. Fan L, Yuvarajan S, Kavasseri R (2010) Harmonic analysis of a DFIG for a wind energy conversion system. IEEE Trans Energy Conver 25(1):181–190, March 2010
3. Lindholm M, Rasmussen T (2003) Harmonic analysis of doubly fed induction generators. In: Proceedings of the 5th international conference on power electronics and drive systems, Nov 2003, vol 2, pp 837–841
4. Xu L, Wang Y (2007) Dynamic modeling and control of DFIG-based wind turbines under unbalanced network conditions. IEEE Trans Power Syst 22(1):314–323
5. Das RK, Pradhan SK, Pati SK (2018) An investigation of harmonics in grid connected DFIG-WT experiencing transmission faults. In: International IEEE conference on technologies for smart city energy security and power (ICSESP-2018), Bhubaneswar, India. https://doi.org/10.1109/icsesp.2018.8376689, Mar 2018
6. Muller S, Deicke M, Doncker RWD (2002) Doubly fed induction generator systems for wind turbine. IEEE Ind Appl Mag 8(3):26–33
7. Fletcher J, Yang J (2010) Introduction to doubly-fed induction generator for wind power applications. University of Strathclyde, Glasgow, UK
8. Abo-Khalil AG, Lee D-C, Jang J-I (2007) Control of back-to-back PWM converters for DFIG wind turbine systems under unbalanced grid voltage. In: Proceedings of IEEE international symposium on industrial electronics, pp 2637–2642
9. Petersen L, Kryezi F, Iov F (2017) Design and tuning of wind power plant voltage controller with embedded application of wind turbines and STATCOMs. IET Renew Power Gener 11 (3):216–225
10. Izadpanahi N et al (2017) DFIG-based wind turbine using STATCOM for improvement performance. In: 2017 14th International Conference on Electrical Engineering/Electronics, Computer, Telecommunications and Information Technology (ECTI-CON). IEEE

11. Pradhan SK, Srivastava RK (2013) Analysis of PMSG under stator turn to turn fault using winding function theory. In: International conference on condition assessment techniques organised by IEEE DEIS Kolkata, India, pp 116–121. https://doi.org/10.1109/catcon, 6737483, 6–8 Dec 2013
12. Liao Y, Ran L, Putrus GA, Smith K (2003) Evaluation of the effects of rotor harmonics in a doubly-fed induction generator with harmonic induced speed ripple. IEEE Trans Energy Convers 18(4):508–515
13. Kheshti M, Kang X, Song G, Jiao Z (2015) Modeling and fault analysis of doubly fed induction generators for Gansu wind farm application. IEEE J Mag 38:52–64

Dictionary Design for Block-Based Intra-image Compression

Arabinda Sahoo[1(✉)] and Pranati Das[2]

[1] Department of ECE, ITER, Siksha 'O' Anusandhan Deemed to be University, Bhubaneswar, Odisha, India
[2] Department of Electrical Engineering, Indira Gandhi Institute of Technology, Sarang, Odisha, India

1 Introduction

An effective dictionary can lead to efficient sparse approximation. The dictionary can be fixed, generated from pre-defined sets of functions or designed from the learning sets. The fixed dictionary is efficient for sparse approximation of certain classes of signals, but quite disappointing for others. The shortcomings of the fixed dictionary have paved the way for another versatile dictionary called learned dictionary. The performance of block-based intra-image compression scheme depends on how well the residuals obtained from intra-prediction are encoded. Transform-based intra-compression method uses fixed DCT dictionary for encoding of prediction residuals. However, DCT dictionary is not suited for efficient encoding of prediction residuals with complex and non-periodic characteristics.

In recent years, various learned dictionaries have been developed by researchers for image compression. Bryt and Elad [1] designed a dictionary based on K-SVD algorithm for facial image compression. Zepeda et al. [2] proposed an iteration-tuned dictionary (ITD). Zhan et al. [3] proposed a dictionary for compression of complex synthetic aperture radar (SAR) images. Shao et al. [4] proposed a dictionary for fingerprint compression. Zhu et al. [5] proposed a dictionary for surveillance image compression. Most of the above methods are related to some specific class of images. These dictionaries are not suitable for compression of prediction residuals.

Recently, Sahoo and Das [6] developed a dictionary-based image compression for general arbitrary images. Experimental result shows the compression efficiency using this scheme is better than other standard methods but it is computationally complex. This motivated us to design a dictionary for block-based intra-image compression to reduce the computational complexity.

This paper presents an algorithm to design an over-complete residual dictionary, denominated as residual dictionary, suitable for encoding prediction residuals. The proposed over-complete residual dictionary is embedded in block-based intra-image compression framework to encode the residuals [6]. Comparative analysis demonstrates that the proposed dictionary design method yields superior performance in terms of compression efficiency and computational complexity as compared to other standard methods.

M. N. Mohanty and S. Das (eds.), *Advances in Intelligent Computing and Communication*, Lecture Notes in Networks and Systems 109,
https://doi.org/10.1007/978-981-15-2774-6_27

The rest of the paper is organized as follows. Section 2 introduces the basics of dictionary learning. Section 3 presents the proposed algorithm to design an over-complete residual dictionary. Results and comparative analysis are discussed in Sect. 4. Finally, Sect. 5 presents the conclusion.

2 Dictionary Design

The central idea in dictionary learning is to train a dictionary based on set of training signals until a well-suited specialized final dictionary is obtained that can be used in fixed form to perform efficient sparse representation of other signal sets of the same class. The learned dictionary has been shown to be very well suited for sparse approximation.

2.1 Background

Let y represent a signal set given by:

$$y = [y_1, y_2, y_3, \ldots, y_K] \in R^{M \times K} \tag{1}$$

Dictionary learning aims to find a dictionary $D = [d_1, d_2, d_3, \ldots, d_N] \in R^{M \times N}$ with $M < N$, such that each y_K can be sparsely represented by atoms of D. The sparse representation of each y_K is summarized in the matrix $X = [x_1, x_2, x_3, \ldots, x_N] \in \mathbb{R}^{N \times K}$, where x_K is sparse representation vector of y_K. The minimization problem with sparsity constraint T_0 can be represented as:

$$\min_{D,X} \|Y - DX\|_2 \text{ subject to } \forall i, \|x_i\|_0 \leq T_0 \tag{2}$$

In dictionary learning, the first step finds the coefficients with constant D. Then, in second step the atoms of dictionary D are updated by keeping fixed coefficients. These two steps continue for a certain number of iterations or until a desired approximation error is reached. The first step of the problem is usually solved with sparse decomposition algorithm such as MP [7] and OMP [8]. The second step of the problem, dictionary update, can be solved by various update techniques. The K-SVD algorithm [9] is very efficient and often used as an offline preprocessing step to design the dictionary.

3 Proposed Residual Dictionary Design

In block-based intra-compression method [10], the prediction residual is encoded. In intra-compression method, the prediction residuals are transformed, quantized and entropy-encoded. It uses fixed DCT dictionary. The fixed DCT dictionary is particularly well suited for representing prediction residuals with periodic texture patterns.

However, DCT dictionary is not suited for efficient encoding of prediction residuals with complex and non-periodic characteristics. In this section, we will describe the detailed algorithm to design an over-complete residual dictionary.

The proposed residual dictionary design consists of three main steps: intra-prediction, generation of the training samples and dictionary learning.

3.1 Intra-prediction

Let I denote the original image block, and I_P is its predicted block for best mode N. The prediction residual image block is represented as:

$$I_R = I - I_P \tag{3}$$

After intra-prediction, the best mode and the corresponding prediction residues of each block are generated for encoding.

3.2 Generation of the Training Samples

In this step, prediction residuals I_R generated from a wide variety of images are used as training samples to train the dictionary. Prediction residuals from each mode are randomly chosen as training set to train the residual dictionary.

3.3 Dictionary Learning

Let $I_T = [I_{R1}, I_{R2}, I_{R3}, \ldots, I_{RK}] \in \mathbb{R}^{n^2 \times K}$, where $I_{R1}, I_{R2}, I_{R3}, \ldots, I_{RK}$ represents K training vectors of training set I_T. Residual dictionary $D_R = [d_{R1}, d_{R2}, d_{R3}, \ldots, d_{RN}] \in \mathbb{R}^{n^2 \times N}$ with $N > n^2$ aims to sparsely represent each I_{RK} by linear combination of few atoms of D_R. The minimization problem to find the best dictionary D_R with sparsity constraint T_0 is represented by:

$$\min_{D_R, X} \| I_T - D_R X \|_2 \text{ subject to } \forall i, \| x_i \|_0 \leq T_0 \tag{4}$$

The overall problem can be represented as a joint optimization:

$$D_R = \arg\min_{D_R, \mathbf{x}} \sum_{i=1}^{k} \left(\|I_{Ri} - D_R x_i\|_2^2 + \lambda \|x_i\|_0 \right) \tag{5}$$

This optimization problem can be solved by splitting the problems into two parts:

1. With fixed D_R, compute sparse representation X (sparse coding step).
2. With fixed X, update atoms of D (dictionary update step).

Sparse coding:

With fixed D_R, it can be represented in k distinct problem of the form

$$\min_{X_i} \|I_{Ri} - D_R x_i\|_2^2 \text{ subject to } \forall i, \|x_i\|_0 \leq T_0 \tag{6}$$

This problem is adequately addressed by the OMP sparse decomposition algorithm.

Dictionary update step:

K-SVD algorithm is used to update the dictionary atoms. Let d_{Rk} be the one column of the residual dictionary that needs to be updated, then the error matrix without atom d_{Rk} is represented as:

$$E_k = I_T - \sum_{j \neq k} d_{Rj} x_T^j \tag{7}$$

SVD decomposes $E_k = U \Delta V^T$. The updated atom for d_{Rk} is the first column of U, i.e., $d_{Rk} = U_1$. This step repeats for all atoms or columns of D_R. The above two steps are repeated until the sparsity constraint is reached, i.e., $\forall i, \| x_i \|_0 \leq T_0$. *Algorithm 1* summarizes the steps of residual dictionary design.

Algorithm 1: Residual dictionary design

Input: Training prediction residual $I_T = [I_{R1}, I_{R2}, I_{R3}, \ldots, I_{RK}] \in R^{n^2 \times K}$
Output: $D_R = [d_{R1}, d_{R2}, d_{R3}, \ldots, d_{RN}] \in \mathbb{R}^{n^2 \times N}$
1: initialization: randomly initialize $D_R \in \mathbb{R}^{n^2 \times N}$ with normalized columns
2: repeat sparse coding stage and codebook update stage until halting criteria is satisfied
Sparse coding stage:
For $i = 1$ to k do
3: solve $\min_{x_i} \| I_{Ri} - D_R x_i \|_2^2$ subject to $\forall i, \| x_i \|_0 \leq T_0$ using OMP sparse decomposition algorithm
Codebook update stage:
For $k = 1$ to N do
4: Compute the overall representation error matrix
$E_k = I_T - \sum_{j \neq k} d_{Rj} x_T^j$
5: Apply SVD which decomposes $E_k = U \Delta V^T$
6: Update atom d_{Rk} by first column of U, i.e., $d_{Rk} = U_1$
7: Until halting criterion $\| I_{Ri} - D_R x_i \|_2^2 \leq \delta$ is *true* $\forall i$

3.4 The Compression Scheme

The block diagram of the residual dictionary-based compression process is shown in Fig. 1 [6].

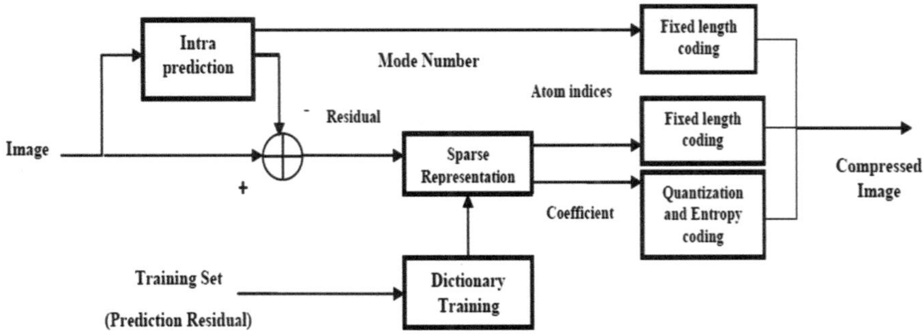

Fig. 1. Detailed block diagram of the compression process

In this scheme, after intra-prediction, the prediction mode number N of each block is transmitted for encoding and the corresponding prediction residue I_R is transmitted for sparse representation. The coefficients and position of nonzero elements of solution vector obtained from sparse representation are encoded.

4 Experimental Results

In order to evaluate the performance of proposed dictionary, experiments are conducted on several test images taken from the Berkeley segmentation database [11, 12]. Test images with different features are selected for experiment. The experiments concentrate on compression performance and computational complexity. The implementation is done in MATLAB.

4.1 Experimental Setup

In the experiment, a set of 45,000 prediction residual blocks, 5000 from each mode, are randomly chosen as training set to train the residual dictionary. Residual dictionary D of size 64×512 is selected. Our proposed residual dictionary design algorithm is employed for dictionary training. Sparsity constraint $T_0 = 4$ and 100 iterations are selected as stopping condition of the algorithm.

Our proposed residual dictionary D and OMP algorithm are used for sparse representation of prediction residual I_R. Error $\delta = 20$ and $T_0 = 4$ are selected for OMP.

After sparse representation, fixed-length codes are used to encode mode number and position. Four bits are used to encode mode number. Positions of nonzero elements are encoded with $\log_2 k$ bits, where k is the number of atoms in dictionary. The coefficient is uniformly quantized and entropy-coded similar to JPEG to form the final bit stream.

4.2 Compression Performance Comparison

We compared the image compression performance of our dictionary with intra-image compression scheme using fixed DCT dictionary and JPEG 2000. The objective evaluation of our proposed method is accomplished by assessment of PSNR [13]. Table 1 shows comparison of PSNR for test image at low bit rate as well as high bit rate. It is evident from Table 1 that the proposed method gives rise to a PSNR improvement up to 0.35 dB in comparison with JPEG 2000 and 0.6 dB in comparison with intra-image compression scheme using DCT.

Table 1. Compression performance comparison of proposed dictionary with intra-image compression scheme using fixed DCT dictionary and JPEG 2000

Images	PSNR (dB) at bit rate 0.2 bpp			PSNR (dB) at bit rate 1 bpp		
	DCT	JPEG 2000	Proposed dictionary	DCT	JPEG 2000	Proposed dictionary
Barbara	26.92	27.30	27.61	36.94	37.16	37.53
Sailboat	28.88	29.13	29.54	36.38	36.82	37.02
Baboon	24.74	25.25	25.46	30.22	30.66	30.82
Couple	28.12	28.49	28.81	36.47	36.80	37.11
Hill	29.44	29.89	30.22	36.25	36.43	36.85
Jet plane	31.56	31.93	32.29	41.61	41.89	42.21
Lena	32.87	33.03	33.47	40.23	40.41	40.83
Lighthouse	28.11	28.40	28.83	38.23	38.43	38.84
Peppers	32.13	32.50	32.89	38.12	38.39	38.73
Average	29.19	29.55	29.90	37.16	37.44	37.77

4.3 Computational Complexity Comparison

In order to evaluate the computational complexity of our dictionary, we compared our method with the dictionary-based image compression scheme proposed by Sahoo and Das [6]. Both the methods differ due to sparse representation of image block. Both the methods are applied over 100 residual image blocks of size 8×8. It is observed that for error $\delta = 20$ around 12 dictionary elements are required using the method proposed by Sahoo and Das [6] for sparse representation of residual image blocks, whereas using our method around 4 dictionary elements are required for sparse representation of residual image blocks. Note that sparse representation using our dictionary is around 3 times faster compared to other one.

5 Conclusion

In this paper, we presented an algorithm to design an over-complete residual dictionary suitable for block-based intra-image compression. The performance of proposed image compression method is compared with intra-image compression scheme using fixed DCT dictionary and state-of-the-art JPEG 2000 compression standard. The computational complexity is also compared with other methods. Simulation results demonstrate that the proposed image compression method outperforms other standard methods in terms of compression efficiency and computational complexity.

References

1. Bryt O, Elad M (2008) Compression of facial images using the K-SVD algorithm. J Vis Commun Image Representation 19(4):270–282
2. Zepeda J, Guillemot C, Kijak E (2011) Image compression using sparse representations and the iteration-tuned and aligned dictionary. IEEE J Sel Topics Signal Process 5(5):1061–1073
3. Zhan X, Zhang Y, Huo C (2013) SAR image compression using multiscale dictionary learning and sparse representation. IEEE Geosci Remote Sens Lett 10(5):1090–1094
4. Shao G, Wu Y, Yong A, Liu X, Guo T (2014) Fingerprint compression based on sparse representation. IEEE Trans Image Process 23(2):489–501
5. Zhu J-Y, Wang Z-Y, Zhong R (2015) Dictionary based surveillance image compression. J Vis Commun Image R 31(7):225–230
6. Sahoo A, Das P (2017) Dictionary based image compression via sparse representation. Int J Electr Comput Eng 7(4):1964–1972
7. Mallat S, Zhang Z (1993) Matching pursuit with time-frequency dictionaries. IEEE Trans Signal Process 41(12):3397–3415
8. Pati YC, Rezaiifar R, Krishnaprasad PS (1993) Orthogonal matching pursuit: recursive function approximation with applications to wavelet decomposition. In: Proceedings of the twenty-seventh Asilomar conference on signals, systems and computers, pp 40–44
9. Aharon M, Elad M, Bruckstein A (2006) K-SVD: an algorithm for designing overcomplete dictionaries for sparse representation. IEEE Trans Signal Process 54(11):4311–4322
10. Turkan M (2012) Novel texture synthesis methods and their application to image prediction and image inpainting, Ph.D. Thesis
11. (Online). Available:www.eecs.berkeley.edu/Research/Projects/CS/vision/grouping/segbench/
12. (Online). Available: http://sipi.usc.edu/services/database/
13. Wang Z, Bovik AC, Sheikh HR, Simoncelli EP (2004) Image quality assessment: from error visibility to structural similarity. IEEE Trans Image Process 13(4):600–612

Performance Analysis of Fractional Order Low-pass Filter

Kumar Biswal[1(✉)], Madhab Chandra Tripathy[2], and Sanjeeb Kar[3]

[1] School of Electronics Engineering, Kalinga Institute of Industrial Technology (DU), Bhubaneswar, India
kumar.biswalfet@kiit.ac.in
[2] Department of Instrumentation & Electronics Engineering, CET, BPUT, Bhubaneswar, India
mctripathy@cet.edu.in
[3] ITER, SoA, Deemed to be University, Bhubaneswar, Odisha, India
sanjeebkar@soa.ac.in

1 Introduction

Fractional order differential and integral equations are the generalization of conventional integral and differential equations that can be used for modeling the real world we live in. The integer-order models were used due to the absence of solution methods for fractional differential equations. At present, many methods have been developed for realizing the applications based on fractional derivative and integration [1–3]. Fractional calculus covers all non-integer numbers including fractions and irrational numbers. The integer-order filters have integer coefficients in its mathematical representation which leads to some design criterion that makes it difficult to obtain sharp cutoff frequency [4–7]. The exact design requirements can be satisfied by fractional order filters, but in conventional integer-order filters this was not possible. This fractional concept adds a little bit of safety in the design specifications in comparison with the integer-order filters [8]. In fractional order filters, the circuit design specifications with tuning parameters can be easily achieved. For example, the roll-off frequency can be varied and set to any desired slope which is only possible in case of fractional order filters. Hence, fractional order filters provide better design flexibility as compared to integer-order filters [9–14]. This paper is organized as follows. Section 2 deals with model, transfer function and response of fractional order low-pass filter. Section 3 illustrates the time response of single-stage fractional low-pass filter. Similarly, Sect. 4 deals with frequency response of single-stage fractional order low-pass filter. Finally, conclusion is presented in Sect. 5.

M. N. Mohanty and S. Das (eds.), *Advances in Intelligent Computing and Communication*, Lecture Notes in Networks and Systems 109,
https://doi.org/10.1007/978-981-15-2774-6_28

2 Fractional Order Low-Pass RC Filter

Low-pass filters allow low-frequency signals to pass through and attenuate high-frequency signals through it. The cutoff frequency (ω_c) locates the passband ($\omega < \omega_c$) and the stop-band ($\omega > \omega_c$) in a complete frequency response. A low-pass filter allows signal of low frequency to pass and gradually attenuates high signals having frequencies higher than the cutoff frequency.

Low-pass filters have a wide range of applications in analog and digital filters. The filter is characterized by its cutoff frequency and rate of roll-off frequency. So, the value of attenuation frequencies which are generally higher than cutoff frequency is determined by the order of filter [13, 14]. The circuit model for the single-stage fractional order low-pass filter is presented in Fig. 1.

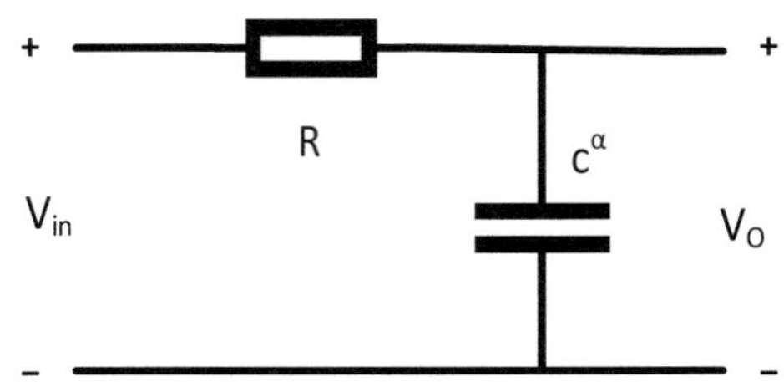

Fig. 1. Circuit model of a single-stage fractional order low-pass filter

The characteristic equation for the circuit is given by an expression:

$$\frac{V_O}{R} + C^{\propto}\frac{d^{\propto}V_O(t)}{dt^{\propto}} = \frac{V_{in}}{R}$$

Taking Laplace transform, the transfer function of the circuit is

$$H(s) = \frac{V_O(S)}{V_i(S)} = \frac{\tau}{s^{\alpha} + \tau};$$

Here, $\tau = \frac{1}{RC^{\alpha}}$ is a constant (i.e., time constant).

H(*s*) represents the transfer function expression of the (integer) first-order low-pass filter. From the above transfer function, the magnitude *H*(*w*) and phase $\emptyset(w)$ of the (fractional) first-order low-pass filter [15–18] are given by

$$|H(w)| = \frac{\tau}{\sqrt{w^{2\alpha} + \tau^2 + 2w^{\alpha}\cos\frac{\alpha\pi}{2}}} \quad (1)$$

$$\emptyset(w) = -\tan^{-1}\left[\frac{w^{\alpha}\sin\left(\frac{\alpha\pi}{2}\right)}{\tau + w^{\alpha}\sin\left(\frac{\alpha\pi}{2}\right)}\right] \quad (2)$$

The step response is expressed as

$$v_o(t) = 1 - E_{\alpha,1}(-\tau t^{\alpha}) \tag{3}$$

The impulse response is expressed

$$h(t) = \tau^{\alpha-1} E_{\alpha,\alpha}(-\tau t^{\alpha}) \tag{4}$$

$$|H(w)| = \frac{1}{\sqrt{1 + (w\mathrm{RC})^2}} \tag{5}$$

$$\emptyset(w) = -\tan^{-1}(w\mathrm{RC}) \tag{6}$$

Equations (5) and (6) represent the transfer function expressions for magnitude and phase of (integer) first-order low-pass filter.

3 Time Response of a Single-Stage Low-Pass Filter

In fractional order filter, the time domain response defines the different measuring parameters such as the transfer function, magnitude, phase, stability, peak overshoot (M_p), settling time (t_s) and rise time (t_r) [6, 14]. The step response and impulse response of single-stage fractional low-pass filter are plotted for the above parameters with different orders (α) of a fractional capacitor.

Settling time: It is the time required for the response of a system to be within the fraction of the steady-state value and to stay there.

Rise time: It is the time taken for the response of a system to go from 10% to 90% of the steady-state value at its first instant.

Peak time: It is the time taken for the response of a system to reach the peak or maximum value of overshoot.

The time at which the system's output attains its maximum value is expressed as [19],

$$t_p = \frac{0.131(\propto -0.255)^2}{(\propto -0.921)w_c} \tag{7}$$

The unit step response of single-stage fractional order low-pass filter is shown in Fig. 2, and its response is plotted against order α. This figure can be used for comparing the responses of fractional order and integer-order filters. A standardized value of time constant, i.e., RC = 1, is assumed for all simulations.

An impulse is a short duration signal that starts from zero to a maximum value and returns to zero again in a short time. The filter's impulse response is obtained by applying an impulse input and studied with order α of filter as shown in Fig. 3.

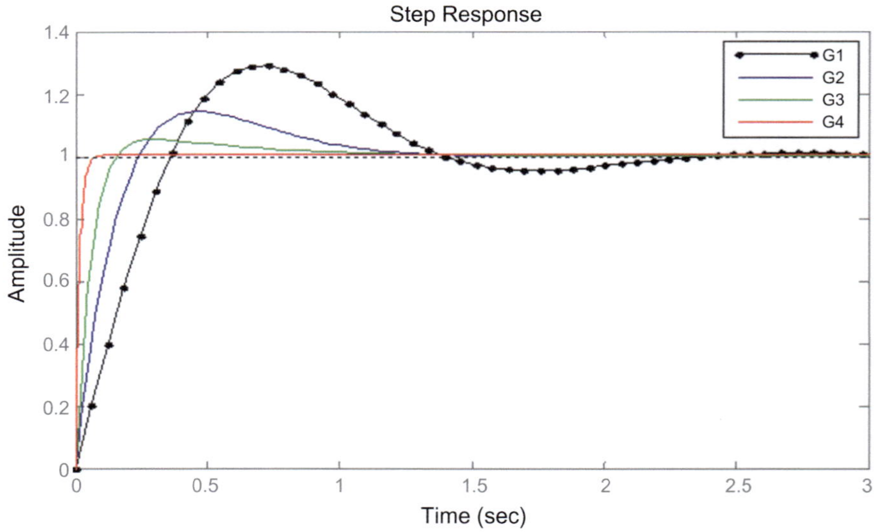

Fig. 2. Unit step response of single-stage low-pass filter where order **a** $G1 = 0.1$, **b** $G2 = 0.4$, **c** $G3 = 0.6$, **d** $G4 = 0.8$

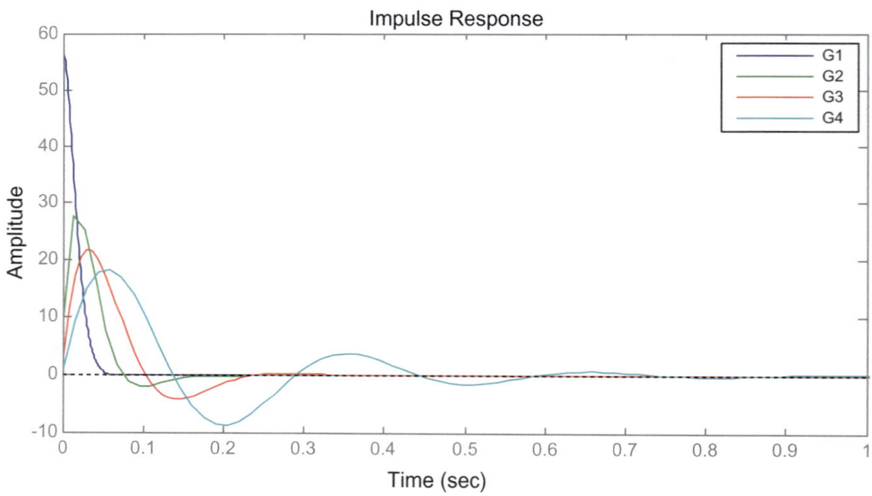

Fig. 3. Impulse response of a single-stage low-pass filter where order **a** $G1 = 0.1$, **b** $G2 = 0.4$, **c** $G3 = 0.6$, **d** $G4 = 0.8$

Furthermore when α varies from 0 to 1, the circuit's behavior approaches toward an equivalent second-order system as observed in Fig. 3. The different measuring specifications such as rise time, settling time, peak overshoot and peak time are given in Table 1 by changing order α of fractional capacitor.

Table 1. Time domain response specification of a fractional low-pass filter

α	Rise time	Settling time	Peak overshoot	Peak time
0.1	0.0331	0.0566	0.475	0.141
0.2	0.0483	0.0810	1.377	0.177
0.3	0.0709	0.4751	2.929	0.238
0.4	0.1021	0.7999	5.381	0.010
0.5	0.1361	1.0019	9.101	0.367
0.6	0.1780	1.1159	14.210	0.468
0.7	0.2257	1.2067	20.771	0.594
0.8	0.2781	2.1160	28.789	0.728
0.9	0.3351	3.4006	38.277	0.486
1	0.3968	5.1004	48.353	1.063

From Figs. 4, 5 and 6, the circuit's response like percent overshoot, rise time and settling time is studied for different values of α. So by using the results of figures shown below, an approximate fractional capacitor can be chosen to meet the desired parameters given in Table 1.

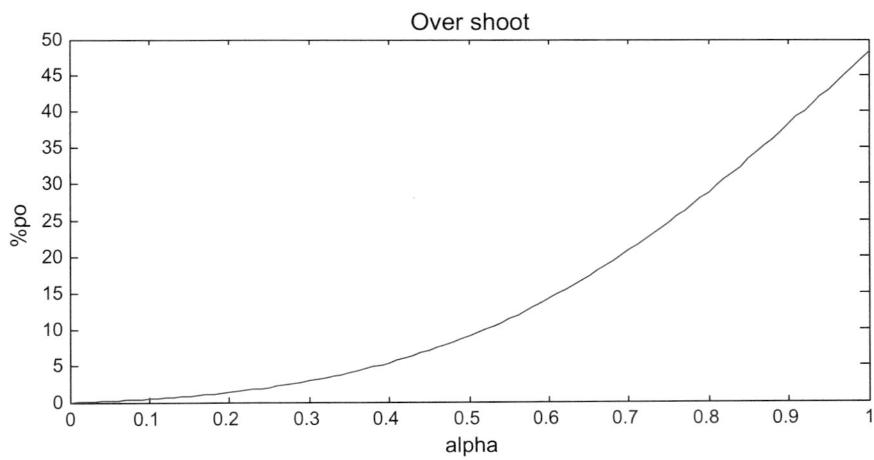

Fig. 4. Percent overshoot versus alpha (α) for single-stage fractional low-pass filter

Figure 4 shows the percentage overshoot versus order of the fractional capacitor α used in the low-pass filter. From Fig. 4, it is observed that overshoot decreases when order of the fractional capacitor α decreases. It is maximum in classical capacitor.

Figure 5 indicates the variation of settling time by varying the order of the fractional capacitor. The settling time is maximum at $\alpha = 1$, i.e., when normal capacitor is used. This shows the circuit output stabilizes quickly when fractional capacitor is used.

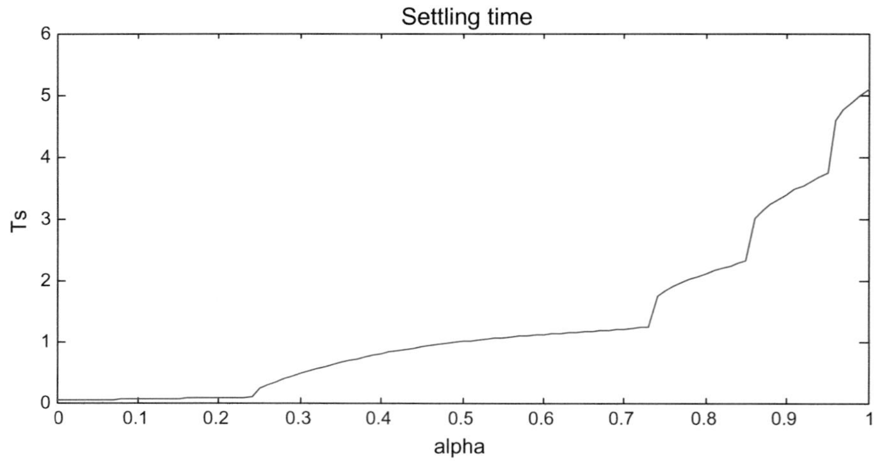

Fig. 5. Settling time versus alpha (α) for single-stage fractional low-pass filter

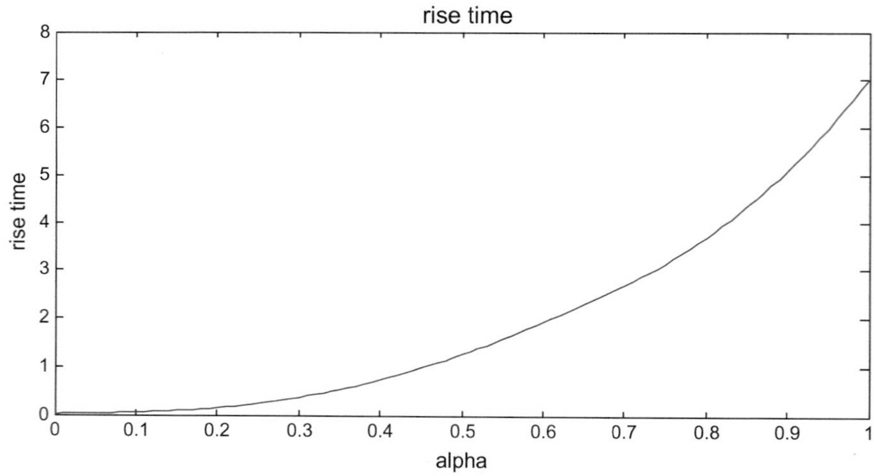

Fig. 6. Rise time versus alpha for single-stage fractional low-pass filter

Figure 6 shows rise time versus order of the fractional capacitor α. It is clear from the figure that rise time decreases when fractional capacitor replaces the classical capacitor. This indicates that fractional circuits have less delay and high speed, and thus the speed of response has low delay time.

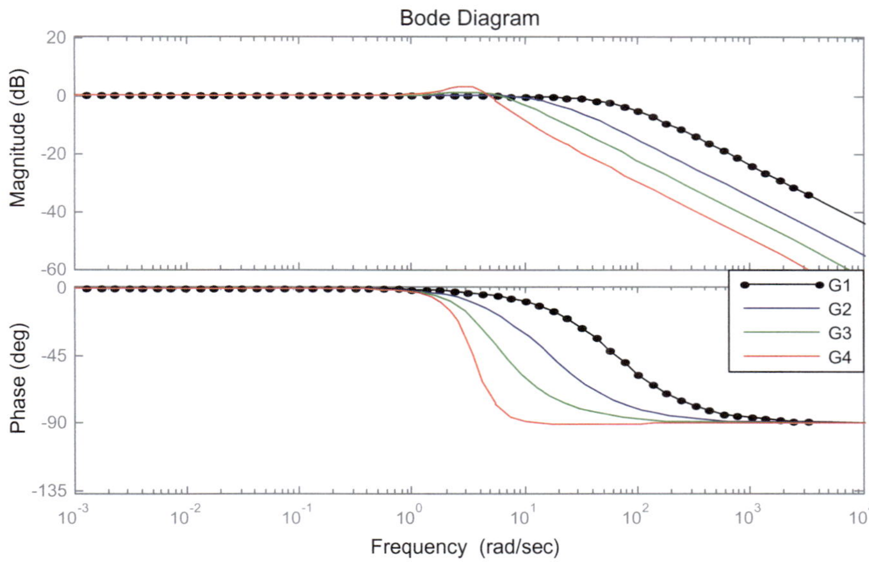

Fig. 7. Bode plot of single-stage fractional low-pass filter where order **a** $G1 = 0.1$, **b** $G2 = 0.4$, **c** $G3 = 0.6$, **d** $G4 = 0.8$

4 Frequency Response of a Single-Stage Low-Pass Filter

The Bode plots for magnitude and phase of a fractional order filter with order α are shown in Fig. 7; it is observed that the slope of magnitude plot in the stopband region varies proportionally to α. As observed from Fig. 7, the asymptotic approximation becomes inaccurate for very small values of α, whereas the asymptotic part in the high-frequency region has greater slope which increases with increase of α. Hence, the cutoff frequency becomes sharper, which leads to an increase in speed of response obtained in frequency domain response as shown in Fig. 7.

The Bode plot for phase is also shown in Fig. 7 which indicates for small value of α, the phase variation is almost linear over a narrow frequency range. From the figure, it is also observed that as α increases, the phase plot moves toward saturation state having asymptotic values of 0° and 90° for low and high frequencies, respectively.

5 Conclusion

In this article, the time domain and frequency domain responses of a RC low-pass fractional order filter are discussed. First, the fundamentals of magnitude and phase characteristics have been worked out as fundamental for the fractional order filter. Secondly, from the filter's response the amplitude–frequency characteristics, phase–frequency characteristics and cutoff frequency are analyzed in detail with respect to time constant τ and order α that shows better flexibility of the fractional order filter.

The filter behavior is studied and represented in figures by changing the order α. Finally, it is concluded that the speed of response increases with decrease in order α of a filter. It is also observed that the magnitude and phase change almost linearly with the fractional order in the region of stopband.

References

1. Debnath L (2003) Recent applications of fractional calculus to science and engineering. Int J Math Math Sci 54:3413–3442
2. Ortigueira MD (2008) An introduction to the fractional continuous-time linear systems, the 21st century systems. IEEE Circ Syst Mag 8(3):19–26
3. Elwakil AS (2010) Fractional-order circuits and systems: an emerging interdisciplinary research area. IEEE Circ Syst Mag 10(4):40–50
4. Machado JT (2013) Fractional generalization of memristor and higher order elements. Commun Nonlinear Sci Numer Simul 18:264–275
5. Moreles MA, Lainez R (2017) Mathematical modelling of fractional order circuit elements and bio-impedance applications. Commun Nonlinear Sci Numer Simul 46:81–88
6. Zahra WK, Hikal MM, Bahnasy TA (2017) Solutions of fractional order electrical circuits via Laplace transform and nonstandard finite difference method. J Egypt Math Soc 25:252–261
7. Sierociuk D, Podlubny I, Petráš I (2013) Experimental evidence of variable-order behavior of ladders and nested ladders. IEEE Trans Control Syst Technol 21:459–466
8. Walczak J, Jakubowska A (2015) Analysis of resonance phenomena in series RLC circuit with super capacitor. In: Analysis and simulation of electrical and computer systems, Lecture Notes in Electrical Engineering, vol 324, Springer, Heidelberg, pp 27–34
9. Rousan AA, Ayoub NY, Alzoubi FY, Khateeb H, Al-Qadi M, Hasan (Qaseer) MK, Albiss BA (2006) A fractional LC-RC circuit, Fractional calculus and applied analysis. Int J Theor Appl 9(1):33–41
10. Obeidad A, Gharaibeh M, Al-Ali M, Rousan A (2011) Evaluation of a current in resistor, Fractional calculus and applied analysis. Int. J Theor Appl 14(2):247–250
11. Gómez-Aguilar JF, Escobar-Jiménez RF, Olivares-Peregrino VH, Taneco-Hernández MA, Guerrero-Ramírez GV (2017) Electrical circuits RC and RL involving fractional operators with bi-order. Adv Mech Eng 9:1–10
12. El-Sayed MA, Nour HM, Raslan WE, El-Shazly ES (2012) Fractional parallel RLC circuit. Alexandria J Math 3:11–23
13. Radwan AG, Salama KN (2012) Fractional-order RC and RL circuits. Circ Syst Signal Process 31(6):1901–1915
14. Jakubowska A, Walczak J (2016) Analysis of the transient state in a series circuit of the class R-L_{β}-Cα. Circ Syst Signal Process 35:1831–1853
15. Hu Y, Luo Y, Lu Z (2008) Analytical solution of the linear fractional differential equation by Adomian decomposition method. J Comput Appl Math 215:220–229
16. Gorenflo RR, Kilbas AA, Mainardi F, Rogosin SV (2014) Mittag-Leffler functions, related topics and applications. Theory and applications. Springer, Heidelberg, New York, Dordrecht, London
17. Garrappa R (2015) Numerical evaluation of two and three parameter Mittag-Leffler functions. SIAM J Numer Anal 53(3):1350–1369
18. Garrappa R (2014) The Mittag–Leffler function, MATLAB Central File Exchange. File ID: 48154
19. Podlubny I (1999) Fractional differential equations. Academic Press, London

Chronic Disease Risk (CDR) Prediction in Biomedical Data Using Machine Learning Approach

Lambodar Jena[1(✉)], Soumen Nayak[1,2], and Ramakrushna Swain[3]

[1] Department of Computer Science and Engineering, Siksha O Anusandhan (Deemed to Be University), Bhubaneswar, India
lambodarjena@soa.ac.in
[2] Department of Computer Science and Engineering, Indian Institute of Technology (ISM), Dhanbad, India
[3] Department of Computer Science and Engineering, Silicon Institute of Technology, Bhubaneswar, India

1 Introduction

Day by day, large amount of data are collected due to the rapid growth in technology and science. Particularly in medical science, huge amount of medical databases are created to assist the healthcare system [1]. Now, a major field of research is management of large scale of data and efficiently extraction of knowledge from large databases [2]. In the recent era, machine learning has become an emerging field to be used on bioinformatics and complex studies of medical science. Large volumes of data are yet to be explored and mined to gather knowledge. The aim of this work is to process the chronic kidney disease dataset for prediction of chronic kidney disease of a large population. CKD is a major health issue in many societies because it gradually decreases the function of kidney. Based on the symptoms identified in patients, it predicts the sick ones [5]. In this work, genetic search algorithm is used for feature selection and three machine learning classifiers are used for classification. The classifiers are namely naive Bayes, J48, and decision table. For better prediction and diagnosis of chronic kidney disease, first the classifiers are used and compared with each other. Then, subsequently the features are reduced from the dataset by using genetic search algorithm and again classification results of each classifier are compared with each other.

2 Related Work

Many authors have implemented different machine learning techniques for classification and future prediction.

Dhamodharan [1] predicted risk in liver disease by using a number of machine learning classifiers, and he found that Naïve Bayes becomes suitable algorithm with high prediction accuracy. Solanki [2] has used J48 and random tree algorithms for the

M. N. Mohanty and S. Das (eds.), *Advances in Intelligent Computing and Communication*, Lecture Notes in Networks and Systems 109,
https://doi.org/10.1007/978-981-15-2774-6_29

classification of sickle cell disease and found that random tree performs better. Breast cancer diagnosis and prognosis have been done by Joshi et al. [3] using some rules of classification. They found that logistic model tree (LMT) algorithm identify more correctly (76%) the healthy and sick patients. However, the disease leukemia risk has been predicted by David et al. [4]. They have implemented a number of classifiers like KNN, J48, random tree, and Bayesian n/w and found that Bayesian performs well. Heart disease risk prediction has been done by Vijayarani and Sudha [5] by using different types of classifiers. Kumar [6] has used alternating decision trees for dengue fever diagnosis. In healthcare sector, Durairaj and Ranjani [7] have used and compared the efficiency of various types of data mining algorithms. They have worked on the prediction and diagnose of life killing diseases. Sugandhi et al. [8] used weka to analyze cataract patient's database. Yasodha and Kannan [9] analyzed diabetic patient database using different methods namely network, Bayes tree, J48, and random tree. Different classification techniques such as radial basis function, Bayes network, decision tree, and pruning algorithms are compared by Bin and Yau [10] for breast cancer. Jena et al. [11] worked on chronic kidney disease dataset by comparing different classification techniques such as SVM, naive, conjunctive rule, MLP, J48, and decision table. Also, Jena et al. [12] further worked on the prediction of kidney disease using different algorithms and feature selection techniques. Jena et al. [13] have also worked on the prediction of human depression using apriori algorithm.

3 Methodology

3.1 Preliminary

Classification comes under supervised learning technique. Large scales of population of records are classified by a classification model. There is a mapping of the data into predefined groups by the model. Based on the attribute value, new classes will be defined by classification model. A measure in terms of classification accuracy is considered. To carry out the experiments, we have obtained the biomedical dataset on chronic kidney disease with 400 instances and 24 attributes.

3.2 Classifiers Used

Classifiers are compared with each other to note the similarity and dissimilarity present in each parameter. In machine learning, classifiers are divided into various types:

- Bayes
- Functions
- Lazy
- Rules
- Tree-based classifiers, etc.

A balance mixture of classifiers, namely naive Bayes, J48, and decision table, from different groups has been chosen.

3.3 Performance Indicators

The five performance indicators namely accuracy, error rates such as mean absolute error (MAE) and root mean square error (RMSE), kappa statistics and ROC value are considered to measure the performance of all the classifiers.

4 Result and Discussion

The experiments have been conducted on the dataset using all mentioned classifiers. The result before applying feature selection procedure and the result after applying feature selection are recorded in Table 1. The result of feature selection based on chronic kidney disease dataset using genetic search algorithm is shown below. The attributes selected by genetic search after 100 iterations are 2, 3, 4, 6, 10, 12, 13, 14, 15, 16, 17, 19, 20, 22, 23, 24 = 16. That is, the selected attributes are {bp, sg, al, rbc, bgr, sc, sod, pot, hemo, pcv, wbcc, htn, dm, appet, pe, ane} from the dataset. The attributes which are reported by genetic search as irrelevant are neglected here. Taking relevant attributes selected by genetic search into consideration is the objective of our further study. That is, after attribute selection we have also applied all the three classification algorithms to see their behavior for various parameters shown in Table 1.

Table 1. Result of accuracy, kappa statistics, MAE, RMSE, and ROC

Experiment 1: Classification result before feature reduction					
Algorithm	**Accuracy (%)**	**Kappa statistics**	**MAE**	**RMSE**	**ROC**
Naïve Bayes	95.5	0.8962	0.0478	0.2047	1
J48	99	0.9786	0.0225	0.0807	0.999
Decision table	99	0.9786	0.1815	0.2507	0.992
Experiment 2: Classification result after feature reduction using genetic search					
Algorithm	**Accuracy (%)**	**Kappa statistics**	**MAE**	**RMSE**	**ROC**
Naïve Bayes	97	0.937	0.0334	0.1608	1
J48	99	0.9786	0.0225	0.0805	0.999
Decision table	98.75	0.9732	0.163	0.2331	0.99

5 Analysis of the Performance Indicators (Parameters)

5.1 Classification Accuracy

The accuracy tells that the estimated result is how much away from the actual value.

$$\text{Accuracy} = \frac{\text{TP} + \text{TN}}{P + N}$$

where
TP = True positive
TN = True negative

P = TP + FN (total positive)
N = FP + TN (total negative).

Any classifier with high accuracy will be more suitably chosen for prediction task. Accuracy result of the algorithm varies from dataset to dataset. The accuracy of an algorithm changes if we change the dataset. From experiment 1 in Table 1, it is found that J48 and decision table algorithm give the same accuracy result, i.e., 99%, whereas naïve Bayes classifier gives classification accuracy of 95%. All algorithms produce more than 90% of classification accuracy. Hence, it is concluded that J48 and decision table performs well for CKD dataset by taking all features of the dataset into consideration. However after feature selection using genetic search algorithm, it is found that J48 gives high accuracy result (= 99%) comparing to all other algorithms. Hence, it is concluded that J48 performs well in both the cases for the biomedical dataset to determine chronic kidney disease, which is shown in Fig. 1.

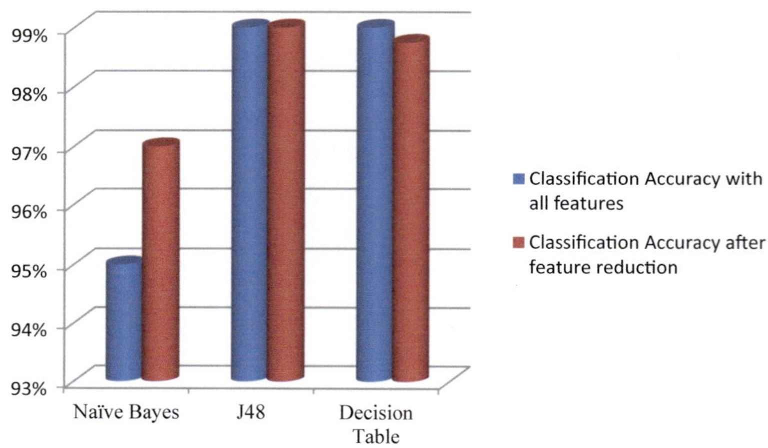

Fig. 1. Classification accuracy analysis

But from Table 1, it is observed that the accuracy of naive Bayes classifier algorithm is increased by 2%, whereas the accuracy of decision table is getting reduced mildly. Other algorithms, i.e., J48 have the same classification accuracy in both the experiments.

5.2 Kappa Statistics

In both the experiments, classification with J48 has better kappa statistics value with 0.9786 than other classifiers shown in Fig. 2. From Table 1, it is interesting to note that the values of kappa statistics, MAE, and RMSE for the classifiers are also changed. J48 and decision table has same kappa statistics value in comparison to naïve Bayes in the first experiment.

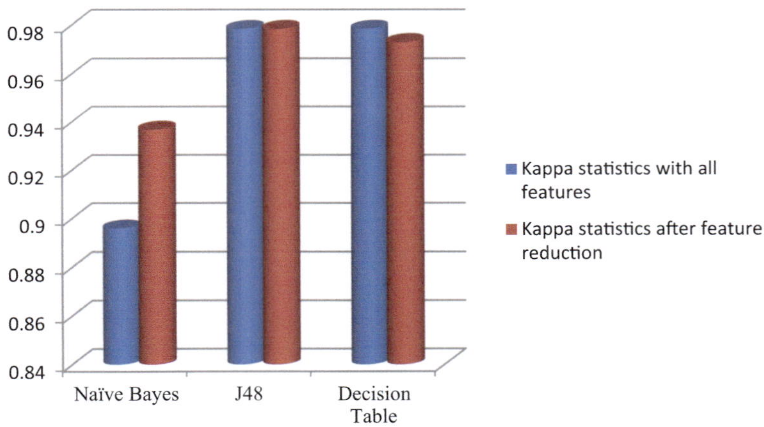

Fig. 2. Kappa statistics analysis

5.3 Error Rate in Terms of Mean Absolute Error (MAE)

The classifier J48 achieves minimum MAE value = 0.0225 that remains the same in both the experiments shown in Fig. 3. Thus in case of J48 classifier, the calculated result is closer to the true value of the chronic kidney disease criteria. But it is also observed from Table 1 that the MAE value varies slightly from experiment 1 to experiment 2 for naïve Bayes and decision tree algorithms.

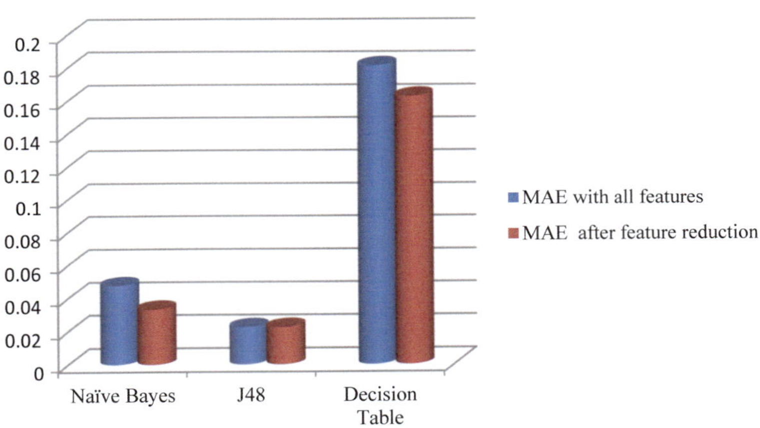

Fig. 3. Mean absolute error (MAE) analysis

5.4 Error Rate in Terms of Root Mean Square Error (RMSE)

From Table 1, it is found that J48 classifier results in minimum error rate (i.e., 0.0807 with all features and 0.0805 after feature reduction) compared to other algorithms, naïve Bayes and decision tree shown in Fig. 4. Here, also it is observed that in Table 1, the error rate (RMSE) value gets reduced in case of both naïve Bayes and decision table as the number of features is reduced.

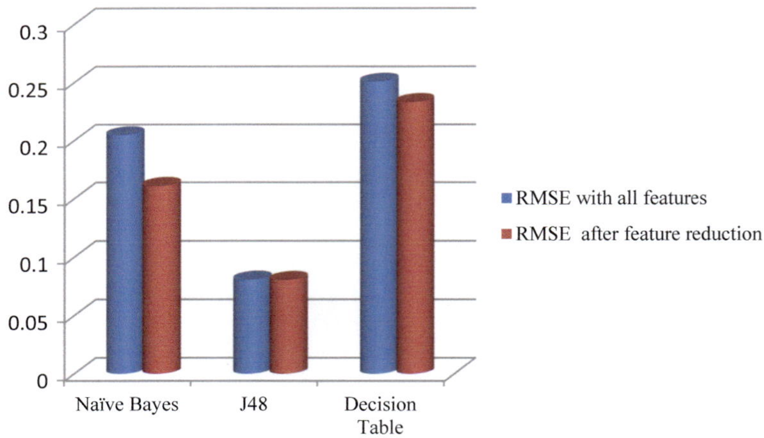

Fig. 4. RMSE analysis

5.5 ROC Value

From Fig. 5 and Table 1, it is seen that the ROC value of algorithms is approximately the same in both experiments except decision table algorithm. In decision table algorithm, the ROC value is slightly reduced as the numbers of features are reduced.

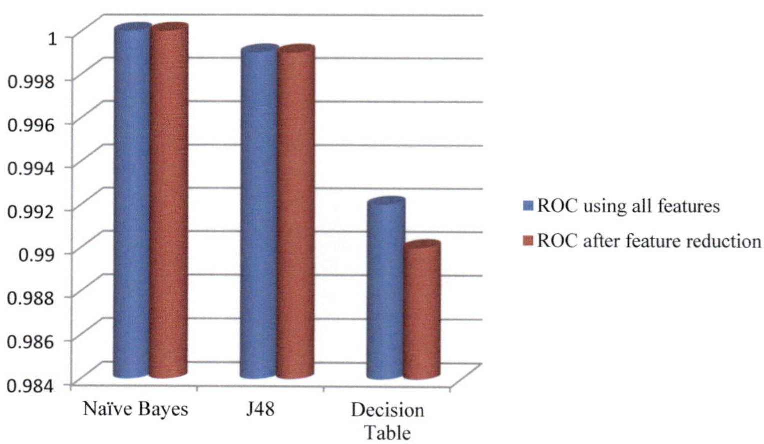

Fig. 5. Value of ROC

From Figs. 1, 2, 3, 4, and 5, it is observed that in all the cases, J48 classifier's performance is quite better than all other classifiers being used for our experiment.

6 Conclusion

The major focus of this work is to analyze the complex usage of the classifiers to predict the risk in chronic disease accurately. Here, three classifiers have been used to determine the target class accurately. They have been used in two typical scenarios; one with all attributes of the dataset and second after reducing the number of attributes. In the experiments, we observe that J48 algorithm performs better with high accuracy of 99% and hence gives better performance in prediction. But the prediction accuracy level of naïve Bayes and decision table is fluctuating with the increase of 2% in naïve Bayes and decrease of 0.25% in case of the later one.

References

1. Dhamodharan S (2014) Liver disease prediction using Bayesian classification, Special Issues. In: 4th National conference on advance computing, application technologies, May 2014
2. Solanki AV (2014) Data mining techniques using WEKA classification for sickle cell disease. Int J Comput Sci Inf Technol 5(4):5857–5860
3. Joshi J, Rinal D, Patel J (2014) Diagnosis and prognosis of breast cancer using classification rules. Int J Eng Res General Sci 2(6):315–323
4. David SK, Saeb AT, Al Rubeaan K (2013) Comparative analysis of data mining tools and classification techniques using WEKA in medical bioinformatics. Comput Eng Intell Syst 4 (13):28–38
5. Vijayarani S, Sudha S (2013) Comparative analysis of classification function techniques for heart disease prediction. Int J Innov Res Comput Commun Eng 1(3):735–741
6. Kumar MN (2013) Alternating decision trees for early diagnosis of dengue fever. arXiv preprint arXiv:1305.7331
7. Durairaj M, Ranjani V (2013) Data mining applications in healthcare sector a study. Int J Sci Technol Res IJSTR 2(10)
8. Sugandhi C, Ysodha P, Kannan M (2011) Analysis of a population of cataract patient database in WEKA tool. Int J Sci Eng Res 2(10), October 2011
9. Yasodha P, Kannan M (2011) Analysis of population of diabetic patient database in WEKA tool. Int J Sci Eng Res 2(5), May 2011
10. Bin Othman MF, Yau TMS (2007) Comparison of different classification techniques using WEKA for breast cancer. In: 3rd Kuala Lumpur international conference on biomedical engineering 2006. Springer, Heidelberg, pp 520–523, Jan 2007
11. Jena L, Kamila NK (2015) Distributed data mining classification algorithms for prediction of chronic- kidney-disease. Int J Emerg Res Manage Technol 4(11):110–118. ISSN 2278-9359

12. Jena L, Swain R (2017) Chronic disease risk prediction using distributed machine learning classifiers. In: Proceedings of IEEE International Conference on Information Technology (ICIT)-2017. doi:https://doi.org/10.1109/icit.2017.46, 978-1-5386-2924-6/17 © 2017 IEEE
13. Jena L, Kamila NK (2014) A model for prediction of human depression using apriori algorithm. In: IEEE Xplore Digital Library proceedings, pp 240–244. ISBN 978-1-4799-4152-0

A Novel Approach to Detection of and Protection from Sybil Attack in VANET

Binod Kumar Pattanayak[1(✉)], Omkar Pattnaik[2], and Sasmita Pani[2]

[1] Department of Computer Science and Engineering, Institute of Technical Education and Research, Siksha 'O' Anusandhan Deemed to be University, Bhubaneswar, Odisha 751030, India
binodpattanayak@soa.ac.in

[2] Department of Computer Science and Engineering, Government College of Engineering, Keonjhar, Odisha 758002, India

1 Introduction

The research community nowadays focus on the concept of mobile ad hoc network (MANET) which is implemented in road and transport communication where vehicles act as nodes, and it forms a vehicular ad hoc network (VANET). It is also regarded as a subset of MANET. Vehicular ad hoc network (VANET) was created in October 2002 by the Federal Communications Commission (FCC). The main objective of this concept is to provide safety on roads and secure driving. To create a VANET environment, some parameters are necessarily required like onboard unit (OBU) and a roadside unit (RSU) that must be installed along the roads. Dedicated short-range communications (DSRC) protocol is used to communicate between OBU and RSU. Again this protocol is modified later, and a new protocol is formed called as wireless access in vehicular environment (WAVE). It is based on IEEE 802.11p group. The architecture of VANET supports vehicle-to-vehicle (V2V) and vehicle-to-infrastructure (V2I) interchanges associated with RSU and OBU.

Security is necessarily required for VANET due to its wireless communications. So, there are different types of attacks possible in VANET. **Sybil attack** is one of the harmful attacks among them. In this attack, multiple identities are created by the attacker, thereby trying to forge or steal new identities from the neighboring vehicles. The attacker performs various malicious operations with the help of Sybil attack. An illusion is created by the attacker by sending messages to different vehicles from different identities. So, there occurs a jamming in the network which enforces the driver of vehicles to divert from the original route.

Generally, it affects two major areas like routing and voting and reputation systems as well. It may appear as a malicious node at more than one locations at a time in geographical routing protocols [1]. Also, it may disrupt the cluster head and disrupt the valuable information in cluster-based routing protocols [2]. But in the second case, a voting process is performed and the vehicle's exact location is monitored and the misbehaving node is identified. When a malicious vehicle performs Sybil attack, all messages come from a particular location that circulates the false information in the entire network. So, the network may be unable to communicate with neighbor nodes.

M. N. Mohanty and S. Das (eds.), *Advances in Intelligent Computing and Communication*, Lecture Notes in Networks and Systems 109,
https://doi.org/10.1007/978-981-15-2774-6_30

The rest of the paper is organized as follows. Section 2 describes the attack model along with the assumptions. Section 3 covers the way Sybil attack works. Section 4 includes the methods of protection from Sybil attack in VANET. Section 5 concludes the paper.

2 Attack Model and Assumptions

In this part, the Sybil attack model is defined, thereby presenting some assumptions on the system and the attacker's probable actions.

Attack Model:

In wireless networks, the new neighboring nodes are discovered by the mobile nodes to whom the beacon packets are periodically broadcast in the network. In VANET, the vehicle is to be considered as a node. Sybil attack is based on the idea of multiple identities being produced by a malicious node [3]. Also, the identities and positions are determined by the beacon packets. A malicious node has the potential to create multiple identities without the knowledge of legitimate nodes. A malicious driver may have additional identity information because either he has borrowed identity from another driver or might be stealing identity from neighbors. The main objective of Sybil attack detection is that only one identity is assigned to each physical node [4].

In this paper, we assume a malicious node to be a physical node which produces multiple identities, and correspondingly, the Sybil nodes are the fabricated identities produced by the malicious node.

Assumptions:

The assumptions are focused here on the basis of Sybil attack detection and attacker's actions.

Attack Detection:

In this paper, we have assumed the following assumptions about detection of Sybil attack in VANET. (i) The Sybil attack is launched by the individual drivers (vehicles), which is the basic threat to the system. We also assume that the Sybil attack may be created by multiple malicious vehicles. (ii) The RSUs are deployed separately along the road, and to maintain the authentication of infrastructure, an electronic license plate (ELP) is issued to whole network [5]. (iii) We focus on the same radio module called received signal strength indicator (RSSI) such as DSRC which is based on radio-frequency (RF) communication strategy [6]. (iv) Finally, we assume that all the vehicles must be equipped with GPS devices along with digital maps since the accurate position is determined by GPS technology. But in real-world scenario, we cannot rely on roadside base stations present in RSU for the whole network.

Attackers' Actions:

The following are the possible actions conducted by the attacker.

(i) Compromised RSBs: The attacker may be successful in compromising the RSBs (part of RSU) which are a semi-trusted parties. The detection of compromised RSBs can be canceled by the DMV (Department of Motor Vehicle). But, there is still a chance for the attacker to retrieve the information stored in the RSBs. So, the attacker uses its technique to recover the amount of information quickly from the RSBs.

(ii) Announce false messages: It is also regarded *as* false data injection. The false message can be signed by a vehicle before it is broadcast in the network. The message can also be signed by a pseudonym that is certified by CA. So, it is impossible to detect the message. When the number of vehicles is more than the number of attackers in a network, the voting scheme is used. But, if the false identities are generated sufficiently by the attacker, then the Sybil attack occurs.

(iii) Sybil attack: It refers to a situation wherein the attacker may pretend to be multiple vehicles by virtue of using multiple pseudonyms. When multiple pseudonyms are used to sign the message by a vehicle, the Sybil attack is formed in VANET. The privacy of a vehicle is compromised when the vehicle is identified from a group of pseudonyms it uses. Therefore, RSBs and vehicles tend to sign multiple messages in attacker's pseudonyms [7].

3 How Sybil Attack Works

The concept of Sybil attack was first introduced by Douseur [8]. This is a threat to functionality of VANET. The attacker uses multiple identities in order for sending messages to other nodes that belong to the network. The identities of other nodes may be spoofed by a node called as malicious node that is also regarded as the Sybil attacker. Sybil nodes are those nodes whose identities are spoofed by the Sybil attacker [9]. An illusion is created in the network, wherein message is sent to the vehicles that an accident or a traffic jam has already taken place, and thus, other vehicles are forced to change their routes and tend to move along the attacker's route. Also, the false information may be injected by the Sybil attacker into the network. The attacker also tries to disrupt the warning message sent by the actual vehicle in the network and fulfill its target. So, this may put the life of passengers in danger [10].

The different types of possible Sybil attacks are given in Fig. 1.

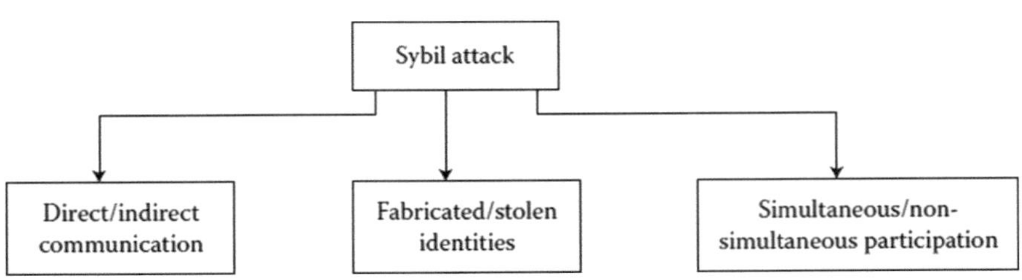

Fig. 1. Different forms of Sybil attack

In Fig. 1, it is clear that Sybil attack classification is carried out on the basis of the type of communication, identity, and their participation in the network [11]. These features are explained clearly in the following paragraphs.

(i) **Communication category**: When a Sybil node receives a radio message from an honest node, the message may be listened by a malicious node present in the network. Along the same path, the malicious devices actually send messages but not the Sybil nodes. In direct communication, the legitimate nodes communicate with all the Sybil nodes which are created by malicious devices. But in indirect communication, the malicious node helps the legitimate nodes to enter the area of Sybil nodes.
(ii) **Identity category**: In this category, a new Sybil identity is created by an attacker and that is generated randomly with a 32-bit integer which is a fabricated identity or the identities of neighboring nodes are stolen by the attacker.
(iii) **Participation category**: The malicious nodes create multiple Sybil identities which can participate simultaneously in the attack. In this category at a time, one identity is used for each node, and thus, an observation takes place in the network that many times a particular node may join or leave the network.

4 Protection Methods for Sybil Attack in VANET

In VANET, the protection mechanism can be classified into three different types that are: (i) public key cryptography method, (ii) methods based on position verification, and (iii) resource testing method.

Public Key Cryptography Method:

Public key cryptography is a standard method to detect Sybil attack, and also, it provides authentication mechanism [12]. The concept of a symmetric cryptography is used to develop the security solution which is based on the combination of digital certificates with signatures.

In this technique, a CA (central authority) is present that is responsible for issuing the certificates, thereby maintaining the hierarchy. A secure channel is used to communicate between CAs and monitor the issued certificates which are used by every signed message. In VANET, the signed information sent by vehicles may be verified by the receiver as demonstrated in Fig. 2. In this technique, it is verified if a message is valid or invalid. The messages which are attached with certificates are to be considered as valid ones, and the messages having no certificates are ignored. So, the Sybil attacks can be prevented. At a time only one certificate is assigned to each node, and to maintain better privacy, it is necessary to change the certificates from time to time. But in VANET, due to rapid changes in mobility, it is difficult to implement PKI. But to maintain privacy, the key pair (public/private) must be authenticated by the CA. But, it may not be useful to check the identity of a vehicle [13].

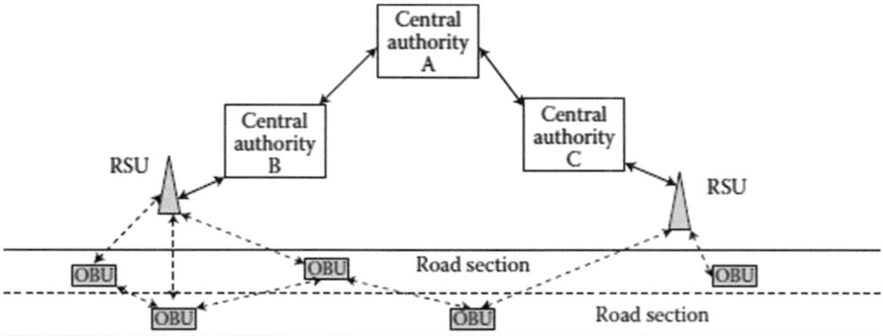

Fig. 2. Hierarchy of central authority

Position Verification Method:

In vehicular environment, the concept of sensors, GPS has also important role in creating a node (vehicle). It may also be possible to tamper the sensors on the part of an attacker and send false information to other vehicles present in the network. To achieve effective moving of vehicles in the network accurate positions of vehicle are essentially required. So it is an important task for all the nodes to maintain accurate position in the network. This is very much important for geographic routing. In this routing scheme, accurate positions of all nodes are essential because intermediate nodes are responsible for transmitting messages to the destination based on their locations.

The method of verification of a secure position was first addressed by the authors in [14], called as **verifiable multilateration** approach. To fix the position of the vehicle, a trustworthy network is built by the base stations which measures the sending and receiving information sequentially in time. In this approach, a node neither delays the reply time nor advances the time. So, the nearer distance is fixed for communication.

The second proposal is given by authors in [15] where the concept of threshold values is used to determine the correct position of the node by receiving information from other nodes. In this technique, nodes rely on GPS rather than base station scheme to determine their positions. Beacon messages are used for sending and receiving location information of nodes and also for verifying it through specified checks.

There are 4 types of check strategies based on the above concept, which are given below:

- Acceptance range threshold;
- Mobility grade threshold;
- Maximum density threshold;
- Map-based technique.

In **Acceptance Range Threshold**, the verified node obtains a maximum threshold value which determines their communication ranges in the network. So, it helps to detect wrong positions of nodes.

In **Mobility Grade Threshold**, the last node position is known, and thus, if any false information about changing of position is detected, then an extreme speed is defined for a hub. In order to communicate among the nodes, signal messages are used.

The idea about **Maximum Density Threshold** is to accommodate a specific region where maximum number of vehicles can be physically present.

In map-based technique, the correct location of the vehicle is to be determined.

In order to detect Sybil attack based on the position, the parameters like collision avoidance, resource availability, emergency alert, and traffic conditions can be useful. A VANET may be affected by inserted bogus packets, dropping packets, replying packets, and modifying existing packets which are created by the malicious attacker [15, 16].

There are two efficient schemes for localization like spectrum-based and spectrum-free techniques. In spectrum-based technique, in order to locate vehicle's position, the estimated distance between a receiver and transmitter must be calculated. Some of the techniques are: time of arrival (TOA), angle of arrival (AOA), and received signal strength indicator (RSSI). But distance-free localization technique helps in gathering other position information. In order to achieve more accuracy in localization, distance-based method must be adopted [17].

According to Yan et al., a new method has been developed for Sybil attack in which vehicles can listen GPS coordinates of other vehicles and also visualize the neighbor's vehicle with the help of radar transmission range. It is also called as a virtual vehicle eye. Therefore, it is easy to track the actual location of neighboring vehicle and malicious vehicles are separated from others [18].

The limitation of this method is that when an existing vehicle position is acquired by a target vehicle, then they are in the radar network of established vehicle [19, 20]. Such a scenario is depicted in Fig. 3.

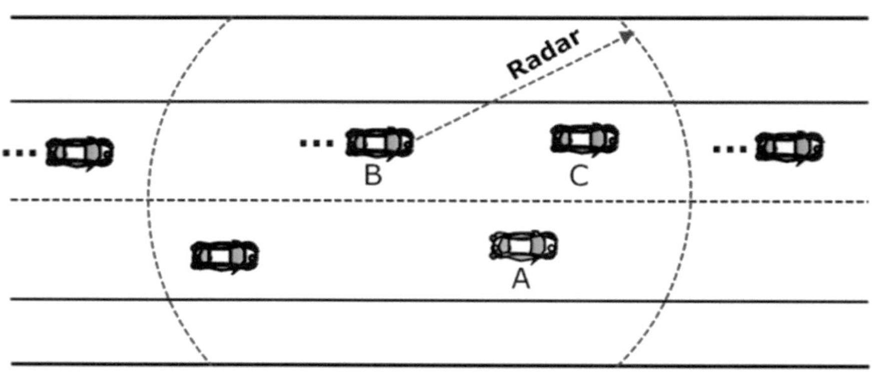

Fig. 3. A potential Sybil attack

As shown in Fig. 3, A gets *C*'s position to be Lc. *A* claims to have victimized node *B* that its position is Lc, and its ID is IDa. *B* recognizes a vehicle that is located at Lc and then concludes that it is the location of *A* [15].

Resource Testing Method:

In this technique, assumption is generated in user's mind for limited resources of each physical entity. The concept of radio resource testing specifies that a message is broadcast by a node to all its neighbor nodes and tries to listen the response message by channel which is selected randomly.

The nodes which do not response the broadcast message positively in the same channel are assumed as Sybil entities, and the Sybil attack is detected. Otherwise the neighbor node response in the same channel [21].

The resources which can be shared between malicious vehicles and Sybil entities are memory, IP, and computational recourses. So it is possible to generate for Sybil attack conditions.

According to researcher opinion, this approach may not be suitable for VANET. The malignant vehicle can produce numerous integrity which does not match with the identity of the vehicle present in the network. So they are registered separately in other list [17].

The main objective of resource testing method is to detect the Sybil attack and check the fake identities from the attacker. But in practical scenario to obtain sufficient fake identities for a attacker is not a difficult task. So this method may not be implemented in VANET [20, 22].

5 Conclusion and Future Scope

In this paper, we propose a method to detect and a protection mechanism for Sybil attack in vehicular communication. Also, we have assumed the attacker's action in different scenarios. In this case, we prefer position verification method as an efficient technique to protect from Sybil attack in real-world environment. Extensive work is still required in the future. Also, we focus on the acceptance range threshold technique of position verification method to detect the wrong position of more number of Sybil nodes in VANET. In future, the performance of our method can be examined using NS2 and traffic simulators called SUMO and MOVE for real-world environment.

References

1. Karlof C, Wagner D (2003) Secure routing in wireless sensor networks: attacks and countermeasures. Ad hoc Netw J 1:293–315
2. Sood M, Vasudeva A (2013) Perspectives of Sybil attack in routing protocols of mobile ad hoc network. Comput Netw Commun (NetCom) 131:3–13
3. Newsome J, Shi E, Song D, Perrig A (2004) The Sybil attack in sensor networks: analysis & defenses. In: Third international symposium on information processing in sensor networks, IPSN 2004, pp 259–268
4. Pouyan AA, Alimohammadi M (2014) Sybil attack detection in vehicular networks. Comput Sci Inf Technol 2(4):197–202

5. Hubaux JP, Capkun S, Luo J (2004) The security and privacy of smart vehicles. IEEE Secur Priv Mag 2(3):49–55
6. 5.9 GHz DSRC. http://grouper.ieee.org/groups/scc32/dsrc/index.html
7. Yu B, Xu C-Z, Xiao B (2013) Detecting Sybil attacks in VANETs. J Parallel Distrib Comput. Available online 9 Feb 2013
8. Douceur JR (2002) The Sybil attack. In Proceedings of the international workshop on peer to peer systems, March 2002, pp 251–260
9. Saini M, Kumar K, Bhatnagar KV (2013) Efficient and feasible methods to detect Sybil attack in VANET. Int J Eng Res Technol 6:431–440
10. Grover J, Gaur MS, Laxmi V (2011) Sybil attack in VANETs detection and prevention. Published in security of self-organizing networks. Printed in the United States of America on acid-free paper-10 in 2011
11. Newsome J, Shi E, Song D, Perrig A (2004) The Sybil attack in sensor networks: analysis and defenses. In: Proceedings of international symposium on information processing in sensor networks, April 2004, pp 259–268
12. Raya M, Hubaux JP (2007) Securing vehicular ad hoc networks. J Comput Secur 39–68
13. Khalili A, Katz J, Arbaugh WA (2003) Towards secure key distribution in truly ad-hoc networks. In: Proceedings of IEEE workshop on security and assurance in ad-hoc networks
14. Martucci LA, Kohlweiss M, Anderson C, Panchenko A (2008) Self-certified Sybil-free pseudonyms. In: WiSec'08: proceedings of the first ACM conference on wireless network security, New York, NY, USA, ACM Press, pp 154–159
15. Yan G, Olariu S, Weigle MC (2008) Providing VANET security through active position detection. Comput Commun 31(12):2883–2897
16. Pathre A, Agrawal C, Jain A (2013) Identification of malicious vehicle in vanet environment. J Glob Res Comput Sci 4:30–34
17. Isaac T, Zeadally S, Camara JS (2010) Security attacks and solutions for vehicular ad hoc networks. Commun IET 4(7)
18. Samara G, A-Salihy WAH, Sures R (2010) Security Analysis of Vehicular Ad Hoc Networks (VANET). In: IEEE second international conference on network applications, protocols and services
19. Ibrahim K (2011) Data aggregation and dissemination in vehicular ad-hoc networks. Doctoral dissertation, Old Dominion University, Norfolk, Virginia
20. Rahbari M, Jamali MAJ (2011) Efficient detection of Sybil attack based on cryptography in VANET. Int J Netw Secur Its Appl (IJNSA) 3:185–195
21. Zhou T, Choudhury RR, Ning P, Chakrabarty K (2011) P^2 DAP-Sybil attacks detection in vehicular ad hoc networks. IEEE J. Sel Areas Commun 29(3)
22. Levine BN, Shields C, Margolin NB (2006) A survey of solutions to the Sybil attack. University of Massachusetts, Amherst
23. Hao Y, Tang J, Cheng Y (2011) Cooperative Sybil attack detection for position based applications in privacy preserved VANETs. In: IEEE communications society in the IEEE Globecom 2011 proceedings

Design of Monochromatic Photonic Filter in Near-Infrared Region Using Plane Wave Expansion Method

K. P. Swain[1(✉)], Sangram Kishore Mohanty[2], P. S. Das[1], and G. Palai[1]

[1] Department of Electronics and Communication Engineering, GITA, Bhubaneswar, India
[2] Department of ECE, ICE, Bhubaneswar, India

1 Introduction

PWE is customarily used technique in the photonics fraternity to solve Maxwell equation for finding the dispersion relation. Generally, it uses eigenvalue problem to solve the Maxwell equation. Many literatures [1–10] implement this method, especially in the field of photonics. In [1], PWE is used to compute the band structure by considering material dispersion, whereas paraxial beam propagation is demonstrated in [2] using PWE. Photonic band gap (PBG) is calculated using PWE method in [3–5] to design a 2-D band-pass filter for C-band WDM application, band-pass filter using dual-mode square lattice and band-pass filter using dual square ring, respectively. Using PWE, a detail calculation of defect mode in PCF is proposed in [6] and in [7], a detail investigation is proposed using PWE to design add/drop filter. A PWE-based model for laser beam propagation is also proposed in [8] in anisotropic medium to calculate the optical field distribution and in [9], PWE is reported to study the electric field distribution in MSOF, whereas doping concentration is investigated in [10] in an optical thyristor.

Photonics is nowadays found many applications to design the different kinds of filters as described in [11–13]. An agile RF photonic notch filter is demonstrated experimentally in [11], where silicon Bragg grating waveguide is implemented. A tunable photonic single notch filter in microwave region is proposed and demonstrated experimentally in [12] which is based on phase modulation technique and in [13], tunable photonic band-pass filter based on Brillouin is presented.

Monochromatic photonic filters with some real-life application are designed and implemented in [14–17] where silicon grating structure is used as background material. PWE simulation method is carried out to study the reflectance curve of the filter. In the present work, monochromatic filters are designed in near-infrared region using PWE simulation method, where the width of SiO and Si decide the allowable signal wavelength. Simulations are made for the three commonly used ranges like 800–900, 1300–1400 and 1500–1600 nm and a relation is also established between the width of the grating materials and the allowable signal wavelength.

M. N. Mohanty and S. Das (eds.), *Advances in Intelligent Computing and Communication*, Lecture Notes in Networks and Systems 109,
https://doi.org/10.1007/978-981-15-2774-6_31

2 Proposed Grating Structure

The design of monochromatic photonic filter using silicon grating structure is shown in the figure where SiO and Si materials are used alternatively to form total 68 layers with a defect at 18th and 52nd position. Defect or imperfection in material science plays a vital role to change the physical and mechanical property in the crystalline materials. Here, defect is introduced deliberately as a line defect to obtain the characteristics of an ideal filter. Here, air replaces Si material at the 18th and 52nd layer whose refractive index is 1. In the proposed structure, the combination of refractive index of the materials (SiO and Si), the width of the materials (taken as t′ for SiO and t″ for Si) and overall grating length decide the allowable signal wavelength in the range. Helmholtz equation which is the representation of Maxwell equation is used to calculate the reflectance of grating structure. The final equation can be obtained by solving the Helmholtz equation with help of Crammer's rule as given below (Fig. 1).

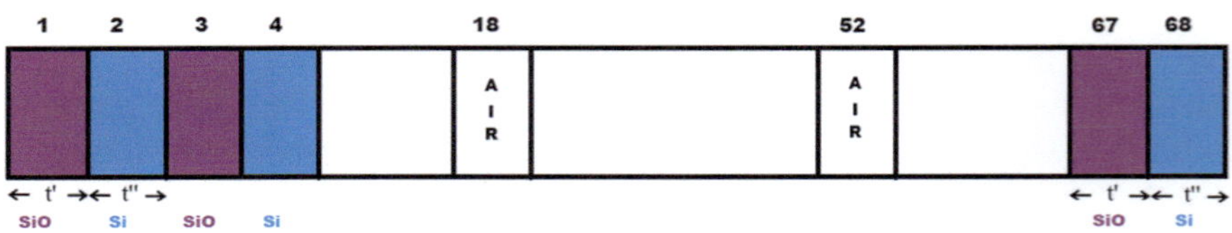

Fig. 1. Grating structure of proposed structure

$$\lambda = 2\left(n' \cdot t' + n'' \cdot t''\right)$$

where

n' Refractive index of SiO (taken as 1.913)
n'' Refractive index of Si (taken as 3.6726)
t' Width of SiO (varies depending upon the allowable wavelength)
t'' Width of Si (varies depending upon the allowable wavelength).

3 Simulation Result and Discussion

PWE simulations are carried out to get the reflectance characteristics of monochromatic filters for NIR. Basically, simulations are implemented maximum times for three ranges like 800–900, 1300–1400 and 1500–1600 nm but only two reflectance curve are displayed in Figs. 2, 3 and 4 for each range.

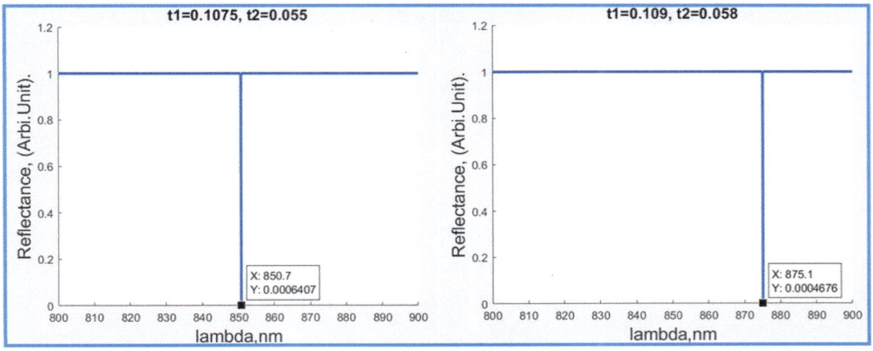

Fig. 2. Reflectance characteristics curve for the allowable signal wavelength 850.7 and 875.1 nm

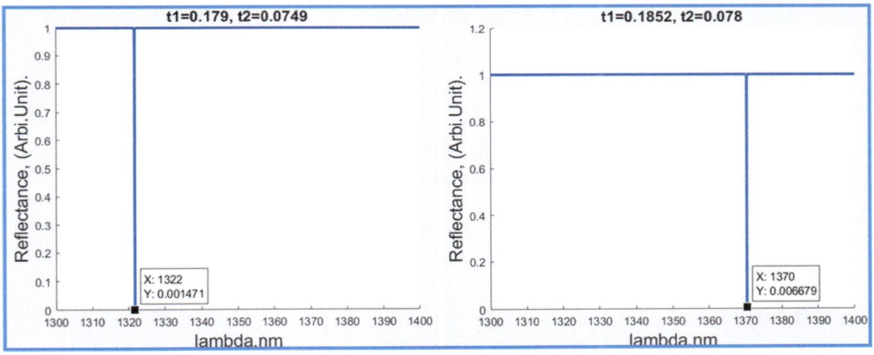

Fig. 3. Reflectance characteristics curve for the allowable signal wavelength 1322 and 1370 nm

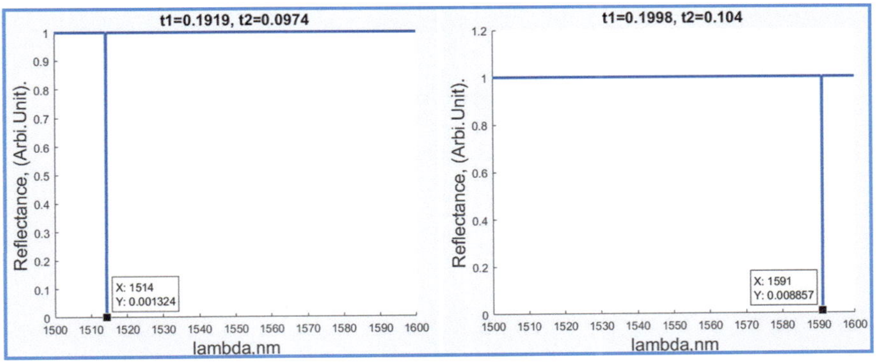

Fig. 4. Reflectance characteristics curve for the allowable signal wavelength 1514 and 1591 nm

Considering the reflectance characteristics curve (Fig. 2) for the allowable wavelength of 850.7 nm and 875.1 nm in the 800–900 nm range, the reflectance curve is approached to zero for the allowable signal 850.7 nm and 875.1 nm, whereas one for rest part of the signal in the range. So, it clear that this particular design is used to allow only 850.7 nm for the width of SiO (t') is equal to 0.1075 μm and Si (t'') is equal to 0.055 μm while rejecting completely other signals within the range. Similarly, Figs. 3 and 4 are representing the reflectance characteristics of 1322 nm and 1370 nm in the range 1300–1400 nm and 1514 nm and 1591 nm in the range 1500–1600 nm range, respectively. In the entire reflectance curves, which are obtained from PWE simulations, *X*-axis represents the range of the signals in nanometer (nm), whereas reflectance is represented in Arb. Units by *Y*-axis.

After getting all the simulation results for each range, one interesting facts is revealed that the width of the SiO and Si is varied almost linearly with the increasing allowable signal wavelength in the range as shown in Fig. 5, 6 and 7. So, any researcher can easily find the width parameters of SiO and Si by applying simple mathematics or any machine learning algorithm.

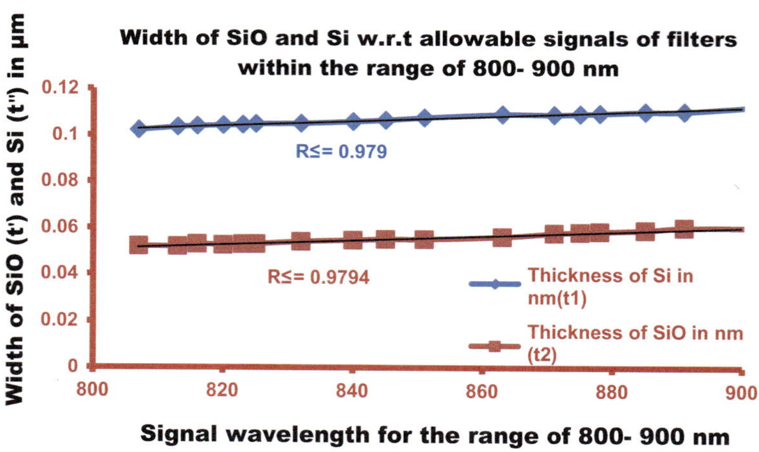

Fig. 5. Width of SiO and Si w.r.t. allowable signal of range 800–900 nm

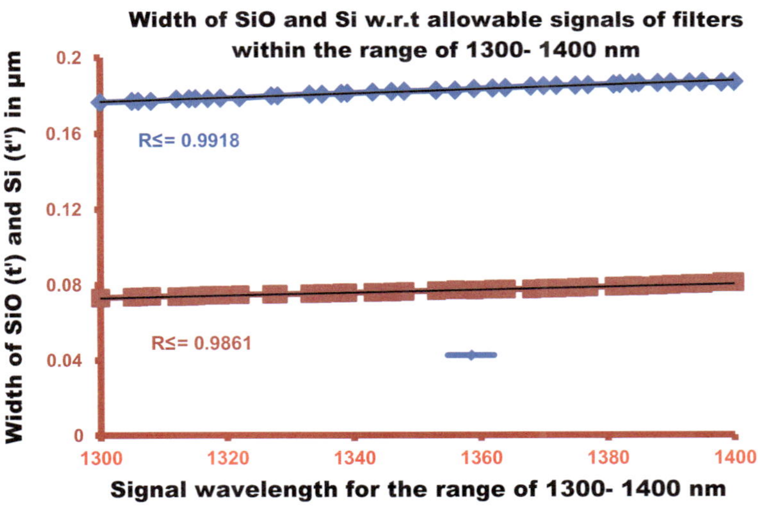

Fig. 6. Width of SiO and Si w.r.t. allowable signal of range 1300–1400 nm

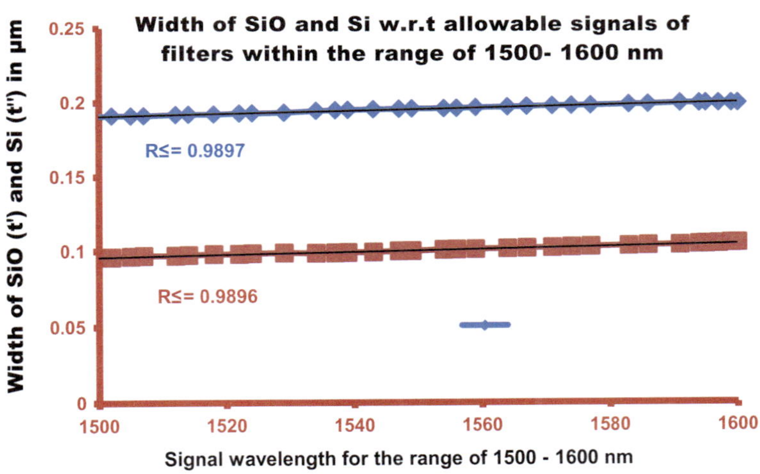

Fig. 7. Width of SiO and Si w.r.t. allowable signal of range 1500–1600 nm

4 Conclusion

Monochromatic photonic filters in the near-infrared region are successfully designed using PWE simulation method, where the width of waveguide structure (SiO and Si) plays a vital role to select the allowable signal. Aside this, the allowed or disallowed by the waveguide is lucidly investigated through the analysis of the reflectance characteristics pertaining to the wavelength of the signal. It is also divulged that the increasing allowable signal wavelength is varied almost linearly with the width of the waveguide structures.

References

1. Guryev IV, Sukhoivanov IA (2007) Plane wave expansion method with considered material dispersion. In: 2007 9th international conference-The Experience of Designing and Applications of CAD Systems in Microelectronics (CADSM). IEEE, pp 23–24
2. Shuklal BK, Patel RH (2008) Simulation of paraxial beam propagation using plane wave expansion method. In: 2008 International Conference on Recent Advances in Microwave Theory and Applications (ICM). IEEE, pp 652–656
3. Robinson S, Nakkeeran R (2011) Two dimensional photonic crystal ring resonator based bandpass filter for C-band of CWDM applications. In: 2011 National Conference on Communications (NCC). IEEE
4. Kamiji Y, Nagaoka N, Chun-Ping C, Anada T, Jui-Pang H (2012) A novel photonic crystal bandpass filter using degenerate modes of point-defect microcavity for terahertz communication system. In: 2012 Asia Pacific Microwave Conference Proceedings (APMC). IEEE, pp 583–585
5. Sharma A, Inaniya PK (2015) Dual square ring with 3 × 3 dielectric rods structure based band pass filter using two dimensional photonic crystal. In: International Conference on Computer Science and Network Technology (ICCSNT). IEEE pp 45–49
6. Zhu Y, Zhuang Y, Shi X (2014) An improved algorithm of photonic crystal fibers' defect mode based on the plane-wave expansion and supercell method. In: 2014 International Symposium on Computer, Consumer and Control (ISCCC). IEEE, pp 322–324
7. Kuzma A, Uherek F, Skriniarova J, Kuzma A, Uherek F (2016) Photonic crystal based add/drop filters for sensing. In: 2016 Photonics North (PN). IEEE
8. Pilgun Y, Smirnov E (2016) Plane-wave expansion based modelling of laser beam propagation in anisotropic medium. In: 2016 IEEE 7th International Conference on Advanced Optoelectronics and Lasers (CAOL). IEEE, pp 139–141
9. Das PS, Jena D, Palai G (2016) PWE approach to MSOF for beam splitting application. Optik 127:10228–10231
10. Panda A, Mishra CS, Palai G (2016) PWE approach to optical thyristor for investigation of doping concentration. Optik 127:4831–4833
11. Burla M, Bazargani HP, St-Yves J, Shi W, Chrostowski L, Azana J (2014) Frequency agile microwave photonics notch filter based on a waveguide Bragg grating on silicon. In: 2014 international topical meeting on Microwave Photonics (MWP) and the 2014 9th Asia-Pacific Microwave Photonics Conference (APMP). IEEE, pp 392–394
12. Xu E, Yao J (2015) Frequency- and Notch-depth-tunable single-notch microwave photonic filter. IEEE Photo Technol Lett 1–4
13. Mahendra A, Magi E, Choudhary A, Liu Y, Marpaung D, Eggleton BJ (2018) High performance, low noise figure Brillouin-based tunable microwave photonic bandpass filter. In: 2018 international topical Meeting on Microwave Photonics (MWP). IEEE
14. Swain KP, Palai G, Moharana JK (2017) Analysis for '101' channels of MUX/DEMUX using grating SOI Structure at sub nanometer scale. Optik 129:78–82
15. Swain KP, Palai G, Moharana JK (2019) Optical filter based electrical control vis-à-vis cloud data: a new hybrid optoelectronics device for embedded application. Optik Int J Light Electron Opt 178:964–969
16. Swain KP, Palai G, Moharana JK (2019) Design and implementation of opto-electro decoder using photonic structure: a new application of Li-fi vis-a-vis optical embedded system. Optik Int J Light Electron Opt 178:658–663
17. Swain KP, Palai G (2018) Photonic structure for embedded application: realization of optical filter based locking system. Optik Int J Light Electron Opt 169:344–349

Modeling, Analysis, and Control of Vehicle Suspension System Based on Linear Quadratic Regulator

Akshaya Kumar Patra[1(✉)], Alok Kumar Mishra[1], Anuja Nanda[1], Narendra Kumar Jena[1], and Bidyadhar Rout[2]

[1] Department of EEE, ITER, S'O'A University, Bhubaneswar 751030, India
narendrajena@soa.ac.in
[2] Department of EEE, Veer Surendra Sai University of Technology, Burla 768017, India

1 Introduction

In recent times, many researchers in the field of vehicle suspension (VS) system dynamics have devoted to arrive at an optimal solution with a compromise between vehicle handling, ride comfort, and stability. There is a need for better approach today as the above problems are very much evident in the modern vehicle cases. Specific to the large sedan and luxury cars, even if excellent ride qualities are achieved, it is limited to acquiring adequate handling behavior. With reference to sports vehicles, although it is provided with very good handling capability, but fails to provide desirable ride quality. There are many options in between for variations in the designing stage of the vehicle manufacturers to meet the customer needs. Designing point of view, passenger comfort and vehicle control are the two primary objectives to be considered. Road disturbances such as bumps or potholes are to be handled not to sacrifice the passengers comfort. At the control stage, these factors are generally considered either through keeping the vehicle body from rolling and pitching excessively, or by maintaining good contact between the tire and the road.

Nowaday's hydraulic dampers (shock absorbers) and springs are extensively used for vehicle suspensions. As a principle, these are charged with the job of absorbing bumps, minimizing the vehicle's body motions during acceleration, braking and turning of the vehicle, and keeping the tires in contact with the road surface. However, from the design point view, these objectives are contradictory to each other to achieve all simultaneously at their optimum level.

The spring and the damper are the two essential components in VS design. The spring design mostly depends on the weight of the vehicle. The damper design is based on the suspensions placement on the compromise curve and so, it is essential to be perfectly chosen to make the optimal vehicle performance for any type of vehicle. For ideal performance, the damper should act such that passengers isolate from low-frequency road disturbances and absorb high-frequency road disturbances. High damping is essential to achieve for best isolation of passengers from low-frequency disturbances.

M. N. Mohanty and S. Das (eds.), *Advances in Intelligent Computing and Communication*, Lecture Notes in Networks and Systems 109,
https://doi.org/10.1007/978-981-15-2774-6_32

However, it is desirable to design a high-damping system, on the other hand it degrades the high-frequency absorption rate. In other way, by providing low damping the damper offers adequate high-frequency absorption in terms of scarifying low-frequency isolation. To meet these contradictory objectives, it is essential to design and focus on automotive suspensions without compromising any of the factors mentioned above. As a solution to the above, three types of suspensions can be improved. These are passive, fully active, and semi-active-type of suspensions. The spring and damper are the two basic components of the conventional passive suspension. Both the components are considered and fixed at the design stage. The suspension stores energy in the spring. Later it dissipates energy through the damper.

It is very much needed for further research to develop robust control algorithms to enhance the performance of VS system. This in turn enhances the vehicle capabilities to handle the aforementioned issues. An ideal design of VS system needs to achieve many performance characteristics such as: (1) control of body movement, (2) control of suspension movement and force distribution. Performance point of view, the VS should able to isolate the body for comfort against the road impact and inertial disturbances. These are generally associated with cornering and braking or acceleration of the vehicle system [1]. As discussed above, many performance objectives are conflicting in nature, that to enhance one the other factor degrades. Considering all the objectives, as a goal for the designing of a suspension system is difficult to meet [2]. Minimization of vertical force to the passengers can be achieved by minimizing the vertical vehicles body acceleration of the suspension. Another factor which plays a vital role in passengers comfort is the optimal contact between wheel and road surface and this is essential in various driving conditions in order to maximize safety factor [3]. Among few design in the past, the system presented in [4] based on unconstrained optimizations for passive suspension (PS) system case is widely accepted and used. This successfully performs to achieve the desirability of low suspension stiffness, reduced unsprung mass (UM), and an optimum damping ratio for the better controllability. As the PS system performs satisfactorily some extent, it is considered in many applications for the VS system. However, both the suspension spring and damper do not supply energy to the suspension system. They only control the motion of the vehicles body and wheel by limiting the suspension velocity. This is computed according to the rate specified by the designer. To overcome this issue, the active suspension (AS) system is considered as an efficient option for this application. The AS systems have the ability to dynamically respond to changes in the road profile. It is due to this fact that it can supply energy to produce relative motion between the body and the wheel. Sensors are provided in suspension system to measure the parameters dynamically. The parameters such as body velocity, suspension displacement, wheel velocity, and wheel and body acceleration are sensed and computed for the controller as input parameters [5]. An AS can be thought as the integration of the passive components to actuators that supply additional forces. These additional forces are computed by a feedback control law based on the input data from the sensors fixed to the vehicle. For modeling the real-time dynamic conditions, the uncertainties due to system design and other external disturbances are needed to be considered for the controller. This motivates for a better control design to increase the robustness and controllability under uncertainties and disturbances.

During the past three decades, so many control strategy techniques are suggested and tested by absorbing the shocks due to a rough and bumpy road in case of VS system. Time-discrete and switching PID control strategy is implemented in VS problems with variable control gains based on the measured suspension variables [6, 7]. However, the optimal gain parameter setting, a lesser range of robust control, and need of change of gain setting with varying conditions are the major limitations to limit the real-time application of these controllers. Among other projected robust control algorithms applied for limiting the oscillation and velocity of the VS system are fuzzy-logic control [8], fuzzy-PID control [9], genetic algorithm [10], neural network [11], neuro-fuzzy (NF) control [12], linear quadratic regulator (LQR) [13], H-infinity control [14], and sliding mode (SM) control [15]. However, even if these control techniques are implemented effectively by absorbing the shocks due to the rough and bumpy road in case of VS system with enhanced accuracy and damping of oscillation, still fail to handle various constraints and random change found in a suspension environment. These approaches are not fully insensitive to the disturbances and the uncertainties of the model in spite of the improved performance. Hence, optimal control parameters setting for still better performance and for avoiding slow response following road disturbance (road impact), the current work suggests an alternative novel technique based on the LQR concept. Application of the suggested approach to control the oscillation and velocity in vehicle suspension system results to ensure a better robust controller in comparison to other contemporary well-established approaches under both harmonized and incompatible uncertainties.

2 Problem Formulation and Modeling

2.1 System Overview

The overall model of VS system is depicted in Fig. 1. The road disturbance $w(t)$ and $v(t)$ are reflected as the process disturbance and the sensor noise, respectively, in this study. The controller receives information about the oscillation of the VS system as input to provide the optimal control force $u(t)$, and it is applied in between the UM and the sprung mass (SM) to reduce the relative motion between them.

2.2 Modeling of VS System

In the year 1958, the concept of VS system was emerged into limelight. It has been found in the later period that the vibration suppression capability is limited in case of traditional PS and semi-active suspension systems. This leads to focus on an AS system with additional control forces as a better alternative to the above drawbacks to suppress the oscillations and used in modern vehicle industry. The major factor that differentiates the active suspension system performance wise is the ability to inject energy into vehicle dynamic system via actuators unlike dissipates energy in case of conventional suspension system (SS). To produce a desirable control force to handle the variety of road disturbances (road impact) in real-time applications, actuators are placed in between the UM and the SM in AS system. The most challenging task of designing an AS system is to enhance the ride comfort by absorbing the shocks due to the rough and bumpy road. The force actuator in the case of the AS system is capable to add and also dissipates energy from the system. This results an increase in the ride comfort and vehicle handing due to the capability of the suspension system to regulate the vehicle altitude and to eliminate the adverse effects of braking and vehicle roll during cornering and braking. The schematic model of the VS system is illustrated in Fig. 1.

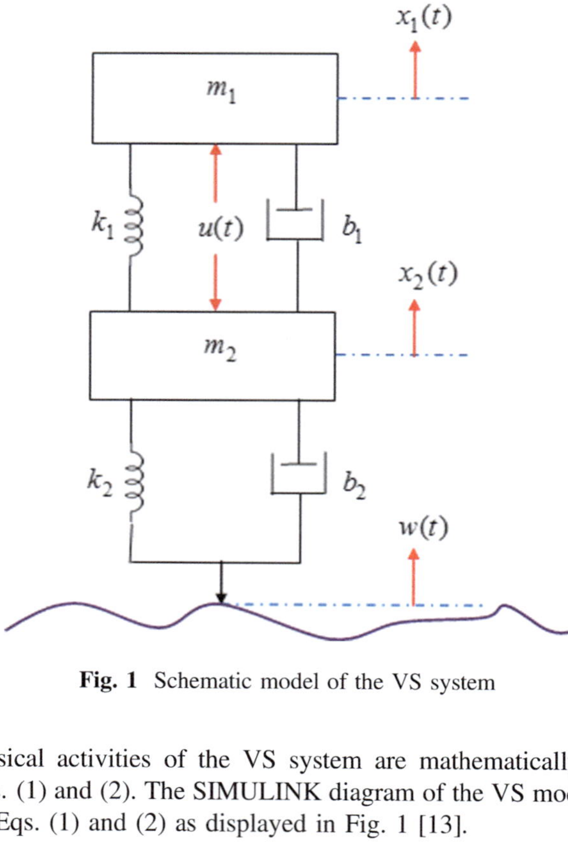

Fig. 1 Schematic model of the VS system

All the physical activities of the VS system are mathematically expressed and specified by Eqs. (1) and (2). The SIMULINK diagram of the VS model is established with respect to Eqs. (1) and (2) as displayed in Fig. 1 [13].

$$m_1\left(\frac{\mathrm{d}^2x_1(t)}{\mathrm{d}t^2}\right) = -b_1\left(\frac{\mathrm{d}x_1(t)}{\mathrm{d}t} - \frac{\mathrm{d}x_2(t)}{\mathrm{d}t}\right) - k_1(x_1(t) - x_2(t)) + u(t) \tag{1}$$

$$\begin{aligned} m_2\left(\frac{\mathrm{d}^2x_2(t)}{\mathrm{d}t^2}\right) &= b_1\left(\frac{\mathrm{d}x_1(t)}{\mathrm{d}t} - \frac{\mathrm{d}x_2(t)}{\mathrm{d}t}\right) + k_1(x_1(t) - x_2(t)) \\ &\quad + b_2\left(\frac{\mathrm{d}v(t)}{\mathrm{d}t} - \frac{\mathrm{d}x_2(t)}{\mathrm{d}t}\right) + k_2(v(t) - x_2(t)) - u(t) \end{aligned} \tag{2}$$

2.3 Linearization of VS System

Figure 1 shows the non-linear model of VS system, and its simplified form is shown in Fig. 1. The state-space equation of the VS model with $w(t)$ and $v(t)$ can be expressed as [13]:

$$\left.\begin{array}{l}\dot{x}_m(t) = A_m x_m(t) + B_m u(t) + B_d w(t) \\ y(t) = C_m x_m(t) + D_m u(t) + v(t)\end{array}\right\} \quad (3)$$

where, $w(t)$, $x_m(t)$, $v(t)$, and $y(t)$ are the road disturbance, state variable, sensor noise, and measured output of the VS system, respectively. The state variable $x_m(t)$ consists of $x_1(t)$, $x_2(t)$, $\dot{x}_1(t)$, and $\dot{x}_2(t)$. The A_m, B_m, C_m, D_m, and B_d represent the matrices of the state-space model of VS system. The state-space matrices are found out from the linearization of the system dynamics nearby the reference point. The detailed derivation is described in the literature [13]. In the current study, the initial reference such as the oscillation and the velocity of the suspension system are taken as zero. The VS model is linearized surrounding the reference point for the design of robust controllers to control the oscillation and the velocity of the suspension system within the stable range.

2.4 Performance Analysis of VS System

There are four poles in VS system. One of them lies in right-hand side of the complex plane. As a result, the system becomes unstable. This needs the design of an adaptive controller for improving the stability of the system by means of shifting the poles into the left-hand side of the complex plane. The VS system in the open loop form is depicted in Fig. 1.

The VS system has two inputs and two outputs. The control force $u(t)$ and road disturbance $w(t)$ are the two inputs of the VS system. The oscillation and velocity of the SS are the two outputs of the VS system. An uncontrolled oscillation and velocity of the suspension system are being observed owing to the application of 0.1 m impulsive road disturbance on the unsprung mass at the simulation time of 1.0 s. The irregular oscillation and velocity of the suspension system are illustrated in Fig. 2a, b. Figure 2a, b illustrates the unstable dynamics under various model uncertainties and disturbances. These unstable dynamics can be reduced by applying the suitable control algorithms. In this case, the oscillation of the suspension system is the most essential outcome needs to be controlled within a stable range through suitable control techniques, and velocity of the suspension system is analyzed in order to view the motion trajectory.

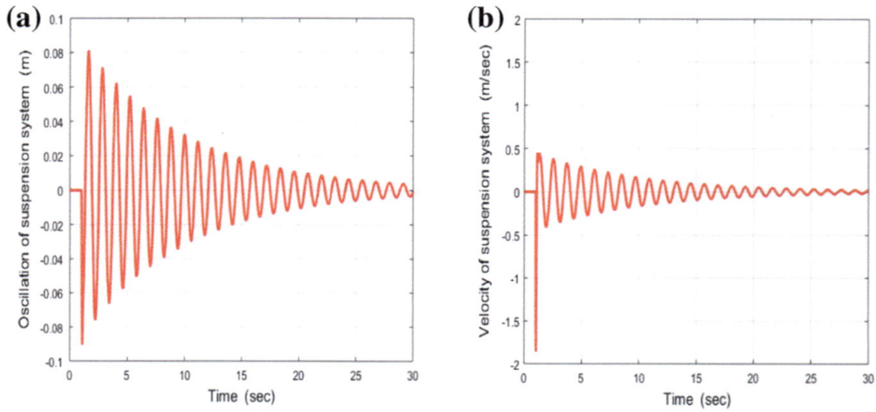

Fig. 2 **a** Response of oscillation of the SS with road impact; **b** response of velocity of the SS with road impact

3 Control Algorithm

The control specifications such as settling time t_s, steady-state error e_{ss}, maximum overshoot O_{Max}, and maximum undershoot U_{Max} are evaluated and examined with proper validation of the controller action.

3.1 LQR Design

The linearized model of the VS system as discussed in Sect. 2.3 has been taken for the formulation of the suggested control algorithm to regulate the oscillation and the velocity of the suspension system. For accomplishing an upgraded performance and the adjustment of controller parameters of the suggested LQR, it is integrated to the linearized model of the VS system as illustrated in Fig. 3. The linearized model of the VS system with $w(t)$ and $v(t)$ is formulated as [16–19]:

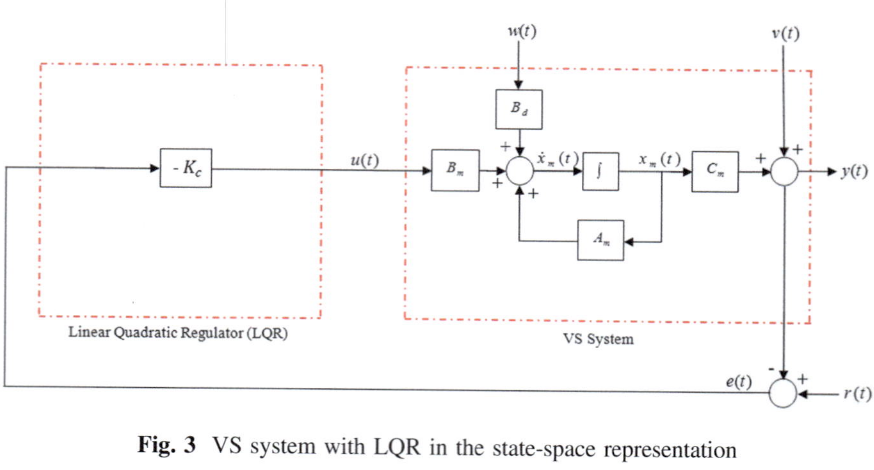

Fig. 3 VS system with LQR in the state-space representation

$$\frac{dx_m(t)}{dt} = A_m x_m(t) + B_m u(t) + B_d w(t) \tag{4}$$

$$y(t) = C_m x_m(t) + v(t) \tag{5}$$

$$u(t) = -K_c e(t) \tag{6}$$

Figure 3 shows the linearized model of the diabetes patient with the LQR gain K_c. The evaluation procedures of K_c is mentioned in detail in Sect. 3.1.1. The evaluation of the TF of the LQR is described in Sect. 3.1.2.

3.1.1 *LQR Gain* K_c

The K_c is evaluated with respect to the system dynamics for the minimum value of the quadratic performance index j as specified in Eq. (7).

$$j = \int_0^\infty \left[x_m(t)^T Q_1 x_m(t) + u(t)^T R_1 u(t)\right] dt \tag{7}$$

where, $Q_1 = C_m^T q_1 C_m$. The Q_1, q_1, and R_1 are denoted as the positive semi-definite weighted matrix of state, the intensity of a weighted matrix of state, and the positive definite weighted matrix of input, respectively. The K_c is estimated for the minimum value of the quadratic performance index as specified in Eq. (8).

$$K_c = R_1^{-1} B_m^T \Pi_k \tag{8}$$

Π_k is the solution of the Controller Algebraic Reccati Equation (CARE). The CARE is specified as follows:

$$A_m^T \Pi_k + \Pi_k A_m - \Pi_k B_m R_1^{-1} B_m^T \Pi_k + C_m^T q_1 C_m = 0 \tag{9}$$

3.1.2 TF of LQR

The TF of suggested controller $K(s)$ is estimated with the use of K_c. The $K(S)$ of LQR is specified as follows:

$$K(s) = K_c (sI_n - A_m + B_m K_c)^{-1} B_m \tag{10}$$

The packed matrix notation of $K(S)$ is specified as follows:

$$K(s) = \begin{bmatrix} A_m - B_m K_c & B_m \\ K_c & 0 \end{bmatrix} \tag{11}$$

For the design of suggested LQR, the optimal values of control parameters are evaluated with help of MATLAB and represented as in the Table 1.

Table 1 Optimal values of control parameters

Q_1	R_1
$1000 * C_m^T * C_m$	1

4 Result and Discussions

The performance of the closed loop VS system with LQR is described in detail in this section. The suggested control approach is compared with other popular control algorithms to justify its enhanced performance.

4.1 Performance Analysis of VS System with LQR

In this section, all physical activities of the closed loop VS model with suggested LQR are examined under different conditions and the huge deviation of road disturbance. The time domain response of the oscillation and the velocity of the suspension system with 0.1 m impulsive road disturbance at the simulation time of 1.0 s are displayed in Fig. 4a, b. The output results clearly specify the suspension system attains the zero oscillation and zero velocity with less settling time and also attains the balance position where the system is absolutely steady. To achieve the enhanced system response, the required control force $u(t)$ is generated by the suggested LQR.

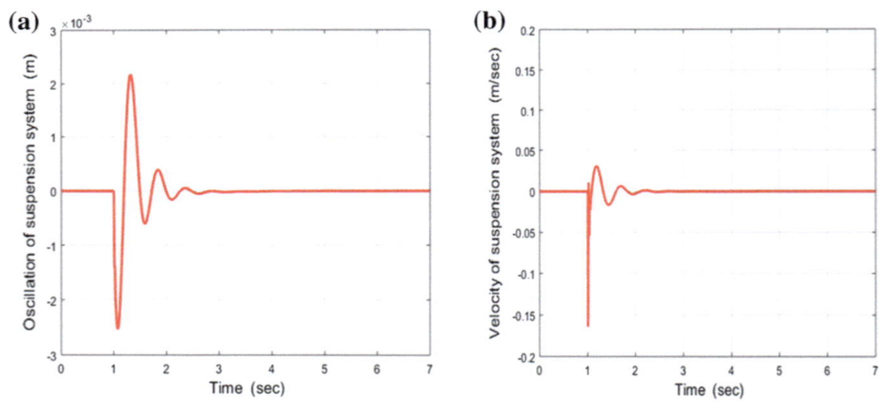

Fig. 4 **a** Response of oscillation of the suspension system with road impact based on LQR; **b** response of velocity of the suspension system with road impact based on LQR

4.2 Comparative Study

The proposed LQR control approach is compared with other popular control approaches such as PID, fuzzy, NF, LQR, H_∞, and SM to justify its enhanced performance as a controller. Figure 4a illustrates the effect of road impact in the oscillation of the suspension system with the suggested LQR approach. Table 2 presents a comparative analysis with respect to t_s (sec), $O_{\text{Max}}(m)$, $U_{\text{Max}}(m)$, noise (%), and $e_{ss}(\%)$. The effect of road disturbance in the oscillation of the suspension system applying different control approaches such as PID, fuzzy, NF, LQR, H_∞, and SM is also documented in Table 2 based on the references, and [6, 8, 12–14] and [15], respectively. Similar working conditions are followed with the same level of sensor noise in all control techniques application for comparison.

Table 2 Comparative result analysis related to oscillation of the suspension system

Controller	PID [6]	Fuzzy [8]	NF [12]	LQR [13]	H_∞ [14]	SM [15]	LQR (Proposed)
Road disturbance (m)	0.1	0.1	0.1	0.1	0.1	0.1	0.1
t_s (sec)	4.2	4.1	3.7	3.4	3.1	3.0	2.3
$O_{\text{Max}}(m)$	0.0050	0.0049	0.0047	0.0045	0.0043	0.0040	0.0022
$U_{\text{Max}}(m)$	0.0042	0.0045	0.0046	0.0042	0.0041	0.0048	0.0026
Noise (%)	10	10	5	5	5	5	1
e_{ss} (%)	0	0	0	0	0	0	0

The oscillations of the suspension system under 0.1 m impulsive road disturbance are tested. The corresponding results are presented for the various control approaches along with the proposed LQR with respect to time domain specifications such as $O_{\text{Max}}(m)$, $U_{\text{Max}}(m)$, and t_s(sec). The results signify the better controllability of the LQR. The simulation results also demonstrate the high noise and chattering elimination capability with high robustness for the proposed approach. Overall, by looking to the above comparative analysis, the findings of suggested approach advantages are the higher accuracy and stability, more robustness, high noise and chattering elimination capability, and better capability to handle uncertainty under various conditions, and huge deviation of road disturbance.

5 Conclusions

In this manuscript, a novel control strategy LQR is proposed based on self-tuned approach. To justify its enhanced performance, it has been applied and tested to control the oscillation and the velocity of the suspension system in the vehicle. In suggested LQR, self-tuned algorithm is utilized to enhance the control performance. The comparative results clearly reflect the suggested LQR is arrived at better performance than the other control approaches such as PID, fuzzy, NF, LQR, H_∞, and SM with respect to stability, reliability, and robustness under various abnormal conditions and disturbances. The better performance of the suggested approach in terms of improved accuracy and stability, enhanced robustness, high noise elimination capability, and better ability to control uncertainty justifies its real-time application.

References

1. Wright PG (1984) The application of active suspension to high performance road vehicles. In: Microprocessors in fluid engineering IMechE conference publications
2. Fateh MM, Alavi SS (2009) Impedance control of an active suspension system. J Mechatron 19:134–140
3. Lin JS, Kanellakopoulos I (1995) Nonlinear design of active suspension, In: 34th IEEE Conference 1995, vol 17, pp 45–59
4. Alleyen A, Hedrick JK (1995) Nonlinear adaptive control of active suspensions. IEEE Trans Control Syst Technol 3(1):845–860
5. Esmailzadeh E, Taghirad HD (1997) Active vehicle suspensions with optimal state feedback control. J Mech Sci 200(4):1–18
6. Kumar MS (2008) Development of active suspension system for automobiles using PID controller
7. Talib MHA, Darns IZM (2013) Self-tuning PID controller for active suspension system with hydraulic actuator. In: 2013 IEEE Symposium on Computers & Informatics (ISCI), April 2013, pp 86–91
8. Pekgökgöz RK, Kısa M (2010) Actıve suspension of cars using fuzzy logic controller optimized by genetic algorithm. Int J Eng Appl Sci 2(4):27–37
9. Yang Q, Yuan C (2008) Fuzzy-PID control on sem-i active air suspension. Trans Chin Soc Agric Mach 39(9):24–29
10. Du H, Lam J, Sze KY (2003) Non-fragile output feedback H1 vehicle suspension control using genetic algorithm. Eng Appl Artif Intell 16:667–680
11. Wang YJ (2007) Analysis of vehicle suspension control using neural networks. Northeastern university, Shenyang
12. Aldair AA, Wang WJ (2012) A neurofuzzy controller for full vehicle active suspension systems. J Vib Control 18(12):1837–1854
13. ElMadany MM (1999) Linear quadratic Gaussian control of a quarter-car suspension. Veh Syst Dyn 32(6):479–497
14. Kaleemullah, M., Hasbullah, F.: Design of robust H∞, fuzzy and LQR controller for active suspension of a quarter car model, In 2011 4th International Conference on Mechatronics (ICOM), (2011, May) 1–6, IEEE
15. Li H, Liu H (2013) Adaptive sliding-mode control for nonlinear active suspension vehicle systems using T-S fuzzy approach. IEEE Trans Indus Electron 60(8):3328–3338
16. Patra AK, Rout PK (2018) Backstepping sliding mode Gaussian insulin injection control for blood glucose regulation in TIDM patient. J Dyn Sys Meas Control 140(9):091006
17. Patra AK (2018) Backstepping model predictive controller for blood glucose regulation in type-I diabetes patient. IETE J Res. https://doi.org/10.1080/03772063.2018.1493404
18. Petra AK (2017) Adaptive continuous-time model predictive controller for implantable insulin delivery system in type I diabetic patient. Optimal Control Appl Meth 38(2):184–204
19. Patra AK, Rout PK (2017) Adaptive sliding mode Gaussian controller for artificial pancreas in TIDM patient. J Process Control 59:13–27

Performance and Evaluation of Different Kernels in Support Vector Machine for Text Mining

Ashish Kumar Mourya, ShafqatUlAhsaan, and Harleen Kaur(✉)

Department of Computer Science and Engineering, School of Engineering Sciences and Technology, JamiaHamdard, New Delhi, India

1 Introduction

The significance of utilization of an electronic data has increased in recent years. In recent years, approximately 75% of digital data is in the form of text or multimedia form [1]. This tremendous amount of data is not only unstructured and unclassified but also contains unless data, useful data, heterogeneous data, and scientific data. Currently, more than 35% of corporation is working with this huge data. Text mining and analysis is similar to data mining, excluding data mining techniques applied on relational data, whereas text pre-processing works on unstructured or semi-structured data [2]. In text mining or analysis, we extracted the useful information from heterogeneous or multimedia data by separating the plain text and heterogeneity and analyzing the classification frequency or correlation. Machine learning technique such as SVM or DT, the possibility of the high frequency result even when the amount of data is heterogeneous or enormous, is confirmed. Data mining is liable for pulling out of implied, potential data and liable for explicitly data [3]. Machine learning techniques and algorithms designed to analyzed these explicitly data know the configuration of numbers, and they are pre-programmed to process it with precision. We have used SVM for predicting the performance and analyzing the multimedia in text mining.

SVM algorithm is also known to be very powerful in terms of accomplishing performance on the numerical data. SVM technique was introduced to solve the linear and binary classification problem of two classes [4]. Now, it is currently implemented in any kind of problem [9]. We have now used them in classification as well as regression problem. It is broadly used in lots of applications like gene classification, time series prediction, image segmentation, and face recognition. SVM is successfully applied on different classification tasks that take statistical data or heterogeneous data. On the other hand, artificial neural network techniques are biologically inspired machine learning methods designed like a human brain process the information [5]. Its critical structure allows knowledge practically with any function beneficially with adequate hidden layers of processing nodes with training data. SVM and neural network are normally O (n3) or at best quadratic ally complex with more examples.

SVM falls under the category of supervised machine learning. SVM classifier performs classification by example within multiple data classes. SVM algorithms are used in the field of computational biology on a large number of problems, like

M. N. Mohanty and S. Das (eds.), *Advances in Intelligent Computing and Communication*, Lecture Notes in Networks and Systems 109,
https://doi.org/10.1007/978-981-15-2774-6_33

detection of homology in protein, analyses of gene expressions and tandem mass, etc. SVMs work on the principle of structural risk minimization [6]. SVMs are universal learning systems [7]. SVM, within a multi-dimensional feature space, works on training samples of multiple classes in order to get a best possible hyper-plane that separates the two classes. The training samples that fall on hyper-plane that is closest to each other are called support vectors, and the minimum distance between the support vectors is called margin. The result with the maximum margin is considered as optimal, and the error rate for this result is very low. SVM, by means of a kernel function, projects vectors into a higher-dimensional input space and obtains the best possible hyper-plane that performs the distinction of data classes via optimization function. Let consider pattern $\boldsymbol{x}$; the corresponding estimated label $\hat{\boldsymbol{Z}}$ is given by equation (1).

$$\hat{Z} = \text{sign}[f(x)] = \text{sign}\left(\sum_{i=1}^{n} a_i y_i K(x_i, x) + c\right) \tag{1}$$

Kernel functions or tricks give a principled means to characterize the data and thus depict assumptions from different variety of data. The kernel methods while deploying multiple kernel functions correspond to a collection of varied data objects and characterize the relationship among any pair of data objects. For example, that defines a set of multiple data objects that can be adequately and sufficiently represented by means of '$m * n$' matrix of pair-wise kernel standards.

While dealing with heterogeneous data, SVM makes use of different types of kernel functions that perform efficiently on the data objects and must be combined numerically. In [8], the authors used the SVM approach in their study to combine gene data and phylogenetic profiles in an un-weighted mode, by deploying numerical operators to axis the SVM within dataset correlations surrounded by features, whereas do not take into account the correlations among various datasets [9]. Un-weighted sums of kernels have also been used successfully in the forecasting of protein relations.

Researchers proposed a new learning method known as multi-kernel learning (MKL), which is an ensemble of different kernels contained by the SVM classifier itself [10]. These methods combine a single optimization procedure, which concurrently finds the solution for classification followed by SVM and weights on the individual facts types in the heterogeneous dataset.

2 Background

In [11] this paper, a comparison is made among multilayer perceptron and convolutional network and concluded that convolutional networks perform better as compared to multilayer perceptron. The experimental results are performed on Czech newspaper text dataset. Jadon and Sharma [12] deployed the SVM algorithm for document classification. In their approach, they used SVM to categorize the text documents automatically. They worked on Reuters-corpus-text document dataset and obtained 86% accuracy while applying SVM algorithm. Devika et al. [13] applied the sentiment-based classifier to analyze and determine the relevant news content. Their dataset consists of 250 news documents in the form of text labeled with multiple sentiments on which classification is done. Arundthati [14] explained that Naïve Bayes classifier is used for text classification. They first applied the approach in linear way and then hierarchically. The hierarchical method results to be more effective and improved the efficiency and accuracy of the classifier.

Purohit [15] has compared multiple types of sentiment analysis methods like rule-based approach, learning classifier approach, and lexicon-based approach. In their work, the parameters that are being highlighted are accuracy, performance, and efficiency, irrespective of advantages and disadvantages of these techniques. Tilve and Jain [16] used the decision tree for the assessment of dataset related to educational background. The dataset constitutes the name, marks, enrollment number, and grades of students. A number of algorithms have been deployed like random forest, REPTree, Hoeffding, and J48. It has been evaluated that the decision tree algorithms show a descent increment in performance while dealing with student's dataset.

Kumar and Verma [17] went through different research works and used different datasets for sentiment analysis. In their work, a number of algorithms such as association rule-based decision tree and Naïve Bayes classifiers have been used. A comparison is performed among these algorithms. The algorithm proposed by Purohit A. et al. makes use of Porter stemmer, association rule, a priori algorithm, Naïve Bayes classification which showed a nice performance and attained a classification accuracy of 75%. Vaghela et al. [18] compared the proposed algorithms, Naive Bayes and vector space model, while deploying them on newsgroup datasets. The dataset constitutes 50 categories of news text data. The experimental results show that Naive Bayes classifier has best accuracy. Kharde and Sonawane [19], Kalarikkal and Remya [20] explored feature selection and submitted that if it is done reasonably, the classification algorithms would perform better specially in terms of accuracy. In their study, a comparison among lexicon and machine learning algorithms has been performed and came to the point that machine learning algorithms are domain dependent. Supervised learning classifiers have been used in the comparative study, and the authors explored that supervised learning models are easy to train and need much less maintenance. Supervised machine learning models acquire high accuracy in comparison to lexicon-based models.

Das et al. [21] performed a detailed review and study of supervised and untrained methods of learning by means of some estimation metrics. They focus on challenges and applications of sentiment analysis. A comparative study has been carried out on multiple classification algorithms such as Naive Bayes, SVM, logistic regression, and random forest. Shrivatava et al. [22] mainly focused on processing the datasets using machine learning techniques like support vector network, principal component analysis, and maximum entropy. They used qualitative training set for better implementation and outcome. Chavan et al. [23] developed an application that allows users to gather data from social media Twitter handle, analyze the data, and visualize it by means of pie charts and tables. Medhat et al. [24] deployed an efficient method to sort out the structure of tweets with the help of SVM and attained a precision of 70.5%.

Senthilkumar and Paulraj [25] proposed sentiment classification using multiple techniques of machine learning, particularly **Naïve** Bayes, SVM, and decision trees.

They used these classifiers to carry out the text categorization. In their study, the authors came to know that SVM classifier performs better in terms of accuracy as compared to other classifiers as **Naïve** Bayes and decision trees. In addition to accuracy, SVM is the best candidate to adjust the parameter settings automatically. Moraes et al. [26] studied and came up with a detailed summary regarding utilization of sentiment classification via feature selection.

The survey highlighted those techniques of sentiment classification that are very effective and depicted how to choose features that play a key role in improving the performance of sentiment classification. Kang et al. [27] presented a quality of classification techniques that can be deployed in data mining. In this paper, a couple of classification techniques like SVM, k-nearest neighbor classifier, Bayesian networks, neural network, and decision tree are reviewed. Abbasi et al. [28] classified the reviews in text form for positive and negative sentiments. The authors evaluated the support vector network and artificial neural networks on the bases of document-level sentiment analysis.

The empirical results presented that the performance of artificial neural networks generates superior results as compared to SVM classifiers. Terachi et al. [29] presented an innovative improved **Naïve** Bayes algorithm. The proposed algorithm performed better and showed a descent improvement in comparison to basic **Naïve** Bayes algorithm when trained on bigram and unigram features. The proposed algorithm expected very powerful than SVM and **Naïve** Bayes. [30] evaluated the **Naïve** Bayes and maximum entropy classification models for Twitter data, and the main objective of their study is the performance evaluation. The exploratory results that are being carried out in their study clearly proved that **Naïve** Bayes models are very proficient as compared to maximum entropy model.

3 Experiments and Results

In our experiments, we have compared the performance of SVMs using polynomial and RBF kernels using text mining. For performing experiment, we have taken dataset from UCI data repository. The name of dataset is the Reuters-21578 R8 dataset. The data has various documents whose outcome classes belong to eight of the ten most frequent document classes of Reuters' articles. We have split the entire data into two parts. The first part is training dataset, which contains 5485 documents, and the second part contains 2189 documents for testing purpose.

The problem of text classification consists of text example, training example, and newly derived example. Newly derived examples consist of predictor feature only, and testing examples are used to weigh up the models; therefore, they have both predictor and target attributes. The objective of text classification is to construct a model as a function of the values of the attributes using training examples or documents that would then forecast the target values of the new documents as accurately as possible.

The proposed study has applied different corpus tools on the raw data as shown in Fig. 1. The raw data contains various stop words such as the, and, is, and for. This process is known as pre-processing. There are many libraries available in *R* package for performing various tasks on text classification. Caret and tm are used in this study. Before evaluating the model performance, we have tested the cross-validation using tenfold methods. This method uses 10 time repeated values for evaluating the model. Thereafter, we have applied three different kernel tricks with SVM. The linear SVM has achieved 0.91 training accuracy, whereas 0.89 testing accuracy. The radial basis function has tremendous performance over any kind of data as shown in Fig. 2. These kernel tricks perform good on text as well as numeric data. The outcomes of all three kernel tricks have been shown in Table 1.

Fig. 1. Word cloud of the raw data

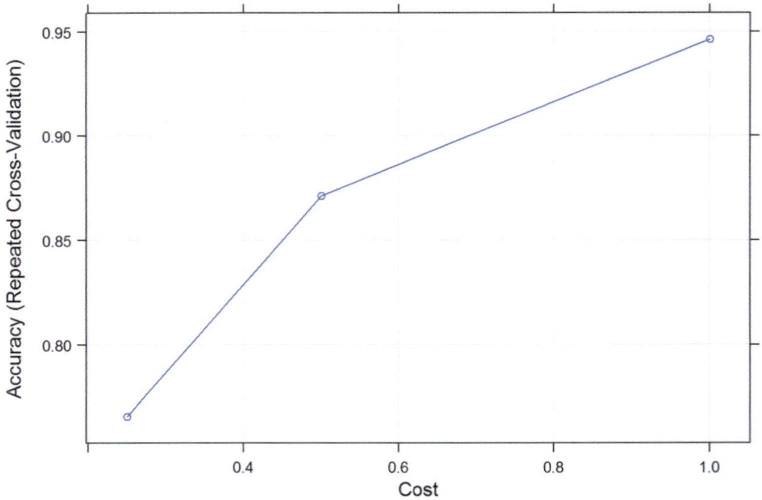

Fig. 2. Accuracy plot with RBF kernel

Table 1. Performance table

Kernels	Training accuracy	Testing accuracy
Linear kernel	0.91	0.89
Polynomial kernel	0.86	0.90
Radial basis function	0.94	0.91

4 Conclusion

SVM is a subfield of machine learning. SVM can be applied in both classification and regression problem. This paper discusses how different kernel functions that can be implemented and applied in conjunction with the SVMs in the automatic text classification problem. Several models for text analysis are represented; in most of them, text is determined by a deposit of vectors in a higher-dimensional space. The use of SVM classifier in combination with some text pre-processing techniques and kernel function assortment is illustrated by some experiments, conducted in a limited datasets obtained from the different frequent classes of the datasets. In this experiment, SVM repeatedly achieves good performance on the text analysis and categorization task. SVM eliminates the need for feature space due to its ability to generalize in high-dimensional vector space. The proposed study shows that radial basis function has the highest tendency to classify data.

References

1. Kaushik A, Naithani S (2015) A study on sentiment analysis: methods and tools. Int J Sci Res 4(12). ISSN (Online): 2319-7064
2. Gupta V, Lehal GS A survey of text mining techniques and applications. J Emerg
3. Duda RO, Hart PE, Stork DG (2001) Pattern classification. Wiley, New York
4. Bishop CM (2006) Pattern recognition and machine learning. Springer, New York
5. Marsland S (2009) Machine learning. An algorithmic perspective. Chapman and Hall/CRC Press, Boca Raton
6. Mohri A, Rostamizadeh A, Talwalker A (2012) Foundations of machine learning. The MIT Press, Cambridge
7. Tapia E, Bulacio P, Angelone L (2012) Sparse and stable gene selection with consensus SVM-RFE. Pattern Recogn Lett 33(2):164–172
8. Ruiz R, Riquelme JC, Aguilar-Ruiz JS (2006) Incremental wrapper-based gene selection from microarray data for cancer classification. Pattern Recogn 39(12):2383–2392
9. Huerta R, Vembu S, Muezzinoglu MK, Vergara A (2012) Dynamical svm for time series classification. Pattern recognition. Springer, Berlin, Germany, pp 216–225
10. Fatima S, Srinivasu B (2017) Text document categorization using support vector machine. Int Res J Eng Technol (IRJET) 4(2):141–147
11. Hui JLO, Hoon GK, Zainon WMNW (2017) Effects of word class and text position in sentiment-based news classification. Procedia Comput Sci 124:77–85
12. Jadon E, Sharma R (2017) Data mining: document classification using Naive Bayes classifier. Int J Comput Appl 167(6):13–16
13. Devika MD, Sunitha C, Ganesh A (2016) Sentiment analysis: a comparative study on different approaches. Procedia Comput Sci 87:44–49
14. Arundthati A (2017) Assessment of decision tree algorithm on student's recital. Int Res J Eng Technol (IRJET) 4
15. Purohit A, Atre D, Jaswani P, Asawara P (2015) Text classification in data mining. Int J Sci Res Publ 5(6):1–7
16. Tilve AKS, Jain SN (2017) A survey on machine learning techniques for text classification. Int J Eng Sci Res, Technol
17. Kumar R, Verma R (2012) Classification algorithms for data mining: a survey. Int J Innovations Eng Technol (IJIET) 1(2):7–14
18. Vaghela VB, Jadav BM, Scholar ME (2016) Analysis of various sentiment classification techniques. Int J Comput Appl 140(3):0975–8887
19. Kharde V, Sonawane P (2016) Sentiment analysis of twitter data: a survey of techniques. arXiv preprint arXiv:1601.06971
20. Kalarikkal S, Remya PC (2015, June) Sentiment analysis and dataset collection: a comparative study. In: 2015 IEEE International Advance Computing Conference (IACC). IEEE, pp 519–524
21. Das TK, Acharjya DP, Patra MR (2014, January) Opinion mining about a product by analyzing public tweets in Twitter. In: 2014 international conference on computer communication and informatics. IEEE, pp 1–4
22. Shrivatava A, Mayor S, Pant B (2014) Opinion mining of real time twitter tweets. Int J Comput Appl 100(19)
23. Chavan GS, Manjare S, Hegde P, Sankhe A (2014) A survey of various machine learning techniques for text classification. Int J Eng Trends Technol 15(6)
24. Medhat W, Hassan A, Korashy H (2014) Sentiment analysis algorithms and applications: a survey. Ain Shams Eng J 5(4):1093–1113

25. Senthilkumar D, Paulraj S (2015, March) Prediction of low birth weight infants and its risk factors using data mining techniques. In: Proceedings of the 2015 international conference on industrial engineering and operations management, pp 186–194
26. Moraes R, Valiati JF, Neto WPG (2013) Document-level sentiment classification: an empirical comparison between SVM and ANN. Expert Syst Appl 40(2):621–633
27. Kang H, Yoo SJ, Han D (2012) Senti-lexicon and improved Naïve Bayes algorithms for sentiment analysis of restaurant reviews. Expert Syst Appl 39(5):6000–6010
28. Abbasi A, Chen H, Salem A (2008) Sentiment analysis in multiple languages: feature selection for opinion classification in web forums. ACM Trans Inf Syst (TOIS) 26(3):12
29. Terachi M, Saga R, Tsuji H (2006, October) Trends recognition in journal papers by text mining. In 2006 IEEE international conference on systems, man and cybernetics, vol 6, pp 4784–4789

High Gain Slot Antenna by Using Artificial Magnetic Conductor

Prakash Kumar Panda[1(✉)] and Debalina Ghosh[2]

[1] Department of ECE, ITER, SOA Deemed to be University, Bhubaneswar, India
prakashpanda@soa.ac.in

[2] School of Electrical Sciences, IIT Bhubaneswar, Bhubaneswar, India
deghosh@iitbbs.ac.in

1 Introduction

With the increase demand in communication technology, requirement of efficient antenna system has also been increased. The need of high gain and high directive antennas is highly essential in satellite communication and personal communication system. Besides these features, antennas also need to maintain low profile behaviour [1]. In the evolution of research on increasing gain and directivity of planar monopole or dipole antenna, metamaterial plays an important role in transforming the bidirectional radiation pattern into directional radiation pattern.

Metamaterial, an artificial periodic structure, is formed by placing metallic patches above a grounded dielectric slab. When this structure reflects an incident wave with a 0° refection phase, then it is referred to as artificial magnetic conductor (AMC), and basically the phase reflected by an AMC surface varies with frequency. This surface behaves like a reflector when placed behind a planar monopole or patch antenna [2]. The radiation pattern of a monopole or dipole antenna is omni-directional in nature. If such antennas are supported by a PEC ground plane or a reflector, then it radiates in a single direction [3, 4]. But the antenna must be $\lambda/4$ distance apart from the ground plane for avoiding antenna impedance mismatch. To overcome this problem, AMCs are used in such antennas for increasing bandwidth, gain and directivity instead of PEC ground plane [5–7].

In this work, a planar slot antenna having omni-directional radiation pattern has been placed above the AMC surface for obtaining directional pattern. The gain of the slot antenna is restricted to only 2–3 dBi due to omni-directional pattern. By transforming this radiation pattern to directional, the gain is increased to 8.38 dBi. Also, the impedance bandwidth of the antenna does not change due to the AMC surface. The AMC surface benefits to yield an in-phase image current in the distinct locality of the antenna. But conversely, PEC creates a current which is reversed with the antenna current. Section 2 describes the design techniques of AMC surface. Section 3 suggests the slot antenna and designs its combination with AMC surface. Section 4 describes the simulation results, and Sect. 5 concludes the overall research work.

M. N. Mohanty and S. Das (eds.), *Advances in Intelligent Computing and Communication*, Lecture Notes in Networks and Systems 109,
https://doi.org/10.1007/978-981-15-2774-6_34

2 Design Methodology of Artificial Magnetic Conductor

The reflection phase characteristics and the boundary condition applied to the unit cell are depicted in Fig. 1a, b, respectively. FR4 material having relative permittivity (ε_r) of 4.4 and tangent loss (tan δ) of 0.02 is chosen as the substrate for the unit cell. Square-type unit cell is considered for wide bandwidth and easy manufacturing [8]. The substrate's length and width are 19.3 mm, and the length and width of the patch are 18.76 mm. The ground plane size is similar to the unit cell substrate. The unit cell substrate height is considered as 3.2 mm. Periodic boundary condition is given on the unit cell structure, and a normal plane wave is incident on it to find the phase characteristics. The reflecting phase bandwidth of the unit cell is found to be 2.3–2.54 GHz (bandwidth in the range of ±90°) [8, 9], and it crosses 0° at 2.39 GHz.

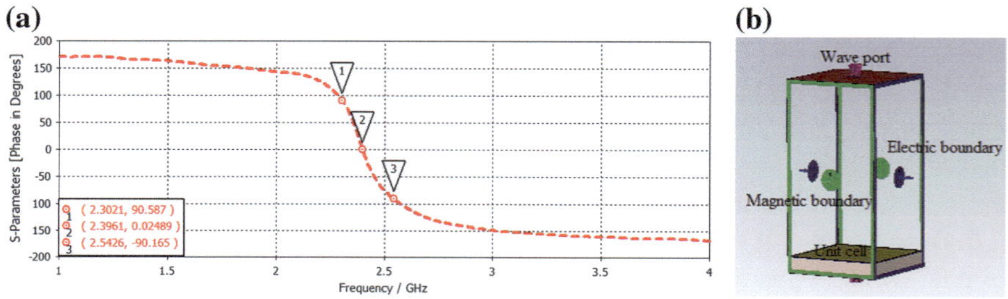

Fig. 1. **a** Phase diagram. **b** Boundary condition applied in unit cell

3 Design of Slot Antenna and Its Integration with the AMC Surface

Figure 2 shows the geometrical configuration of a coplanar folded slot antenna. It is based on RO4003C having a 0.8 mm thick and occupies an area of 44.3 × 55 mm^2. The relative permittivity and the loss tangent of RO4003C are 3.38 and 0.0027, respectively. Coplanar waveguide with 50Ω characteristic impedance is fed to the antenna. The resonating frequency of the antenna is adjusted by the length of the stub added in the slot or aperture created on the radiating patch. After a simulation study on varying the stub length and finally choosing its length b = 35.6 mm, the antenna resonates at a centre frequency of 2.4 GHz (Fig. 3).

The CPW-fed slot antenna is employed on an AMC consisting of 4 × 4 unit cells. The separation distance among the slot antenna and the artificial magnetic conductor is maintained to be 10 mm for proper impedance matching of the antenna. If the antenna will be placed very close to the AMC surface, then due to the high electromagnetic interference among antenna and AMC surface mismatch the antenna input impedance. So after a parametric analysis, this separation distance is found to be 10 mm for proper impedance matching and a suitable gain pattern. The simulated return loss parameter of the slot antenna and the integration with artificial magnetic conductor is shown in

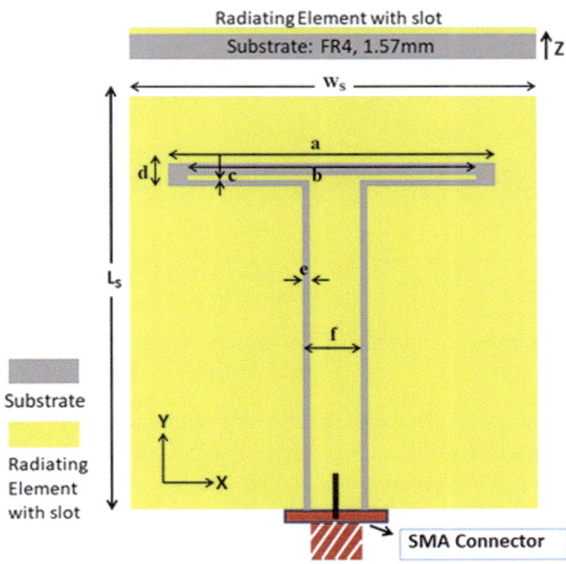

Fig. 2. Geometrical dimensions of slot antenna. [$L_S = 44.3$, $W_S = 55$, $a = 47$, $b = 35.6$, $c = 0.35$, $d = 2.06$, $e = 0.21$, $f = 3.1$, $Z = 0.8$] (all dimensions in mm)

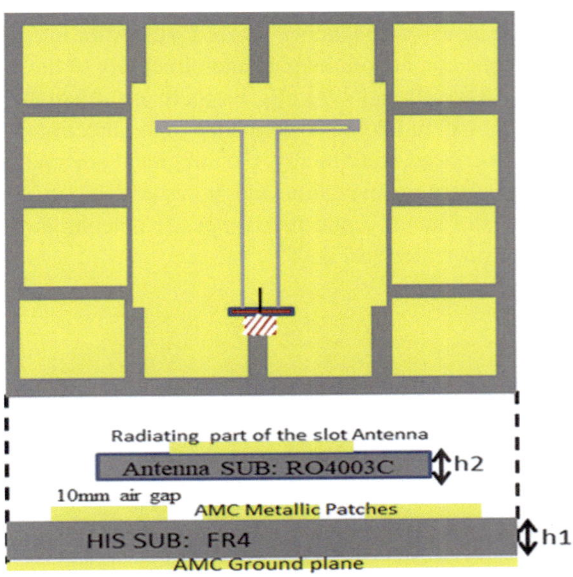

Fig. 3. Slot antenna over AMC surface: **a** top view; **b** side view with $h1 = 3.2$ mm and $h2 = 0.8$ mm

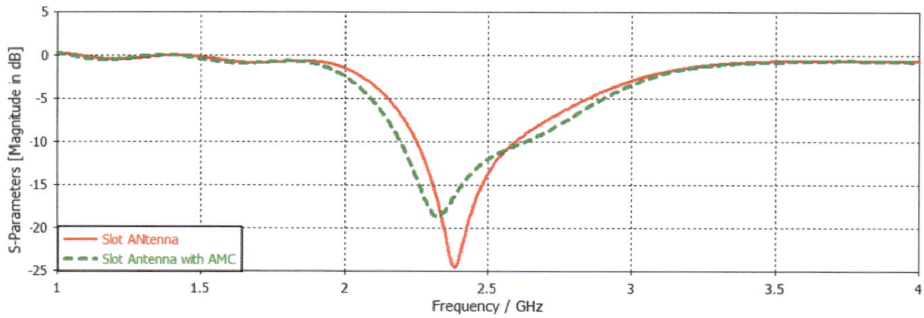

Fig. 4. Return loss parameter of slot antenna and the integrated structure

Fig. 4. The S parameter results between the slot antenna and its integration with the AMC surface look almost similar. The slight change in the result is due to the mutual coupling effect between the antenna and AMC surface.

4 Simulation Results

The simulated results for gain and directivity of the slot antenna along with the composite antenna are displayed in Figs. 5 and 6 correspondingly. The following results are taken at the resonating frequency of 2.4 GHz. After integrating the antenna with the artificial magnetic conductor, the gain and directivity of the composite antenna have been increased by 4.25 dB and 4.99 dBi, respectively. Also, the simulated results verify that the bidirectional radiation is converted into directional after adding the antenna with the AMC surface. Additionally, the integrated antenna is characterized in the form of radiation pattern in elevation and horizontal plane. The elevation and horizontal plane of the slot antenna and the composite antenna at the resonating frequency of 2.4 GHz are presented in Fig. 7.

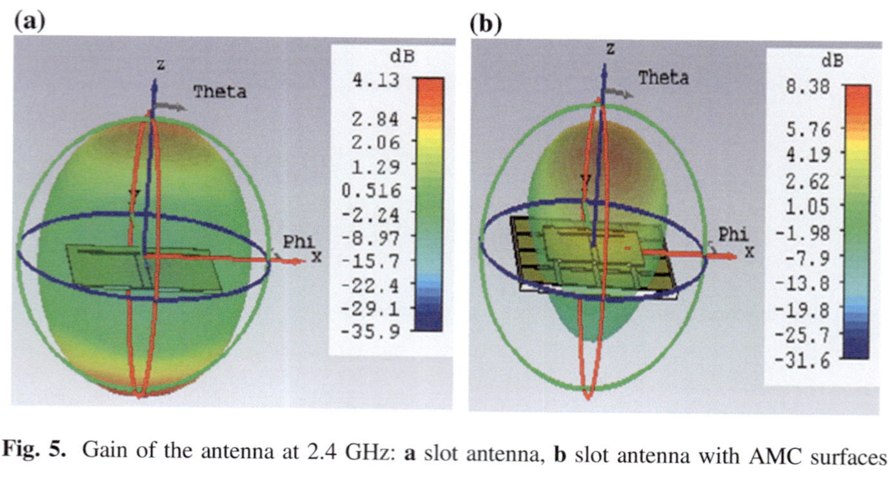

Fig. 5. Gain of the antenna at 2.4 GHz: **a** slot antenna, **b** slot antenna with AMC surfaces

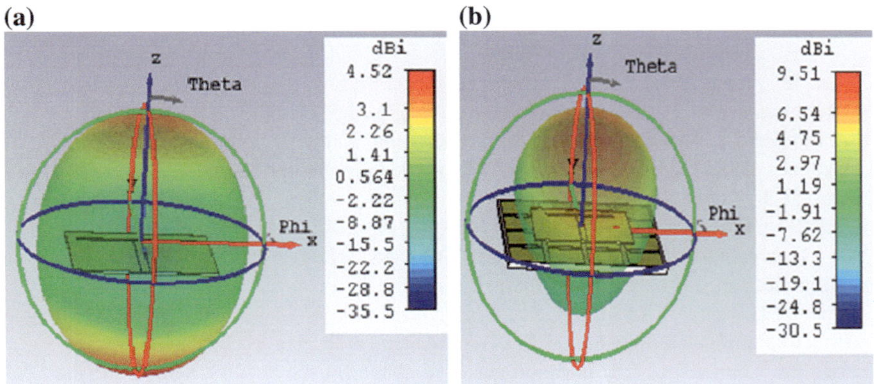

Fig. 6. Directivity of the antenna at 2.4 GHz: **a** slot antenna, **b** slot antenna with AMC surfaces

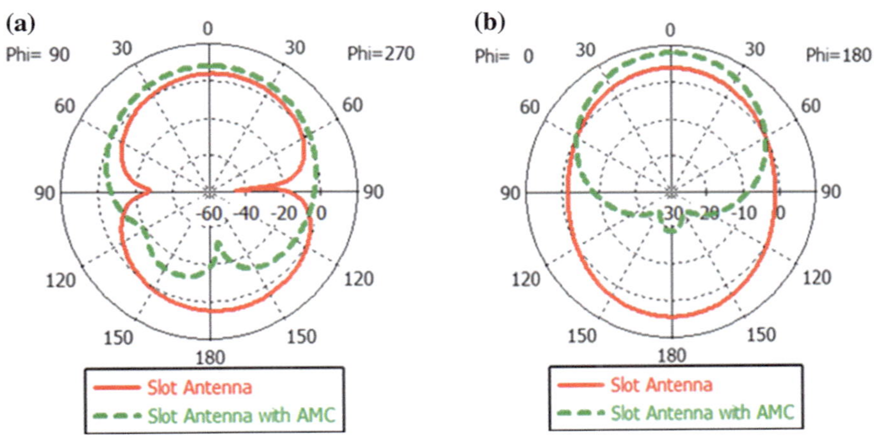

Fig. 7. **a** E-plane radiation pattern. **b** H-plane radiation pattern

5 Conclusions

This work proposes a slot antenna combined with artificial magnetic conductor surface by which bidirectional radiation behaviour of the slot antenna has been transformed into directional radiation pattern. This phenomenon also provides very low profile behaviour of the antenna and confirms an increased gain and directivity of 8.38 dB and 9.5 dBi at the resonant frequency, i.e. 2.4 GHz. Overall, this analysis concludes that slot antenna with AMC surface is good for wireless system (at ISM band) for its high gain and directivity. Further, the antenna possesses low profile characteristics.

References

1. Zaker R, Abdipour A (2010) A very compact ultrawide band printed omnidirectional monopole antenna. IEEE Antennas Wirel Propag Lett 9:471–473
2. Sievenpiper D, Zhang L, Broas RFJ, Alexopolous NG, Yablonovitch E (1999) High impedance electromagnetic surfaces with a forbidden frequency band. IEEE Trans Microwave Theor Tech 47(11):2059–2074
3. Kraus JD (1940) The corner—reflector antenna. Proc IRE 28(11):513–519
4. Uda S, Mushiake Y (1954) Yagi-Uda Antenna. The Research Institute of Electrical Communication, Japan, Tohoki University
5. Zhang Y, Hagen JV, Younis M, Fischer C, Wiesbeck W (2003) Planar artificial magnetic conductors and patch antennas. IEEE Trans Antennas Propag 51(10):581–583
6. Panda PK, Ghosh D (2018) Isolation and gain enhancement of patch antennas using EMNZ superstrate. AEU-Int J Electron Commun 86:164–170
7. Panda PK, Ghosh D (2019) Wideband bow-tie antenna with increased gain and directivity by using high impedance surface. Int J RF Microwave Comput Aided Eng 29(3):1–9
8. Foroozesh A, Shafai L (2011) Investigation into the application of artificial magnetic conductors to bandwidth broadening, gain enhancement and beam shaping of low profile and conventional monopole antennas. IEEE Trans Antennas Propag 59(1):04–19
9. Aminian A, Yang F, Rahmat-Samii Y (2003) In-phase reflection and EM wave suppression characteristics of electromagnetic band gap ground planes. In: Proceedings of the IEEE antennas and propagation society international symposium, vol 4, pp 430–433, June 2003

Electrical Energy Storage: A Great Business Ahead

Renu Sharma(✉), Satish Choudhury, Bibekananda Jena, and Sonali Goel

Department of Electrical Engineering, ITER Siksha 'O' Anusandhan (Deemed to be University), Bhubaneswar, India
{renusharma, satishchoudhury, bibekanandajena}@soa.ac.in

Abbreviations

GW	Giga watt
MW	Mega watt
kW	Kilo watt
UPS	Uninterruptible power supply
LCOS	Levelized cost of storage
ISOs	Independent system operators
BESS	Battery energy storage system
VRF	Vanadium redox flow
PCS	Power control system
BMS	Battery management system
PER	Primary frequency response
AGC	Automatic generation control
FFR	Fast frequency response
PFR	Primary frequency response
BTW	Behind-the-meter
PV	Photovoltaic
MISO	Midcontinent independent system operator
CAISO	California Independent System Operator Corporation
BESS	Battery energy storage systems
RegD	Dynamic regulation signal
RegA	Traditional regulation signal
FERC	Federal Energy Regulatory Commission
PHS	Pumped hydro storage
DOD	Depth of discharge

1 Introduction

Energy storage is one of the fastest growing areas in the Indian utility industry and rapidly becoming the mainstream. About 202 GW of grid-connected storage systems are available globally of which 135 GW is pumped hydro and 65 GW is UPS systems

M. N. Mohanty and S. Das (eds.), *Advances in Intelligent Computing and Communication*, Lecture Notes in Networks and Systems 109,
https://doi.org/10.1007/978-981-15-2774-6_35

and rest about 2677 MW being new storage technologies [1]. Policymakers are using mandates to encourage storage investments. National Mission on Transformative Mobility and Battery Storage approved by Cabinet Mission have approved for setting up a National Mission on Transformative Mobility and Battery Storage for driving sustainable and holistic mobility initiatives. The Phased Manufacturing Program valid for 5 years till 2024 to support setting up of a few large-scale, export-competitive integrated batteries and cell-manufacturing giga plants in India for confine production of the entire electric vehicle value chain. Several new application areas for batteries have opened up that require large-scale batteries for storage and grid balancing applications. Flow batteries are suitable for energy storage applications requiring long hours of backup, ranging from a few hours to several days. As power (in kW) and energy capacity (in kWh) are separated, the battery capacity (or discharge time) can simply be enlarged by filling more liquid into the tanks and the power capacity can be increased by increasing the battery stacks. Flow batteries usually offer long life and stable capacity, but low energy density makes them more suitable for use in stationary applications.

A large number of battery technologies are available in the market. The conventional battery technology like the lead–acid battery is still good for residential and industrial applications. The advantages in this type of battery is the low cost ($100–200/kWh), it has no major safety issues and can deep discharge. The drawbacks, on the other hand, are a short life (300–1000 cycles, i.e., 2–3 years of life), low specific energy (35–50 Wh/kg), limited depth of discharge (up to 50%), thermal runway (performance and life degrades as battery heats up and as temperature changes), and disposal issues, as lead is toxic. The selection of batteries depends entirely upon the application and one should consider various factors such as capital cost, LCOS, lifetime, efficiency, depth of discharge, footprint, etc., for selecting the right battery technology. The Li-ion battery is a market leader for its portable applications and the electric vehicle. Lead–carbon batteries are an improved version of lead–acid batteries and provide long life and fast charging/discharging Li-ion batteries provide higher energy density (80–240 Wh/kg) are compact in size, fast charging, and hence are best for portable applications and for electric vehicles. However, Li-ion and lead–acid batteries are designed to discharge for 3–4 h only. For storage applications requiring long hours of backup, Li-ion and lead–acid batteries are not suitable. Further, safety issues like fire hazard associated with the Li-ion batteries are still a concern over their use in large-scale storage. Flow batteries are suitable for large-scale energy storage applications requiring long hours of backup and heavy-duty cycles. Among them, vanadium redox battery is one of the suitable candidates for renewable integration.

Vanadium redox flow (VRF) battery is one of the proven and commercially available flow battery technologies ideals for applications in micro-grids, grid balancing, electric vehicle charging stations, etc., worldwide, about 1 GWh of VRF batteries has been installed and is in operation. Since vanadium at different oxidation states is stored in two separate tanks, there is no detrimental effect of cross mixing of the electrolyte. The battery stacks are made of a graphic electrode and polymer membranes, and no components are consumed or degraded during the operation. Therefore, the lifetime of the battery is more than 20 years (>10,000 cycles) and provides a stable performance (no capacity decay) over its lifetime price wise, VRF batteries are available \$1000/kWh for small to medium systems and \$500/kWh for MWh level systems. Due to the long life of VRF batteries, the leveled cost of energy storage is the cheapest compared with other battery types. The major supply of vanadium currently comes from China, and due to the recent rebar policy in China the price of vanadium is a commodity and can be recovered 100% from the used battery, a leading model is being developed to eliminate the upfront cost burden to developers and users.

Zinc bromine hybrid flow batteries are emerging as a suitable technology for residential, telecommunication, and other energy storage markets. This battery is good for deep discharging and provides stable performance for 10 years. However, the use of halide chemistry and the need for routine maintenance of this battery are the drawbacks. In large-scale energy storage applications, technology requiring frequent and long duration maintenance also adds extra cost and necessitates the redundancy plan. In terms of price, zinc–bromine batteries are available at \$1000/kWh.

Iron chromium and all-iron flow batteries are other types of flow batteries available at low cost, but they have limited installations and their long-term performance in the field has to be evaluated. Sodium sulfur batteries are also available for large-scale energy storage applications. They have a high energy density, high roundtrip efficiency, and long life, but they need high temperature (>300 °C) for operation and have a safety risk associated with sodium. Rechargeable zinc–air batteries are also emerging as a low cost option at 4100/kWh battery for energy storage with long hours of discharge capability, high energy density (450 Wh/l), and good performance of over 10,000 discharge hours. The drawbacks of this battery are low-power density, relatively short life, low roundtrip efficiency, and the requirement of preventive maintenance. The overall analysis about different type of batteries is given in Fig. 1.

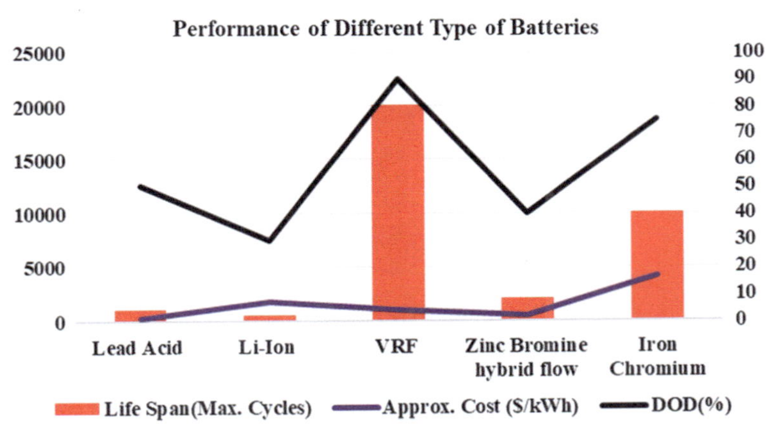

Fig. 1. Performance analysis of different type of batteries in terms of cost, life span, and DOD

In terms of cost of energy storage, one should consider not only the battery or module cost but also the cost for power control system (PCS), battery management system (BMS), balance of the plant, fire suppression system, air-conditioning, etc.

2 Storage System

Storage can provide more value to a system or to infrastructure that is highly stressed. For example, systems that are running up against minimum generation constraints on thermal units and curtailing wind and solar, or distribution feeders that are hitting thermal overloads, or island systems where the rate of change of frequency can be very high, may all get more value from storage than systems that are less stressed. Storage can provide system flexibility, fast reserves, etc., but the value of these is greater in very constrained systems. Other infrastructure and/or modifications to planning and operations may also provide the same benefits that storage provides, so it is also a matter of least-cost mitigation options. Basic application of storage system is shown in Fig. 2. It shows the interconnection between different sources of electricity and the existing utility grid [2]. Out of many sources shown in the figure, energy storage device, i.e., battery is one of the vital components to provide bridge between source and the sink [3]. It also plays a key role at the time of power outage from different existing sources and in the remote places where electricity is unreachable.

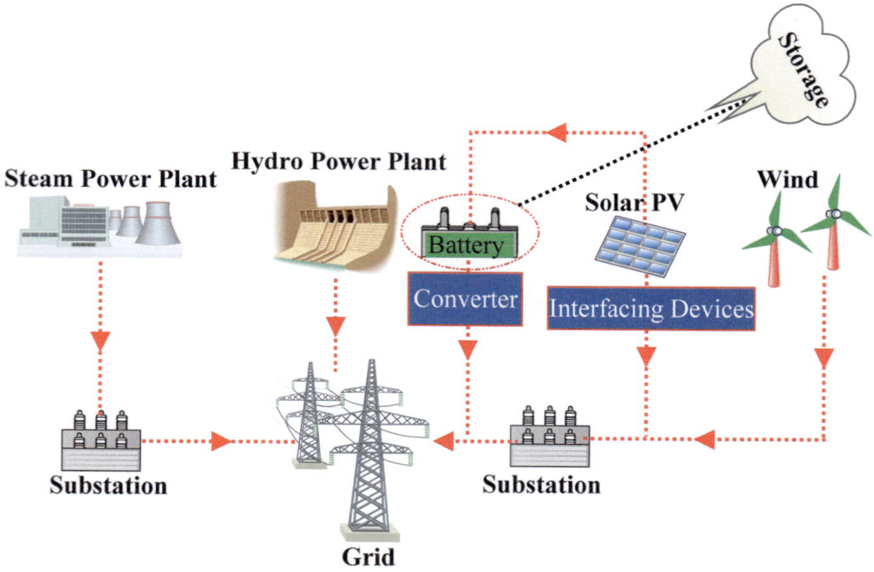

Fig. 2. Interconnection of different sources of electricity with utility grid

2.1 Solar and Wind Integration

The variability and uncertainty of solar and wind create challenges for utilities and ISOs that need to maintain system reliability. A large number of potential options have been discussed in literature for mitigating the impacts of solar and wind on the grid [4, 5]. Storage has been considered an option, but cost recovery for storage to integrate wind is that the high forecast errors (over—or under—forecasts) [6]. To take advantage of that price spread, storage needed to better forecasts than the wind forecasts.

Storage can provide a better match for solar integration than for wind integration [7]. First, solar is much more predictable and storage owners easily know when to charge and discharge, unlike with wind. Second, solar integration is much more about time-shifting of large amounts of energy because the midday solar peak is so pronounced, whereas there tends to be some wind at all hours of the day, on average.

2.2 Where Is Storage Most Effective?

Storage can be implemented at various levels on the power system:

- Bulk power system/transmission level—providing energy arbitrage, avoiding transmission congestion or losses
- Distribution feeder level—in addition to the above, storage can also provide more location-specific services such as deferring distribution upgrades.
- Behind-the-meter (BTM)—some of the above benefits may be accessible to BTM storage. Storage can also provide services such as reducing demand. Charges or managing energy bills with time-of-use rates.

The further downstream storage is, the more benefits it can theoretically provide [8]. However, the further downstream storage is, the less accurately the markets/incentives/rates reflect the physics of power flows. For example, the volumetric and bundled rates for most residential customers reflect very little about the actual cost of service at any given time. In another example, the rates for commercial customers may include a demand charge or a time-of-use charge but still are likely to ignore the impacts of that commercial customer on ancillary services.

Additionally, the further downstream storage is, the less it may be able to monetize the various value streams. BTM customers are unlikely to have a large enough system, and the telemetry and control sophistication to participate in energy, capacity or ancillary service markets. For example, while a BTM system may defer upgrades of the distribution system, it cannot monetize this value.

2.3 Ownership

Ownership of the storage asset constrains the applications that the storage unit provides. On the bulk power system level, there may be issues as to whether storage is treated as a generator or as a load. As a load, it fits into demand-side management markets but perhaps not when it acts as a generator.

BTM systems can bring qualitative benefits that are not economic. For example, consumers that already own distributed PV systems may wish to add storage to increase their usage of self-generated electricity, thus increasing their self-sufficiency. While these benefits may not make economic sense, they can be important drivers in decision-making.

BTM systems are most likely to be under-utilized. Part of this is the lack of sophistication of the end-users; part is the lack of access to the markets that would allow the owners to better utilize the systems. For example, in providing demand charge reduction, there will be many hours when the unit could be providing regulation and still have time to charge to prepare for offsetting the demand charge. This means there are opportunities for third parties to step in, aggregate these systems and install the telemetry and controls needed to bid services into the markets.

2.4 Market Rules

Some market rules limit the ability of storage to fully monetize its benefits. For example, in MISO, storage resources that have one hour or less duration can provide regulation, but if it does provide regulation, it cannot simultaneously provide spin, supplemental reserve or ramping reserve. That limits the value stream that storage can access. In CAISO, storage is treated as a generator that produces both positive and negative energy.

2.5 Co-location with Solar and Wind

Many utility-scale battery projects are co-located with new or existing wind and solar plants. This allows the battery to leverage significant interconnection infrastructure. Additionally, storage paired with solar or wind plants can be eligible for federal investment tax credits, state incentives, and/or subsidized loans. Despite co-location with a wind plant and capability of a BESS may be operated independently of the wind plant. In PJM, this is frequently done to prioritize BESS use the highly valued fast regulation service.

3 Energy Arbitrage

Even though energy arbitrage (changing when prices are low and discharging when prices are high) is the main application that most people consider for energy storage, it is difficult to make a business case for new storage with arbitrage as the sole application. First, the current spate of low natural gas prices does not drive frequently high price spreads that drive arbitrage opportunities. Additionally, price forecast errors tend to be worse when arbitrage opportunities are best (prices are volatile), making bidding less certain.

As mentioned earlier, storage in energy arbitrage applications should be able to provide high value to grids with high solar penetrations. Unlike wind, the predictability of solar means that high solar penetrations will drive much more predictable price signals. The negative prices in the day-ahead market due to the very high levels of solar are an indication of this predictability.

A final issue for storage is determining when to bid on as a generator and when to bid on as a load. Because storage owners do not have the full information that the ISO has, e.g., transmission congestion, it is difficult for storage owners to know how to bid the storage and storage owners may attempt to schedule the storage in a way that is not beneficial to the grid. Incorporating storage scheduling into the ISO's security-constrained unit commitment in ISOs that allow this, such as PJM, can essentially give the ISO control over the resource to optimize operation.

Storage is one of many options for flexibility on the grid. Other option also holds promise for rapid adoption that provides time-shifting of the load. Dynamic pricing is being considered as rate restructuring is debated in many states, partly driven by shortcomings of net metering and partly driven by advanced metering capabilities. To some extent, dynamic pricing will compete with storage to provide the same arbitrage value to the system. Essentially, dynamic pricing could drive arbitrage storage out of the bulk power system and into the BTM market segment.

Another emerging technology option is transportation electrification. Electric vehicles (EVs) with vehicle to grid (V2G) technology essentially look like storage with limited runtimes. But even without V2G technology, EVs provide one part of the storage arbitrage value, that of providing load, and if EV charging is controlled, this may reduce curtailment of excess wind or solar.

4 Frequency Regulation

One of the most promising applications for storage technologies is frequency regulation. Frequency regulation (also known as secondary reserves or automatic generation control) tends to be the highest value reserve service. While conventional storage technologies such as pumped hydro storage, fit well in this application, there is a "fast" frequency regulation application for which new storage technologies (such as batteries and flywheels) are ideal.

4.1 Tuning the Fast Regulation Signal

In 2009, PJM had collaborated with AES on the use of lithium battery storage to provide higher performance regulation. At first, they tried to use a fast ramping hydrogenation-type signal. However, lithium-ion batteries tend to have a limited energy capacity, typically 4:1 power to energy ratio. The limited energy capacity was not compatible with this signal, which had tended to push the battery to its top or bottom limit and pin it there. PJM designed the dynamic control signal (RegD) with this type of limited energy capacity in mind. The target time for the RegD signal to converge to zero was 5 min and 95% of the time, it converges to zero in less than 15 min [9].

4.2 Performance-Based Regulation

In 2011, The Federal Energy Regulatory Commission (FERC) Order 755 directed system operators to "pay for performance" for providing frequency regulation and grid stabilization services. Under this order, resources that can more accurately follow an operator's regulation signal should be rewarded. PJM was the first ISO in the USA to implement performance-based regulation. Regulation resources are compensated for capability (reserve MWs), performance (mileage), and lost opportunity cost ($/MW of lost revenue in the 5-min interval). Performance of resources determines whether a resource can bid on the traditional regulation signal (RegA) or the faster RegD signal. Additionally, fast regulation provides greater benefits: PJM has found that the first MW of RegD offsets 3 MW of RegA, but that this benefit declines as more resources provide RegD, and that above 60% fast resources, there is no additional benefit [10].

Performance-based regulation has enabled PJM to reduce regulatory requirements. Previously, PJM required 1% of the on or off-peak load forecast, depending on the time of day. After introducing RegD, the total regulation requirement ratcheted down over subsequent months to its current level of 0.7%. PJM's control performance standard metrics demonstrate a continued performance in regulating area control error.

4.3 Impact on Prices

Regulation prices increased from $13.75/MWh prior to performance-based regulation to $38.75/MWh when RegD was introduced partly because the old framework did not include lost opportunity cost as part of the compensation for regulation. If this lost opportunity cost is included, then PJM's overall regulation cost remains similar to its cost prior to performance-based regulation.

4.4 Storage Market Impacts

PJM's performance-based regulation, including the fine-tuning of the RegD signal for the limitations of battery storage, has made PJM the leading battery storage market in the USA. Battery storage projects in PJM's interconnection queue went from 9 MW in 2012 to 534 MW in 2015. The share of regulation capability provided by batteries increased from 16% in 2014 to 28% in 2015. At the same time, the share of regulation credits earned by batteries increased from 12 to 21%. There is a concern; however, that the RegD market is saturated because RegD is over-procured due to an inconsistency in how RegD substitutes for RegA in the market design [11].

5 Demand Charge and Bill Management

The fastest growing sector for energy storage is the BTM sector which grew from 6 MW in 2014 to 35 MW in 2015. The ability of BTM storage, especially BESSs, to reduce demand charges and manage energy bills, has driven the commercial sector to be the fastest growing storage segment in the USA over the last two years [12].

This application is heavily dependent on the characteristics of the particular load and the rate structure. Finding storage technologies that fit the needs of the load and rates are also critical.

Energy storage systems for demand charge reduction and bill management may have relatively low utilization factors. For example, if the system's primary application is to help meet energy use during expensive afternoon and evening hours, the system may have all night and morning (far more hours than necessary) to ensure a full charge. There is thus an opportunity for this system to provide other grid benefits during that downtime. However, these systems tend to be too small to qualify to be in the market and the technical and financial hurdles to participate in the market tend to be too great.

6 Fast Frequency Response

New storage technologies such as batteries, flywheels, and supercapacitors, may be capable of providing a very fast response to frequency deviations. With increased penetrations of non-synchronous generators and decreased levels of synchronous generators that provide inertia, the system may have more challenges maintaining frequency stability. If a large generator trip offline, a power system with lower inertia will have a faster drop in frequency. To recover the system, frequency responsive generation is needed to increase output. Conventional generators providing primary frequency response are relatively slow to respond (several seconds), and may be too late to help a system that has very low inertia. The FFR is a new service in some regions (required or likely to be required in Ireland, UK, ERCOT) that can provide a very fast response (less than one second) to help recover frequency. Additionally, simultaneous have shown that in some system conditions (low inertia), 1 MW of FFR can provide the same level of support to the system as twice the amount of conventional primary frequency response (PFR) [10]. This means that if some of the frequency response is FFR, the total requirement for frequency response could be lower.

While it is too soon to know what prices will be for this service and how large of a market there will be, there are relatively few resources that can provide this service, depending on requirements for response speed, and so storage will likely have a large role to play.

7 Life Cycle Costing

A cost analysis [13] is made to explain how the storage device is economical as compared to the existing sources and how it leads to great business ahead. A small community has taken into consideration for the life cycle costing analysis. Let 100 numbers of people are staying in the community and they are accessing electricity from a hydel plant. Let the capital cost of consumption of electricity in hydel plant is about ₹50,000 + 5000 * l, where it includes installation cost, transmission cost, cost of the line, and losses. 'l' denotes the length of the line in km. For lifetime of 25 years and taking a present worth factor of 7.84, the annual life cycle cost (ALCC) of hydel system is calculated in (1).

$$\mathrm{ALCC_{hydel}} = \frac{50000 + 5000 * l}{7.86} \approx 6378 + 636 * l \tag{1}$$

Similarly, if the same community wants to install a solar PV system with energy storage device it costs approximately 200 per watt, where the availability of solar insolation is expected to be 5 kWh/m^2/day. For 'x' kW of consumption, the capital cost of solar PV system will be 200x. So, taking a life span of 25 years and present worth of 7.84 the ALCC of PV system is calculated in (2).

$$\mathrm{ALCC_{pv}} = \frac{200x}{7.84} \approx 25.5x \tag{2}$$

Annual energy generation in hydel plant is calculated in (3).

$$E_{\mathrm{hydel}} = 1\,\mathrm{kW} \times 24\,\mathrm{h} \times 365\,\mathrm{days} \times \eta = 1 \times 24 \times 365 \times 0.3 = 2628\,\mathrm{kWh} \tag{3}$$

$$\begin{aligned} \text{Cost per unit consumption} &= \frac{\mathrm{ALCC_{hydel}}}{E_{\mathrm{hydel}}} = \frac{6378 + 636 * l}{2628} \\ &= 2.42 + 0.242 * l \end{aligned} \tag{4}$$

Annual energy consumption in PV system is calculated in (5).

$$\begin{aligned} E_{\mathrm{pv}} &= \frac{x}{1000}\,\mathrm{kW} \times 5\,\mathrm{kWh/m^2/day} \times 365\,\mathrm{days} \\ &= \frac{x}{1000} \times 5 \times 365 = 1.825x \end{aligned} \tag{5}$$

$$\text{Cost per unit consumption} = \frac{\mathrm{ALCC_{pv}}}{E_{\mathrm{pv}}} = \frac{25.5x}{1.825x} = ₹14 \tag{6}$$

To compare between two systems, breakeven point can be calculated as in (7).

$$\begin{aligned} 2.42 + 0.242 * l = 14 \\ \Rightarrow l \approx 48\,\mathrm{km} \end{aligned} \tag{7}$$

It can be concluded from the above analysis that if your community is about 48 km distance from the hydel system then it is more beneficial to use such sources of electricity, whereas if your location is more than 48 km from hydel station then it is more economical to use PV system with battery storage facility. In this paper, the cases of remote area where electricity is unreachable and the area, where the power outage is more frequent are taken into consideration. The analysis shows the storage system (PV along with battery) can be better replacement for the sources of electricity for such type of community and the business can be more propagated in such commodity.

8 Conclusion

Storage has the potential to provide a large number of values to the power system. However, different actors can access and monetize different portions of that overall value stream, and even that varies depending on the ISO or jurisdiction. Existing PHS that was installed in the past years to integrate the nuclear fleet has earned revenue by stacking these benefits for years, but it is difficult to recover costs on new utility-scale storage by stacking benefits, at least in the current low natural gas price paradigm.

The main commercial markets for storage are utility-scale BESSs offering fast regulation reserves and BTM BESSs for demand charge/bill management. FFR is likely to be a new niche market for utility-scale BESSs. As rules evolve, to allow aggregators into the market and to allow different market participants to access more of the value stream, storage growth should accelerate. And as rates are restructured to send more granular signals to end-users, dynamic pricing is likely to drive much more BTM storage down to the residential level.

Acknowledgements. The assistance provided by the authorities of Siksha 'O' Anusandhan (Deemed to be University), Bhubaneswar are highly acknowledged.

References

1. Smart grid handbook for regulators and policy makers. Nov 2017
2. Vazquez S, Lukic SM, Galvan E, Franquelo LG, Carrasco JM (2010) Energy storage systems for transport and grid applications. IEEE Trans Ind Electron 57(12):3881–3895
3. Kalaiselvam S, Parameshwaran R (2014) Thermal energy storage technologies for sustainability: systems design, assessment and applications. Elsevier
4. GE Energy (2010) Western wind and solar integration study. NREL/SR
5. GE Energy Consulting (2012) Hawaii solar integration study
6. Gharehpetian GB, Agah SM (eds) (2017) Distributed generation systems: design, operation and grid integration. Butterworth-Heinemann
7. Mills A, Wiser R (2014) Strategies for mitigating the reduction in economic value of variable generation with increasing penetration levels. LBNL-6590E
8. Fitzgerald G, Mandel J, Morris J, Touati H (2015) The economics of battery energy storage. RMI, Boulder, CO
9. Benner S, PJM (2015) A brief history of regulation signals at PJM
10. PJM (2013) Performance based regulation: year one analysis
11. Monitoring Analytics (2016) LIC, state of the market report for PJM January through June, Aug. 11, 2016
12. GTM Research/Energy Storage Association (2016) U.S. energy storage monitor 2015 year in review
13. Zakeri B, Syri S (2015) Electrical energy storage systems: a comparative life cycle cost analysis. Renew Sust Energy Rev 1(42):569–596

Embedded Entropy-Based Registration of CT and MR Images

Sunita Samant, Subhaluxmi Sahoo(✉), and Pradipta Kumar Nanda

Image & Video Analysis Laboratory, Department of ECE,
S O A, Deemed to be University, Bhubaneswar, India

1 Introduction

Multimodal image registration is a methodology of spatial mapping between different images acquired by different diagnostics processes. Multimodal image registration is among the most challenging problems in medical imaging. In recent years, different registration methods have been proposed. One of the methodologies is the measure similarity in image registration is maximizing the MI. Registration of multimodal images using MI is proposed by Maes et al. [1] and for registration, where no assumptions were made and no boundary constraints were imposed. Subsequently, Chen and Varshney [2] and Juni et al. [3] have addressed the artifacts using the interpolation method. Woo et al. [4] have proposed a new registration technique incorporating spatial and geometric information to high resolution images. Mohanalin et al. [5] have proposed an innovative idea to increase the fidelity of multimodal image registration using Tsallis entropy. Khadar and Hamza [6] have proposed a Jensen–Tsallis divergence along with histogram-based modality transformation, and Li et al. [7] have proposed Jensen–Arimoto divergence. But, to take care of number of bins used for the computation of mutual information, a minimum MSE histogram-based estimation method has been proposed [8]. Han et al. [9] have proposed a non-rigid image registration technique that is based on a weighted Horn-type regularizer. Another registration approach that has been used in [10] involves finding structural information of the images to register. In our work, we have proposed a new approach for registering noisy images. Our registration scheme for noisy images has been developed on the notion of mutual information to find out the optimal registration parameter. The 2D joint histogram of the noisy MR and CT image is assumed to be degraded which would result uncertainty in finding mutual information and optimal registration parameter. So instead of registering the noisy images, we have created embedded images for registration process to take care of noisy situation. We have considered noisy MR-T1 image and CT images from retrospective image registration evaluation (RIRE) database and noisy T2 and T1 from brain web database. MR image is considered to be noisy one, so Gaussian kernel filter is applied to smooth out the noise component. We have obtained structural representations of the images by calculating embedded entropy of the images and then input images are replaced by the embedded entropy images for the registration. Then, maximization of mutual information determines the registration parameter. The proposed method has been tested for 5, 10, 20 dB SNRs of MR-T1 (RIRE) and MR-T2 (brain web) data set.

M. N. Mohanty and S. Das (eds.), *Advances in Intelligent Computing and Communication*, Lecture Notes in Networks and Systems 109,
https://doi.org/10.1007/978-981-15-2774-6_36

2 Background on Entropy-Based Image Registration

Viola [11] has introduced a notion of image registration based on mutual information (MI). The motive of registration is to compute a transformation that will align two or more images captured with different techniques at different time. Consider two images $\boldsymbol{I_f}$ and $\boldsymbol{I_r}$ as the floating and reference images, respectively. The registration of these images is then formulated using mutual information. The estimation of the transformation (*T*) which will align the floating image *F* with reference image *R* is obtained by maximizing the mutual information over transformation *T*,

$$T^* = \arg\max \mathrm{MI}(F, R) \tag{1}$$

MI of two random variables *F* and *R* with marginal probability and joint probability mass function is defined by

$$\mathrm{MI}(F, R) = p_{\mathrm{FR}}(f, r)\log_2 \frac{p_{\mathrm{FR}}(f, r)}{p_{\mathrm{F}}(f)p_{\mathrm{R}}(r)} \tag{2}$$

MI measures the degree of interdependence between *F* and *R* in terms of the distance measured between joint distribution $P_{F,R}\ (f,\ r)$ and distribution associated with $P_F(f)$ and $P_R(r)$. Entropy can be interpreted as a measure of uncertainty, variability or complexity. The MI related to entropies is stated as

$$\mathrm{MI}(F, R) = H(F) + H(R) - H(F, R) \tag{3}$$

where $H(F)$, $H(R)$ and $H(F,\ R)$ are the marginal and joint entropies of the random variable *F* and *R*, respectively.

3 Embedded Entropy Images

The similarity between images can be interpreted as whether intensity changes occurring at similar locations. So, the intensity changes or the bound for a lossless compression in an image can be quantified by the entropy measures.

Here, we have introduced a new structural representation of the image by embedding Arimoto entropy and Tsallis entropy. In the following, Arimoto and Tsallis entropies are defined.

3.1 Arimoto Entropy

For any discrete random variable $X(x_1,\ x_2, \ldots,\ x_N)$ with its distribution function $P(p_1,\ p_2,\ \ldots,\ p_N)$, i.e., Arimoto entropy is given by

$$H_{\mathrm{A}}(X) = \frac{q}{q-1}\left[1 - \left(\sum_{i \in X}\sum_{j \in X} p_{ij}^{q}\right)^{1/q}\right] \tag{4}$$

where 'q' is the Arimoto's parameter. $p_{i,j}$ is the joint probability of the current pixel with the corresponding neighborhood. Arimoto entropy can be viewed as generalization of Shannon's entropy. Arimoto entropy is pseudo-additivity and non-negative in nature.

3.2 Tsallis Entropy

For any discrete random variable $X(x_1, x_2,\ldots, x_N)$ with its probability distribution $P(p_1, p_2,\ldots, p_N)$, i.e., Tsallis entropy is given by

$$H_{\mathrm{T}}(X) = \frac{q}{q-1}\left[1 - \sum_{i \in X}\sum_{j \in X} p_{ij}^{q}\right] \tag{5}$$

It is extended version of Shannon's entropy and generalization of Boltzmann–Gibb's entropy. It contains a 'q' parameter which is a real and represents the system's non-extensivity.

3.3 Embedded Entropy

The Arimoto and Tsallis parameters play a vital role for obtaining the structural information of the image. The different values of the parameters result in sub-extensive or super-extensive property. Therefore, we have introduced the function to embed Arimoto and Tsallis entropy.

$$\mathrm{EE}(X) = (1+q)H_{\mathrm{T}}(X) + (1-q)H_{\mathrm{A}}(X) \tag{6}$$

3.4 Embedded Entropy Image Generation

In order to generate entropy image for image registration, we have enforced a second-order entropic measure to image. To avoid the spatial ambiguities, each region is processed through Gaussian filter. For the original image with a size of MN, the neighborhood of size b is extracted for each image pixel i in l. In each region, Gaussian kernel filter is applied. Then, we have calculated the embedded entropy. This embedded entropy gives the statistical dependency between current pixel and its neighborhood pixels, and calculation of entropy value depends upon the number of bins used. Higher the bin size, more finer the histogram is, which gives more information about the image and which tends to more accurate result. But, at the same time higher bin size also requires more storage space, which leads to increase in redundancy in the joint histogram. So, we have limited the bin size to 10 by using global normalization. The normalized image is used for the estimation of the second-order entropy for all the pixels.

Due to local intensity changes, there is a similarity between current pixel intensity and its neighborhood intensity. And second-order entropy technique addresses changes in intensity in an image region locally when the second-order entropy values of these regions are minimized. In this proposed work, we have considered MR-T1 image as shown in Fig. 1a. The corresponding entropy images have been computed using Arimoto entropy, Tsallis entropy and embedded entropy as shown in Fig. 1b–d.

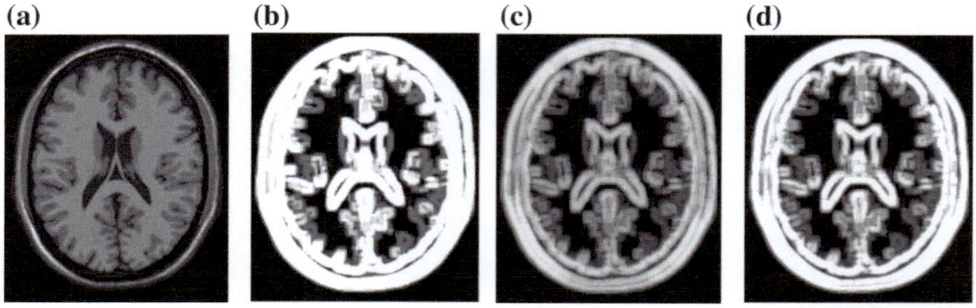

Fig. 1. **a** MR-T1 image, **b–d** entropy images using Arimoto entropy, Tsallis entropy and embedded entropy, respectively

4 Embedded Entropy Images-Based Registration

Registration of images has been achieved from the mutual information of the embedded entropy images.

Our registration process has been represented in Fig. 2. In our work, we first used the embedded entropy (EE) operator which indicates the structural changes in the images on floating image I_f and reference image I_r. Then, joint histogram of the entropy images is computed, and then the marginal and joint probability distributions have been calculated.

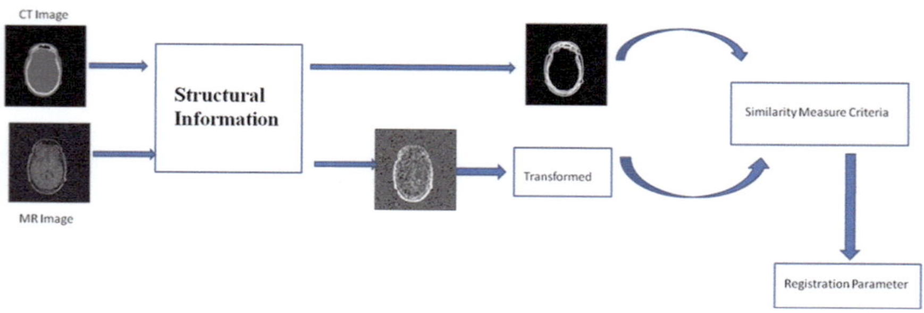

Fig. 2. Block diagram of proposed registration algorithm

Therefore, the entropy image registration is formulated as follows:

$$t^* = \arg\max \mathrm{MI}(\mathrm{EE}(I_R(r)), \mathrm{EE}(I_F(t(x)))) \quad (7)$$

where MI is the similarity measure criteria given as

$$\mathrm{MI}(F, R) = p_{FR}(f, r)\log_2 \frac{p_{FR}(f, r)}{p_F(f)p_R(r)} \quad (8)$$

and transformation function 't' maps the correspondence of the pixels x of the floating image I_F in the reference image I_R.

5 Pseudo Codes of the Algorithm

a) Calculation of embedded entropy images:
Input: I_f(Floating image) and I_r(Reference image), b=size of neighborhood, standard deviation = σ= 1.0, mean = μ = 0.0
Output: Ie_f, Ie_r, IG_f = gaussian(I_f, sigma, mu), IG_r = gaussian(I_r, sigma, mu), I_f = normalization (IG_f), I_r = normalization (IG_r)
for **i= 1 to MN in I (both floating and reference image)** do
P = neighborhood region(i,b)
e_Arimoto = second-orderArimotoEntropy(P)
e_Tsallis = second orderTsallisEntropy(P)
Ie (i) = (1+alpha) * e_Tsallis + (1-alpha) *e_Arimoto
end for
b) Calculation of MI of Ie_r & Ie_f
c) Determination of registration (α) from the computed MI

6 Results and Discussions

For the registration process, we have considered CT and MR-T1 images of RIRE database and MR-T1 and MR-T2 images of brain web database. In order to reveal the efficiency of our proposed method, Gaussian noise has been added to MR images at different SNR label. SNR can be defined as $SNR_{dB} = 10\log_{10}SP/NP$, where SP is signal power and NP is noise power.

In our paper, we have considered MR-T1 as the floating image and CT image as reference image from RIRE database for registration. MR-T1 image has been degraded by 2, 5, 10 and 20 dB SNRs as shown in Fig. 3b–e, respectively. We have extracted the structural images corresponding to CT and noisy MR-T1 images with 2, 5, 10 and 20 dB SNRs as shown in Fig. 3f–i, respectively. Then, the MI curve of the structural CT and MR images with 20 dB SNR has been computed from the joint histogram, and the

MI plot is shown in Fig. 4a. Maximum value of MI is found to be −6°, and the registration parameter alpha (rotational angle) that is estimated has been found to be at −6°. Subsequently, noise content has been increased in MR-T1 image to have of 10 and 2 dB SNRs. The mutual information for 10 dB and 2 dB SNR has been shown in Fig. 4b, c, respectively. The estimated registration parameter (alpha) using proposed method is found to be −4°. For the noise-free case, this alpha (registration) parameter has been found to be −4°. We can observe that the efficiency of our proposed method increases with increase in noise content. We have compared our proposed method with two existing methods as shown in Fig. 5. As we can see from the figure, the estimated errors do not change up to 10 dB SNR. MI, SSD method diverged beyond 5 dB SNR; whereas, the estimated error showed a converging trend beyond 5 dB SNR. It has been found that the performance of existing methods decreases as increase in noise. But, the proposed method gives better result with increase in noise content. We have also tested our proposed methods with MR-T1 and MR-T2 from brain web database. We have obtained the noisy MR-T2 images for 20, 10, 5 and 2 dB as shown in Fig. 6a–e. The corresponding structural images of MR-T1 and noisy MR-T2 are shown in Fig. 7a–e. The performance of our method has been found to be better than the other exiting method for 2 and 10 dB as shown in Fig. 8a–c for 10 dB, 5 dB and 2 dB SNR, respectively. The estimate errors of proposed method and existing methods have been found out as shown in Fig. 9.

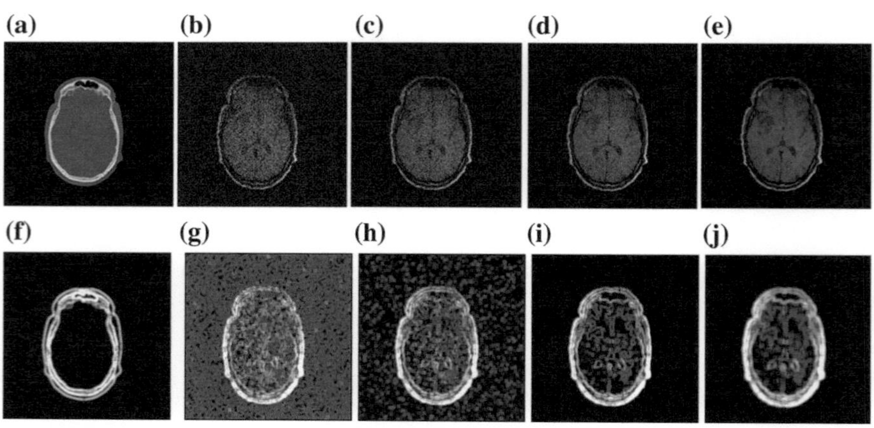

Fig. 3. **a–e** CT, noisy MR-T1 images with 2 dB, 5 dB, 10 dB and 20 dB SNR, respectively, **f–j** embedded entropy image of **a–e,** respectively

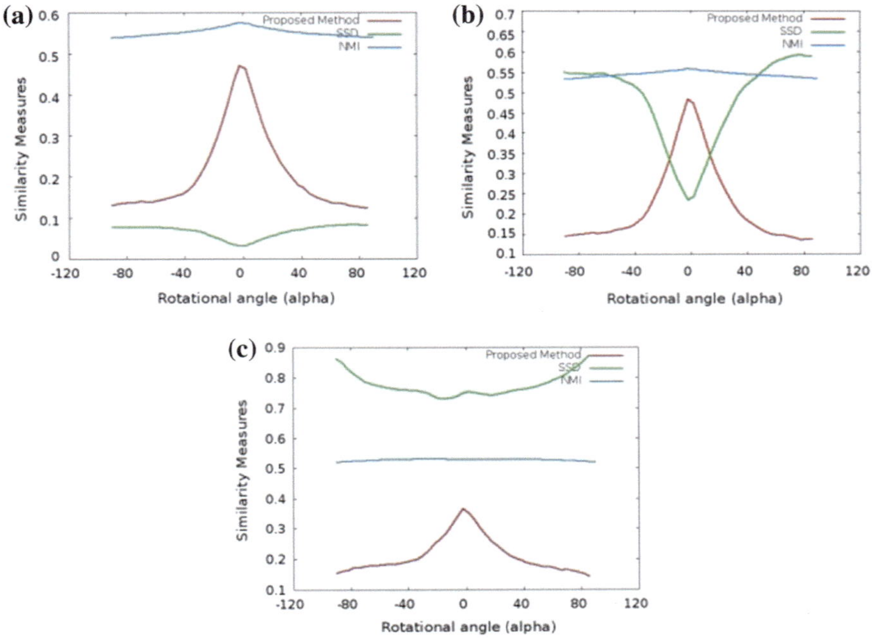

Fig. 4. Estimation of similarity metric for CT and noisy MR-T1 with **a** 20 dB and **b** 10 dB SNR **c** 2 dB SNR, respectively

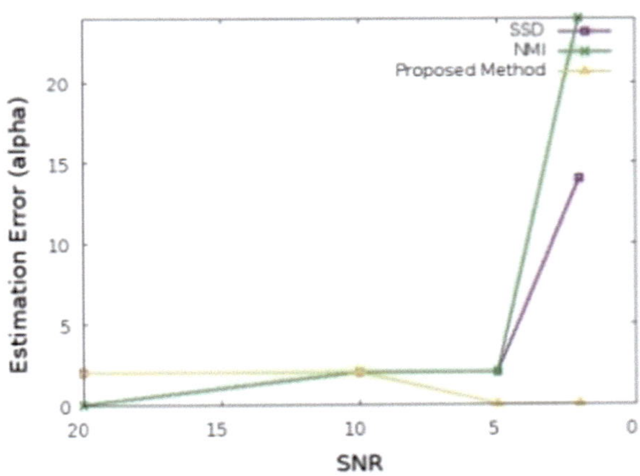

Fig. 5. Error plot for registration parameter (rotational angle)

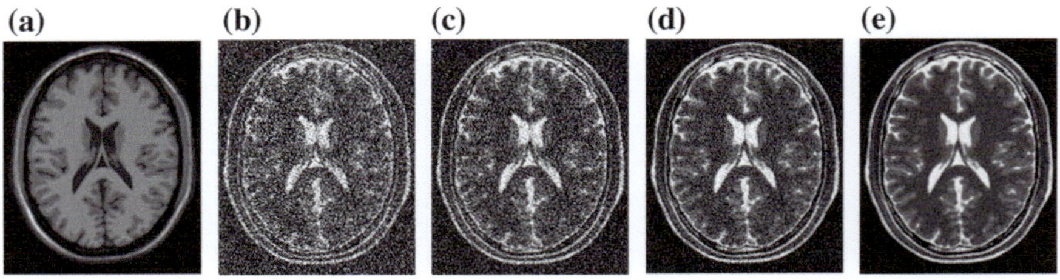

Fig. 6. **a** MR-T1, **b–e** MR-T2 with 2 dB, 5 dB, 10 dB and 20 dB SNR, respectively (brain web database)

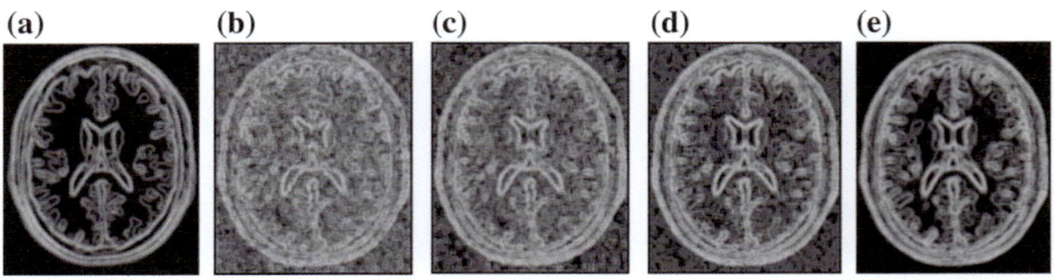

Fig. 7. **a–e** Embedded entropy images for MR-T1 and MR-T2 with 2 dB, 5 dB, 10 dB and 20 dB SNR, respectively (brain web database)

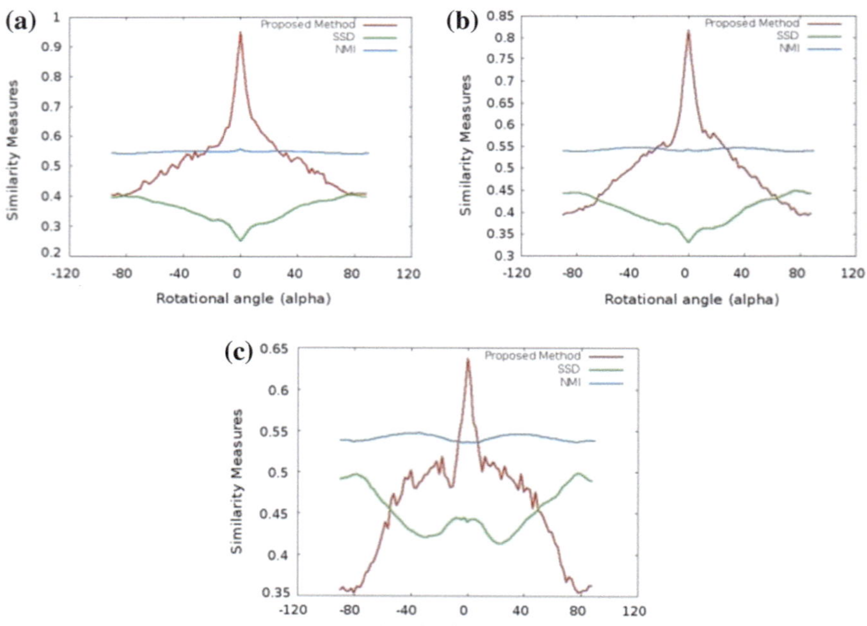

Fig. 8. Estimation of similarity metric for MR-T1 and noisy MR-T2 with **a** 10 dB SNR, **b** 5 dB SNR and **c** 2 dB SNR, respectively

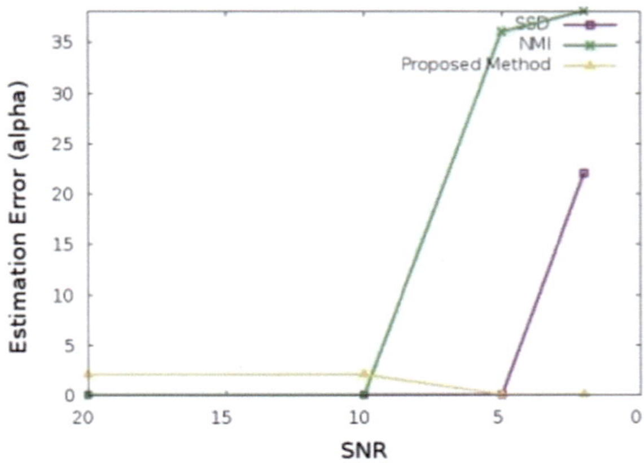

Fig. 9. Error plot for registration parameter (rotational angle)

7 Conclusions

In this work, we have presented an embedded entropy image-based multimodal image registration. The proposed method is based on the statistical relation between a pixel and its neighborhood in an image region. The embedded entropy, which is computed from this joint histogram, gets affected with increase in noise content. However, the proposed method achieved better results as compared to other existing methods with high noise levels in terms of estimation error.

References

1. Maes F, Vandermeulen D, Suetens P (1997) Medical image registration using mutual information. IEEE Trans Med Imaging 16(2):187–198
2. Chen H, Varshney PK (2003) Mutual information-based CT-MR brain image registration using generalized partial volume joint histogram estimation. IEEE Trans Med Imaging **22**(9):1111–1119
3. Li J, Cong R, Jin L, Wei P (2008) A medical image registration method based on weighted mutual information. In: 2nd International conference on bioinformatics and biomedical engineering, Shanghai, pp 2549–2552
4. Woo J, Stone M, Prince JL (2015) Multimodal registration via mutual information incorporating geometric and spatial context. IEEE Trans Image Process 24(2):757–769
5. MohanalinB, Kalra PK, Kumar N (2010) An automatic image registration scheme using Tsallis entropy. Biom Sig Process Control 5(4):328–335
6. Mohammed Khader A, Hamza B (2012) An information-theoretic method for multimodality medical image registration. Expert Syst Appl 39(5):5548–5556
7. Li B, Yang G, Shu H, Coatrieux JL (2014) A new divergence measure based on Arimoto entropy for medical image registration. In: 22nd International conference on pattern recognition, Stockholm, pp 3197–3202

8. Hacine-Gharbi A, Deriche M, Ravier P, Harba R, Mohamadi T (2013) A new histogram-based estimation technique of entropy and mutual information using mean squared error minimization. Comput. Electr. Eng. 39(3):918–933
9. Han Y, Feng X-C, Baciu G (2013) Local joint entropy based non-rigid multimodality image registration/ Pattern Recogn. Lett. 34(12):1405–1415
10. Wachinger Christian, Navab Nassir (2012) Entropy and Laplacian images: structural representations for multi-modal registration. Med Image Anal 16(1):1–17
11. Viola P, Wells III WM (1997) Alignment by maximization of mutual information. Int j comput vis 24(2):137–154

Power Factor Corrected SEPIC and Cuk Converter—A Comparison

Alok Kumar Mishra[1(✉)], Akshaya Kumar Patra[1], Ramachandra Agrawal[1], Lalit Mohan Satapathy[1], Shekharesh Barik[2], Samarjeet Satapathy[1], and Jnana Ranjan Swain[1]

[1] EEE Department, ITER, S'O'A (Deemed to be University), Bhubaneswar 751030, India
lalitsatapathy@soa.ac.in
[2] CSE Department, DRIEMS (Autonomous), Cuttack 754022, India

1 Introduction

According to the international standards and for the best power transfer and utilization, power factor correction has become a necessity. To get a DC output voltage we use a rectifier and a parallel capacitor [1, 2] as shown in Fig. 1a. The input voltage and current are shown in Fig. 1b. Figure 1b depicts a pulse-shaped input current with much harmonics content causing a very poor power factor. The IEEE and IEC are some international entities to define or standardize the permissible limits of the harmonic content in-line current such as IEEE 519 and IEC 61000-3-2 [3]. The design of an AC to DC power converter to overcome these power quality issues like obtaining unity power factor (UPF) at the AC input mains and a close regulation of the DC output voltage has been discussed in [4–7]. So, for small power uses in single-phase supplies, DC-DC converters are used which are switch-based and controlled accordingly to ensure high power factor (HPF) at the mains side. The main motive is to emulate a resistive circuit when seen from mains side and improve power factor. Theoretically, there exist three families of non-isolated power factor correction topologies which are buck, boost and buck-boost topology [8, 9]. The buck topology is generally used when low output voltage is required but has high-frequency commuted current at input. Due to commuted nature, it exhibits a discontinuous nature, and hence should be connected with high-speed recovery circuit. The major drawback faced by buck topology is the inclusion of a filter of high-frequency range within the diode bridge and the power source [10]. The filtering inductor causes a soft variable current in the input end in the topology of a boost converter. This topology provides high output voltage and leads to over-voltage stress at the switches [11]. The third family is buck-boost topology and finds its applications in power factor correction circuits. It includes Cuk and SEPIC converter [12, 13]. The output voltage polarity makes these two converters different from each other. In case of Cuk converter, the position of free-wheeling diode and the inductor is reversed from that of the SEPIC converter to obtain reversed output voltage polarity [14, 15]. This paper presents two different topologies (namely SEPIC and Cuk)

M. N. Mohanty and S. Das (eds.), *Advances in Intelligent Computing and Communication*, Lecture Notes in Networks and Systems 109,
https://doi.org/10.1007/978-981-15-2774-6_37

type power factor correction (PFC) circuit to improve the input-side power factor of the diode bridge rectifier, lower the THD factor with decreased ripple contained voltage at output [16]. The entire modeling and the design scheme of input power factor (IPF) rectified AC to DC power converters are carried out in the MATLAB/Simulink environment.

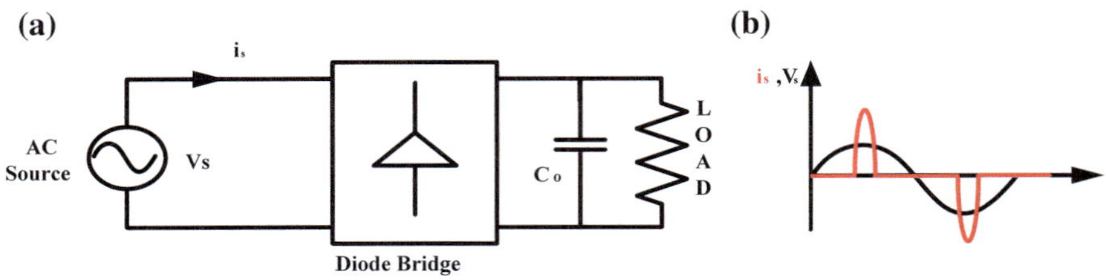

Fig. 1. **a** AC-DC rectifier, **b** source voltage and current waveform

2 Circuit Configuration and Design Equation

To improve the power factor at the input and to get a regulated voltage at the output, we have to interface a DC to DC (SEPIC or Cuk) converter in between the bridge rectifier and the load as shown in Fig. 2 (SEPIC) Fig. 3 (Cuk), respectively. By adopting these topologies, a low current ripple can found at input for a low-level DC-bus voltage, which is rare in case of conventional buck/boost converter. With adequate coupling coefficient design in a Cuk or SEPIC converter, the same magnetic core can accommodate the output and input inductors. In both the converter, when switch is on inductor L_1 and L_2 stores energy, inductor current increases linearly and output capacitor supply power to the load and diode is reverse biased. When switch is off diode gets forward biased and stored energy of the inductor is supplied to the load.

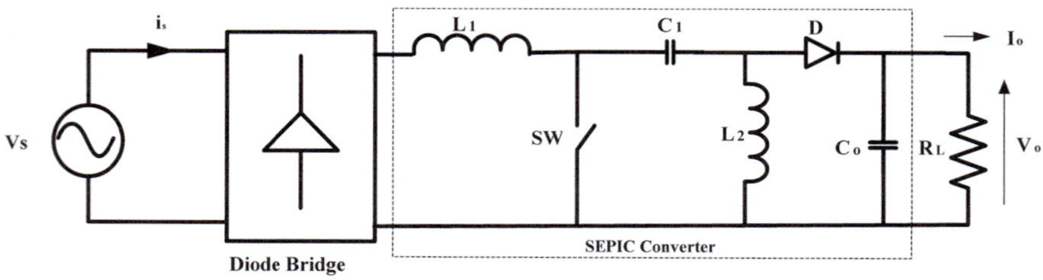

Fig. 2. SEPIC converter topology for power factor correction

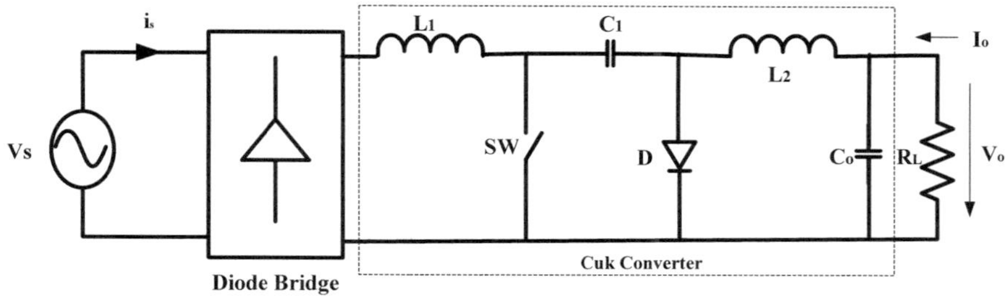

Fig. 3. Cuk converter topology for power factor correction

Table 1. Designed equation for SEPIC and Cuk converter

Parameters	SEPIC	Cuk
L_1	$\frac{v_s * d}{\Delta i L_1 * f_s}$	$\frac{v_s * d}{\Delta i L_1 * f_s}$
L_2	$\frac{v_s * d}{2 * \Delta i L_2 * f_s}$	$\frac{v_s * d}{\Delta i L_2 * f_s}$
C_1	$\frac{i L_2 * d}{\Delta v_{c_1} * f_s}$	$\frac{I_0 * (1-d)}{\Delta v_{c_1} * f_s}$
C_o	$\frac{P_o}{4\pi * f_s * V_o * \Delta v_o}$	$\frac{v_s * d}{8 * f_s^2 * L_2 * \Delta v_o}$

The different parts of all the converter system are modeled using basic equations and all equations are summarized in Table 1. Where v_s is the rms source voltage, d is the duty cycle, ΔiL represents inductor current ripple, Δv_c, Δv_o are the voltage ripple of the output voltage and capacitor voltage, respectively, f_s is the switching frequency. P_o is the output power, V_o is the output voltage, I_o is the output current.

3 Different Methods of Power Factor Correction

There are two objectives of power factor correction: 1. To get a regulated voltage at the output. 2. The wave shape of current at the input must be a sine wave. To achieve the first objective, we have to use a feedback loop at the output. There are two methods to achieve the second objective. First method is called "multiplier approach" and the second is called "voltage follower approach". In multiplier approach, a feedback loop of input current is used to govern the DC to DC power converter to function as an input voltage programmed current sink as shown in Fig. 4. In this paper, multiplier approach is used, where multiplier approach control is further sub-divided into four different methods for generating the gate pulse for the SEPIC and Cuk converter, such as:

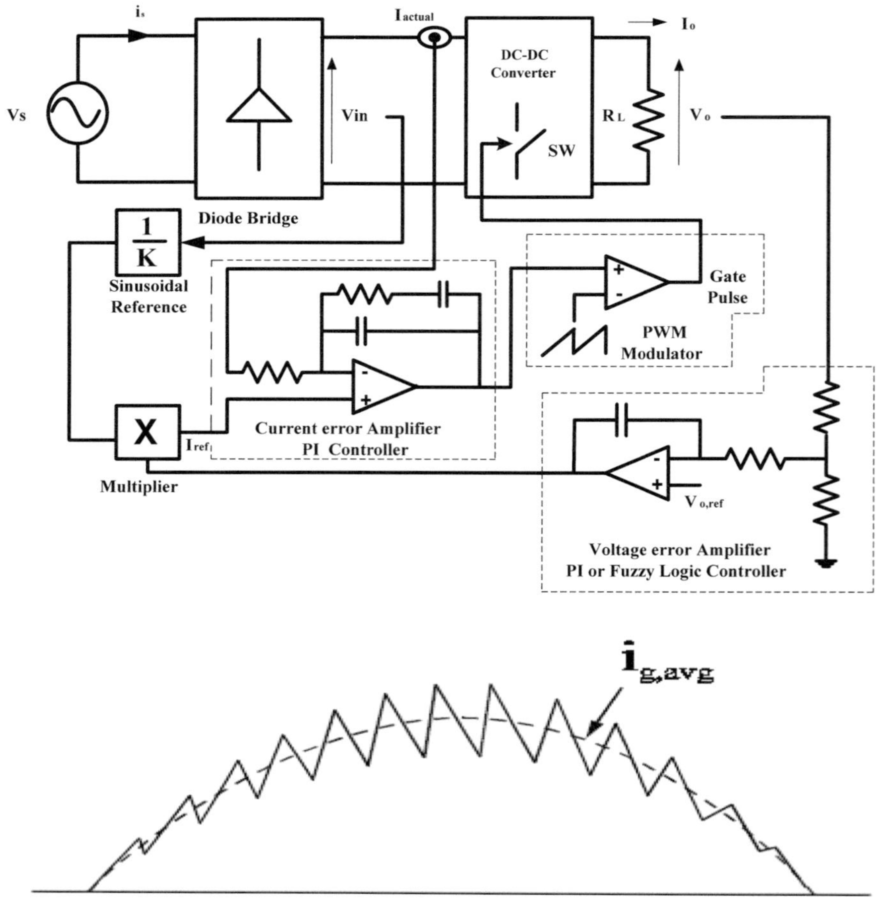

Fig. 4. Multiplier approach with average current control technique

1. Hysteresis current control
2. Peak current control
3. Average current control
4. Borderline current control.

In this paper, the average current control method is considered which allows a better input current waveform, which is depicted in Fig. 4. In this technique, the sensed inductor current is to be filtered with the help of a current error amplifier and the output of the same drives the PWM modulator. Hence, the error between input current Ig and its reference gets minimized with the help of inner current loop. The reference current is generated by a voltage error amplifier (PI or FLC). Due to the PWM modulator, average current control method gives a constant switching frequency because of current filtering commutation noises gets eliminated. This method does not need any compensation ramp. Inductor current must have to be sensed which is the demerits of this method.

4 Controller Used for Output Voltage Regulation

In this paper, two different controllers are used to regulate the output voltage namely PI controller and fuzzy logic controller and the performance of the same is compared.

4.1 PI Controller

The name itself proportional-integral controller is a type of control loop mechanism adopted in continuous modulated operation of control. A PI controller repeatedly calculates the error in between a set point (SP) value and measured process value(PV) and get a correction accordingly in PI base. In real-time operation, it accordingly provides an accurate and respective change of correction to a control function. For example, furnace temperature control that it applies a derivative term to effectively correct the error despite a huge change.

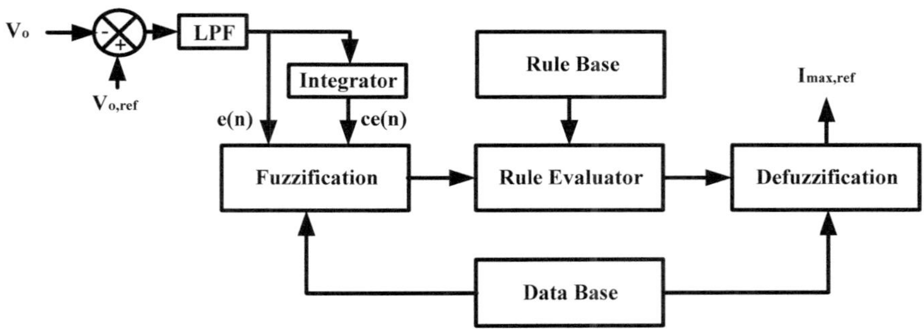

Fig. 5. Basic block diagram of FLC

4.2 Fuzzy Logic Controller

The basic block diagram of the proposed FLC for output voltage regulation of SEPIC and Cuk converter is shown in Fig. 5.

There are two separate meanings of fuzzy logic. Generally, fuzzy logic is the further application of multivalued logic and is also known as logic system. We can also say, fuzzy logic is same as fuzzy sets theory that which relates to collection of objects with unsharp boundaries and the membership is a point of degree. There is a specific object which degree of membership in a given set which can vary between the ranges 0–1 in fuzzy set theory. Fuzzy logic is based on sound quantitative and also deals with imprecise information and data. In mathematical theory, the values of fuzzy variable are expressed by proper English language. Error in voltage of a power converter can be defined in linguistic variables like negative big (NB), negative medium (NM), negative small (NS), zero (ZE), positive small (PS), positive medium (PM), positive big (PB), and each variable can be defined by varying triangular membership function. Seven fuzzy levels were chosen and were defined by fuzzy set library which value of the error is e and change in error is de. The larger the number of fuzzy levels, the higher the input resolution. We know that a rule is n dimensional and n is the number of variable included in the rule. The sum of rules is known as rule *R*. FIS editor edits the input and output variables, which are *e*, *de* and output. After editing, we have to design the membership function for each variable. The final step involves of writing rules in rule editor using the rule given in Table 2.

Table 2. Fuzzy rules for closed loop control of SEPIC and Cuk converter

*e**de*	NB	NM	NS	ZE	PS	PM	PB
NB	NB	NB	NB	NB	NM	NS	ZE
NM	NB	NB	NB	NM	NS	ZE	PS
NS	NB	NB	NM	NS	ZE	PS	PM
ZE	NB	NM	NS	ZE	PS	PM	PB
PS	NM	NS	ZE	PS	PM	PB	PB
PM	NS	ZE	PS	PM	PB	PB	PB
PB	ZE	PS	PM	PM	PB	PB	PB

5 Results and Discussion

To investigate the performance of the proposed system, a Simulink model of a Cuk and SEPIC converter used for power factor correction is developed as shown in Figs. 6 and 7, respectively. The designed parameter used in simulation for both SEPIC and Cuk converter is given in Table 3. The system performance is investigated in six steps.

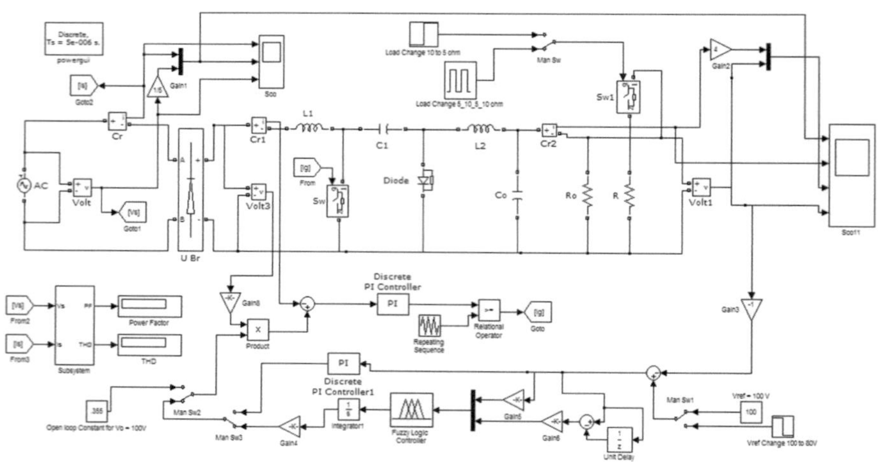

Fig. 6. Simulink model of Cuk converter

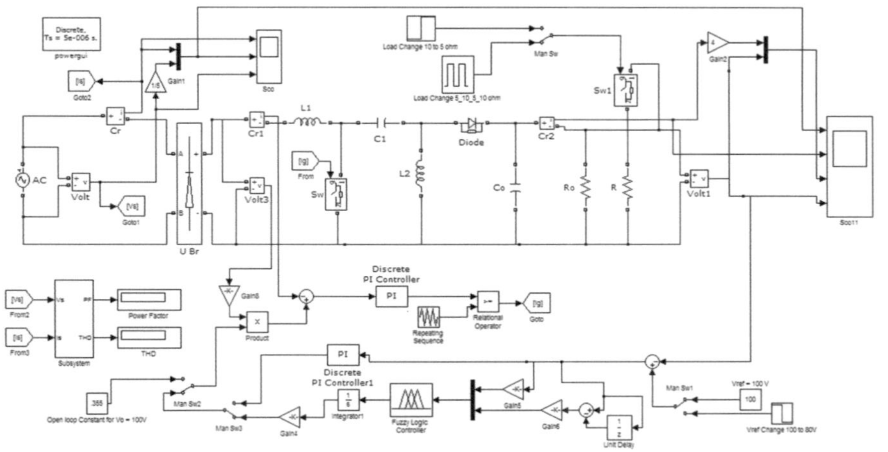

Fig. 7. Simulink model of SEPIC converter

Table 3. Designed parameter of the system

Parameters	Values
Input voltage RMS value	120 V
Supply frequency	50 Hz
L_1, L_2	6 mH, 10 mH
C_1, C_o	10, 10 mF
Load power	1 kW
Reference output voltage	100 V
Switching frequency	40 kHz
Voltage and current ripple	0.05
Duty cycle	0.45

Step: 1 when none of these converters is connected to the system. Step: 2 when Cuk and SEPIC converters are connected to the system with inner current loop only and without the output voltage feedback loop. Step: 3 when Cuk converter is connected to the system with PI controller in the feedback loop. Step: 4 when Cuk converter is connected to the system with FLC in the feedback loop. Step: 5 when SEPIC converter is connected to the system with PI controller in the feedback loop. Step: 6 when SEPIC converter is connected to the system with FLC in the feedback loop. To convert AC to DC, we generally connect a bridge rectifier and filter capacitor across the load as shown in Fig. 1, the simulation result of the same along with its harmonic spectrum is shown in Fig. 8a–d.

To improve the performance of the input current and output voltage, Cuk or SEPIC converter is used as discussed in Figs. 3 and 4. Various waveforms at steady state in open loop are shown in Fig. 9 (Fig. 9a–c for SEPIC and Fig. 9d–f for Cuk). From

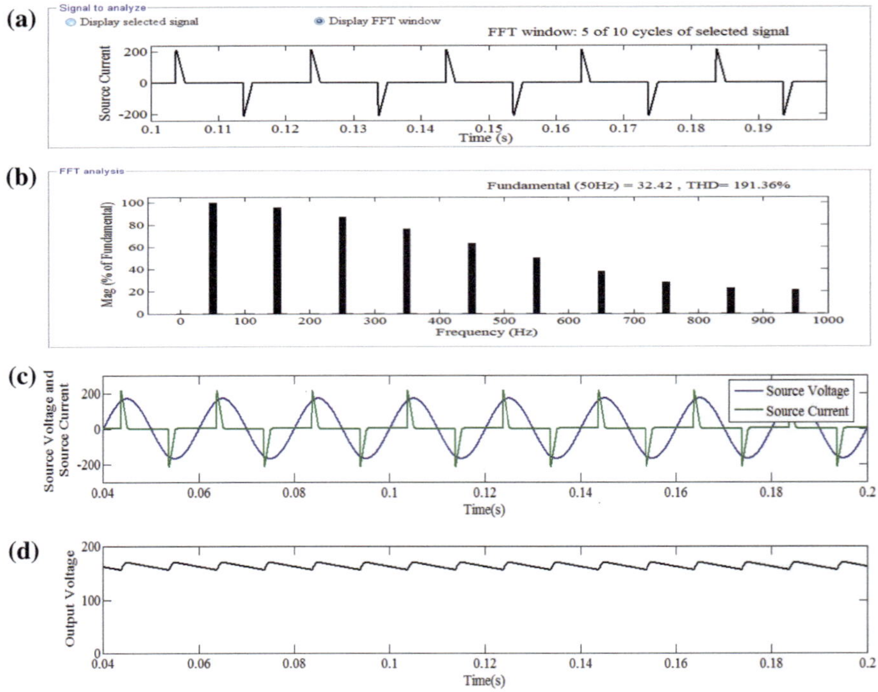

Fig. 8. Simulation results of the proposed system without any converter

(Fig. 9b or e), it is clear that when load is increased at t = 1 s, load voltage is not a regulated one. To get a regulated output voltage and improved input current wave, PI or FLC is used in the output feedback loop. The dynamic response of the system is shown in Figs. 10a–f and 11a–f for Cuk converter, Figs. 12a–f and 13a–f for SEPIC converter, respectively, for PI and fuzzy logic controller in the outer loop for both the converter. To test the validity of the PI and FLC, a load change and reference voltage change are created at t = 1 s and at t = 2 s, respectively. With this dynamic change, the regulated output voltage is obtained which can be clearly seen from Figs. 10e, 12e and 11e, 13e, respectively, for PI and FLC. Various performance parameters obtained in simulation are given in Table 4 for comparison.

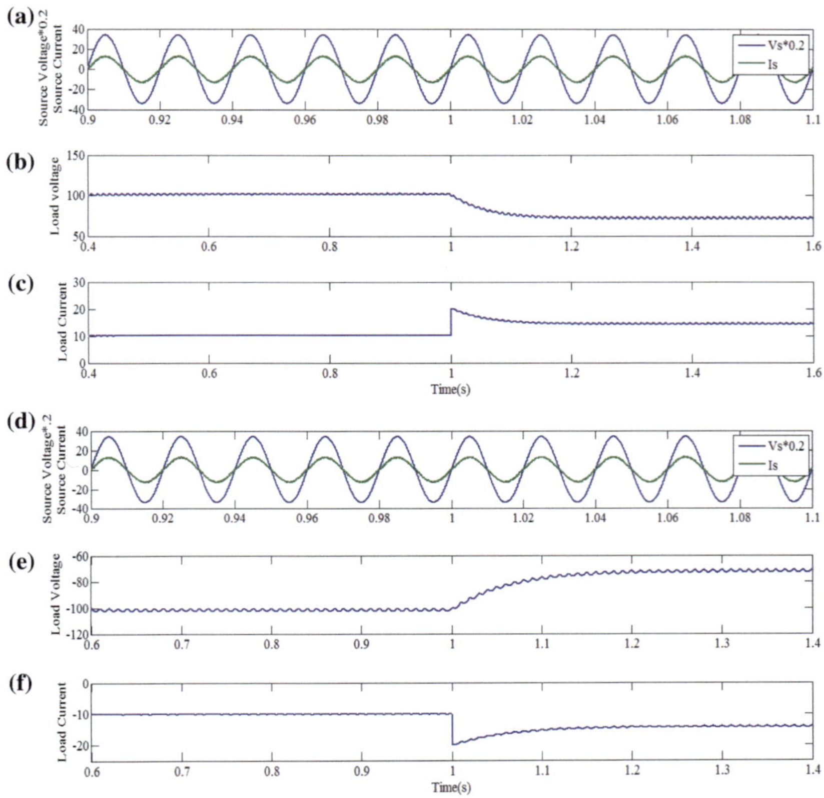

Fig. 9. Simulation results of the system in open loop (**a–c** for SEPIC and **d–f** for CUK)

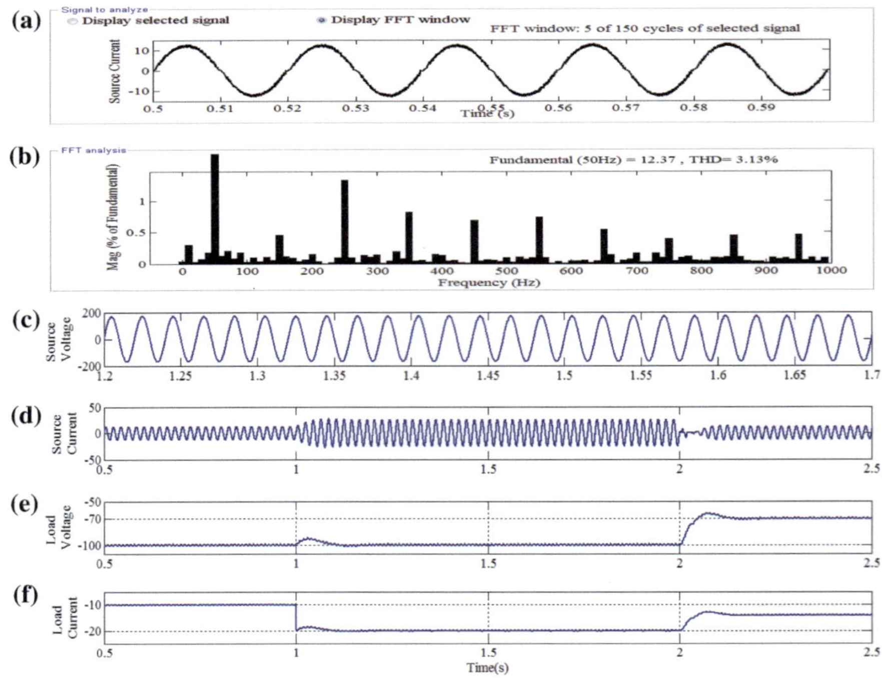

Fig. 10. Simulation results of PI Controlled Cuk converter

6 Conclusion

The design, modeling and simulation of Cuk and SEPIC converter for power factor correction purposes considered in MATLAB/Simulink environment. The outcomes of the simulation work provided low THD of the supply current with upgraded AC mains power factor and reduced output voltage ripple. Comparing with PI and FLC used in the output feedback loop, FLC gives better result in terms of THD of supply current, IPF and output voltage ripple. From the obtained result, it is clear that fuzzy logic

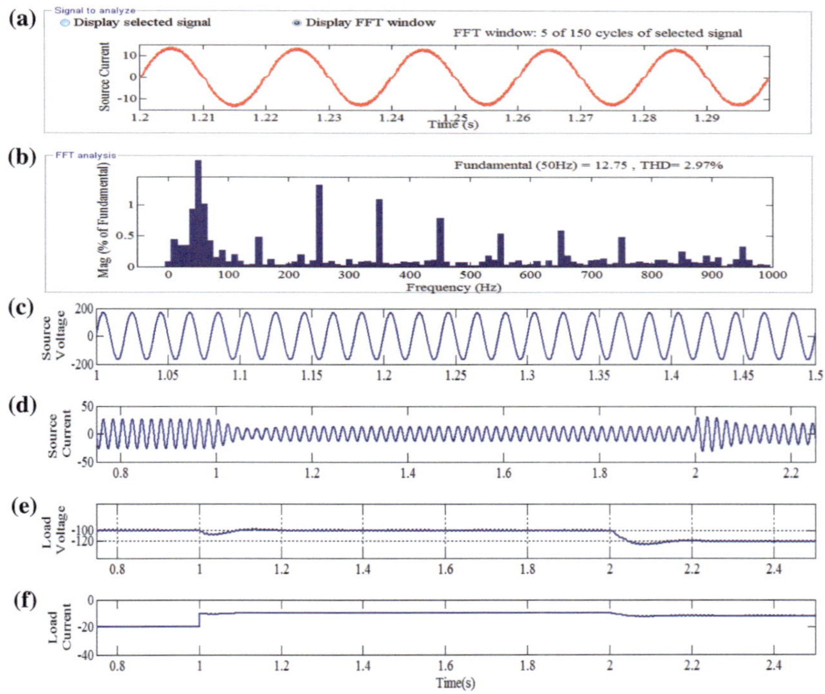

Fig. 11. Simulation results of fuzzy logic controlled Cuk converter

controlled Cuk converter gives better result when load disturbance and reference change occur. The prototype of the proposed Cuk converter can be developed that would be attempted as future work.

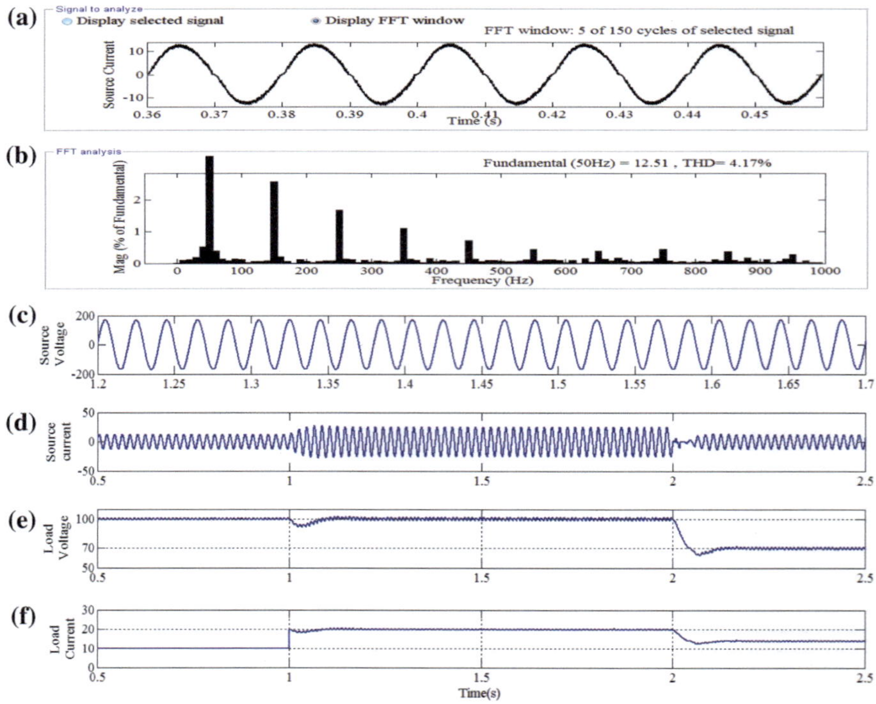

Fig. 12. Simulation results of PI controlled SEPIC converter

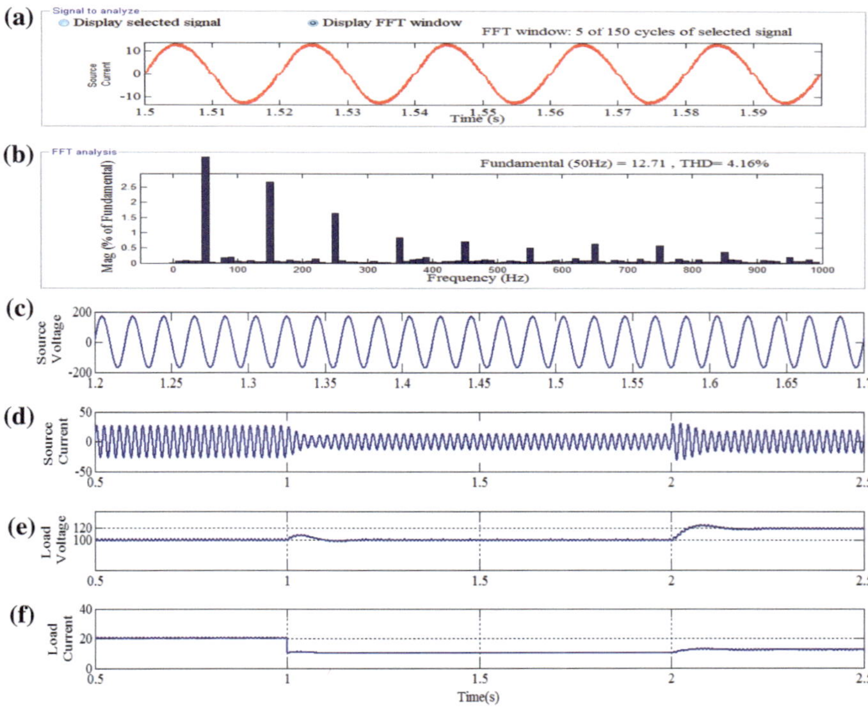

Fig. 13. Simulation results of fuzzy logic controlled SEPIC converter

Table 4. Performance parameters of the system under different condition

Performance parameter / Different condition	THD (%)	Power factor	Output voltage ripple (%)
Without any converter	191.36	0.4479	9.2
Open loop SEPIC converter	4.37	0.9986	2.0
Open loop Cuk converter	4.70	0.9986	2.2
PI controlled SEPIC converter	4.17	0.9983	2.1
PI controlled Cuk converter	3.13	0.9990	2.1
Fuzzy logic controlled SEPIC converter	4.16	0.9997	1.5
Fuzzy logic controlled Cuk converter	2.97	0.9999	1.8

References

1. Mishra AK, Pathak MK, Das S (2011) Isolated converter topologies for power factor correction—a comparison. In: IEEE International conference on energy, automation and signal 28 Dec 2011, pp 1–6
2. Kuiyuan W (2006) The comparison and choice of several power factor correction methods. In: IEEE vehicle power and propulsion conference, Sep 2006. IEEE, pp 1–5
3. 519–2014 IEEE recommended practice and requirements for harmonic control in electric power systems, 11 June 2014, pp 1–29
4. Pressman I (1991) Switching power supply design. McGraw-Hill, New York
5. Singh B, Singh BN, Chandra A, Al-Haddad K, Pandey A, Kothari DP (2003) A review of single-phase improved power quality AC-DC converters. IEEE Trans Ind Electron 50(5): 962–981
6. Batarseh I Power electronics circuits, 3rd edn. Wiley
7. Richard R (1994) The fundamentals of power factor correction. Int. J. Elect. Eng. Educ. 31:213–229
8. Dah D, Ki SK (2013) Light-load efficiency improvement in buck-derived single-stage single-switch PFC converters. IEEE Trans Power Electron 28(5):2105–2110
9. Liu X, Xu J, Chen Z et al (2015) Single-inductor dual-output buck–boost power factor correction converter. IEEE Trans Ind Electron 62(2):943–952
10. Kanaan HY, Al-Haddad K, Fnaiech F (2004) Switching-function-based modeling and control of a SEPIC power factor correction circuit operating in continuous and discontinuous current modes. In: IEEE international conference on industrial technology, vol 1, no 8, Dec 2004, pp 431-437
11. Umamaheswari MG, Uma G (2013) Analysis and design of reduced order linear quadratic regulator control for three phase power factor correction using Cuk rectifiers. Electr Power Syst Res 1(96):1–8
12. Umamaheswari MG, Uma G, Vijitha SR (2012) Comparison of hysteresis control and reduced order linear quadratic regulator control for power factor correction using DC–DC Cuk converters. J Circ Syst Comput 21(1):1250002
13. Umamaheswari MG, Uma G, Isabella LA (2014) Analysis and design of digital predictive controller for PFC Cuk converter. J Comput Electron 13(1):142–154
14. Poorali B, Adib E (2016) Analysis of the integrated SEPIC-flyback converter as a single-stage single- switch power-factor-correction LED driver. IEEE Trans Ind Electron 63(6): 3562–3569
15. Melo PF, Gules R, Ribeiro EF et al (2010) A modified SEPIC converter for high-power-factor rectifier and universal input voltage applications. IEEE Trans Power Electron 25(2): 310–321
16. Sudhakarababu C, Veerachary M (2004) Zeta converter for power factor correction and voltage regulation. In: 2004 IEEE region 10 conference TENCON 2004, vol 500. IEEE, pp 61–64

An Efficient Classifier-Based Approach for Early Arrhythmia Detection with Feature Reduction Using Ranker Search Algorithm

Monalisa Mohanty[1], Asit Kumar Subudhi[1](✉), Pradyut Kumar Biswal[2], and Sukanta Sabut[3]

[1] Department of Electronics and Communication, Institute of Technical Education & Research, SOA Deemed to be University, Bhubaneswar, India
asitsubudhi@soa.ac.in
[2] Department of Electronics Engineering, IIIT Bhubaneswar, Bhubaneswar, India
[3] School of Electronics Engineering, KIIT Deemed to be University, Bhubaneswar, Odisha, India

1 Introduction

The VF and the VT are hazardous cardiac chaos, which requires quick response and request of high-energy shock in the heart area [1]. Due to VF, there occurs an improper electrical movement in the ventricles, and hence, the heart is no longer an effective pump.

The authors have used the temporal and spectral domain approach to discriminate ventricular arrhythmias [2]. The wavelet transforms (WT), support vector machines (SVMs), and neural network approaches are used for differentiating several ventricular arrhythmias [3–5]. The spectral and temporal features were combined with the morphological characteristics of ECG signal for the detection of VF and VT conditions [6]. The SPRT, i.e., sequential probability ratio test, and TCI, i.e., threshold crossing interval methods, have been used in time domain for precise classification of the VF/VT condition. Zhang et al. [7] have described a complexity measure approach for the detection of VT/VF conditions by changing the ECG signal into 0–1 string.

Various algorithms and methods have been proposed for recognition of the cardiac arrhythmias which are required for efficient classification of VT and VF arrhythmia rhythms. The detection performance of VF conditions may be improved using combined features extracted from ECG signal. The current work aims at recognition and classification of VT/VF arrhythmia conditions with the help of spectral, statistical, and temporal domain features. For a 5 s window length, the de-noised signals were subjected for feature extraction. The extracted features have been ordered and ranked with the help of correlation attribute evaluation along with a ranker search method. The arrhythmia conditions were classified with the help of ordered features using J48 decision tree algorithm, SVM, MLP, and by CVR which has been discussed thoroughly in Sect. 2. The evaluation of results with different parameters and comparison with other time–frequency domain methods has been done and is explained in Sect. 3.

M. N. Mohanty and S. Das (eds.), *Advances in Intelligent Computing and Communication*, Lecture Notes in Networks and Systems 109,
https://doi.org/10.1007/978-981-15-2774-6_38

Section 4 corresponds to the conclusion of the entire work that has been used for classification the ventricular arrhythmia effectively.

2 Methodologies

The proposed block diagram of ventricular arrhythmia detection process is as shown in Fig. 1.

Fig. 1. Proposed block diagram of ventricular arrhythmia detection process

2.1 Database

Two different databases of PhysioNet repository, i.e., CUDB and VFDB, were considered. Data of 35 recordings with eight-minute duration of single-channel ECG signal from CUDB database along with data of 22 recordings with thirty-five-minute duration of two-channel ECG signal from VFDB database were taken for feature extraction of ECG signal. A total 57 records and 78 complete duration annotated ECG channels were used for the study. A suitable window length of 5 s is considered for feature extraction methods.

2.2 Preprocessing of ECG Signal

With a sampling frequency of 250 Hz, the preprocessing of acquired ECG signals was done to reduce high-frequency noise, to suppress baseline wandering effect, and to eliminate power line interference. Thus, the steps of preprocessing are:

(1) High-pass filtering with a f_c = 1 Hz to reduce baseline wander;
(2) Second-order Butterworth low-pass filtering with a cutoff frequency of f_c = 30 Hz to reduce high-frequency noise; and
(3) Notch filtering with a f_c = 60 Hz to eliminate power line interference.

2.3 Feature Extraction

A total nineteen number of features were extracted in statistical, spectral, and temporal domains for precise classification of the ventricular arrhythmia. The weighted and ranked features are given as LKG, TCSC, A3, AB, C1, CO, C3, FSMN, FB, VAL, MEA, CROSS, TCI, A2, A1, KUR, HURST, SKW, and PERM.

2.4 Feature Selection

The feasibility of extracted parameters was measured using a feature selection technique called "correlation attribute evaluation method." The selected features are further ranked according to their respective weightage value using a ranker search method.

Correlation Attribute Evaluation (CAE) [8].

This algorithm estimates the significance of a feature by computing the correlation between the class and the feature and is conferred as:

$$C_{\text{zc}} = \frac{n\overline{C_{\text{zi}}}}{\sqrt{n + n(n-1)\overline{C_{\text{ii}}}}}$$

where

C_{zc} indicates the correlation between the class variable and the summed feature subsets.

n indicates the no. of subset features.

C_{zi} indicates the average correlations of the subset features and the class variable, and C_{ii} indicates the average inter-correlation between subset features.

2.5 Classifiers

Four different classifiers have been used in this section. The ranked and ordered features are fed as the input to the J48 decision tree, SVM, multilayer perceptrons, i.e., MLP and classification via regression classifiers, i.e., CVR, for precise classification of NSR, VT, and VF arrhythmia rhythms.

2.5.1 J48 Decision Tree Algorithm

J48 decision tree algorithm is a classifier which classifies a specified data point using a top-down approach [9]. The node having the maximum information gain has been carefully chosen as the test attribute. For a given attribute A, the information gain (IG) at any node is calculated as:

$$\text{IG}(S, A) = \text{Entropy}(S) - \sum_{v \epsilon \text{Values}(A)} \left(\frac{|S_{\text{V}}|}{|S|} \text{Entropy}(S) \right) \quad (1)$$

where S denotes the set of instances; $|S|$ denotes its cardinality; S_v denotes the subset of S for which attribute A has value v.

2.5.2 Support Vector Machine

SVM is usually a supervised machine learning algorithm which has been used for both classification and regression tasks [5]. Let us consider a training data set with N pairs, i.e., $(x_{1,}y_1), (x_{2,}y_2), \ldots, (x_{N,}y_N)$ with $x_i \in \Re^p$ and $y_i \in \{-1, 1\}$.

Let us assume the hyper-plane is given as $\{x : f(x) = x^T\beta + \beta_0 = 0\}$, where β is a unit vector,

$$\boldsymbol{G}(\boldsymbol{x}) = \mathbf{sign}\left[\boldsymbol{x}^T\boldsymbol{\beta} + \boldsymbol{\beta}_0\right]\boldsymbol{C} \tag{2}$$

2.5.3 Multilayer Perceptron

The multilayer perceptron calculates a single output from number of real-valued inputs [10]. It can be given mathematically as:

$$Y = \varphi\left(\sum_{i=1}^{n} v_i a_i + b\right) = \varphi\left(V^T A + b\right) \tag{3}$$

where

V represents the weight vector.
A represents the input vector.
b represents the bias.
φ represents the activation function.

2.5.4 Classification via Regression

It is used for performing classification using regression methods [11]. It generates a binary class along with one regression model corresponding to each class.

2.6 Performance Measures

The typical statistical indices used for performance evaluations such as sensitivity, i.e., S_e, specificity, i.e., S_p, precision, i.e., PPV, and accuracy, i.e., Acc, are given as:

$$S_\mathrm{e} = \frac{\mathrm{TP}}{\mathrm{TP} + \mathrm{FN}}$$

$$S_\mathrm{p} = \frac{\mathrm{TN}}{\mathrm{TN} + \mathrm{FP}}$$

$$\mathrm{PPV} = \frac{\mathrm{TP} + \mathrm{TN}}{\mathrm{TP} + \mathrm{FP}}$$

$$\mathrm{Acc} = \frac{\mathrm{TP} + \mathrm{TN}}{\mathrm{TP} + \mathrm{FP} + \mathrm{TN} + \mathrm{FN}}$$

where TP is true positive, TN is true negative, FN is false negative, and FP is false positive.

3 Results

A total 19 number of features were extracted from the de-noised ECG signals. The ordered and ranked features are given as LKG, C1, C3, CROSS, FSMN, A1, A2, A3, VAL, MEA, SKW, KUR, CO, FB, AB, HURST, PERM, TCI, and TCSC. The evaluation was done by correlation attribute evaluator by computing the correlation between the attribute and the class. Table 1 demonstrates the ordered ranking of weighted features, and the selected features are given as LKG, TCSC, A3, AB, C1, CO, C3, FSMN, FB, VAL, MEA, CROSS, TCI, A2, A1, KUR, HURST, SKW, and PERM.

Table 1. Ranking of the selected features using correlation attribute evaluation technique along with ranker search method

Rank	Feature name	Weightage
1	LKG	0.6809
2	TCSC	0.5642
3	A3	0.5261
4	AB	0.4564
5	C1	0.4250
6	CO	0.4111
7	C3	0.3962
8	FSMN	0.3640
9	FB	0.2208
10	VAL	0.2206
11	MEA	0.2206
12	CROSS	0.2206
13	TCI	0.2056
14	A2	0.1827
15	A1	0.1754
16	KUR	0.1604
17	HURST	0.1072
18	SKW	0.0441
19	PERM	0.0390

The performance valuation of J48 algorithm with different features combinations at a confidence factor, i.e., CF = 1, was done, and the finest classification was obtained for all 19 features together with a CF = 1 as shown in Table 2. The maximum accuracy rate was found to be 97.7317% with Se = 92.7822%, Sp = 98.1123%, and PPV = 96.0063%, respectively. Figures 2 and 3 correspond to the accuracy with different feature combinations and the comparative analysis of evaluated parameters, respectively.

Table 2. Performance valuation of J48 algorithm with different feature combinations at CF = 1

Feature combinations	TP	FP	FN	TN	ACC	SE	SP	PPV
1	4214	902	902	9330	88.25	62.67	89.65	79.17
1, 2	4233	881	881	9347	88.52	64.51	90.01	72.95
1, 2, 3	4402	712	712	9516	90.72	71.13	92.40	81.61
1, 2, 3, 4	4402	712	712	9516	90.72	71.13	92.40	81.61
1, 2, 3, 4, 5	4457	657	657	9571	91.44	73.21	93.02	82.16
1, 2, 3, 4, 5, 6	4449	665	665	9563	91.33	73.07	92.96	81.42
1, 2, 3, 4, 5, 6, 7	4607	507	507	9721	93.39	79.25	94.62	86.75
1, 2, 3, 4, 5, 6, 7, 8	4662	452	452	9776	94.11	81.39	95.17	89.20
1, 2, 3, 4, 5, 6, 7, 8, 9	4801	313	313	9915	95.92	87.73	96.72	91.93
1, 2, 3, 4, 5, 6, 7, 8, 9, 10	4850	264	264	9964	96.56	89.22	97.19	93.65
1, 2, 3, 4, 5, 6, 7, 8, 9, 10, 11	4812	302	302	9926	96.06	87.44	96.78	92.77
1, 2, 3, 4, 5, 6, 7, 8, 9, 10, 11, 12	4812	302	302	9926	96.06	87.44	96.78	92.77
1, 2, 3, 4, 5, 6, 7, 8, 9, 10, 11, 12, 13	4854	260	260	9968	96.61	89.11	97.22	93.87
1, 2, 3, 4, 5, 6, 7, 8, 9, 10, 11, 12, 13, 14	4873	241	241	9987	96.86	89.72	97.39	94.54
1, 2, 3, 4, 5, 6, 7, 8, 9, 10, 11, 12, 13, 14, 15	4844	270	270	9958	96.48	88.27	97.08	94.02
1, 2, 3, 4, 5, 6, 7, 8, 9, 10, 11, 12, 13, 14, 15, 16	4847	267	267	9961	96.52	88.39	97.13	94.01
1, 2, 3, 4, 5, 6, 7, 8, 9, 10, 11, 12, 13, 14, 15, 16, 17	4847	267	267	9961	96.52	88.89	97.16	93.35
1, 2, 3, 4, 5, 6, 7, 8, 9, 10, 11, 12, 13, 14, 15, 16, 17, 18	4934	234	234	10,102	96.98	90.48	97.51	94.33
All 19 features	**4940**	**174**	**174**	**10,054**	**97.73**	**92.78**	**98.11**	**96.01**

Bold indicates the best results as compared to the other feature combinations

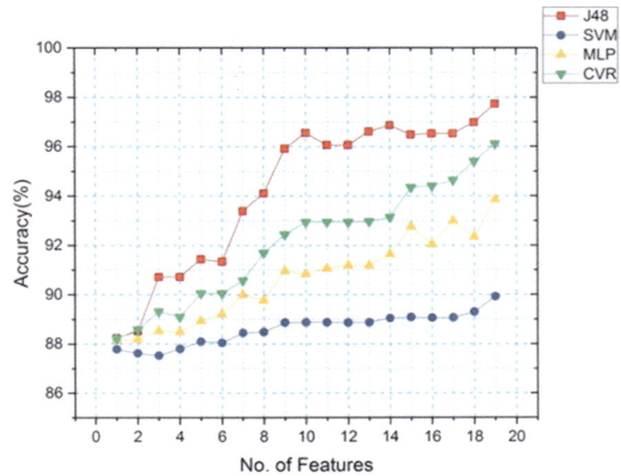

Fig. 2. Results of accuracies for different feature combinations using four classifiers

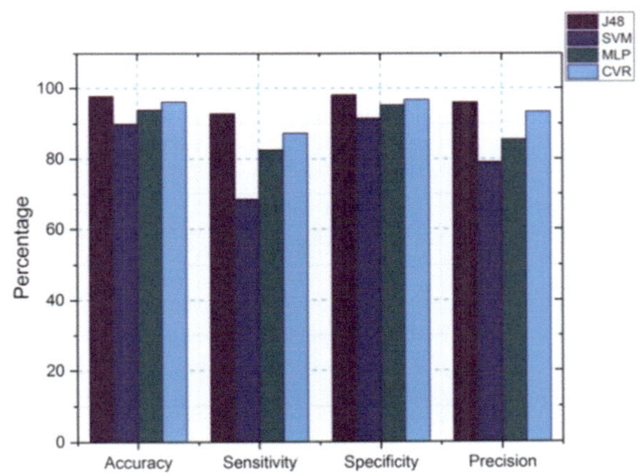

Fig. 3. Comparison of evaluated parameters of four different classifiers

4 Conclusion

Precise detection and accurate classification of arrhythmia rhythms are extremely vital in order to provide external defibrillation. A group of 19 features in statistical, spectral, and temporal domain such as LKG, TCSC, A3, AB, C1, CO, C3, FSMN, FB, VAL, MEA, CROSS, TCI, A2, A1, KUR, HURST, SKW, and PERM was considered for a 5 s window length. The extracted features were subjected to a feature selection procedure called correlation attribute evaluation to evaluate the prejudiced properties of the extracted ECG features. The weighted and ranked features are combined together for the J48 decision tree algorithm to provide a robust classifier for ventricular arrhythmia. The results in this study show that the use of J48 decision tree algorithm significantly improves the capability for the detection of life-threatening arrhythmias as compared to the SVM, MLP, and CVR classifiers.

References

1. Jekova I (2007) Shock advisory tool: detection of life-threatening cardiac arrhythmias and shock success prediction by means of a common parameter set. Biomed Sig Process Control 2:25–33
2. Li Q, Rajagopalan C, Clifford G (2014) Ventricular fibrillation and tachycardia classification using a machine learning approach. IEEE Trans Biomed Eng 61:1607–1613
3. Clayton RH, Murray A, Campbell RWF (1994) Recognition of ventricular fibrillation using neural networks. Med Biol Eng Comput 32:217–220
4. Khadra L, Al-Fahoum AS, Al-Nashash H (1997) Detection of life-threatening cardiac arrhythmias using the wavelet transformation. Med Biol Eng Comput 35:626–632
5. Alonso-Atienza F, Morgado E, Fernandez-Martınez L, Alberola AG, Rojo-Alvarez JL (2012) Combination of ECG parameters with support vector machines for the detection of life-threatening arrhythmias. Comput Cardiol 39:385–388
6. Krasteva V, Jekova I (2005) Assessment of ECG frequency and morphology parameters for automatic classification of life-threatening cardiac arrhythmias. Physiol Meas 26:707–723
7. Zhang XS, Zhu YS, Thakor NV, Wang ZZ (1999) Detecting ventricular tachycardia and fibrillation by complexity measure IEEE Trans Biomed Eng 46:548–555
8. Karegowda AG, Manjunath AS, Jayaram MA (2010) Comparative study of attribute selection using gain ratio and correlation based feature selection. Int J Inf Tech Know Man 2:271–277
9. Panigrahi R, Borah S (2018) Rank allocation to J48 group of decision tree classifiers using binary and multiclass intrusion detection datasets. Procedia Comput. Sci. 132:323–332
10. Raji G, Chandra SCSV (2017) Long-term forecasting the survival in liver transplantation using multilayer perceptron networks. IEEE Trans Syst Man. Cybern 47:2318–2329
11. Hou C, Nie F, Yi D, Wu Y (2013) Efficient image classification via multiple rank regression. IEEE Trans Image Proc 22:340–352

The Fractional Order PID Controller Design for BG Control in Type-I Diabetes Patient

Akshaya Kumar Patra[1(✉)], Anuja Nanda[1], Santisudha Panigrahi[2], and Alok Kumar Mishra[1]

[1] Department of EEE, ITER, S'O'A University, Bhubaneswar 751030, India
[2] Department of CSE, ITER, S'O'A University, Bhubaneswar 751030, India

1 Introduction

As per the World Health Organization report, one among the widespread diseases is diabetes mellitus and is resulted due to the malfunctioning of the pancreas. This reduces insulin sensitivity affecting the normoglycaemic range of BG concentration (70–120 mg/dl) in a healthy human being. At present, numerous research projects are undertaken by several researchers to get rid of this problem by devising advanced medical equipment like automated MID. Till date, BG concentration is manually controlled to adopt the open-loop control strategy. The hypoglycemic or hyperglycemic conditions may evolve due to the difficulties in handling the internal system changes and external disturbances by implementing the control loop technique. The development of implanted artificial pancreases (AP) enabling the adequate dose of insulin delivery proportionate to the sensor measurement in the patient's body may provide the means to incorporate the closed loop control strategy. Figure 1 describes a closed loop patient model with an AP. The AP consists of glucose sensor, MID, and controller. The sensor measures the BG concentration of the human body continuously and sends signal to the controller for generating the desired control actions. The control signal generated by this controller also depicts the association of the additional model uncertainties and disturbances. Thereafter, according to the control signal $u(t)$, the optimal insulin dose is infused into the patient's venous blood by MID to achieve the *normoglycaemic* range of BG concentration [1, 2].

To determine an optimal solution for the AP like building an appropriate model of the complex BG regulatory system, a number of obstacles and challenges such as the effects of non-linear behavior, time-dependent dynamics, presence of several sources of disturbance, uncertainty, and lack of glucose sensing are to be faced. Additionally, the challenges and constraints related to control, specifically for BG regulation controller design are considerable glucose measurement delay, insulin absorption delay after being injected, irreversible action of insulin, meal detection and estimation, model parameter variation, asymmetric risk of extreme BG concentration variations and time-based control needs, etc. [1]. Despite of technical progress and considerable development on aforesaid issues, substantial improvement is still required in the control algorithm. Assessment of the glucose excursions following the insulin dose adjustment needs a controller for BG regulation in AP and many authors suggested the PID controller as a viable solution [3, 4]. However, due to glucose sensing time delay,

M. N. Mohanty and S. Das (eds.), *Advances in Intelligent Computing and Communication*, Lecture Notes in Networks and Systems 109,
https://doi.org/10.1007/978-981-15-2774-6_39

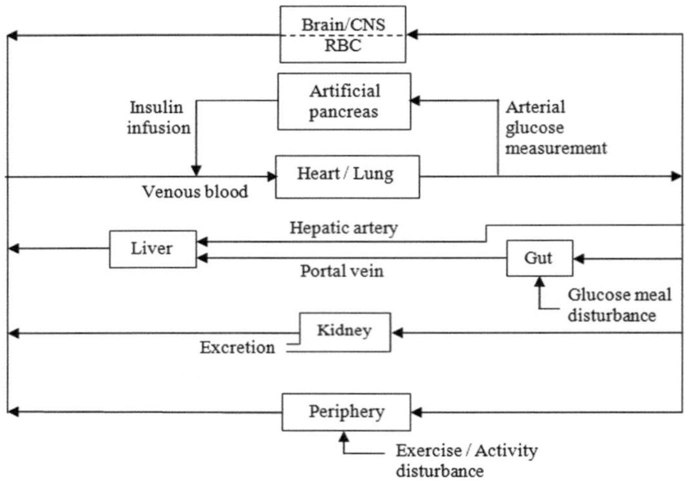

Fig. 1. Compartmental model of TIDM patient with AP

insulin action and non-variable gain parameters, the desired performance, assuring high accuracy, reliability, and robustness could not be achieved. The fuzzy control [5, 6], LQG control [7], *H*∞ control [8–10], Sliding Mode (SM) control [11–14], and model predictive (MP) control [15, 16] are some of the well proven controllers to deal with the BG regulation issues. The BG control in the diabetic patients within the *normoglycaemia* range inculcating the above controllers enhanced the accuracy and robustness to some extent compared to the PID controllers. However, these control algorithms are not fully insensitive to the disturbances and the uncertainties of the model in spite of the improved performance. Hence, optimal control parameters setting for better performance and for avoiding slow response following meal disturbance, the current work suggests an alternative novel technique implementing the fractional order PID approach.

The FOPIDC approach concept leads to an enhanced control performance with respect to robustness and delay of time compensation characteristics to counteract the negative impact of associated errors, disturbances, and uncertainties. The stabilizing control law $u(t)$ is developed based on the fractional order approach. Application of the FOPIDC to control the BG concentration in TIDM patient results to ensure a better robust controller in comparison to other contemporary well-accepted methods under both harmonized and incompatible uncertainties.

This manuscript is organized as follows. Section 2 concisely illustrates the TIDM patient model with mathematical details reflecting to its dynamic characteristics of its GM process. A detailed presentation on how the control technique is formulated and how it is implemented for this problem is presented in Sect. 3. Comparative results of the proposed approach are presented in Sect. 4. The concluding comments are summarized in the Sect. 5.

2 Problem Formulation and Modeling

2.1 System Overview

The compartmental model schematic diagram as depicted in Fig. 1 reveals the GM dynamics of the human body. The proposed controller is tested using the above model. The human body has liver, kidney, periphery, heart/lungs, gut, and brain as different functional parts and is shown as six compartments separately. The blood flow directions or circulation are indicated by the arrow marks. The controller receives the measured arterial glucose value as input to provide the optimal insulin dose to be injected as an output with the help of MID.

2.2 Clinical Background

A human body with a prolonged high arterial BG level that exceeds 144 mg/dl is known to suffer from *hyperglycemia* disease. This type of the clinical disorder is also termed as diabetes mellitus. The insulin deficiency along with its less resistive or insensitiveness independently or combinedly cause *hyperglycaemia* and are some of the major reasons as discussed in [17, 18]. Diabetes mellitus may occur if the blood glucose remains unutilized effectively under any normal human life cycle. Diabetes is classified as type-I and type-II as per the rate of insulin creation in the patient's body. In the type-I diabetic patients, the insulin creation by the pancreatic cells is completely impossible, but in case of the type-II is at a bit smaller rate, which also gets inhibited in due course. Hence, in the present situation, an immediate insulin injection to restore the normal BG level is essential that also prevents allied adverse impacts on the normal operation of any human body. Subject to diverse fooding, fasting, and exercise behaviors, the regulation of the BG levels are not very easy in the real human life cycle.

The external sources like the carbohydrate foods are digested down into glucose at different parts of the digestive system (gut), and then filtered. The filtered glucose is added to the venous blood. Later on, the liver stores this glucose after being transferred from the venous blood as glycogen. When the BG concentration in the venous blood is less than *normoglycaemic* range of glucose concentration, glycogen is re-injected as glucose by the liver into venous blood and is known as the internal glucose source. The glucose utilization and production by the liver is known as the net hepatic glucose balance (NHGB). which is the general phenomena in the glucose metabolism process. The total energy needed by human body organs irrespective of insulin dependency derives from the degree of glucose utilization.

The creation of insulin is due to the β-cells present in the pancreas and thus the BG level is regulated. A normal human with high BG level has two major functions of insulin, firstly to allow the liver for the glucose absorption along with the storing it in the form of glycogen, specifically under the meal intake conditions. As a result, the production of excess 'internal' glucose by the liver and muscles is stopped. Secondly, insulin speeds up the glucose absorption in the muscles and fulfills the peripheral energy needs of the body. But, in case of diabetes patients, the abnormality in the metabolic process dynamics, both the aforementioned functions are found to impair partially or completely. In case of diabetes patient, the cells stop the glucose utilization and internal glucose is produced by the liver, then an uncontrolled BG level is noticed. Later, as the BG level is more than the renal threshold glucose (RTG) value of 162 mg/dl, venous BG in excess is extracted through the kidney.

2.3 Modeling of a TIDM Patient

In present scenario, with due consideration of the dynamics of glucose metabolism process, several simulation models are proposed to control the BG concentration in the TIDM patients [19, 20]. Owing to the simplified structure and desired approximation to the dynamics of human metabolism with reduced error, the model suggested by Lehmann [19, 20] are among the widely accepted ones. The testing and verification of the proposed technique using Lehmann [19, 20] model for BG control are considered in the present study. Figure 1 describes the compartmental set up of diabetic patient with the implanted AP. The controller computes the insulin injection amount to the venous blood at 5 min interval using an integrated implanted insulin pump. Depending on the supporting device and sensing technology, the sampling rate varies when is being applied in the real time domain [21, 22]. The gut compartment input is the meal and the peripheral input is the exercise for integrating and executing the process disturbances. Compartmental modeling method considering the equations of the fast principle has been used to develop the GI interaction process with a MID model of the patient. Figure 1 demonstrates the simulated model diagram of the patient with MID.

2.4 Analysis of Patient Dynamics

The patient model dynamic and characteristic operations are verified under several operating constraints like the actuator and sensor noises, intake amount of carbohydrates, varying exercises, etc. The BG concentration and the insulin dose of the considered patient model with 60 gm meal at 600 min and the exercise for half-an-hour at 1300 min are illustrated in Fig. 2a. The glucose production and consumption rate of liver, and glucose consumption rate by the BG profile organs such as gut compartment, CNS, and peripheral cells are shown in Fig. 2b. The overall BG regulation in the human body is carried out naturally through glucose utilization by the organs, such as peripheral cells (adipose tissues, muscle cells) and the liver that exclusively depend on insulin. The peripheral cells and liver consume the least amount of glucose under the condition of lack of insulin. Under this situation, BG level rises abnormally and go beyond the glucose level of 144 mg/dl. This results in the *hyperglycaemia* problem. The kidney removes part of the glucose from venous blood under the condition when the BG concentration rises up to RTG value. The rate of glucose excretion of the kidney proportionate to the BG level is represented by Fig. 2b. Figure 2a, b illustrate the unstable dynamics under various model uncertainties and disturbances. These abnormal dynamics can be reduced by applying the AP based on suitable control algorithm.

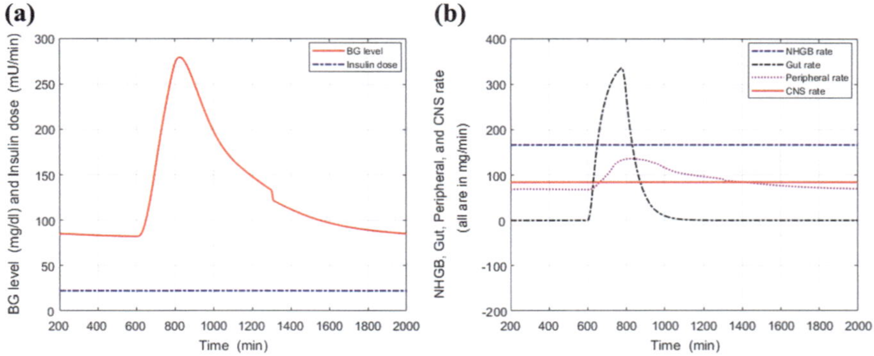

Fig. 2. **a** BG concentration with respect to insulin dose; **b** glucose production and consumption rate by GM organs

3 Control Algorithm

The FOPIDC control algorithm is demonstrated in this section. The closed loop system response with respect to stability, accuracy, and robustness is analyzed. The control specifications such as settling time (min), steady-state error e_{ss} (%), overshoot (mg/dl), and undershoot (mg/dl) are also evaluated and examined with proper validation of the controller actions.

3.1 FOPIDC Design

The TIDM patient model with fractional order $PI^{\alpha}D^{\gamma}$ controller is shown in Fig. 1a. In fractional order $PI^{\alpha}D^{\beta}$ control, the error signal $e(t)$ is used to generate control signal $u(t)$. The transfer function (TF) of the proposed controller can be formulated as [23]:

$$\mathrm{TF} = K_{\mathrm{p}} + \frac{K_{\mathrm{i}}}{s^{\alpha}} + K_{\mathrm{d}}s^{\beta} \quad (1)$$

where, α and β are the fractional orders of the integrator and differentiator of the proposed controller, respectively. K_p, K_i, and K_d are denoted for the proportional, integral, and derivative gains of the proposed controller, respectively. The optimal values of the FOPIDC parameters are considered based on the MATLAB/SIMULINK

Table 1. Optimal values of control parameters

K_p	K_d	K_i	α	β
10	8	0.05	0.7	0.9

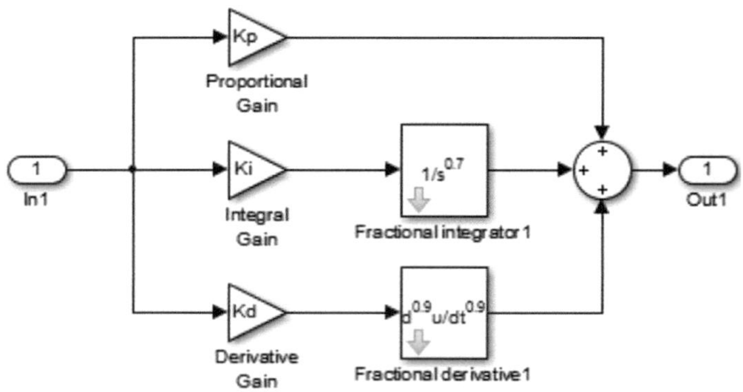

Fig. 3. Simulink diagram of FOPIDC

environment and FOMCON toolbox as represented in Table 1. The structure of FOPIDC is shown in Fig. 3, which is designed based on the Eq. (1).

4 Result and Discussions

Time-domain response of glucose profiles, stability, and robustness of the closed loop model with proposed FOPIDC are described in detail in this section. The proposed control approach is compared with other popular control algorithms to justify its enhanced performance.

4.1 Analysis of Patient Dynamics with FOPIDC

In this section, all glucose profiles of the non-linear patient model with FOPIDC is examined under different operating conditions such as variation of exercise, intake amount of carbohydrate, and noise of sensors and actuators. All glucose profiles like BG level, insulin dose, NHGB rate, gut rate, CNS and peripheral glucose utilization rate in TIDM patient model with proposed controller are illustrated in Fig. 4. Compared to the condition of the uncontrolled process, the result obtained evidences a higher utilization of the plasma glucose by the peripheral cells and liver, those are dependent on the insulin availability and sensitivity. These bring back the BG level to 81 mg/dl with the quick settling time that result in the reduction or prevention of the hyperglycaemia occurrence probability as shown in Fig. 4a. The insulin-independent organ like the CNS, consume the plasma glucose at a constant rate. Figure 4b illustrates the constant rate of glucose utilization by CNS at 84 mg/min. As the BG level is below

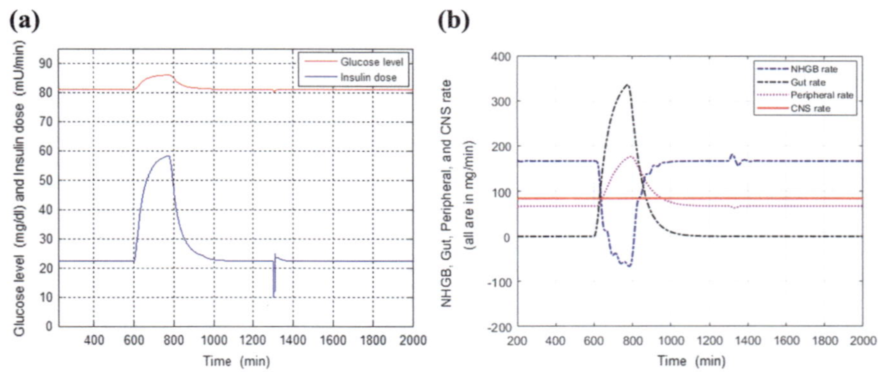

Fig. 4. **a** BG concentration with respect to insulin dose; **b** glucose production and consumption by GM organs

the RTG level, kidney does not extract any glucose and is depicted in Fig. 4b by the constant glucose excretion rate at zero level. By implementing the proposed control techniques to the MID, different operational parameters are evidenced and have the enhanced performances.

4.2 Comparative Study

The suggested FOPIDC approach is compared with other popular control approaches such as PID, fuzzy, $H\infty$, and SM to justify its enhanced performance as the controllers. Figure 4 illustrates the effect of meal disturbance in the BG concentration of the patient with the FOPIDC approaches. Table 2 presents a comparative analysis with respect to settling time (min), peak overshoot (mg/dl), peak undershoot (mg/dl), noise (%), and steady-state error e_{ss} (%). The effect of meal disturbance in the BG concentration of the patient model applying different control approaches such as PID, fuzzy, $H\infty$, and SM is also presented in Table 2 based on the references [4, 5, 10, 14], respectively. Similar working conditions are followed with the same level of actuator and sensor noise in all control techniques application for comparison. The BG concentration in TIDM patient model ingested to 60 gm meal under different controllers is tested. The corresponding results are presented for the various control approaches along with the proposed FOPIDC with respect to control specifications such as overshoots (mg/dl), undershoots (mg/dl), and settling time (min). The results signify the better controllability of the FOPIDC. The simulation results also demonstrate the high noise elimination capability with high robustness for the proposed approach. Overall, by looking to the above comparative analysis, the findings of suggested approach advantages are the higher

Table 2. Comparative result analysis related to the BG concentration

Controller	PID [4]	Fuzzy [5]	$H\infty$ [10]	SM [14]	FOPIDC (Proposed)
Meal intake (mg)	60	60	60	60	60
Insulin infusion rate (mU/min)	59.6	59.2	59.1	59	59.1
Settling time (min)	290	260	255	250	255
Peak overshoot (mg/dl)	5.2	5.3	6.6	6.5	4.4
Peak undershoot (mg/dl)	3.1	2.1	1.2	1.5	1.1
Noise (%)	10	10	5	5	5
e_{ss} (%)	0	0	0	0	0

accuracy and stability, more robustness, high noise elimination capability, and better capability to handle uncertainty under various abnormal conditions and huge variation meal disturbance.

5 Conclusions

In this paper, a novel control strategy FOPIDC is proposed for BG control in TIDM patient model. To justify its enhanced performance, it has been applied and tested to control the BG concentration in patient within *normoglycaemic* range. Initially, the patient is modeled as a ninth-order state-space representation. Then the proposed control approach FOPIDC is designed. The comparative results clearly reflect that the suggested FOPIDC is arrived at better performance than the other control approaches such as PID, fuzzy, $H\infty$, and SM, with respect to stability, reliability, and robustness under various abnormal conditions and disturbances. The related better performance of the suggested approach (FOPIDC) in terms of improved accuracy and stability, enhanced robustness, high noise elimination capability, and better capability to handle uncertainty justify its real-time application.

References

1. Chee F, Fernando T (2003) Closed-loop glucose control in critically ill patients using continuous glucose monitoring system (CGMS) in real time. IEEE Trans Inf Technol Biomed 7(4):43–53
2. Kamath S, Vidyasagar S (2009) Simulation study on closed loop control algorithm of type 1 diabetes mellitus patients. IETE J Res 55:230–235
3. Frederick C, Tyrone L (2003) Expert PID control system for blood glucose control in critically ill patients. Information technology in biomedicine. IEEE Trans 7(4):419–425
4. Sutradhar A, Chaudhuri S (2002) Analysis and design of an optimal PID controller for insulin dispenser system. J Inst Eng (India) 82(2):304–313

5. Ibbini M (2006) The PI-fuzzy logic controller for the regulation of blood glucose level in diabetic patients. J Med Eng Technol 30(2):83–92
6. Singh M, Gupta JRP (2007) A new algorithm-based type-2 fuzzy controller for diabetic patient. Int J Biomed Eng Technol 1(1):18–40
7. Patra AK, Rout PK (2015) An automatic insulin infusion system based on LQG control technique. Int J Biomed Eng Technol 17(3):252–275
8. Chee F, Andrey V (2005) Optimal H∞ insulin injection control for blood glucose regulation in diabetic patients. IEEE Trans Biomed Eng 52(10):1625–1631
9. Yasini S, Sistani MBN (2012) Knowledge-based Closed-loop control of blood glucose concentration in diabetic patients and comparison with H_{∞} control technique. IETE J Res 58:328–336
10. Patra AK, Rout PK (2014) Optimal H-infinity insulin injection control for blood glucose regulation in IDDM patient using physiological model. Int J Autom Control 8(4):309–322
11. Gallardo H, Ana G (2013) High-order sliding-mode control for blood glucose: practical relative degree approach. Control Eng Pract 21(5):747–758
12. Rmileh A, Gabin W (2012) Wiener sliding-mode control for artificial pancreas: a new nonlinear approach to glucose regulation. Comput Methods Programs Biomed 107:327–340
13. Patra AK, Rout PK (2018) 'Backstepping sliding mode Gaussian insulin injection control for blood glucose regulation in TIDM patient. J Dyn Sys Meas Control 140(9):091006–091006-15
14. Patra AK, Rout PK (2017) Adaptive sliding mode Gaussian controller for artificial pancreas in TIDM patient. J Process Control 58:23–27
15. Patra AK, Mishra AK, Rout PK (2018) Backstepping model predictive controller for blood glucose regulation in type-I diabetes patient. IETE J Res. https://doi.org/10.1080/03772063.2018.1493404
16. Patra AK, Rout PK (2017) Adaptive continuous-time model predictive controller for implantable insulin delivery system in Type I diabetic patient. Opt Control Appl Methods 38(2):184–204
17. Guyton JR, Foster RO (1978) A model of glucose-insulin homeostasis in man that incorporates the heterogenous fast pool theory of pancreatic insulin release. Diabetes Care 27(1):1027–1042
18. Barger M, Rodbard D (1989) Computer simulation of plasma insulin and glucose dynamics after subcutaneous insulin injection. Diabetes Care 12(1):725–736
19. Lehmann ED, Deutsch T (1992) Physiological model of glucose–insulin interaction in Type-1 diabetes mellitus. J Biomed Eng 14(3):235–242
20. Lehmann ED, Deutsch T (1998) Compartmental models for glycaemic prediction and decision support in clinical diabetes care: promise and reality. Comput Methods Programs Biomed 56(1):193–204
21. Sperr G (1997) Biosensor research targets medical diagnostics. Medical device and diagnostic industry magazine, Nov 1997
22. Cochin L, Cadwallender W (1997) Analysis and design of dynamic systems, 3rd edn. Addison-Wesley, New York
23. Bingul Z, Karahan O (2018) Comparison of PID and FOPID controllers tuned by PSO and ABC algorithms for unstable and integrating systems with time delay. Opt Control Appl Methods 39(4):1431–1450

Mutual Fund Investment Method Using Recurrent Back Propagation Neural Network

Smruti Rekha Das[1(✉)], Debahuti Mishra[2], Pournamasi Parhi[2], and Prajna Paramita Debata[3]

[1] Department of Computer Science and Engineering, Gandhi Institute for Education and Technology, Baniatangi, Bhubaneswar, Odisha, India
smrutirekhadas@gietbbsr.com
[2] Department of Computer Science and Engineering, Siksha 'O'Anusandhan Deemed to be University, Bhubaneswar, Odisha, India
debahutimishra@soa.ac.in
[3] International Institute of Information Technology, Bhubaneswar, Odisha, India

1 Introduction

Mutual fund is managed professionally. The basic advantages of investing money in mutual fund are the highest level of diversification. Through diversification, it is able to hold many types of securities and diversification also reduces the risk of investing money in securities. In the recent years, a number of methods have been introduced by the researchers and it is found that neural network-based methods have proved its efficiency over a different area of financial market prediction. W-C. Chiang has identified and explored [1] on back propagation neural network (BPNN) to predict net asset value (NAV) of mutual funds. In addition to this, D. C. Indro et al. has applied artificial neural network (ANN) for equity mutual [2] fund prediction. Exploring about the recurrent methods, Manjunath Patel GC et al. proved that in neural network methods, reverse mapping gives [3] effective prediction performance. Various research articles have published, exploring on the efficiency of neural network-based methods, from which, very few are introduced here. By going through the above research, this study was motivated to explore the efficiency of neural network-based methods for predicting mutual fund, as the proper prediction can yield a highest benefit to the investor. Three neural network-based methods such as BPNN, RBPNN and RRBFNN are employed in this paper, as these are the widely accepted models. An empirical comparison based on the mean square error (MSE) results, during training, is established among these prediction models. NAV along with some statistical measures calculated for of SBI Magnum Equity and UTI Equity mutual fund dataset are given as the input to the prediction model, to validate its predicted performance. The prediction methods have to be employed for 1 day ahead, 5 days ahead and 7 days ahead. These three different days ahead are to be predicted by the prediction methods. In addition to this, the prediction models are also validated through various performance measures at the time of testing.

M. N. Mohanty and S. Das (eds.), *Advances in Intelligent Computing and Communication*, Lecture Notes in Networks and Systems 109,
https://doi.org/10.1007/978-981-15-2774-6_40

The rest of the works are organized in the following manner, Sect. 2 focuses on the details of methodologies that have been used in this study. The dataset description and experimental analysis are found in Sects. 3 and 4, respectively. Finally, Sect. 4 concludes the work with future scope.

2 Methodologies

The neural network is developed on the concept of the working principle of brain of human and its nervous system. It is described by consisting an interconnection among each layer, each layer having the number of neurons, the learning algorithm that required for network training and the activation functions. The network is unknown about the prior assumption of the generating process with approximate of any nonlinear continuous function. The training [4–6] of neural network based on a simple concept that whenever the network gives the wrong answer, the weight between the layers is updated with the expectation of getting a minimum error, because of this network's future responses possibly to be correct. Modifying the concept of BPNN, RBPNN is designed, where the network contains a cyclic connection from the output to the hidden layer input, for making a more powerful model than [7, 8] the feed-forward neural network. RBPNN technique has been applied by the resources in many of the research areas. Among different types of feed-forward neural network, RBFNN is able to solve the nonlinear problems. However, RRBFNN is having a major drawback that basically their input to output map is static and to represent the nonlinear data, their efficiency is very [9, 10] limited. It has been observed that both RBFNN and RRBFNN can approximately be used for any nonlinear functions if the structures of any nonlinear functions are constructed properly.

3 Dataset Description

Two mutual fund datasets SBI Magnum Equity and UTI Equity mutual fund are collected. This study considered these two mutual funds for experimentation as UTI Equity is the oldest mutual fund and SBI Magnum is the first non-UTI mutual fund in India. Eleven years of data have selected, with the period of March 1, 2006 to March 1, 2017. In mutual fund only the NAV is available, hence using the NAV the dataset is regenerated by calculating the [11] *mean*, *standard deviation*, *kurtosis* and *skewness*. These statistical measures have generated using the proper mathematical formula. For calculating the statistical measure, 12 is considered as the window size. After regenerating the data from the available NAV, now the dataset is ready for input to the model, in the ratio of 70% data is for training and 30% for is for testing. Table 1 shows the total number of available data and the number of data patterns that are generated during the windowing process. The training patterns and testing patterns are generated accordingly in the ratio of 7:3.

Table 1. Details of mutual fund datasets

Name of mutual fund	Data	Total available data	Total generated patterns	Training patterns	Testing patterns
SBI magnum equity	1-03-2006 to 1-03-2017	3153	3140	2198	942
UTI equity mutual fund	1-03-2006 to 1-03-2017	2676	2663	1864	799

4 Experimental Analysis

The schematic layout of the proposed model is depicted in Fig. 1. At first, two mutual fund datasets such as SBI Magnum Equity and UTI Equity have been collected. Then, in the second step, the dimensionality of the dataset has increased using the statical measures such as *mean*, *standard deviation*, *kurtosis* and *skewness* with 12 window size. Further, the dataset is divided into 70% is for training and 30% is for testing. These training data are set input for all the prediction models such as BPNN, RBPNN and RRBFNN. An empirical comparison is carried out among these predictions used for experimental purpose to get the best predicted outcome.

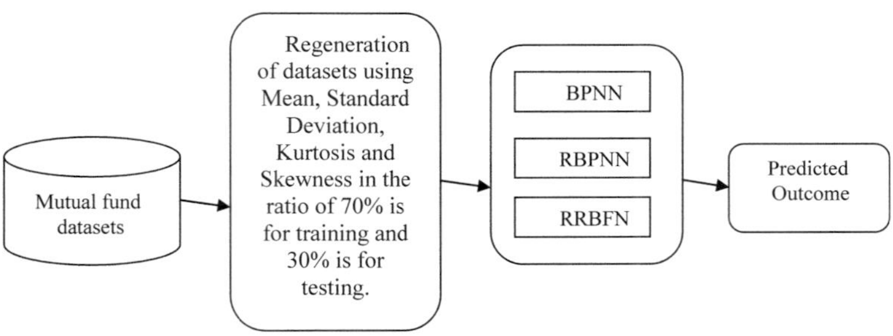

Fig. 1. Abstract view of the proposed model

The simulation result presented in actual versus predicted form for BPNN, RBPNN and RRBFNN is shown in Figs. 2, 3 and 4, respectively. From the figures, it is visualized that in RBPNN, the distance between actual graph and predicted graph is very less.

From the regenerated dataset, five elements such as *NAV*, *mean*, *standard deviation*, *kurtosis* and *skewness* for every experimented day are employed to BPNN, RBPNN and RRBFNN. For all the three prediction models, different parameter values are set according to the specification of the algorithm. 100 number of iterations is well thought out for all the models to make the comparison standardize. Apart from iteration, BPNN needs the number of hidden layers and the number of nodes present in each hidden layer. This study has chosen three numbers of hidden layers and each hidden layer is having five number of nodes. Similarly, for RBPNN, along with the above

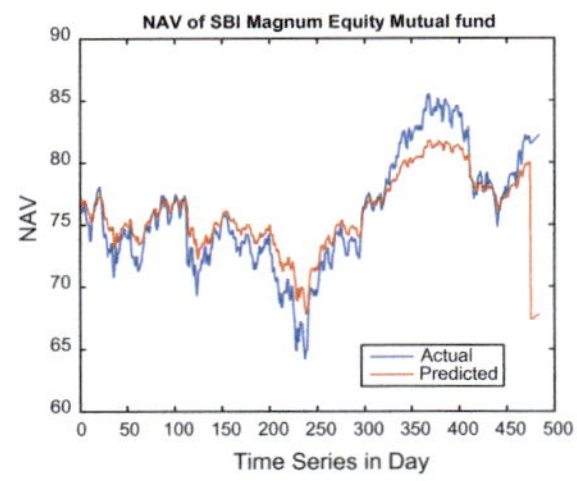

(a) 1 day(SBI)

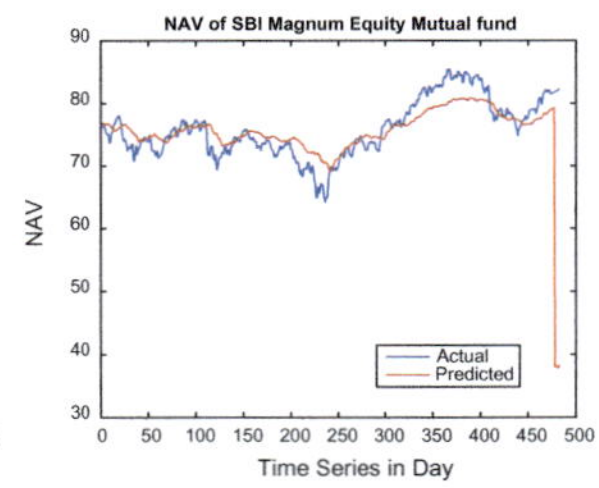

(b) 5 days(SBI)

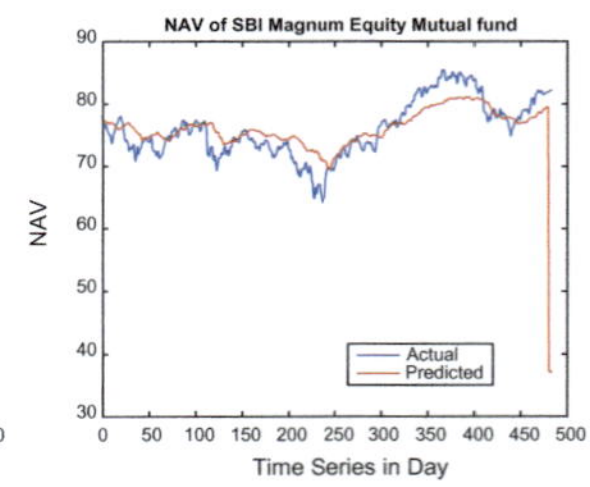

(c) 7 days(SBI)

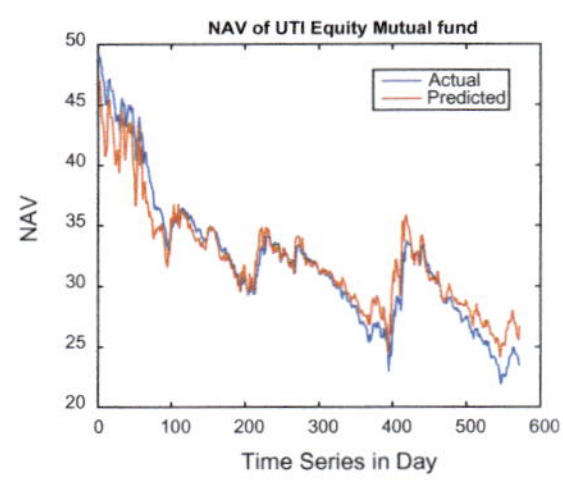

(a) 1 day(UTI)

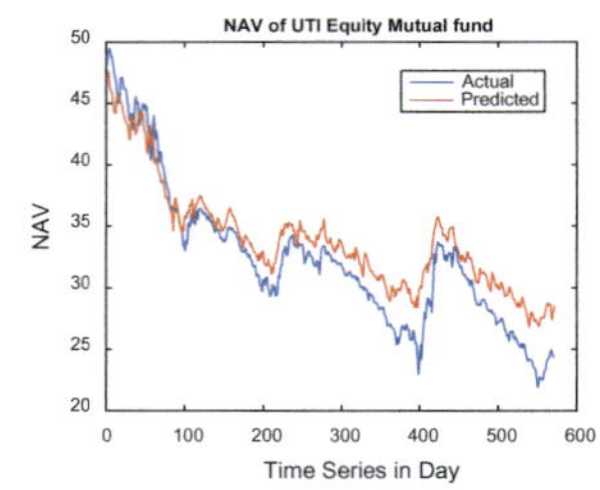

(b) 5 days(UTI)

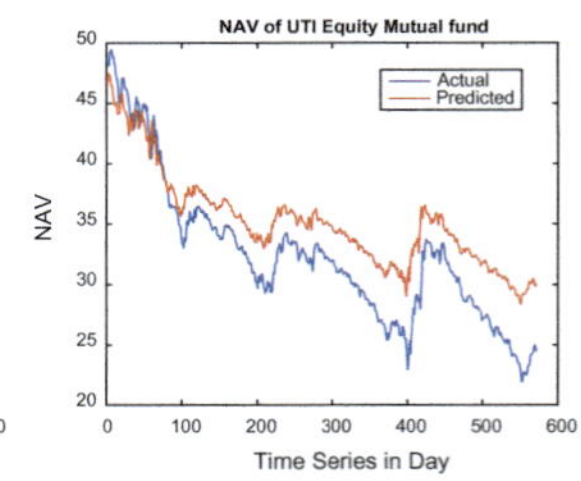

(c) 7 days(UTI)

Fig. 2. NAV of SBI and UTI mutual fund using BPNN

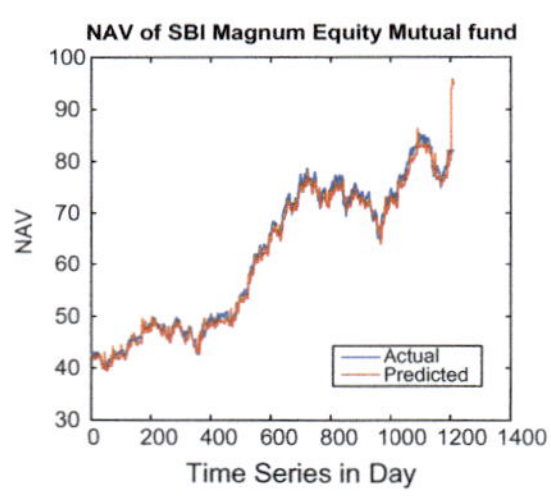

(a) 1 day(SBI)

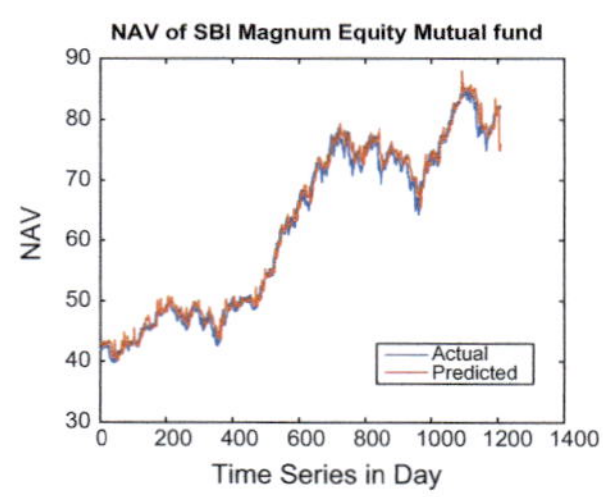

(b) 5 days(SBI)

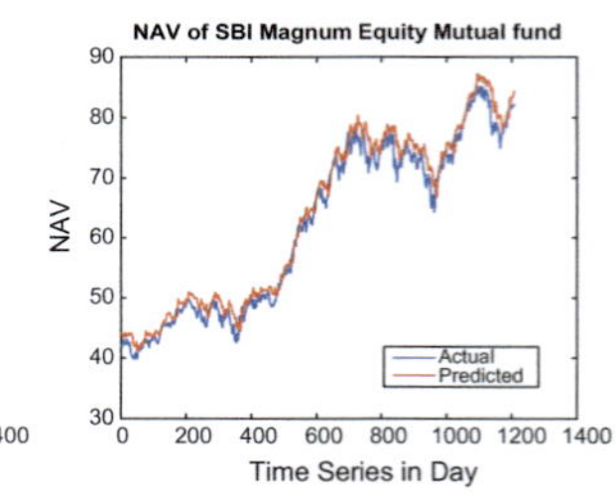

(c) 7 days(SBI)

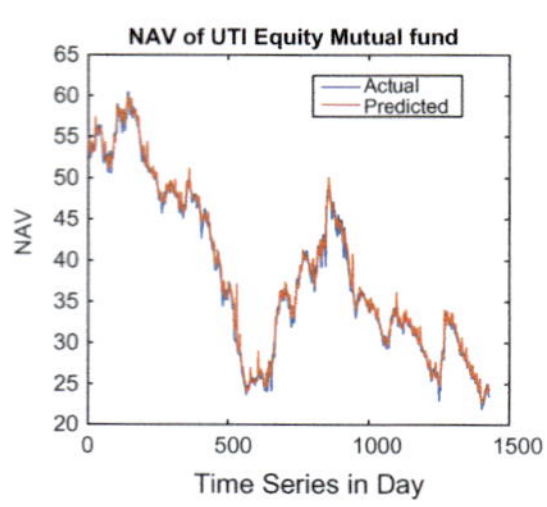

(a) 1 day(UTI)

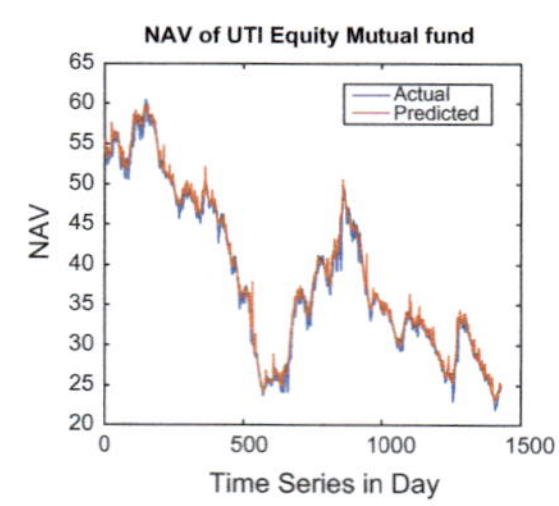

(b) 5 days(UTI)

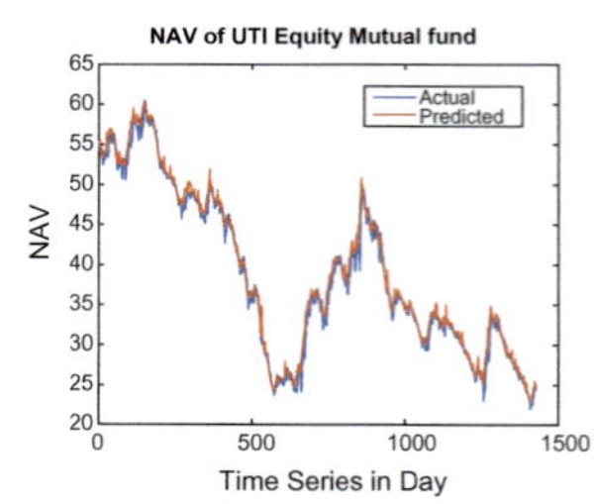

(c) 7 days(UTI)

Fig. 3. NAV of SBI and UTI mutual fund using RBPNN

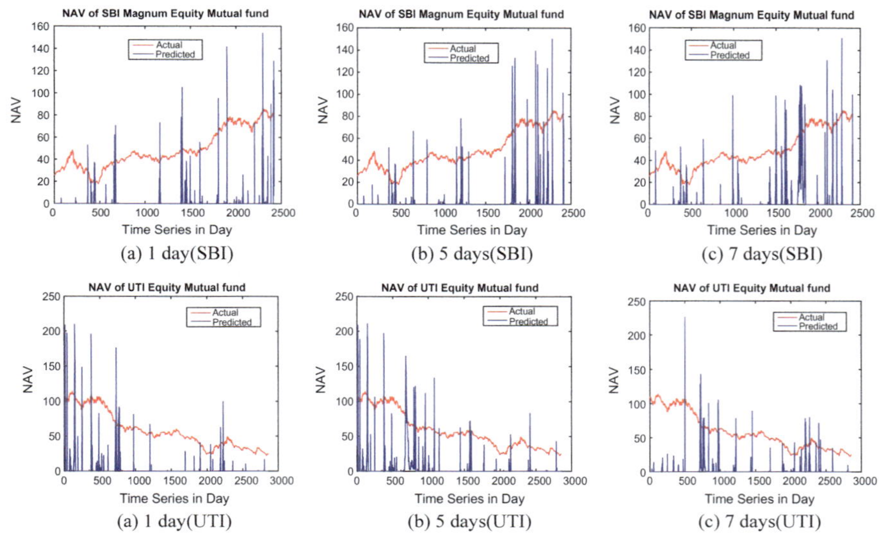

Fig. 4. NAV of SBI and UTI mutual fund using RRBFNN

parameters of BPNN, three other controlling parameters are defined with its suitable value. The value of the threshold is set to 0.001 and for both the learning rate parameter and momentum coefficient 0.01 are set for the experimentation. Further, in RRBFNN, the radial function is the Gaussian function, and the Gaussian function needs two values: One is the value of center *c* and the other is the value of radius *r*. This experimental work fixed 0 for center *c* and 1 for radius *r*. In the training phase, mean square error (MSE) is calculated to train the network. In Table 2, the MSE results of all the models during training are given. From the above table, it can be observed that, for both SBI magnum equity mutual fund and UTI equity mutual fund, RBPNN is performing better than the rest two prediction models. Comparing among the rest two models, it is found that the performance of BPNN is better than RRBFNN. In addition to this, the performance of different days ahead has also examined. The prediction efficiency is very high for the prediction of 1 day ahead than 5 and 7 days. It proves that when the day increases the model performance is decreases. The prediction

Table 2. MSE of BPNN, RBPNN and RRBFNN during training phase

Day(s)		BPNN	RBPNN	RRBFNN
1	SBI	0.8229	0.41571	$1.2769e^{+03}$
5		1.2225	0.45536	$1.2810e^{+03}$
7		1.4436	0.75775	$1.2853e^{+03}$
1	UTI	1.1427	0.091836	$1.8423e^{+03}$
5		1.1675	0.18402	$1.8526e^{+03}$
7		2.2503	0.19421	$1.9752e^{+03}$

methods are validated during the testing phase considering the mean absolute percentage error (MAPE) and root mean square error (RMSE) result. For both the experimented dataset, the validation work is carried out and referring to Table 3, it is noticed that RBPNN is giving less value in the form of error compared to BPNN and RRBFNN.

Table 3. Performance evaluation during the testing phase for 1 day ahead prediction

Methods		MAPE	RMSE
BPNN	SBI	1.56432	1.56432
RBPNN		**0.94251**	**0.98242**
RRBFNN		2.45231	2.18721
BPNN	UTI	1.89522	1.98356
RBPNN		**0.98882**	**0.99231**
RRBFNN		2.01432	2.18164

The bold letter indicates of minimum error. That means RBPNN is giving minimum error than other two models (BPNN and RRBFNN). Minimum error means again maximum accuracy

From the convergence graph depicted in Figs. 5 and 6, it is noticed that RBPNN is converging faster than BPNN and RRBFNN. RBPNN requires only five number of iterations to converge, which is very less compared to the total number of experiments considered here, in this study. By analyzing these convergence graphs, it can be examined that the convergence speed of BPNN is faster than RRBFNN. This convergence graph is plotted for the MSE error during training, considering the data of 1 day ahead to be predicted for both SBI and UTI Mutual fund.

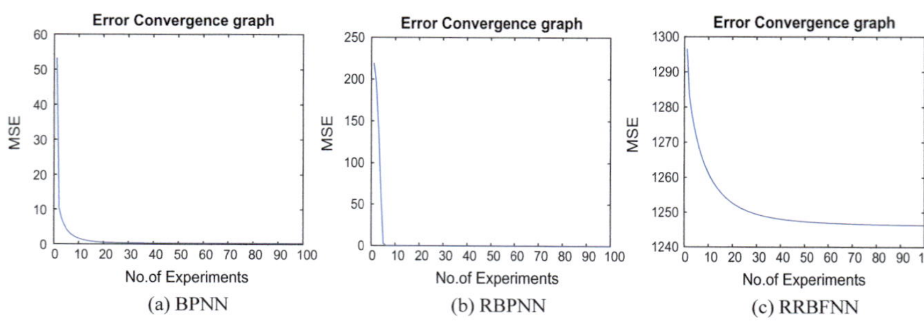

Fig. 5. Error convergence graph of SBI during training for BPNN, RBPNN and RRBFNN

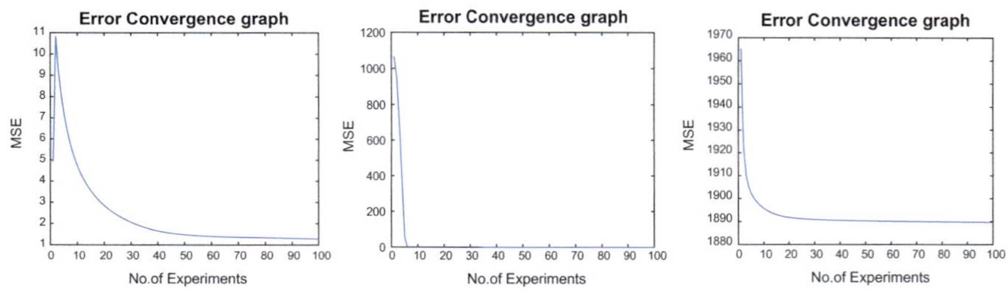

(a) BPNN (b) RBPNN (c) RRBFNN

Fig. 6. Error convergence graph of UTI during training for BPNN, RBPNN and RRBFNN

5 Conclusion

This study has explored the predictive analytics for different time horizon using the base neural network models. So, for these three prediction models such as BPNN, RBPNN and RRBFNN are considered for this experimental work and an empirical comparison has established among these prediction models to assess their performance. Observing the simulated graph and MSE result in training as well as RMSE and MAPE during testing, it is noticed that RBPNN is outperformed over BPNN and RRBFNN. In future, the work can be extended using different optimization techniques.

References

1. Chiang W-C, Urban TL, Baldridge GW (1996) A neural network approach to mutual fund net asset value forecasting. Omega 24(2):205–215
2. Indro DC, Jiang CX, Patuwo BE, Zhang GP (1999) Predicting mutual fund performance using artificial neural networks. Omega 27(3):373–380
3. GC MP, Shettigar AK, Krishna P, Parappagoudar MB (2017) Back propagation genetic and recurrent neural network applications in modelling and analysis of squeeze casting process. Appl Soft Comput 59:418–437
4. Sutheebanjard P, Premchaiswadi W (2010) Stock exchange of Thailand index prediction using back propagation neural networks. In 2010 second international conference on computer and network technology (ICCNT), IEEE, 2010, pp 377–380
5. Khoa NLD, Sakakibara K, Nishikawa I (2006) Stock price forecasting using back propagation neural networks with time and profit based adjusted weight factors. In: International joint conference SICE-ICASE, 2006, IEEE, 2006, pp 5484–5488
6. Lee C-T, Chen Y-P (2007) The efficacy of neural networks and simple technical indicators in predicting stock markets. In: International conference on convergence information technology, 2007, IEEE, pp. 2292–2297
7. Sak H, Senior A, Beaufays F (2014) Long short-term memory recurrent neural network architectures for large scale acoustic modeling. In: Fifteenth annual conference of the international speech communication association, 2014
8. Huang Z, Huang D, Lyu MR, Lok T (2006) Classification based on Gabor filter using RBPNN classification. In: 2006 International conference on computational intelligence and security, IEEE, 2006, vol 1, pp 759–762

9. Han HG, Guo YN, Qiao JF (2017) Self-organization of a recurrent RBF neural network using an information-oriented algorithm. Neurocomputing 225:80–91
10. Han Hong-Gui, Qiao Jun-Fei (2012) Adaptive computation algorithm for RBF neural network. IEEE Trans Neural Netw Learn Syst 23(2):342–347
11. Das SR, Mishra D, Rout M (2019) An optimized feature reduction based currency forecasting model exploring the online sequential extreme learning machine and krill herd strategies. Phys A Statist Mech Appl 513:339–370

Stabilizing and Trajectory Tracking of Inverted Pendulum Based on Fractional Order PID Control

Akshaya Kumar Patra[1](✉), Alok Kumar Mishra[1], Anuja Nanda[1], Dillip Kumar Subudhi[2], Ramachandra Agrawal[2], and Abhishek Patra[2]

[1] Department of EEE, ITER, S'O'A University, Deemed to be University, Bhubaneswar 751030, India

[2] Department of CSIT, ITER, S'O'A University, Deemed to be University, Bhubaneswar 751030, India

1 Introduction

The control of IP is a standard problem among all other problems in the control system engineering field owing to non-minimum phase, nonlinear, and under-actuated characteristics as reported in the references [1–3]. Additionally, the IP system exhibits a significant number of industrial applications like self-balancing two-wheeled vehicles or a kind of sag-way, guided missiles, rockets, intelligent robots, and other crane models [4]. In this present study, an adaptive law of control approach is being tested and analyzed to choose the IP system as it possesses a significant relevance with the control dynamics. During the past three decades, so many control strategy techniques are suggested and tested for control of AP of pendulum within the stable range. Time discrete and switching PID control strategy is implemented in IP problems with variable control gains based on the measured AP of the pendulum [5, 6]. However, the optimal gain parameter setting, a lesser range of robust control and need of change of gain setting with varying conditions are the major limitations to limit the real-time application of these controllers. Among other projected robust control algorithms implemented for limiting the AP of the pendulum are fuzzy [7], linear quadratic regulator (LQR) [8–10], neurocontrol [11], backstepping control [12], passivity control [13], state feedback control [14], H-infinity (H_∞) control [15], sliding mode (SM) control [16], fuzzy-sliding mode (FSM) control [17, 18], and BLQG control [19]. However, even if these control techniques are implemented effectively to control the AP, the pendulum with better accuracy still fails to handle various constraints and random change found in a trajectory of motion in the pendulum. These control techniques are not completely insensitive to the disturbances and the uncertainties of the model in spite of the improved performance. Hence, optimal control parameters setting for better performance and avoiding slow response, the current work suggests an alternative novel technique implementing the fractional order PID approach.

The FOPIDC approach concept leads to an enhanced control performance with respect to robustness and delay of time compensation characteristics to counteract the negative impact of associated errors, disturbances, and uncertainties. The stabilizing

M. N. Mohanty and S. Das (eds.), *Advances in Intelligent Computing and Communication*, Lecture Notes in Networks and Systems 109,
https://doi.org/10.1007/978-981-15-2774-6_41

control law $u(t)$ is developed based on the FOPIDC approach [20–22]. Application of the FOPIDC to control the AP of pendulum results to ensure a better robust controller in comparison with other contemporary well-accepted methods under both harmonized and incompatible uncertainties.

This manuscript is structured as follows. Section 2 concisely illustrates the IP system with mathematical details reflecting to its dynamic characteristics. Also, it clearly demonstrates the simulation execution of the system on MATLAB environment. A detailed presentation on how the control technique is formulated and how it is implemented for this problem is presented in Sect. 3. Comparative results of the proposed approach with other published control techniques and the related analysis are provided in Sect. 4. The concluding comments are summarized in the Sect. 5.

2 Problem Formulation and Modeling

2.1 System Overview

The closed-loop model of the IP is depicted in Fig. 1a. The applied horizontal force $F(t)$ and $v(t)$ is reflected as the process disturbance and the sensor noise, respectively, in this study. The controller receives information about the AP of the pendulum as an input to provide the optimal control force $u(t)$, and it balances the pendulum.

2.2 System Modeling

Figure 1b reflects the cart-pendulum model connected to a flexible cart rail with a free swinging pole. The CP is being controlled by a DC motor. The nonlinear IP modeling through the Newton's law-based mathematical equations has been carried out. It is presumed that the vertical force does not affect the CP, and the CP is disturbed by the horizontal force $F(t)$ based on the operation of the DC motor [23–25]. All the physical activities of the IP system are expressed mathematically and specified by Eqs. (1) and (2). All nomenclature and specifications for IP are shown in Table 1 and Table 2, respectively. The Simulink model of the IP is established with respect to the Eqs. (1) and (2) as displayed in Fig. 1c.

$$(M+m)\frac{\mathrm{d}^2x(t)}{\mathrm{d}t^2} - ml\frac{\mathrm{d}^2\theta(t)}{\mathrm{d}t^2}\cos\theta(t) + ml\left(\frac{\mathrm{d}\theta(t)}{\mathrm{d}t}\right)^2 \sin\theta(t) + b\frac{\mathrm{d}x(t)}{\mathrm{d}t} = F(t) \tag{1}$$

$$(i+Ml^2)\frac{\mathrm{d}^2\theta t}{\mathrm{d}t^2} - mgl\sin\theta(t) = ml\frac{\mathrm{d}^2x(t)}{\mathrm{d}t^2}\cos\theta(t) \tag{2}$$

2.3 Response of IP Model

There are four roots in IP system. One of them lies in right-hand side of the complex plane. As a result, the system becomes unstable.

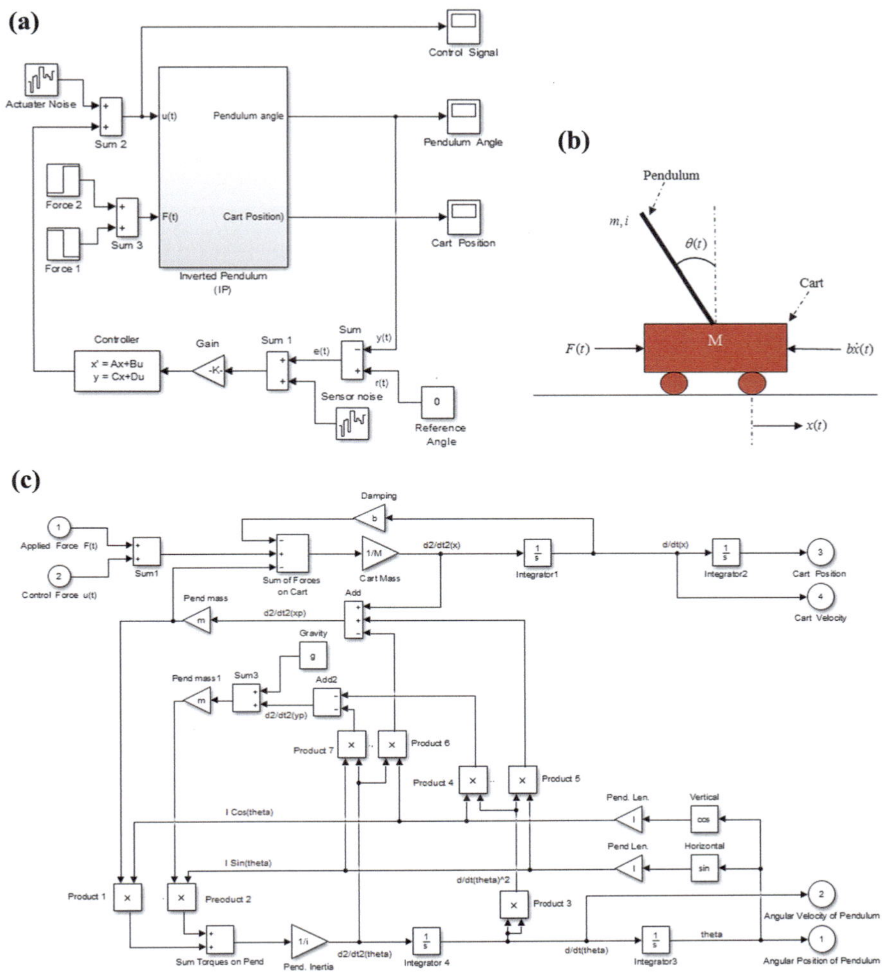

Fig. 1. **a** IP model with LQR; **b** schematic model of the IP system; **c** simulation model of the nonlinear IP system

Table 1. IP model states and parameters

Symbol	Description
$F(t)$	Horizontal force acting on the cart
$b\frac{dx(t)}{dt}$	Frictional force acting on the cart
$x(t), \frac{dx(t)}{dt}, \frac{d^2x(t)}{dt^2}$	Cart position, cart velocity, and cart acceleration, respectively
$\theta(t), \frac{d\theta(t)}{dt}, \frac{d^2\theta(t)}{dt^2}$	AP, AV, and angular acceleration of pendulum, respectively

Table 2. IP model specification values

Symbol	Physical meaning	Value
M	Cart mass	0.5 kg
m	Pendulum mass	0.2 kg
i	Inertia	0.3 kg.m^2
g	Gravitational acceleration	9.8 m/s^2
b	Frictional coefficient	0.1 N.s/m
l	Pendulum length	0.3 m

This needs the design of an adaptive controller for improving the stability of the system by means of shifting the roots into the left-hand side of the complex plane. The IP system Simulinkmodel in the open-loop form is depicted in Fig. 1c. The IP system consists of two inputs and four outputs. The control force $u(t)$ and applied horizontal force $F(t)$ are the two inputs of the IP system. The CP and AP of the pendulum are two outputs of the IP system. An uncontrolled system dynamics such as AP of the pendulum and CP are being observed owing to the application of 1 N impulsive horizontal force $F(t)$ on the cart at the time $t = 1.0\,\text{sec}$. The uncontrolled system dynamics are illustrated in Fig. 2a, b. Figure 2a, b illustrates the unstable dynamics under various model uncertainties and disturbances. The unstable dynamics can be reduced by applying the suitable control techniques. In this case, the AP of the pendulum is the most essential outcome needs to be controlled within a stable range through suitable control techniques, and CP is analyzed in order to view the motion trajectory.

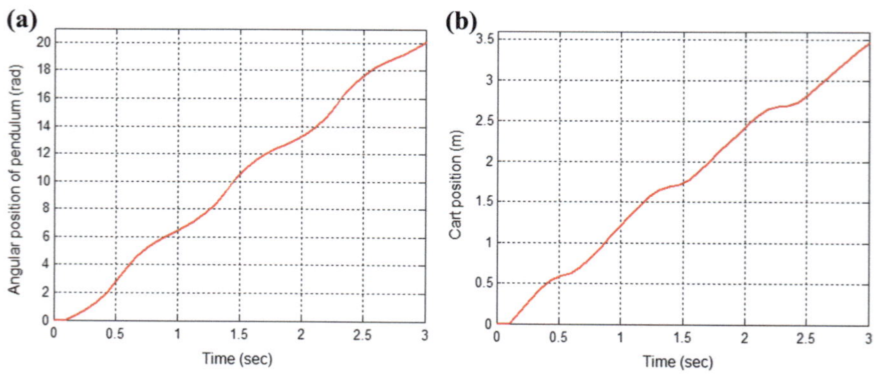

Fig. 2. **a** AP of pendulum with application of $F(t)$; **b** CP with application of $F(t)$

3 Control Algorithm

The FOPIDC algorithm is demonstrated in this section. The closed-loop system response with respect to robustness, accuracy, and stability is analyzed. The control specifications such as settling time t_s, steady-state error e_{ss}, maximum overshoot O_{Max}, and maximum undershoot U_{Max} are also evaluated and examined with proper validation of the controller action.

3.1 FOPIDC Design

The IP model with fractional order $PI^{\alpha}D^{\gamma}$ controller is shown in Fig. 1a. In fractional order $PI^{\alpha}D^{\beta}$ control, the error signal $e(t)$ is used to generate control signal $u(t)$. The transfer function (TF) of the proposed controller can be formulated as [26]:

$$\text{TF} = K_p + \frac{K_i}{s^{\alpha}} + K_d s^{\beta} \tag{3}$$

where α and β are the fractional orders of the integrator and differentiator of the proposed controller, respectively.K_p, K_i, and K_d are denoted for the proportional, integral, and derivative gains of the proposed controller, respectively. The optimal values of the FOPIDC parameters are considered based on the MATLAB/Simulink environment and FOMCON toolbox as represented in Table 3. The structure of FOPIDC is shown in Fig. 3, which is designed based on the Eq. (3).

Table 3. The optimal values of control parameters

K_p	K_d	K_i	α	β
18	6	1.5	0.7	0.9

4 Outcomes and Discussions

The response of the closed-loop IP with proposed FOPIDC is described in detail in this section. The proposed control approach (FOPIDC) is compared with other popular control algorithms to justify its enhanced performance.

4.1 Performance Analysis of IP System with FOPIDC

In this section, all physical activities of the closed-loop IP model with suggested FOPIDC are examined under different conditions and the huge deviation of applied horizontal force. The closed-loop system dynamics with 1 N impulsive horizontal force $F(t)$ at the time of 1.0 s are displayed in Fig. 4a, b. The outcomes clearly specify the pendulum achieves the zero AP with less t_s, and cart also attains the balance position where the system is absolutely stable. To achieve the enhanced system response, the required control force $u(t)$ is generated by the suggested FOPIDC and demonstrated in Fig. 5a.

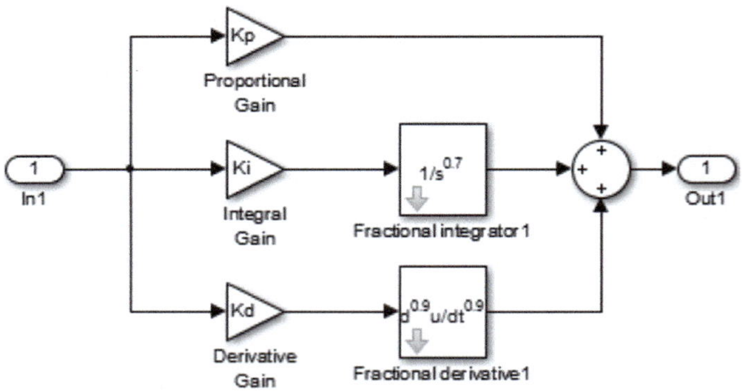

Fig. 3. Simulink diagram of FOPIDC

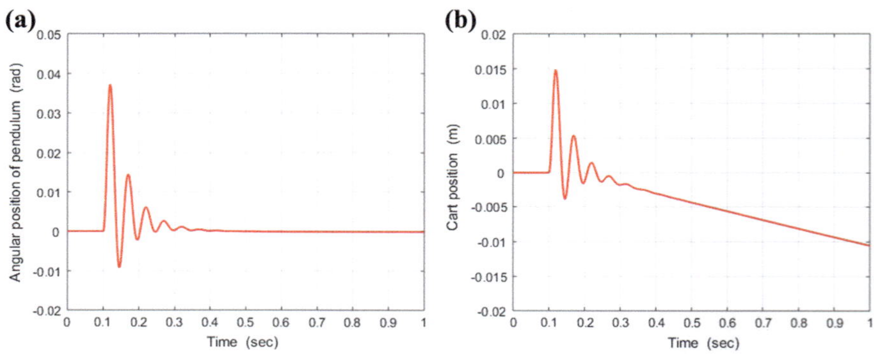

Fig. 4. **a** AP of the pendulum with the application of $F(t)$ based on FOPIDC; **b** CP with the application of $F(t)$ based on FOPIDC

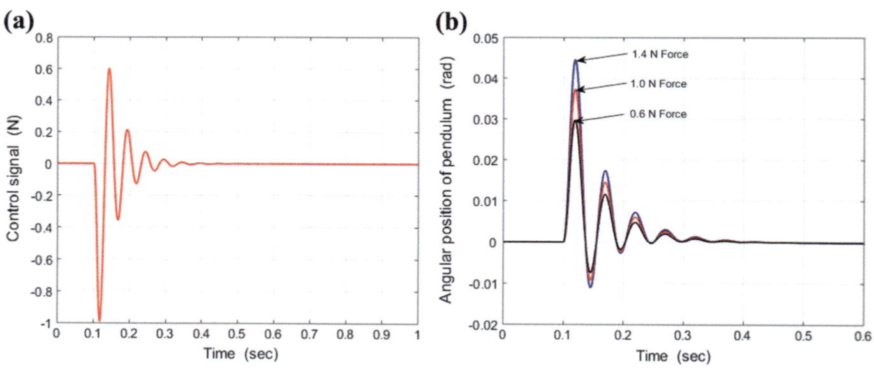

Fig. 5. **a** Controlled signal $u(t)$ of the FOPIDC; **b** AP of the pendulum with a deviation of $\pm 40\%$ applied $F(t)$ based on FOPIDC

4.2 Robustness of the FOPIDC

Figure 5b illustrates the AP of the pendulum with suggested FOPIDC under the huge deviation of applied horizontal force $F(t)$. The time-domain outcomes under huge deviation of applied forces show the enhanced performance of the closed-loop system with FPIDC. Overall in each case, the pendulum achieves finally zero AP with less t_s. As indicated by the outcomes, the suggested FOPIDC robust performance under the huge deviation of applied horizontal forces is much better.

4.3 Comparative Study

The suggested FOPIDC approach is compared with other popular control approaches such as PID, fuzzy, LQR, H_∞, FSM, and BLQG justify its enhanced performance as a controller. Figure 4(a) illustrates the effect of applied force in the AP of the pendulum with the proposed FOPIDC approach. Table 4 presents a comparative analysis with respect to $t_s(\text{sec})$, $O_{\text{Max}}(\text{rad})$, $U_{\text{Max}}(\text{rad})$, noise (%), and $e_{\text{ss}}(\%)$. The effect of applied force in the AP of pendulum applying different control approaches as PID, fuzzy, H_∞, FSM, and BLQG is also demonstrated in Table 4 based on the references [5, 7, 9, 15, 18, 19], respectively. Similar working conditions are followed with the same level of sensor noise in all control techniques application for comparison.

Table 4. Comparative result analysis related to AP of pendulum

Controller	PID [5]	Fuzzy [7]	H_∞ [15]	FSM [18]	BLQG [19]	FOPIDC (Proposed)
Applied force (N)	1	1	1	1	1	1
$t_s(\text{sec})$	2.8	3.0	2.1	1.8	1.9	0.37
$O_{\text{Max}}(\text{rad})$	0.1	0.226	0.107	0.042	0.041	0.037
$U_{\text{Maz}}(\text{rad})$	0.01	0.087	0.045	0.151	0.01	0.008
Noise (%)	10	10	5	5	5	5
$e_{\text{ss}}(\%)$	0	0	0	0	0	0

The AP of pendulum under 1 N impulsive horizontal force is tested. The corresponding results are presented for the various control approaches along with the proposed FOPIDC with respect to time-domain specifications such as $O_{\text{Max}}(\text{rad})$, $U_{\text{Max}}(\text{rad})$, and $t_s(\text{sec})$. The results signify the better controllability of the FOPIDC. The simulation results also demonstrate the high noise and chattering elimination capability with high robustness for the proposed approach. Overall, by looking to the above comparative analysis, the findings of suggested approach advantages are the higher accuracy and stability, more robustness, high noise and chattering elimination capability, and better capability to handle uncertainty under various conditions and huge deviation of road disturbance.

5 Conclusions

The manuscript proposed a novel control strategy (FOPIDC) to balance the pendulum. To justify its enhanced performance, it has been applied and tested to control the system dynamics of IP system within the stable range. Initially, the IP system is modeled as a fourth-order state-space representation. Then, the proposed control approach (FOPIDC) is designed and implemented. The comparative results clearly reflect that the suggested FOPIDC arrives at better performance than the other control approaches such as PID, fuzzy, H_{∞}, FSM, and BLQG with respect to stability, reliability, and robustness under various abnormal conditions and disturbances. The better performance of the suggested approach in terms of improved accuracy and stability, enhanced robustness, high noise and chattering elimination capability, and better ability to control uncertainty justifies its real-time application.

References

1. Iqbal J, Ajwad SA (2016) Automating industrial tasks through mechatronic systems – a review of robotics in industrial perspective. Tehnic ki vjesnik – Technical Gazette 23:917–924
2. Ajwad SA, Iqbal J (2017) Role and review of educational robotic platforms in preparing engineers for industry. Maejo Int J Sci Technol 11:17–34
3. Bettayeb M, Al-Saggaf U (2014) Stabilization of an inverted pendulum-cart system by fractional PI-state feedback. ISA Trans 53:508–516
4. Iqbal J, Khan SG, Ukovic SC (2017) Nonlinear control systems -a brief overview of historical and recent advances. Nonlinear Eng 6:301–312
5. Ghosh A, Subudhi B (2012) Brief paper – robust proportional-integral-derivative compensation of an inverted Cart-Pendulum System: an experimental study. IET Control Theory Appl 6(8):1145–1152
6. Wang C, Liu C, Fu W (2016) Design and simulation of IP system based on the fractional PID controller. In: IEEE 11th Conference on Industrial Electronics and Applications (ICIEA), pp 1760–1764
7. Magana ME, Holzapfel F (1998) Fuzzy-logic control of an inverted pendulum with vision feedback. IEEE Trans Educ 41(2):165–170
8. Ozana S, Hajovsky R (2012) Design and implementation of LQR controller for inverted pendulum by use of REX control system. In: IEEE International Conference on Circuits and Systems. vol 1, pp 343–347
9. Kumar EV, Jerome J (2013) Robust LQR controller design for stabilizing and trajectory tracking of inverted pendulum. Procedia Eng 64:169–178
10. Prasad LB, Gupta HO (2012) Modelling and simulation for optimal control of nonlinear inverted pendulum dynamical system using PID controller and LQR. In: Sixth Asia Modelling Symposium (AMS), pp 138–143
11. Pasemann F (1998) Evolving neurocontrollers for balancing an inverted pendulum. Netw Comput Neural Syst 9:1–4
12. Deng L, Gao S (2011) The design for the controller of the linear IP based on backstepping. In: International conference on electronic and mechanical engineering and information technology (EMEIT). vol 6, pp 2892–2895

13. Jörgl M, Gattringer H (2013) Passivity based control of a cart with inverted pendulum. Appl Mech Mater 332:339–344
14. Žilić T, Pavković D (2009) Modeling and control of a pneumatically actuated inverted pendulum. ISA Trans 48:327–335
15. Lambrecht P, Vander G (1988) H-infinity control of an experimental inverted pendulum with dry friction. IEEE Control Syst Mag 13(4):44–50
16. Wai RJ, Chang LJ (2006) Adaptive stabilizing and tracking control for a nonlinear inverted-pendulum system via sliding-mode technique. IEEE Trans Ind Electron 53:674–692
17. Tao CW, Taur J, Chang J (2010) Adaptive fuzzy switched swing-up and sliding control for the double-pendulum-and-cart system. IEEE Trans Syst Man Cybern B Cybern 40(1): 241–252
18. Chen CS, Chen WL (1998) Robust adaptive sliding-mode control using fuzzy modelling for an inverted-pendulum system. IEEE Trans Ind Electron 45(2):297–306
19. Patra AK, Rout PK (2019) Backstepping linear quadratic gaussian controller design for balancing an inverted pendulum. IETE J Res 1:1–15
20. Khalil HK (2002) Nonlinear systems, 3rd edn. Prentice Hall, Upper Saddle River, New Jersey
21. Patra AK, Rout PK (2017) Adaptive sliding mode Gaussian controller for artificial pancreas in TIDM patient. J Process Control 58:23–27
22. Patra AK, Rout PK (2018) Backstepping sliding mode Gaussian insulin injection control for blood glucose regulation in TIDM patient. J Dyn Sys Meas Control 140(9):091006
23. Irfan S, Razzaq MT, Iqbal J (2018) Advanced sliding mode control techniques for IP: modelling and simulation. Eng Sci Tech Int J. https://doi.org/10.1016/j.jestch.2018.06.010
24. Ronquillo-Lomeli G, Ríos-Moreno GJ (2016) Nonlinear identification of inverted pendulum system using Volterra polynomials. Mech Based Des Struct Mach 44(1):5–15
25. Kajita S, Fujiwara K (2003) Biped walking pattern generation by a simple three-dimensional inverted pendulum model. Adv Robot 17(2):131–147
26. Bingul Z, Karahan O (2018) Comparison of PID and FOPID controllers tuned by PSO and ABC algorithms for unstable and integrating systems with time delay. Optimal Control Appl Meth 39(4):1431–1450

Inter-area and Intra-area Oscillation Damping of Power System Stabilizer Design Using Modified Invasive Weed Optimization

Mohammad Salik[1], Pravat Kumar Rout[1](✉), and Mihir Narayan Mohanty[2]

[1] Department EEE, ITER, Siksha 'O' Anusandhan (Deemed to be University), Bhubaneswar, India
[2] Department of ECE, ITER, Siksha 'O' Anusandhan (Deemed to be University), Bhubaneswar, India

1 Introduction

In power system operation, fault disturbances create low-frequency oscillations that leads to many abnormal situations related to stability and protection. For acquiring a smooth operation, it needs to provide adequate damping to avoid sustained oscillation and may cause later on system outage. From the generator side, PSSs are responsible to provide an adequate damping to these possible oscillations by providing supplementary feedback stabilizing effects to the excitation system. At present times, the conventional PI controller based on lead–lag compensator concept with constant gain parameters is extensively used in the power system utilities [1]. However, this controller fails to offer substantial stability to the system with frequently changing power system environment.

Secondly, these controllers are required to be tuned frequently by an optimal tuning technique to set the right gain parameter values. In past few decades, although many methods are suggested and tested, till this issue is an open research with possibility of better methodology to cope up this problem. Among many successful ideas, the approaches based on modern control theory are prominent and well accepted such as optimal control, adaptive control, variable structure control, and intelligent control for the PSS design [2–5].

Heuristic optimization techniques are successfully applied in many engineering applications to optimally finding the gain parameters and system parameters. The major advantages like simplicity in design, easy to apply, less possibility to be trapped in local minima, non-derivative solution approach, and high possibility to arrive at global minima are the major factors of attraction than the classical techniques. Classical techniques are complex and have high chance of arriving at local minima. With respect to PSS design, recently many heuristic search algorithms are successfully applied such as GA [6], tabu search algorithm [7], and simulated annealing [8]. It has been found that the results are promising and arrive at a better solution. But premature convergence, degradation in efficiency, and reduction in search capability in difficult situations force to apply new evolutionary computation techniques to solve complex nonlinear power system problems.

M. N. Mohanty and S. Das (eds.), *Advances in Intelligent Computing and Communication*, Lecture Notes in Networks and Systems 109,
https://doi.org/10.1007/978-981-15-2774-6_42

A novel evolutionary optimization-based computation technique, known as invasive weed optimization (IWO), has been suggested recently [9]. Basically, it is a new numerical stochastic optimization algorithm. The mathematical formulation and searching strategy are inspired from colonizing weeds. Weeds are plants characterized with its inherent habits like: (1) Vigorous, (2) habitual practices and desirable natural threats, and (3) harvesting of the full grown seeds are challenges in the cultivation. On the other hand, weeds respond according to the climatic changes. Taking this adaptiveness of the weeds into account and efficient search algorithm can be formulated. These optimization techniques titled as invasive weed optimization (IWO) mimic the robustness, adaptation, and randomness of colonizing weeds. Considering the above advantages, the researchers are able to formulate a simple in structure and robust in searching strategy to arrive at global optimum value in many applications. The test results reflect that the solutions from IWO are better in comparison with the other methods.

In this study, the basic problem related to damping of oscillation with PSS design is formulated in terms of an objective function. This objective function is considered to decide the optimum parameter setting through a novel optimization technique based on IWO. The operating constraints are taken care of during the simulation process to arrive at a real-time feasible result. A single-machine infinite bus (SMIB) system has been considered to test the efficacy of the proposed approach. Different test cases with different disturbances and loading conditions are simulated, and the corresponding analysis and nonlinear simulated results are presented to assess the effectiveness of the proposed PSSs. In addition to that, comparative results of the proposed approach are presented to that of GA-based PSS (GAPSS) and conventional approach-based PSS (CPSS).

2 Problem Statement

2.1 System Modeling and Structure of PSS

The small-signal stability model of the PSS-integrated power system can be formulated by a set of nonlinear differential equations and in a generalized form can be expressed as:

$$\mathrm{d}X/\mathrm{d}t = f(X, U) \tag{1}$$

where X denotes the vector of the related state variables defining the characteristic of the system. U denotes the vector of input variables. A fourth-order machine model is considered in this study. The differential equations used for the nonlinear time domain simulation are mathematically represented as follows.

$$\mathrm{d}X/\mathrm{d}t = \omega - \omega_0 \tag{2}$$

$$\mathrm{d}\delta/\mathrm{d}t = (P_\mathrm{m} - P_\mathrm{e})/M \tag{3}$$

$$\mathrm{d}E_q/\mathrm{d}t = \left(E_{fd} - (X_d - X_{dd})I_d - E_q\right)/T_{do} \tag{4}$$

$$\mathrm{d}E_{fd}/\mathrm{d}t = \left(K_a(V_{\mathrm{ref}} - V_t - V_{\mathrm{PSS}}) - E_{fd}\right)/T_a \tag{5}$$

where δ is the rotor angle; ω is the rotor speed; P_{m} is the mechanical power; P_{e} is the electrical power; E_{fd} is the equivalent excitation voltage; d and q are the direct and quadrature axes, respectively; E_q is the internal voltages behind the reactance X_d and X_q; T_a is the regulator time constant; T_{do} is the Time constant of excitation circuit; and K_a is the regulator gain.

The proposed method is applied for designing a PSS in a single-machine infinite bus (SMIB) system and a three-machine power system. Figures 1 and 2 show below the schematic diagram of the SMIB and the three-machine power system, respectively.

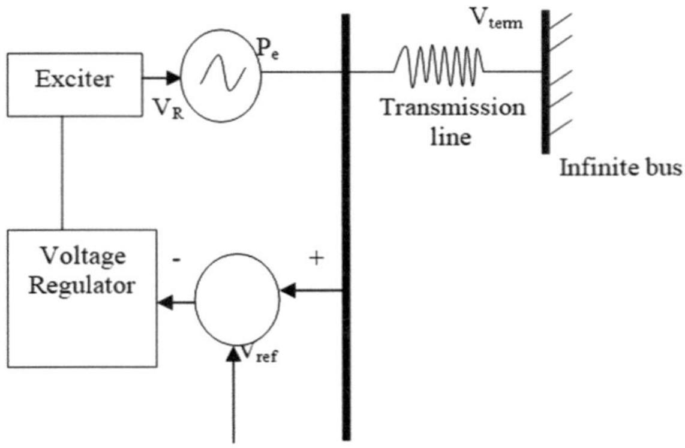

Fig. 1. Single-machine power system

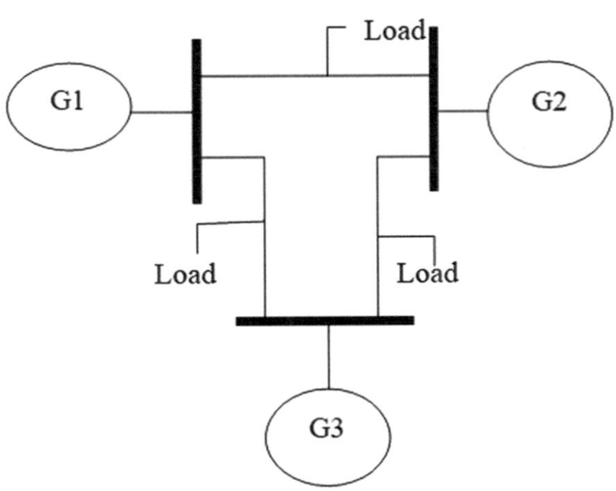

Fig. 2. Three-machine power system

A conventional lead–lag PSS is considered in this test system looking to its wider acceptance in recent times for similar types of study. It can be represented as:

$$V_{\mathrm{PSS}}(i) = K_i \frac{sT_{\mathrm{w}}}{1+sT_{\mathrm{w}}} \frac{(1+sT_1)(1+sT_3)}{(1+sT_2)(1+sT_4)} \Delta w_i \tag{6}$$

where Δw_i is the machine speed deviation from synchronous speed = $(\omega - \omega_{\mathrm{o}})$; V_{PSS} (i) is the PSS output signal at the ith machine; T_{w} is the washout time constant; T_1, T_3 are the lead time constants; and T_2, T_4 are the lag fine constants. The time constants T_{w}, T_2, and T_4 are generally taken from the standard prescribed values for the PSS. However, for the better performance the stabilizer gain, K_i, and time constants T_{1i} and T_{3i} are to be computed and set at their optimized values within the allowed prescribed limits.

2.2 Objective Functions

The PSS is designed with an objective to minimize the power system oscillations through effective damping under a system disturbance to ensure an enhanced power system stability and control. These oscillations are reflected in the system responses as deviations from its normal conditions. The parameters such as power angle, rotor speed, and line power are generally investigated for the analysis and system design. The objective function is formulated according to the deviations occurred for these parameters. In the present study, the two performance indices are considered such as the settling time and overshoots. The objective function in terms of these indices is formulated and considered for the optimum value setting by the searching techniques. These indices can be mathematically represented as:

$$\mathrm{PI}_1 = \sum_{i=1}^{n} \int_{t=0}^{t-t_{\mathrm{sim}}} (t\Delta w) \cdot \mathrm{d}t \tag{7}$$

$$\mathrm{PI}_2 = \sum_{i=1}^{n} \int_{t=0}^{t-t_{\mathrm{sim}}} (\Delta w) \cdot \mathrm{d}t \tag{8}$$

where n is the number of machines and t_{sim} is the simulation time.

Optimize $F(x)$. Subject to

$$K^{\min} \le K \le K^{\max}$$
$$T_1^{\min} \le T_1 \le T_1^{\max}$$
$$T_3^{\min} \le T_3 \le T_3^{\max}$$

The standard limits considered in this study are as follows: The optimized parameters are [0.001 − 50] for K and [0.06 − 1.6] for T_{1i} and T_{3i}. The time constants T_{w1}, T_2, and T_4 are set as 5, 0.05, and 0.055, respectively. The objective function as represented in Eq. (7) is taken to evaluate the fitness value to compute the optimal set of solution vectors in the IWO and GA algorithm. The optimal set of parameters are found out by tuning these parameters within their limits as PSS parameters like $\{K_i, T_{1i}, T_{3i}, i = 1, 2, 3, \ldots n_{\mathrm{PSS}}\}$.

3 Genetic Algorithm

3.1 Overview

The genetic algorithm (GA) is a stochastic-based searching scheme and applied successfully in many real-world applications due to its capability to result in approximate solutions. However, it is found in many cases, fails to arrive at global minima, is trapped at local minima and is stagnant due to saturation effect. The idea of searching in GA algorithm is formulated by the biological principles like survival of fittest and evolution processes like inheritance, crossover, mutation, and selection.

Genetic algorithms are applied in computer simulation for many engineering problems. Like other evolutionary optimization techniques, a set of random solution is formulated named as population. Each member of the population represents one specific solution. This mimics the chromosome structure in biology with genes as variables. Generally, the mathematical representation of solutions is framed in a binary string of 0 and 1 s. The encoding and decoding can be followed in various ways. With every generation represented in simulation as iterations, the solutions vary optimally and better results are obtained by generation after generation by following the major steps like crossover, mutation, and selection. The objective function based on the desirable result to be obtained is formulated, and this function is utilized to find out the fitness value of all the solutions. The fitness vector helps to select the best value considered as an optimal value for the problem it is used for. The stepwise procedure followed in GA algorithm is explained in the next section.

3.2 Details About the Algorithm

3.2.1 Initialization

Initial population is randomly formulated by choosing many solutions as a member of the population. The values are selected within the prescribed range. The size of the population should be chosen suitably according to the complexity and type of the problem. In this study as followed by many authors for similar GA application, here it is taken as 20. Conventionally, even if the size of the population is randomly taken, however it should cover the entire range of the searching apace.

3.2.2 Procedure of Selection

Throughout each succeeding epoch, a proportion of the existing population is preferred by a selection method to breed a new generation. Individual solutions are selected through a fitness-based process by finding the fitness vector by the objective function. For a better selection, in this study the most popular and well-studied selection method based on roulette wheel selection is taken into consideration.

3.2.3 Reproduction

In this step, crossover and mutation operation are followed to generate new population from the existing population. The crossover and mutation process are regulated by crossover probability factor (Cpf) and mutation probability factor (Mpf). In this study, the Cpf and Mpf are considered as 0.8 and 0.25, respectively, according to the standard values taken in many papers. For producing new solution, two parent solutions are chosen from the pool of selected solutions followed by a selection approach based on the survival of fittest concept. The characteristics of parents as found in the child due to inheritance, similarly here also the typical knowledge or factors are shared by the new child solutions from their parent solutions expecting a better generation of solution in comparison. The new solutions are added with the old solutions, and by the selection process new better solutions are selected in proper population size for the next generation.

3.2.4 Termination

In this study, the number of iterations has been taken as a key parameter for the termination of the computation loop. The generation process is allowed to repeat continuously until the termination criteria have been reached. Here, 100 is taken as the maximum number of iterations to be executed.

4 IWO Algorithm

4.1 Overview

Among the stochastic search method, IWO algorithm is one of the novel approaches ever proposed in the literature. This algorithm is formulated based on the natural behavior of weed colonizing and used in the search domain for achieving global optimal point in a complex system. As weeds adopt the environmental changes for proper growth, IWO algorithm is capable to adapt the system complexity for ensuring a complete search in the global search domain. This scheme reproduces seeds for increasing the rate of growth of the population. Implementing the concept of fittest, the colony tends to become dense, as it eliminates the one with lesser fitness and lesser opportunity of life.

4.2 Outline of the Algorithm

4.2.1 Initialization

A population of initial solutions is being spread over the D-dimensional and NP number of population problem space with random positions.

4.2.2 Process of Reproducing the Population Index

An element indexed under the given population is permitted to reproduce elements. The rate of reproduction is indirectly regulated by the maximum and the minimum fitness of the element present in the population. The rate of reproduction of the element through a plant is linearly functioned in this algorithm. Therefore, the rate of reproduction is directly dependent on the fitness factor of an individual plant. Hence, it can be observed in Fig. 3 that the rate of reproduction is always linear and further bounded by the maximum and minimum fitness factor. Eventually, the algorithm ensures the growth of fittest element in the search domain. This forces to increase convergence toward the group's best value. As like other evolutionary algorithms, the fittest element remains in the domain for the reproduction and thus weeds survive to produce in the next generation for better generation. On the other hand, the worst ones are eliminated due to less fitness value from the growth process.

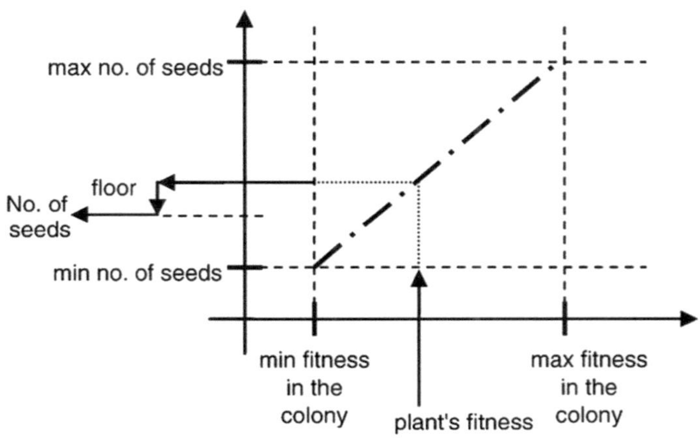

Fig. 3. Reproduction strategy

4.2.3 Spatial Dispersal

This particular part of the manuscript describes the process of randomization and adaptation in the IWO. Reproduced seeds are spread in the search domain, i.e., NP D randomly, maintaining the mean value at zero. As a result, distributed seeds in the search domain maintain an equal density factor. The parent plant is encircled by its own reproduced seeds. Therefore, an uninterrupted search throughout the space is ensured. In contrast, to maintain the speed and accuracy simultaneously, standard deviation (SD) is allowed to vary between the initial and final values σ_{initial} and σ_{final}, respectively. In generation to generation, the procedure converges to the better fitness value and at last arrives at an optimal value considered as the best possible value.

A nonlinear alteration has found to be arriving at better performance. Looking at this idea, the σ is allowed to vary in different iterations as in Eq. (9).

$$\sigma_{\text{iter}} = \frac{(\text{itermax} - \text{iter})^n}{(\text{itermax})^n}(\sigma_{\text{initial}} - \sigma_{\text{final}}) + (\sigma_{\text{final}}) \tag{9}$$

where itermax denotes the maximum number of iterations. The σ_{iter} denotes the standard deviation for the current iteration. Superscript n denotes the weighted average modulation factor.

4.2.4 The Process of Selecting the Fittest Zone of Element

According to the algorithm, if the parent of reproduced seeds is defined as offspring, then it will be eliminated from the space in the next generation of iteration.

This is done by eliminating the low fitness value members. For limiting maximum number of plants in a colony, a competition between the plants is followed based on their fitness value. After consecutive iterations, the number of plants in a colony will reach its maximum due to a combined and fast reproduction. There is a need, and it is expected that the fitter plants have to reproduce more number of new solutions in comparison with the undesirable plants. Once the maximum number of plants (P_{max}) reached the pre-specified number, a mechanism for elimination is followed in terms of comparing their individual fitness values. When the maximum number of weeds in a colony is reached, each weed is allowed to generate seed as specified in reproduction. The produced seeds are then added with the population and allowed to spread over the NP $\times$ D-dimensional search area. However, they ranked together with their parents as a colony of weeds. To avoid the allowed population to be crossed, the lower fitness value members are eliminated. This is necessary to satisfy the constraint that the population should be within the maximum allowable population in a colony. By this process, the plants and offspring are ranked together, and subsequently the ones with better fitness survive as the iteration moves forward. Later, they are only allowed to replicate for producing new and better solutions or members of the population. With a probability to result in a better solution even from the weeds with lower fitness value, all the weeds are allowed to produce new offspring according to their fitness. If they are comparatively better fitness value, they may survive in the colony and be a part of the process in later iterations. However, the population control mechanism is always followed even to their offspring at the end of a given run. This leads to better selection by executing competitive exclusion.

5 Result and Performance Evaluation

For illustration and comparison purposes, a single-machine infinite bus system and a three-machine system have been taken as two test systems as shown in Figs. 1 and 2, respectively. A three-phase ground fault with duration of 0.1 s is considered as a fault to study about the dynamic stability of the system incorporating the effect of PSS on this situation. The test results demonstrate clearly that the proposed IWO-PSS outperforms the GAPSS and CPSS.

Figure 4 shows how quickly the IWO technique optimizes the parameters of PSS with less function value compared to the GA for same type of fault, loading condition, and iterations. Figures 5, 6, and 7 show the control on restoration of bus voltage, the active and reactive power flow in a line in pre- and post-fault condition. Also, the load angle and speed variations of machine are shown in Figs. 8 and 9 in both the techniques. Figures 10, 11, 12, and 13 show the intra-area and inter-area oscillations after the fault and how the damping occurs very quickly due to correct design settings of

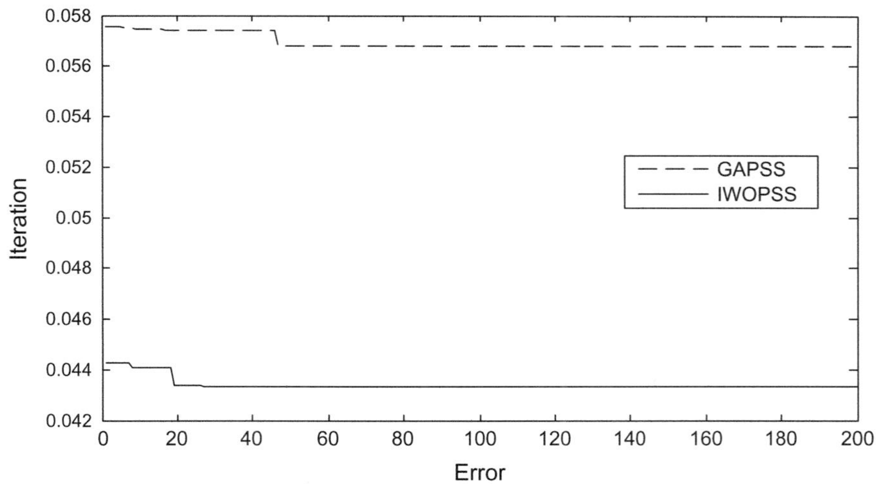

Fig. 4. Convergence characteristics of the IWO and GA

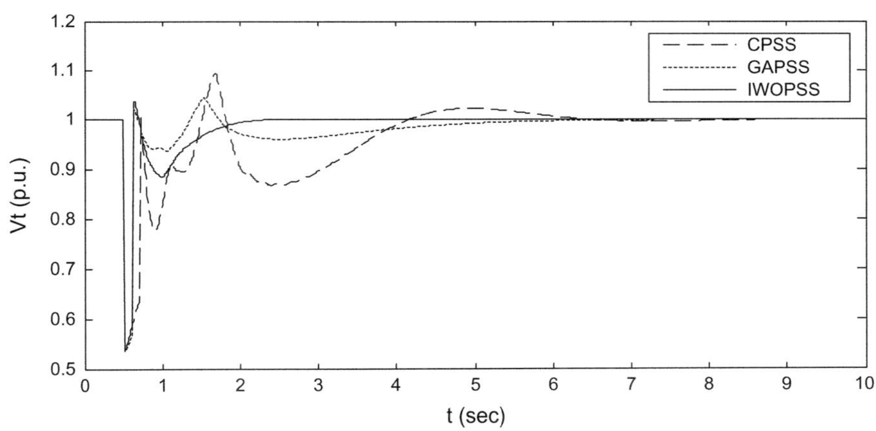

Fig. 5. Terminal bus voltage response at node one

PSS parameters by GA and IWO for a three-multi-machine system. Only selected figures are shown here due to lack of space. Simulation results in different conditions show that the proposed IWO technique is better than the GA.

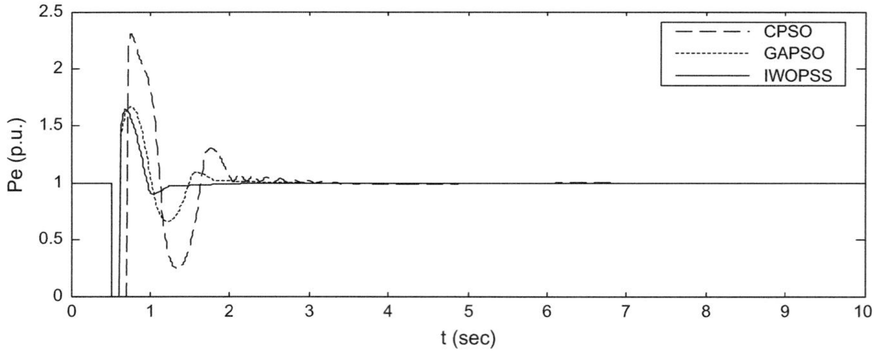

Fig. 6. Active power flow in the transmission line

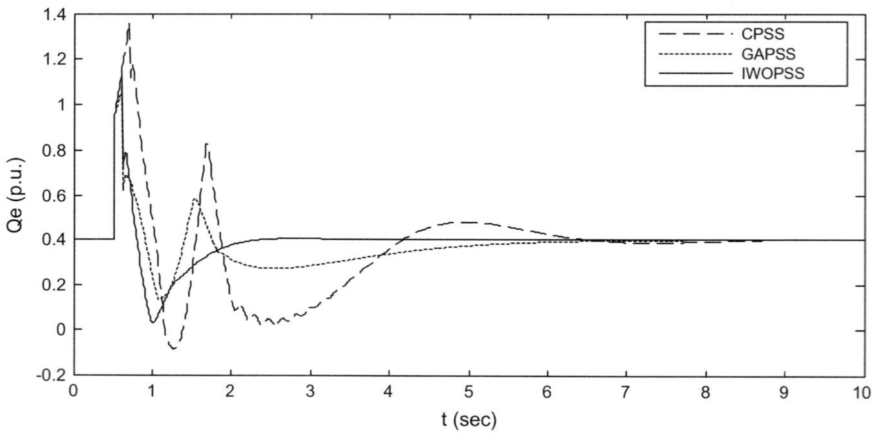

Fig. 7. Reactive power flow in the transmission line

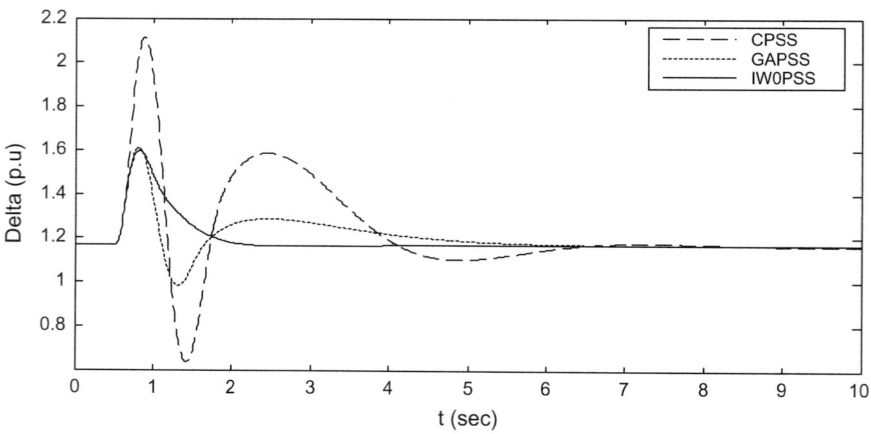

Fig. 8. Rotor angle variation for single-machine case

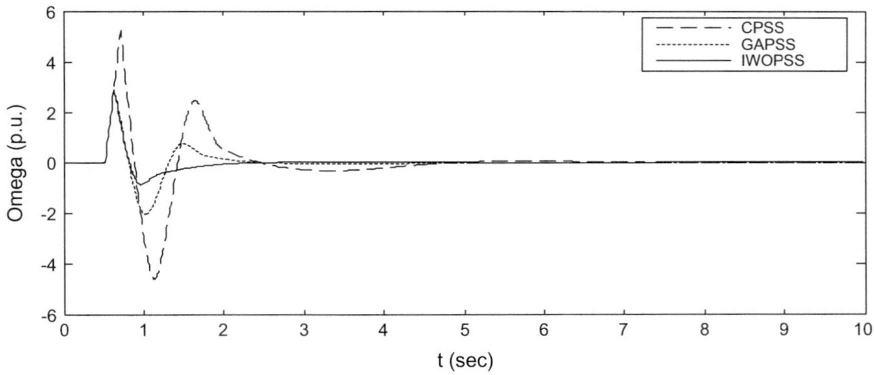

Fig. 9. Speed variations for single machine

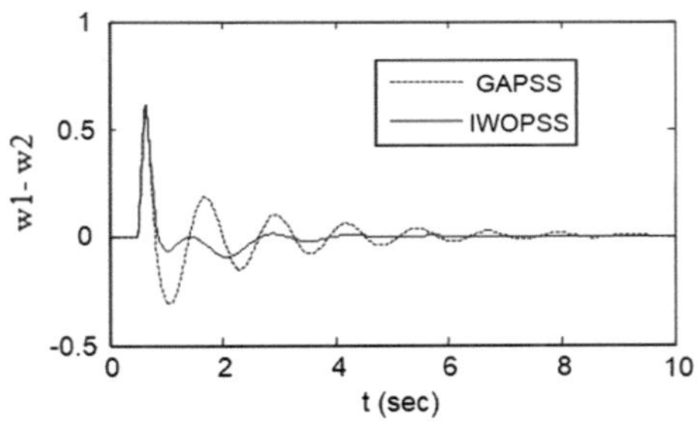

Fig. 10. Speed deviations between generators one and two

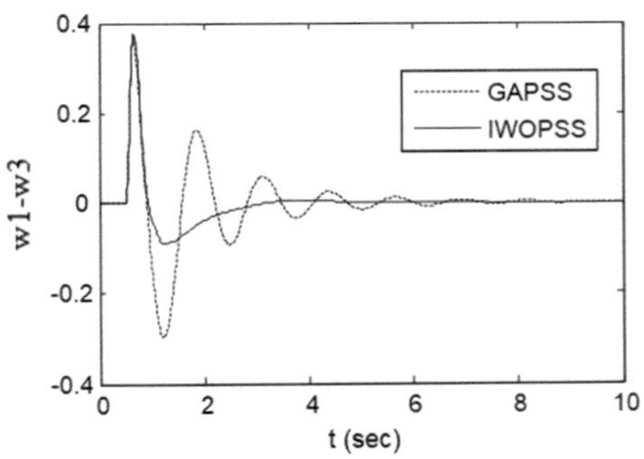

Fig. 11. Speed deviations between generators one and three

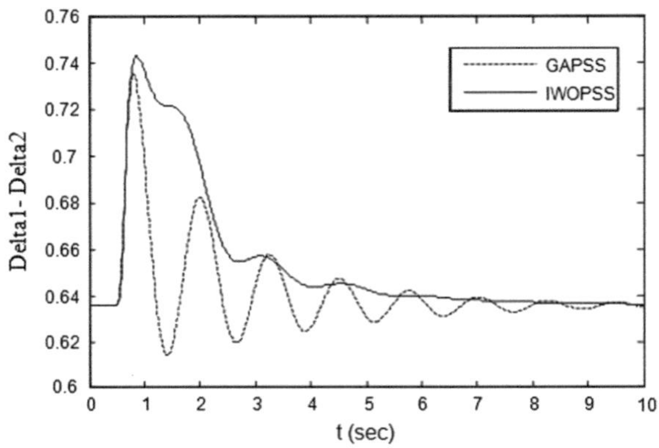

Fig. 12. Rotor angle variations between generators one and two

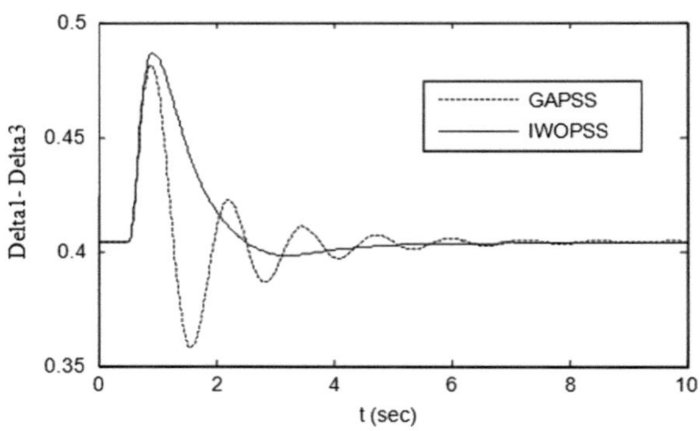

Fig. 13. Rotor angle variations between machines one and three

6 Conclusion

This paper investigates to find out a novel approach IWO for system stability enhancement by quick damping through PSS design with optimal parameter setting. The tuning of power system stabilizers using GA and IWO techniques is investigated under different disturbances and loading conditions in a single-machine and a multi-machine power system. Extensive time domain simulation results have been presented to establish the efficiency and improved dynamic performance of the proposed stabilization techniques. The IWO-PSS technique provides good damping characteristics and outperforms GAPSS. So, the proposed stabilizer bears much potential for practical implementation.

Appendix

For single-machine infinite bus system:

$X_d = 1.5$, $X_{dd} = 0.2$, $X_q = 1$, $T_{do} = 6$, $H = 6$, $X_e = 0.3$, $f = 50$, $K_e = 60$, $T_e = 0.4$, $P = 1$, $Q = 0.4$, $V_t = 1$, $P_{ref} = 1$

Load data: $P_{11} = 0.8$, $Q_{11} = 0.2$

References

1. Anderson PM, Fouad AA (1977) Power system control and stability. Lowa state university press
2. Sauer PW, Pai MA (1998) Power system dynamics and stability. Prentice Hall
3. Larson RV, Swann DA (1981) Applying power system stabilizers, I, II and III. IEEE Trans PAS 100(6):3017–3046
4. IEEE Working Group (1996) Annotated bibliography on power system stability controls 1986–1994. IEEE Trans Power Syst 88(4): 794–804
5. Abido MA (1999) A novel approach to conventional power system stabilizer design using tabu search. Int J Electr Power Energy Syst 21(6): 443–454
6. Kennedy J (1997) The particle swarm: social adaptation of knowledge. In: Proceedings of 1997 IEEE international conference on evolutionary computation. In diaganpolis, IN, pp 303–308
7. Shi Y, Eberhant R (1998) Parameter selection in particle swarm optimization. In: Proceedings 7th annual conference evolutionary program, Mar 1998, pp 591–600
8. Kennedy JE, Shi R (2001) Swarm intelligence. Morgan Raufmann
9. Mehrabian AR, Locus C (2006) A novel numerical optimization algorithm inspired from weed colonization. Ecol Inf, pp 355–366, Elsevier Science

IoT-Based Baby Swing Monitoring, Alerting and Security System

Shuvendra Kumar Tripathy[1(✉)] and Pradyut Kumar Biswal[2]

[1] Trident Academy of Technology, Bhubaneswar, India
[2] Department of Electronics Engineering, IIIT Bhubaneswar, Bhubaneswar, India
pradyut@iiit-bh.ac.in

1 Introduction

This paper presents the smart cradle baby swing prototype model. There is immense issue the working parents are confronting in their day-to-day life. They are not able to pay appropriate time for their infants. Babies and infants need 24 × 7 attention, which is highly difficult for working parents. It is awkward for them to always look out for their infant while doing their work or errands. So, we have structured a basic framework which helps the parents in dealing with their infant. This framework proposes a basic voice identification system which can be connected essentially for planning a gadget with capacity to recognize an infant's cry and naturally turning on baby rest music. This model takes care of the issue of time and vitality of such guardians. The model of the cradle contains humidity and temperature sensor, moisture sensor, load cell, voice recognizer and other electronic gadgets associated with Arduino Uno and Raspberry Pi. The cradle machines are working with coordination of sensors and the processors.

2 Literature Survey

Many child care baby swing frameworks are available; however, dominant part of the framework uncommonly fails due to the absence of proper technology. This paper presents a style of a child observance framework supported by the GSM network. A model is proposed and created which supplies a solid and affordable child recognition framework that may assume a significant job in giving higher child care [1], individuals nowadays are by all accounts consistent in inclination with their work, and they will most likely be unable to screen the tyke constantly. So as to come over this trouble, the youngster checking framework is built up that chips away at the guideline of image processing. The framework comprises of high-goal camera to identify the development of the youngster. Here, the picture preparing is utilized to examine the skin tone of the kid. Arduino controller can store the caught movements. This controller transmits the data to move the robot by utilizing a servo engine. The robot moves alongside the tyke because of the movement caught. Be that as it may, the frameworks'

M. N. Mohanty and S. Das (eds.), *Advances in Intelligent Computing and Communication*, Lecture Notes in Networks and Systems 109,
https://doi.org/10.1007/978-981-15-2774-6_43

reaction time is extremely high. The fundamental intention of this paper is to keep the youngster safe and move the live data by a message to that specific parent cell phone through GSM [2]. The author presents a structure of a baby checking system reliant on the GSM composed. A model is made which gives a reliable and profitable baby watching system that can expect a critical activity in giving better infant youngster care. This system screens fundamental parameters, for instance, body temperature, beat rate, moistness condition, advancement of an infant youngster [3]. The baby's health condition is continuously checked and also communicated with live video in mobile application using IoT [4] as well as wearable technology [5]. The author also proposed a system where PIR sensor, temperature sensor, pressure sensor, voice sensor, light sensor are being used inside the smart room to monitor baby using WSN technology [6]. In 2018, "IoT-based child and women security" paper was recommended to light upon the dislodge youngster away from home. The model is made up of GSM and GPS self-ruling utilizing cell phone [7].

3 Proposed System

From the above study, it is found that the drawback of the existing system is the absence of real-time monitoring and controlling of the baby cradle. The system sends the message signal if baby is crying remotely but not the real-time control. Also, all the systems have used GSM module which needs SIM card, proper network for operation. The major advantage of the proposed system is that it works on wireless sensor network as well as on real-time platform. The block diagram of the proposed system is shown in Fig. 1.

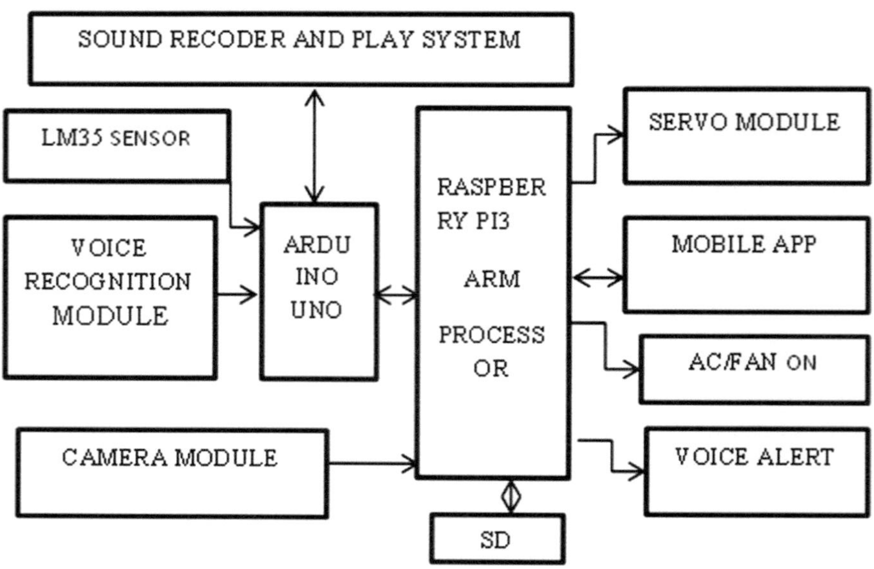

Fig. 1. Block diagram of IoT-based system

4 Working of the Proposed System

The working of the proposed system is explained through the flowchart shown in Fig. 2. In the proposed system, temperature and crying sound of the baby are always monitored through two modules, i.e., temperature sensor and voice module. If baby crying is detected by the voice module, then the cradle automatically starts to swing continuously for 5 min with a recorded audio playing in their parent's tone and one SMS as well as voice alert of "baby crying" will be sent to the parent's mobile.

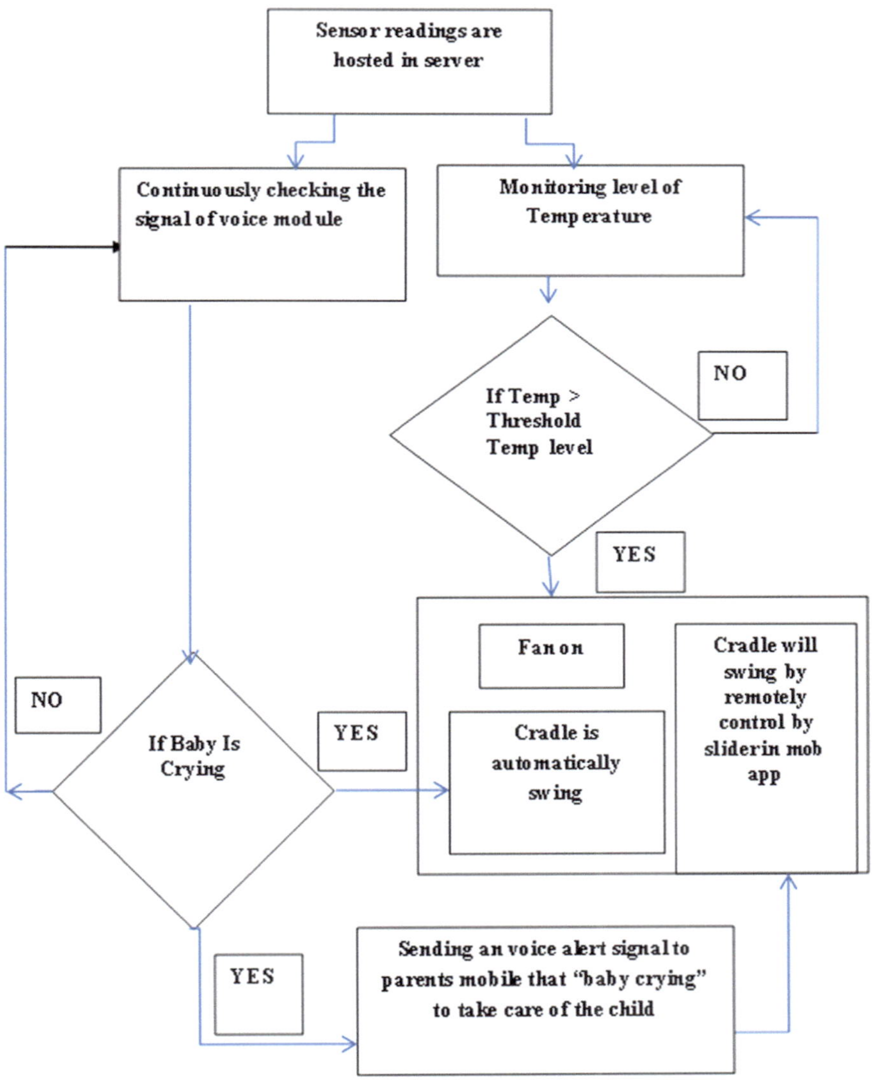

Fig. 2. Flowchart of IoT-based system

Parents also can remotely monitor as well as control the cradle by sliding the slider which is available in the mobile application. Also, when the temperature increases beyond the threshold level, then the fan or the air conditioner automatically turns on without involving any human interference.

Moisture sensor checks the wetness of the bed. If the bed is found to be wet, then a voice alert will be sent to the parent's mobile. Sometimes, it may happen due to busy work inside the home without parent knowledge, and unauthorized person may enter and try to steal the baby. So, to prevent the theft of the baby by the unauthorized person, there is a load cell below the cradle bed so that if the weight decreases below a certain level then a buzzer starts working.

All the sensor and input modules are being connected to Raspberry Pi through Arduino Uno using Firmata software. Arduino will be the interface between all sensor and Raspberry Pi. Camera is directly connected to the Raspberry Pi. All analog data are collected by Arduino Uno, then transferred to the Raspberry Pi through USB port and stored in a database using WSN technology.

A Web application as well as an android-based application is developed through which users (parents) can screen and control the temperature of the room by making fan on or off according to the room temperature where the child is kept and also can control the swing of the cradle through a slider.

5 Implementation of the System

In this proposed framework, Python programming language is used. Python is a translated state programming language for broadly useful programming. Apache server with MySQL database is utilized in this task. The client's parameters gained by Raspberry Pi are put away on a server persistently every 2 min, and the equivalent is transferred to the Web site page which is made utilizing PHP script which makes it comprehensively open. Furthermore, security is given by giving the login id and secret phrase using which the approved individual can login to the page and access client's parameters from anyplace.

The hardware requirements are: Raspberry Pi, Arduino Uno, sound sensor (voice module), sound recorder and play system, NPK sensor, a USB camera and a SD card.

The various softwares that are used to develop the proposed system are: New Out Of Box Software (NOOBS), mobile application, HTML and PHP, MQTT protocol, Firmata protocol, Python and Node-RED.

6 Outcome of the System

The software NOOBS can be installed in the SD card of the Raspberry Pi. Using VNC server, we can access Raspberry Pi window in our desktop. PuTTY is an exceptionally valuable application that is utilized to interface with sequential ports and Secure Shell

(SSH) to Raspberry Pi. A static IP is assigned to the Raspberry Pi through which we can remotely open the Web page.

A Web page as well as mobile app is created to get the real-time information of the child using HTML language and PHP method. Data are stored in the database using MySQL database system. In our case, the Node-RED dashboard window will be opened on the IP address of 192.168.1.102 with the port number 1880. The mobile app view of baby swing monitoring, alerting and security system is shown in Fig. 3.

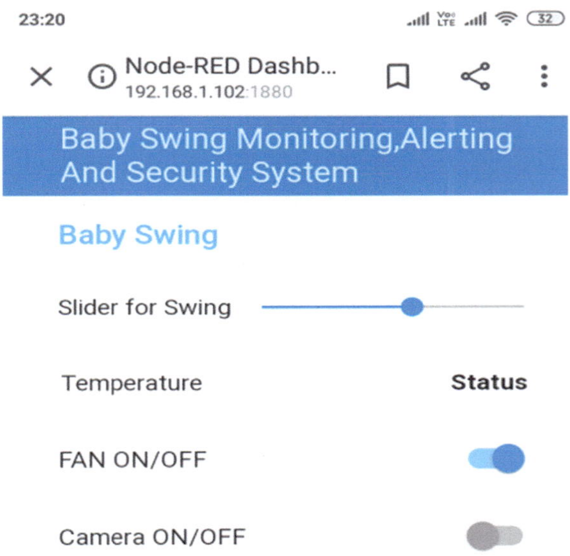

Fig. 3. Mobile app view of the system

Through the slider, the parents can control the cradle rotation by sliding the finger on the mobile screen. As it can be seen from the front screen of the mobile app, the electronic devices in the room can be controlled remotely by the parents depending on the suitability of the situation.

7 Conclusion

In this paper, a framework is proposed and a real-time system is implemented for real-time monitoring, controlling and automation of the baby swing cradle from the remote location. The major advantage of the proposed system is that instead of using a GSM module, a SMS-based notification is designed using IoT technology. The whole system works on WSN technology. The parents can monitor and control various electronic devices that are placed in the room for well-being and comfort of the child. Parents can control the swing of baby cradle by controlling slider which is available in the mobile application; also, a theft alarm system is proposed using a simple and efficient technology. This system can be very much useful for the working parents; even if they are in office, they can take care of their baby.

References

1. Naiknaware S, Gajare D, Vibhute S, Kenjale S, Kulkarni S (2017) Review on baby monitoring system. IJCSN Int J Comput Sci Netw 6(2), Apr 2017
2. Clenisha M, Sandra A, Sajin P, Sam BB (2018) Survey on automated child monitoring system using image processing. Int J Pure Appl Math 118(24)
3. Patil SP, Mhetre MR (2014) Intelligent baby monitoring system. ITSI Trans Electr Electron Eng (ITSI-TEEE), 2014
4. Rajesh1 G, ArunLakshman R, Hari Prasad L, ChandiraMouli R (2014) Baby monitoring system using wireless sensor networks. ICTACT J Commun Technol, Sept 2014
5. Sharma O, Lewis SN, Telang U, Almeida LD, Lewis LDS (2017) Design of a bluetooth enabled health monitoring system for infants using wearable technology. J Adv Res Dyn Control Syst, vol 15
6. Kaur T, Mittal M, Singh H (2018) The baby monitoring room prototype model using IoT. Int J Adv Res Sci Eng 07(6), Apr 2018
7. Budebhai M (2018) IoT based child and woman safety. Int J Comput Sci Mob Comput IJCSMC 7(8), Aug 2018

Effective ECG Beat Classification and Decision Support System Using Dual-Tree Complex Wavelet Transform

Nihar Kar[1], Badrinarayan Sahu[1], Sukanta Sabut[2], and Santanu Sahoo[1(✉)]

[1] Department of ECE, Faculty of Engineering, Siksha 'O' Anusandhan, Bhubaneswar, Odisha 751030, India
santanusahoo@soa.ac.in

[2] School of Electronics Engineering, KIIT, Bhubaneswar, Odisha 751024, India

1 Introduction

In worldwide, the main cause of death is due to cardiovascular diseases. In a recent report by WHO, 17.1 million deaths are claimed due to CVD every year [1]. The irregular electrical activities of the heart are called cardiac arrhythmia. Due to blockage in coronary, artery heart attack occurs. The lack of blood and oxygen supply affects the heart muscle which may die if the blockage is not treated immediately. Change in lifestyle, smoking habits, unhealthy diet, obesity, diabetes, hypertension, use of alcohol may increase the death cause of CVDs in worldwide [2]. Early detection of cardiac arrhythmia may improve the death rate due to cardiac arrest by appropriate analysis of ECG signals. The ECG signal gets corrupted due to the interference such as power line interference, baseline wander, and many other biological artefacts [3]. Several methods such as empirical mode decomposition [4], adaptive filtering [5], mathematical morphology [6], and weighted averaging filter [7] are reported for eliminating these noises. L. Smital et al. have reported wavelet wiener filtering techniques for eliminating noise related to broadband myopotentials (EMG) in ECG signals [8]. In preprocessing stage, the P and T attenuate from noises by smoothing techniques [9]. Proper analysis of ECG signal requires accurate detection of QRS complex which is defined as the depolarization of the right and left ventricles of the human heart and contraction of the large ventricular muscles. Several methods have been reported for the detection of QRS complexes. Discrete wavelet transform (DWT) mostly used to detect the accurate features from ECG signal, where the signal decomposes into different coarseness and coefficients which contains sufficient information of the processed signal in the subbands [10, 11]. Multiscale feature of WT [12] can detect the *P*-, *T*-wave, and QRS complex with a detection error rate 99.8% in the presence of baseline wander. To evaluate the performance of QRS complexes, *P*-wave and *T*-wave, multiresolution wavelet transform with optimum coefficients is used where the author declares an accuracy of 97% for detecting heart rate, 96% for detecting *P*-wave, 95% for QRS complex, and 98% for *T*-wave [13]. Martinez et al. proposed a technique to find positive maximum and negative minimum of peaks using dyadic wavelet [14].

M. N. Mohanty and S. Das (eds.), *Advances in Intelligent Computing and Communication*, Lecture Notes in Networks and Systems 109,
https://doi.org/10.1007/978-981-15-2774-6_44

Extracting the good features is the main key point of signal classification. In a recent work for the detection of QRS complex, Joao et al. [15] present an advanced approach using first derivative, Hilbert transform, wavelet transforms, and adaptive thresholding methods. Author claims 99.5% error rate in the PhysioNet arrhythmia database using moving average computing method for the detection of QRS complexes [16]. In a recent work, using adaptive thresholding-based multiresolution wavelet transform is used to detect the peaks and valleys of a signal, S. Banerjee et al. claimed 99.8% of sensitivity and 99.6% of positive predictivity using MIT-BIH arrhythmia database; however, using PTB ECG database claimed 99.84% and 99.98%, respectively [17]. Based on the power spectrum of decomposition, multiresolution wavelet analysis is used for selecting detail coefficient for the detection of QRS complex with 99.87% of Se, 99.79% of Pp, and detection error rate of 0.34% [18]. Junior et al. in 2016 [19] worked on real-time data to perceive QRS complex using redundant discrete wavelet transform (RDWT) which achieves a detection rate of 99.32% in detecting QRS complexes. Using statistical, geometric, and nonlinear HRV features, Jovic et al., proposed an efficient and accurate classification method for detecting cardiac disorders [20]. Dong et al. [21] proposed a method based on deterministic learning model for efficient ECG beat classification using radial basis function (RBF). Author claims an average accuracy of 97.78 and 97.21% for global and patient-adapting beat classification. Acharya et al. [22] proposed an effective method for automatic detection of myocardial infarction in recorded ECG signals by implementing a convolutional neural network (CNN) algorithm. Using feature selection algorithm, the author claimed an accuracy of 93.53 and 95.22% with and without noise, respectively. Dokur et al. [23] proposed a method which consists of Fourier and wavelet features and classified for ECG beat classification. The features are searched with divergence values and classified 10 types of ECG beat with success of 97% using genetic algorithm-based neural network. Over the last decades, many researchers are working in analysis of ECG signal for automatic detection of cardiac arrhythmias with the use of computer technology. The computer-aided diagnosis system is a mimic to clinician and a reliable tool in the application of detecting cardiac diseases in routine and long-term monitoring of ECG signals. The major steps include noise elimination, detection of *R*-peaks, extraction and formulation of important feature and Classification of cardiac arrhythmias are given in Fig. 1.

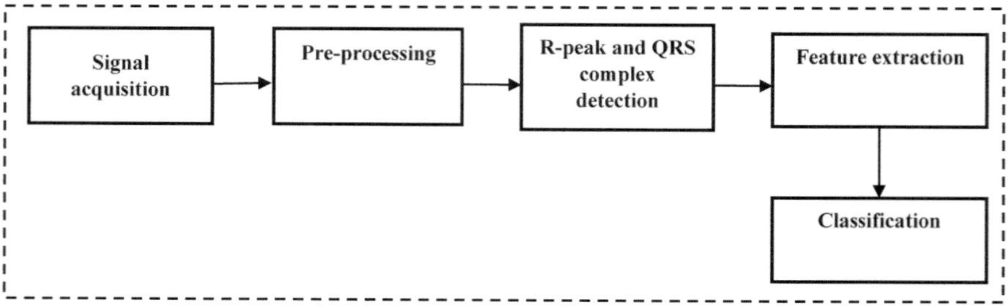

Fig. 1. Overview of ECG beat detection system

In this paper, we have used an application of dual-tree complex wavelet transform to get the accurate features from the processed signal. Signal is first preprocessed by DT-CWT up to four levels and using informative features, and five types of ECG beats were classified from 48 recorded ECG signals taken from MIT-BIH arrhythmia database. Temporal, morphological and statistical are the most important features contain various clinical information are used as an input to the *K*-NN classifier to classify the arrhythmia beats. The performance of the classifier was calibrated in terms of sensitivity (Se), specificity (Sp), and accuracy (Acc).

2 Methodology

2.1 MIT-BIH Arrhythmias Database

In this work, MIT/BIH arrhythmia database is used for classification of arrhythmia beats. The database comprises of 47 records of 2 channel ECG signals each of thirty minutes period obtained from 47 subjects between 1975 and 1979 at the BIH Arrhythmia Laboratory [24]. The database has 25 men of age 32–89 years, and 22 women aged 23–89 years which consists of 116, 137 number of QRS complexes. The ECG signals were sampled at 360 samples per second with 11-bit resolution over a 10 mV range and band pass-filtered at 0.1–100 Hz. For timing and beat class information, the data were verified by independent cardiologist for annotation. Normal (N), left bundle branch block (LBBB), right bundle branch block (RBBB), atrial premature beats (APBs) and premature ventricular contractions (PVCs) beats of data were selected from MIT-BIH arrhythmia database for analysis. The first channel signals taken from database were processed using the algorithm implemented and tested using MATLAB version 2016.

2.2 Dual-Tree Complex Wavelet Transform

The concept of dual-tree complex wavelet transform is introduced in the year 1998 by Kingsbury. Two real DWTs were employed in dual-tree CWT, where the real and imaginary parts of the transform are given by first and second DWT. The poor directional selectivity and shift variance drawbacks of DWT are eliminated by dual-tree CWT. Figure 2 shows that the signal $x(n)$ is implemented using DT-CWT framework where two DWTs are in parallel to the same data. The implementation is much expensive because for an N point signal, it gives $2N$ DWT coefficients [25]. A DT-CWT can be demarcated as:

$$\Psi(l,m) = \Psi(l)\Psi(m) \tag{1}$$

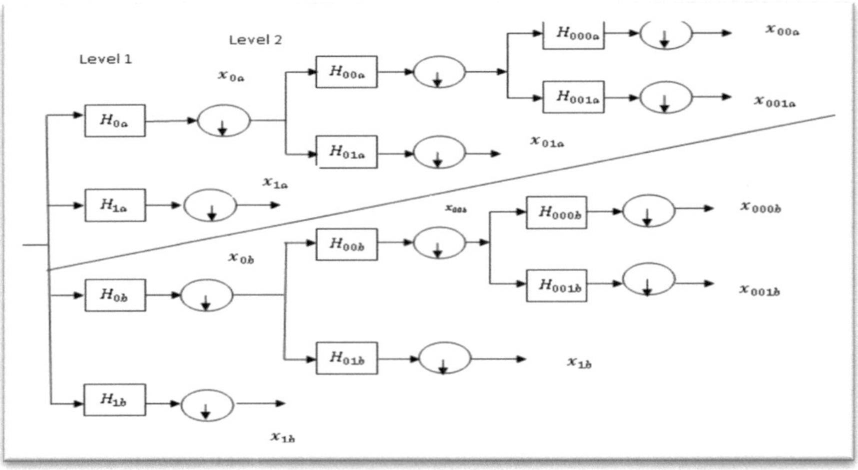

Fig. 2. Dual-tree CWT consistent to three levels

$\Psi(l)$ and $\Psi(m)$ are complex wavelets.

$$\Psi(l) = \Psi_a(l) + j\Psi_b(l) \tag{2}$$

$$(m) = \Psi_a(m) + j\Psi_b(m) \tag{3}$$

where $\Psi_h(xl)$ and $\Psi_h(m)$ are real coefficient of upper and lower filter banks. The overall equation becomes:

$$\Psi(l,m) = [\Psi_a(l) + j\Psi_b(l)][\Psi_a(m) + j\Psi_b(m)] \tag{4}$$

The sets of coefficients used in DT-CWT in level 1 and level 2 are given in Table 1.

Table 1. Filter coefficients used in first and second levels

Level 1				Level 2			
Tree A		Tree B		Tree C		Tree D	
Coeff1	Coeff2	Coeff1	Coeff2	Coeff1	Coeff2	Coeff1	Coeff2
0.000	0.000	0.000	0.000	0.011	0.000	0.011	0.000
0.088	0.011	−0.088	−0.011	0.011	0.000	0.011	0.000
0.088	0.011	0.088	0.011	−0.088	−0.088	−0.088	−0.088
0.696	0.088	0.696	0.088	0.088	−0.088	0.088	−0.088
0.696	0.088	0.696	0.088	0.696	0.696	0.696	0.696
0.088	−0.696	0.088	−0.696	0.696	−0.696	0.696	−0.696
0.088	0.696	−0.088	0.696	0.088	0.088	0.088	0.088
0.011	0.088	0.011	−0.088	−0.088	0.088	−0.088	0.088
0.011	0.088	0.011	−0.088	0.000	0.011	0.000	0.011
0.000	0.000	0.000	0.000	0.000	−0.011	0.000	−0.011

2.3 *K*-Nearest Neighbor (*K*-NN) Classifier

The *K*-NN classifier is the supervised learning process that classifies the signal of unknown instances relating to the known according to the distance/similarity function. The *K*-NN is a robust versatile and classifier that is often used in complex classification problems. In the classification setting, by the appropriate distance function two similar instances are assigned to the same class. Its similarity function is defined according to a distance metric between two data points; the Euclidean distance of the given metric is given by

$$d_{pq} = \sqrt{\sum_{j=1}^{n} \left(x_{pj} - y_{qj}\right)^2} \tag{6}$$

The classifier performance depends on the distance metric and *K* number of neighbors. Training and classification phase of the classifier consist of multidimensional feature space and class labels, respectively. In classification phase, *K* is a constant that assigns a label which is nearest to the query point.

2.4 Feature Extraction

The classification accuracy of the cardiac abnormalities is highly depending upon the information retains in the features. For ECG beat classification, different features were obtained from one cardiac cycle. Accurate detection of R_{loc} could help to determine the features required for the classifier. In this work, we used the features such as temporal, *R-R* interval, and morphological features. The statistical features like mean, median, and variance are also considered for classification. Different features such as *R-R* intervals, *RR*-pre, and *RR*-post are used as temporal features, and QRS duration is the heartbeat features, minimum, maximum, mean, and standard deviation of real and complex wavelet coefficients of DT-CWT which were considered as statistical features used as the input to the classifier for the detection of five types of arrhythmia beats.

3 Results and Discussion

All the recorded signal is initially de-noised with bandpass filters and then processed for signal analysis at preprocessing stage. Due to the irregularity in heartbeats and noises, it is quite a tough work to diagnose the correct arrhythmia conditions of the heart. A dual-tree complex wavelet transform-based method has been proposed in this work to extract discriminant features where the signal is decomposed up to four levels to collect some temporal and morphological features to classify ECG arrhythmia beats. The evaluated result shows the better accuracy and sensitivity in *K*-NN classifier. Figure 3 shows the acquired signal and its preprocessing stages. In Fig. 4, different arrhythmia beats are presented. The measure of performance of the classifier in terms of Se, Sp, Pp, and accuracy is evaluated and presented in Fig. 5. The combined feature set with *K*-NN classifier is tested over tenfold cross validation, and it is observed higher performance is achieved at tenfold with Se (%) of 96.45, Pp (%) of 99.34, Sp (%) of 96.16, and Acc (%) of 98.92 for detecting the arrhythmia beats. The details of result are shown in Table 2.

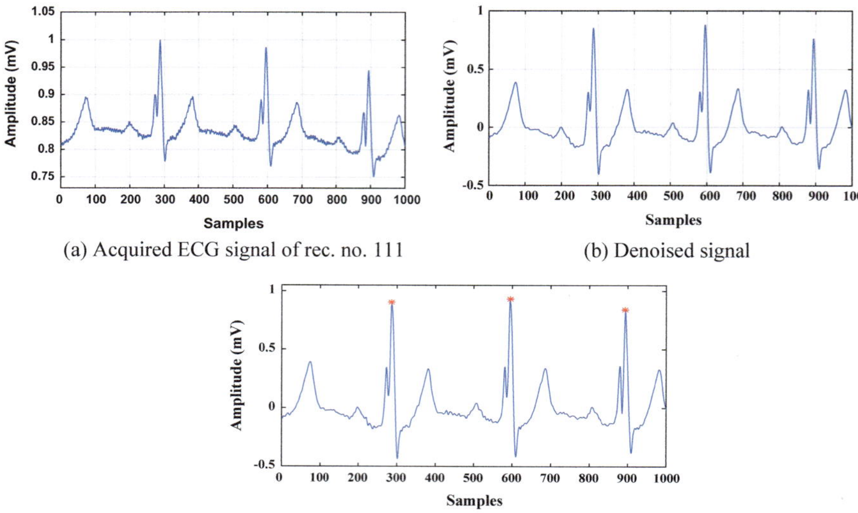

(a) Acquired ECG signal of rec. no. 111

(b) Denoised signal

(c) Detection of R-peaks

Fig. 3. Preprocessing stages of the signal

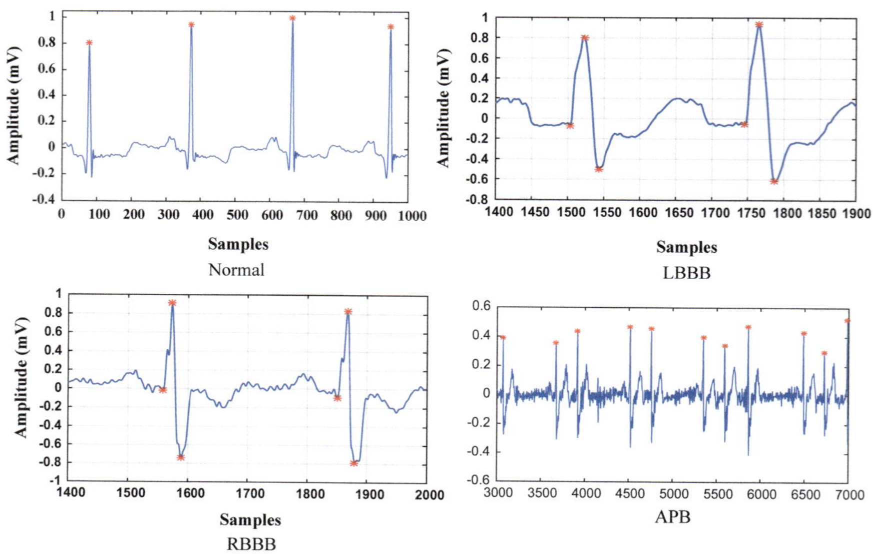

Fig. 4. Types of arrhythmia beats

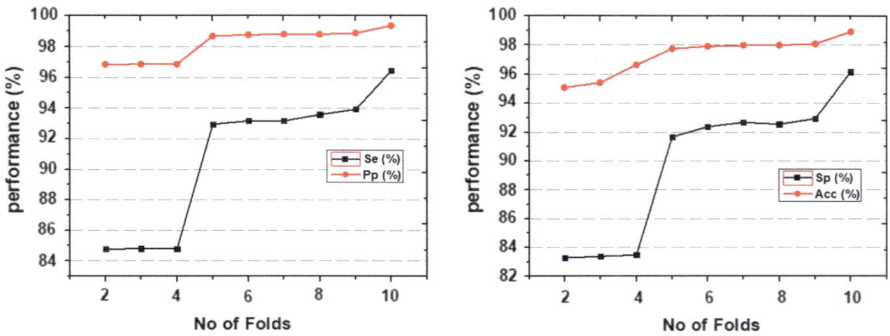

(a) Performance of Se and Pp of the classifier (b) Performance of Sp and Acc of the classifier

Fig. 5. Experimental result of the classifier over tenfold cross validation

Table 2. Classifier performance for ECG beat detection over tenfold cross validation

Fold No.	Se (%)	Pp (%)	Sp (%)	Acc (%)
2	84.76	96.83	83.28	95.09
3	84.80	96.84	83.37	95.42
4	84.79	96.84	83.47	96.64
5	92.95	98.67	91.67	97.75
6	93.17	98.75	92.38	97.91
7	93.17	98.79	92.68	97.98
8	93.58	98.79	92.55	97.99
9	93.94	98.86	92.94	98.09
10	**96.45**	**99.34**	**96.16**	**98.92**

4 Conclusion

This work presents a method for processing and extracting of features from ECG signal using DT-CWT and classifying different heartbeat abnormal conditions using *K*-NN classifiers. With various informative features the projected method confirmed the superiority in terms of higher accuracy for classifying normal, LBBB, RBBB, PVC, and APB beats. The algorithm reached 98.92% of accuracy using *K*-NN classifier. The results suggest that the dual-tree CWT-based features with *K*-NN classifier can be used effectively in the decision-making process of cardiac abnormalities. The machine learning techniques given in this work can be effectively used in diagnosis and decision-making process of accurate and timely detection of abnormal events in the ECG signals so such patients can be treated early in critical care to save life. For real-time ECG signal analysis, a deep learning-based classifier may be used in the future to improve the accuracy and reduce computational time.

References

1. WHO (2015) Cardiovascular diseases (CVDs) (Internet). (Cited 25 Dec 2015). Available from http://www.who.int/mediacentre/factsheets/fs317/en/
2. Aje TO, Miller M (2009) Cardiovascular disease: a global problem extending into the developing world. World J Cardiol 1:3–10
3. Rai HM, Trivedi A, Shukla S (2013) ECG signal processing for abnormalities detection using multi-resolution wavelet transform and artificial neural network classifier. Measurement 46:3238–3246
4. Xing H, Huang M (2008) A new QRS detection algorithm based on empirical mode decomposition. Conf Proc Bioinf Biomed Eng
5. Almenar V, Albiol A (1999) A new adaptive scheme for ECG enhancement. Sig Process 3:253–263
6. Zhang F, Lian Y (2009) QRS detection based on multiscale mathematical morphology for wearable ECG devices in body area networks. IEEE Trans Biomed Circuits Syst 4:220–228
7. Momot A (2009) Methods of weighted averaging of ECG signals using Bayesian inference and criterion function minimization. Biomed Signal Process Control 2:162–169
8. Smital L, Vítek M, Kozumplík J, Provazník I (2013) Adaptive wavelet wiener filtering of ECG signals. IEEE Trans Biomed Eng 60:437–445
9. Kohler BU, Hennig C, Orglmeister R (2002) The principles of software QRS detection. IEEE Eng Med Biol Mag 21:42–57
10. Addison PS, Watson JN, Clegg GR, Holzer M, Sterz F, Robertson CE (2000) Evaluating arrhythmias in ECG signals using wavelet transforms. IEEE Eng Med Biol 19: 104–109
11. Fahoum AS, Howitt I (1999) Combined wavelet transformation and radial basis neural networks for classifying life-threatening cardiac arrhythmias. Med Biol Eng Comput 37:566–573
12. Li C, Zheng C, Tai C (1995) Detection of ECG characteristic points using wavelet transforms. IEEE Trans Biomed Eng 42:21–28
13. Pal S, Mitra M (2010) Detection of ECG characteristic points using multi resolution wavelet analysis based selective coefficient method. Measurement 43:255–261
14. Martinez JP, Almeida R, Olmos S, Rocha AP, Laguna P (2004) A wavelet-based ECG delineator: evaluation on standard database. IEEE Trans Biomed Eng 51:570–581
15. Madeiro JP, Cortez PC, Marques JA, Seisdedos CR, Sobrinho CR (2012) An innovative approach of QRS segmentation based on first-derivative, Hilbert and wavelet transforms. Med Eng Phys 34:1236–1246
16. Chen SW, Chen HC, Chan HL (2006) A real-time QRS method based on moving averaging incorporating with wavelet denoising. Comput Method Programs Biomed 82:187–195
17. Banerjee S, Gupta R, Mitra M (2012) Delineation of ECG characteristic features using multiresolution wavelet analysis method. Measurement 45:474–487
18. Bouaziz F, Boutana D, Benidir M (2014) Multiresolution wavelet-based QRS complex detection algorithm suited to several abnormal morphologies. IET Signal Proc 8:774–782
19. Junior EA, Valentim RADM, Brandao GB (2016) Real-time QRS detection based on redundant discrete wavelet transform. IEEE Latin Am Trans 14:1662–1668
20. Jovic A, Bogunovic N (2011) Electrocardiogram analysis using a combination of statistical, geometric, and nonlinear heart rate variability features. Artif Intell Med 51:175–186
21. Dong X, Wang C, Si W (2017) ECG beat classification via deterministic learning. Neurocomputing 240:1–12

22. Acharya UR, Fujita H, Oh SL, Hagiwara Y, Tan JH, Adam M (2017) Application of deep convolutional neural network for automated detection of myocardial infarction using ECG signals. Inf Sci, pp 190–198
23. Dokur Z, Olmez T, Yazgan E (1999) Comparison of discrete wavelet and Fourier transforms for ECG beat classification. Electron Lett, vol 35
24. Moody GB, Mark RG (2001) The impact of the MIT-BIH Arrhythmia database. IEEE Eng Med Biol Mag 20:45–50
25. Kingsbury N (2010) Complex wavelets for shift invariant analysis and filtering of signal. Appl Comput Harmon Anal 10: 234–253

Classical 2D Face Recognition: A Survey on Methods, Face Databases, and Performance Evaluation

Manoj Kumar Naik and Aneesh Wunnava(✉)

Department of ECE, ITER, Siksha O Anusandhan, Bhubaneswar, Odisha 751030, India
{manojnaik, aneeshwunnava}@soa.ac.in

1 Introduction

The face recognition is a special category of pattern recognition where the input is a face image. The face recognition has started in the early and mid- 1970s, which is used to measure the features between the faces or the face profile. During the 1980s to 1990s, face recognition has treated as not active or not passive. From the early 1990s, face recognition is gaining pace. The reason for the gaining pace in the face recognition: more research project funding for security and surveillance; availability of new generation real-time hardware; the ability of neural networks as a classification and adaptation tolls, etc. So, to get an optimal feature vector in the linear or nonlinear approach is a challenge in the face recognition problem. There is no concrete evidence of the best method to find the optimal feature vectors in the linear face recognition approach.

This section briefly introduces the face recognition. The survey on face recognition methods has stated in Sect. 2. In Sects. 3 and 4, a brief survey on face databases and statistical performance evaluation techniques have been discussed. At last, the conclusion has been drawn in Sect. 5.

2 Face Recognition Methods

The face recognition is one of growing interest due to security reason. The face recognition method based upon the 2D or 3D or combination of 2D and 3D face image. For our limitation, we only discussed the 2D face recognition. To understand the face recognition or face recognition system, we need to investigate various aspects such as classification, classical 2D face recognition methods, face databases, and statistical performance evaluation.

The backbone of any face recognition approach is to find the optimal feature vector using different classification methods or matching the template between the training and testing images. The research on face recognition stated somewhere 1970s, so we are now standing in the fifth decade of FR research, so there are a huge number of

M. N. Mohanty and S. Das (eds.), *Advances in Intelligent Computing and Communication*, Lecture Notes in Networks and Systems 109,
https://doi.org/10.1007/978-981-15-2774-6_45

journal, magazine, technical reports, books, face database, and face recognition systems already developed by the commercial organization, government agencies, research laboratories, and researcher. According to the face recognition review described in [1–6], the methods can broadly classify into two groups. One is a holistic or appearance-based or template-based approach, and the other one analytic or feature-based or geometric-based approach [7].

- *Holistic/Appearance-based/Template-based approach*: In this approach, the feature was extracted by considering the whole face image. These approaches directly operate on the intensity value of the pixel of the facial image. This approach mostly depends on the transform that reduces the high-dimensional feature space to a low-dimensional feature vector because, in high-dimensional space, the probing a face image for recognition is very difficult [8, 9]. The dimensionality reduction makes the face recognition a computationally efficient and traceable.
- *Analytic/Feature-based/Geometric-based approach*: In this approach, the set of geometric features is computed, such as eyes, nose, forehead, and mouth. The face outline and the position of the geometric feature form a feature vector. The location is used to find the geometrical relationship like area, distance, and angle, among the different face images used for the identification or recognition [7, 10, 11].

Mostly all face recognition methods fall on the holistic or analytical approach. The feature extracted using a holistic approach is known as the global feature, and the feature extracted using the analytic approach has referred to as the local feature.

The well-known classical face recognition classification methods are principal components analysis (PCA), linear discriminant analysis (LDA), independent component analysis (ICA), hidden Markov model (HMM)-based, elastic bunch graph matching (EBGM), support vector machine (SVM), and intrinsic discriminant analysis (IDA), etc. The evolution and development of these classification methods have discussed below.

2.1 Principal Component Analysis (PCA)

The feature vector extraction for classification was the backbone of face recognition, where PCA has its significance. The PCA [12] used the well-known Karhunen–Loéve (KL) expression [13–15]. The idea of PCA was first introduced in [16], and then by [17]. Sometimes, the PCA is also known as Hotelling transform [18]. For the pattern recognition problem, the PCA was first time introduced in [19] to represent face images in terms of optimal coordinates. For the first time in [20], try to find the eigenfaces from the face image using Karhunen–Loéve expansion. Then in 1991, Turk and Pentland [21, 22] developed a face recognition system using eigenfaces that can locate and identify face images. A more detailed study on computational and performance aspects of the PCA algorithm has been described in [23, 24].

2.2 Linear Discriminant Analysis (LDA)

The LDA method is a class-specific method proposed by Etemad and Chellappa in [25] in the year 1997 by considering the discriminant analysis [26, 27]. The LDA is also known as Fisher's linear discriminant (FLD) [26]. Then, in the same year 1997, [28] proposed a new face classification image known as Fisherfaces by implementing dimension reduction technique PCA in FLD. One common problem in face recognition is the "small sample size" (SSS) problem, which can be overcome by perturbation or transform modification. To overcome the SSS problem in LDA-based face recognition, a method new LDA (NLDA) has proposed in [29]. Most of the researchers believe that LDA outperforms PCA, but according to [30], PCA can outperform the LDA when the training data set is small. A new direct LDA (DLDA) algorithm has proposed in [31] to overcome the singularity problem in LDA; the diagonalization procedure has modified to make computationally efficient. A new direct fractional-step LDA (DF-LDA) method was proposed in [32], which improves the LDA performance considering the removal of the SSS problem and improving the classification ability. A new algorithm PCA plus LDA has been proposed in [33] that performs LDA in PCA transformed space to overcome the singular problem in LDA. The PCA generally gives linear features, and kernel PCA (KPCA) [34] gives a nonlinear feature. Then, the researcher tries to develop complex LDA-based combine Fisherfaces algorithm known as complex Fisherfaces [35] by fusing the feature extracted from complex LDA and integrated feature of PCA and KPCA. Some more variant LDA algorithm has improved LDA (ILDA) [36], regularized LDA (R-LDA) [37], soft label-based linear discriminant analysis (SL-LDA) [38], uncorrelated linear discriminant analysis (ULDA) [39], orthogonal linear discriminant analysis (OLDA) [39], and perturbation LDA (PLDA) [40] in related extraction of the face feature vector.

2.3 Independent Component Analysis (ICA)

The independent component analysis (ICA) [41, 42] is an extension of the PCA [12], which imposes independence up to the second-order and defines orthogonal directions. The two different architectures have discussed in it: one face images have considered as random variables, and in another one pixel of the face image are treated as random variables. An enhanced independent component analysis coined as Fisher independent component analysis (FIDA) was proposed in [43] by augmenting Fisher linear discriminant (FLD) analysis [26–28] in the generic ICA [42].

2.4 Hidden Markov Model (HMM)

The hidden Markov model (HMM) [44] has also referred to as Markov sources or probabilistic functions of Markov chains [45–48]. The HMM model has successfully applied in the field of speech processing [44]. For the first time, HMM is used for face identification in [49] to extract features and segment the face image automatically. Then in [50], a new face recognition approach based on HMM was discussed, which extracts a two-dimensional discrete cosine transform (2D-DCT)-based feature for face recognition. Again, to reduce the complexity of HMM-based face recognition, the researcher in [51] proposed an HMM-based face recognition, in which the observation vector to characterize the different state of HMM is obtained using Karhunen–Loéve transform (KLT) [15, 19–22, 27].

2.5 Elastic Bunch Graph Matching (EBGM)

The elastic bunch graph matching (ENGM) based face recognition algorithm [52] that implement bunch graph a data structure that serves the general representation of a set of complex Gabor wavelet coefficients [53] known as jets of a small set of individual faces, and also employed the object adapted graphs to find the correspondence between face images. Some more researchers proposed face recognition uses EBGM in [54–56].

2.6 Intrinsic Discriminant Analysis (IDA)

The face image is classified into three components: common facial differences, individuality difference, and interpersonal difference. By considering all these, a new dimensionality reduction technique has proposed as intrinsic discriminant analysis (IDA) in [57]. The face recognition using the IDA algorithm was known as intrinsic faces.

The researcher also presents many more face recognition methods. Some of them are Gabor-Fisher classifier (GFC) [58], kernel eigenfaces [59], kernel Fisherfaces [59], kernel direct discriminant analysis (KDDA) [60], locality preserving projections (LPP) [61, 62], local binary patterns (LBP) [63], locally linear embedding (LLE) [64–66], isometric feature mapping (Isomap) [67], and locality sensitive discriminant analysis (LSDA) [68] algorithms.

3 Face Databases

The human faces are three-dimensional (3D) structures. So, when they record it on the digital camera, it will transform into two-dimensional (2D) coordinates. So, the appearance of the facial image affected by some factors like pose, illumination, facial expression, occlusion, age, and facial hair. To validate a face recognition algorithm, one should consider all these factors. For this reason, the researcher required some standard databases that will be recognized by the researcher community for comparatively evaluate algorithms. Some of these database are Olivetti Research Ltd (ORL) database of faces [69, 70], AR face database [71, 72], face recognition technology (FERET) [73–76], Yale face database [77], extended Yale face database B (B +) [78], the Carnegie Mellon University (CMU) pose, illumination, and expression (CMU-PIE) database [79–81], labeled faces in the wild (LFW) database [82, 83], plastic surgery face database [84], University of Manchester Institute of Science and Technology (UMIST) face database [85], labeled Wikipedia faces (LWF) [86, 87], labeled faces in the wild-a (LFW-a) [88], YouTube makeup (YMU) face image [89], virtual makeup (VMU) face image [89], US Adult Faces Database [90], makeup in the "wild" (MIW) face database [91], Indian movie face database (IMFDB) [92], facescrub database [93], and sibling database [94].

4 Statistical Performance Evaluation

The face recognition algorithm had better been evaluated to study the statistical performance. In the statistical evaluation, for matching of the probe image to the gallery image, one requires a similarity measure. So, the most common algorithm uses a similarity measure as a nearest-neighbor classifier to validate the identity of faces. The similarity measures are the distance function between the feature vectors. Some of these are baseline (L1), Euclidean (L2), angle, Mahalanobis, L1 + Mahalanobis, L2 + Mahalanobis, and angle + Mahalanobis distance [23]. For the performance evaluation, the researcher specifies a CMC curve [75, 76, 95]. Sometimes, the researcher also used the ROC curve for system performance evaluation, which is a special case of the CMC curve [75, 76, 95].

5 Conclusion

The paper briefly discussed the development of the classical 2D-based face recognition method existing in the literature, available databases, and various metrics related to performance evaluation. From the above, it clearly stated that the outperformer classification methods are PCA and LDA for robustness and simplicity. The above methods have been proposed based on the still face databases, so more challenging in the future is how they can be applied to real-time simulations.

References

1. Chellappa R, Wilson CL, Sirohey S (1995) Human and machine recognition of faces: a survey. Proc IEEE 83:705–740
2. Perronnin F, Dugelay J-L (2003) An introduction to biometrics and face recognition. In: IMAGE'2003: learning, understanding, information retrieval
3. Zhao W, Chellappa R, Phillips PJ, Rosenfeld A (2003) Face recognition: a literature survey. ACM Comput Surv 35:399–458
4. Kong SG, Heo J, Abidi BR, Paik J, Abidi MA (2005) Recent advances in visual and infrared face recognition - a review. Comput Vis Image Underst 97:103–135
5. Abate AF, Nappi M, Riccio D, Sabatino G (2007) 2D and 3D face recognition: A survey. Pattern Recogn Lett 28:1885–1906
6. Beham MP, Roomi SMM (2013) A review of face recognition methods. Int J Pattern Recogn Artif Intell 27:1356005
7. Brunelli R, Poggio T (1993) Face recognition: features versus templates. IEEE Trans Pattern Anal Mach Intell 15:1042–1052
8. Beyer K, Goldstein J, Ramakrishnan R, Shaft U (1999) When Is "Nearest Neighbor" Meaningful? In: Beeri C, Buneman P (eds) Database Theory — ICDT'99, vol 1540. Springer, Berlin, Heidelberg, pp 217–235
9. Sirovich L, Kirby M (1987) Low-dimensional procedure for the characterization of human faces. J Opt Soc Am A 4:519–524

10. Bruce V (1999) Identification of human faces. In: 1999 Seventh international conference on (Conference Publication No. 465) image processing and its applications, vol 612, pp 615–619
11. Goldstein AJ, Harmon LD, Lesk AB (1971) Identification of human faces. Proc IEEE 59:748–760
12. Jolliffe IT (1986) Principal cornponent analysis. Springer, New York
13. Karhunen K (1946) Uber lineare methoden in der wahrscheinlichkeits-rechnun. Ann Acad Sri Fennicae ser A1 Math Phys 37
14. Fukunaga K (1972) lntroduction to statistical pattern recognition. Academic, New York
15. Fukunaga K, Koontz WLZ (1970) Application of the Karhunen Loeve expansion to feature selection and ordering. IEEE Trans Comput C-19:311–318
16. Pearson K (1901) On lines and planes of closest fit to systems of points in space. Philosophical Magazine 2\
17. Hotelling H (1933) Analysis of a complex of statistical variables into principal component. J Educ Psychol 24 (1933)
18. Gonzalez RC, Wintz PA (1987) Digital image processing. Addison-Wesley, Reading, MA
19. Watanabe S (1965) Karhunen-Loeve expansion and factor analysis theoretical remarks and applications. In: 4th Prague Conference Information Theory
20. Kirby M, Sirovich L (1990) Application of the Karhunen-Loeve procedure for the characterization of human faces. IEEE Pattern Anal Mach Intell 12:103–108
21. Turk MA, Pentland AP (1991) Face recognition using eigenfaces. In: Proceedings CVPR'91, IEEE computer society conference on computer vision and pattern recognition, pp 586–591
22. Turk MA, Pentland AP (1991) Eigenfaces for recognition. J Cogn Neurosci 3:71–86
23. Moon H, Phillips PJ (2001) Computational and performance aspects of PCA-based face recognition algorithms. Perception 30:303–321
24. Yambor WS, Draper BA, Beveridge JR (2000) Analyzing PCA-based face recognition algorithms: eigenvector selection and distance measures
25. Etemad K, Chellappa R (1997) Discriminant analysis for recognition of human face images. J Opt Soc Am A 14:1724–1733
26. Fisher RA (1936) The use of multiple measurements in taxonomic problems. Ann Eugenics 7:179–188
27. Duchene J, Leclercq S (1988) An optimal transformation for discriminant and principal component analysis. IEEE Trans Pattern Anal Mach Intell 10:978–983
28. Belhumeur PN, Hespanha JP, Kriegman D (1997) Eigenfaces vs. Fisherfaces: recognition using class specific linear projection. IEEE Trans Pattern Anal Mach Intell 19:711–720
29. Chen L-F, Liao H-YM, Ko M-T, Lin J-C, Yu G-J (2000) A new LDA-based face recognition system which can solve the small sample size problem. Pattern Recogn 33:1713–1726
30. Martinez AM, Kak AC (2001) PCA versus LDA. IEEE Trans Pattern Anal Mach Intell 23:228–233
31. Yu H, Yang J (2001) A direct LDA algorithm for high-dimensional data with application to face recognition. Patttern Recogn 34:2067–2070
32. Kostantinos JL, Plataniotis KN, Venetsanopoulos AN (2003) Face recognition using LDA-based algorithms. IEEE Trans Neural Networks 14:195–200
33. Yang J, Yang J-Y (2003) Why can LDA be performed in PCA transformed space? Pattern Recogn 36:563–566
34. Schlkopf B, Smola A, Muller K-R (1998) Nonlinear component analysis as a kernel eigenvalue problem. Neural Comput 10:1299–1319
35. Yang J, Yang J-Y, Frangi AF (2003) Combined fisherfaces framework. Image Vis Comput 21:1037–1044

36. Xiao-Yuan J, Zhang D, Yuan-Yan T (2004) An improved LDA approach. IEEE Trans Syst Man Cybern Part B Cybern 34:1942–1951
37. Lu J, Plataniotis KN, Venetsanopoulos AN (2005) Regularization studies of linear discriminant analysis in small sample size scenarios with application to face recognition. Pattern Recogn Lett 26:181–191
38. Zhao M, Zhang Z, Chow TWS, Li B (2014) Soft label based linear discriminant analysis for image recognition and retrieval. Comput Vis Image Underst 121:86–99
39. Ye J (2005) Characterization of a family of algorithms for generalized discriminant analysis on undersampled problems. J Mach Learn Res 6:483–502
40. Zheng W-S, Lai JH, Yuen PC, Li SZ (2009) Perturbation LDA: learning the difference between the class empirical mean and its expectation. Pattern Recogn 42:764–779
41. Jutten C, Herault J (1991) Blind separation of sources, part 1: an adaptive algorithm based on neuromimetic architecture. Sign Proces 24:1–10
42. Comon P (1994) Independent component analysis, a new concept? Sign Proces 36:287–314
43. Keun-Chang K, Pedrycz W (2007) Face recognition using an enhanced independent component analysis approach. IEEE Trans Neural Networks 18:530–541
44. Rabiner L, Huang B (1993) Fundamentals of speech recognition. Prentice-Hal, Englewood Cliffs, NJ
45. Baum LE, Petrie T (1966) Statistical inference for probabilistic functions of finite state Markov chains. Ann Math Stat 37:1554–1563
46. Baum LE, Eagon JA (1967) An inequality with applications to statistical estimation for probabilistic functions of Markov processes and to a model for ecology, pp 360–363
47. Baum LE, Petrie T, Soules G, Weiss N (1970) A maximization technique occurring in the statistical analysis of probabilistic functions of Markov chains. Ann Math Stat 41:164–171
48. Baum LE (1972) An inequality and associated maximization technique in statistical estimation for probabilistic functions of Markov processes. In: Inequalities III, Academic Press pp 1–8
49. Samaria F, Young S (1994) HMM-based architecture for face identification. Image Vis Comput 12:537–543
50. Nefian AV, Hayes MH, III (1998) Hidden Markov models for face recognition. In: Proceedings of the 1998 IEEE international conference on acoustics, speech and signal processing, vol 2725, pp 2721–2724
51. Nefian AV, Hayes MH, III (1998) Face detection and recognition using hidden Markov models. In: Proceedings 1998 International Conference on Image Processing, ICIP 98, vol 141, pp 141–145
52. Wiskott L, Fellous JM, Kuiger N, von der Malsburg C (1997) Face recognition by elastic bunch graph matching. IEEE Trans Pattern Anal Mach Intell 19:775–779
53. Daugman JG (1988) Complete discrete 2D gabor transform by neural networks for image analysis and compression. IEEE Trans Acoustics Speech Signal Proces 36:1169–1179
54. Kela N, Rattani A, Gupta P (2006) Illumination invariant elastic bunch graph matching for efficient face recognition. In: CVPRW'06 Conference on Computer Vision and Pattern Recognition Workshop, pp 42–42
55. Pervaiz AZ (2010) Real time face recognition system based on EBGM framework. In: 2010 12th International conference on computer modelling and simulation (UKSim), pp 262–266
56. Hanmandlu M, Gupta D, Vasikarla S (2013) Face recognition using elastic bunch graph matching. In: 2013 IEEE applied imagery pattern recognition workshop: sensing for control and augmentation, pp 1–7
57. Wang Y, Wu Y (2010) Face recognition using Intrinsicfaces. Pattern Recogn 43:3580–3590
58. Chengjun L, Wechsler H (2002) Gabor feature based classification using the enhanced fisher linear discriminant model for face recognition. IEEE Trans Image Proc 11:467–476

59. Yang M-H (2002) Kernel Eigenfaces vs. kernel fisherfaces: face recognition using kernel methods. In Proceedings of the fifth IEEE international conference on automatic face and gesture recognition, IEEE computer society, p 215
60. Lu J, Plataniotis KN, Venetsanopoulos AN (2003) Face recognition using kernel direct discriminant analysis algorithms. IEEE Trans on Neural Networks 14:117–126
61. He X, Shuicheng Y, Yuxiao H, Niyogi P, Hong-Jiang Z (2005) Face recognition using Laplacianfaces. IEEE Trans Pattern Anal Mach Intell 27:328–340
62. He X, Niyogi P (2003) Locality preserving projections. In: advances in neural information processing systems
63. Ahonen T, Hadid A, Pietikainen M (2006) Face description with local binary patterns: application to face recognition. IEEE Trans Pattern Anal Mach Intell 28:2037–2041
64. Saul LK, Roweis ST (2003) Think globally, fit locally: unsupervised learning of low dimensional manifolds. J Mach Learn Res 4:119–155
65. Roweis ST, Saul LK (2000) Nonlinear dimensionality reduction by locally linear embedding. Science 290:2323–2326
66. Ridder DD, Duin RPW (2002) Locally linear embedding for classification. Delft University of Technology
67. Samko O, Marshall AD, Rosin PL (2006) Selection of the optimal parameter value for the Isomap algorithm. Pattern Recogn Lett 27:968–979
68. Cai D, He X, Zhou C, Han J, Bao H (2007) Locality sensitive discriminant analysis. In: Proceedings of the 20th international joint conference on Artifical intelligence, Morgan Kaufmann Publishers Inc., Hyderabad, India, pp 708–713
69. http://www.cam-orl.co.uk/facedatabase.html
70. Samaria FS, Harter AC (1994) Parameterisation of a stochastic model for human face identification. In: Proceedings of the second IEEE workshop on applications of computer vision, pp 138–142
71. Martinez AM, Benavente R (1998) The ar face database
72. Li B, Li J, Tang K, Yao X (2015) Many-objective evolutionary algorithms: a survey. ACM Comput Surv 48:1–35
73. Phillips PJ, Hyeonjoon M, Rizvi SA, Rauss PJ (2000) The FERET evaluation methodology for face-recognition algorithms. IEEE Trans Pattern Anal Mach Intell 22:1090–1104
74. Vázquez D, Fernández-Torres MJ, Ruiz-Femenia R, Jiménez L, Caballero JA (2018) MILP method for objective reduction in multi-objective optimization. Comput Chem Eng 108:382–394
75. Phillips PJ, Wechsler H, Huang J, Rauss PJ (1998) The FERET database and evaluation procedure for face-recognition algorithms. Image Vis Comput 16:295–306
76. Rizvi SA, Phillips PJ, Hyeonjoon M (1998) The FERET verification testing protocol for face recognition algorithms. In: Proceedings third IEEE international conference on automatic face and gesture recognition, pp 48–53
77. http://cvc.yale.edu/projects/yalefaces/yalefaces.html
78. Wang H, Jin Y, Yao X (2017) Diversity assessment in many-objective optimization. IEEE Trans Cybern 47:1510–1522
79. Sim T, Baker S, Bsat M (2002) The CMU pose, illumination, and expression (PIE) database. In: Proceedings of fifth IEEE international conference on automatic face and gesture recognition, pp 46–51
80. Sim T, Baker S, Bsat M (2003) The CMU pose, illumination, and expression database. IEEE Trans Pattern Anal Mach Intell 25:1615–1618
81. Sedarous S, El-Gokhy SM, Sallam E (2017) Multi-swarm multi-objective optimization based on a hybrid strategy. Alexandria Eng J 57(3):1619–1629

82. Huang GB, Ramesh M, Berg T, Learned-Miller E (2007) Labeled faces in the wild: a database for studying face recognition in unconstrained environments
83. Meza J, Espitia H, Montenegro C, Giménez E, González-Crespo R (2017) MOVPSO: vortex multi-objective particle swarm optimization. Appl Soft Comput 52:1042–1057
84. Singh R, Vatsa M, Bhatt HS, Bharadwaj S, Noore A, Nooreyezdan SS (2010) Plastic surgery: a new dimension to face recognition. IEEE Trans Inf Forensics Secur 5:441–448
85. http://images.ee.umist.ac.uk/danny/database.html
86. Hasan MK, Pal CJ (2011) Improving alignment of faces for recognition. In: 2011 IEEE international symposium on Robotic and Sensors Environments (ROSE), pp 249–254
87. Hasan MK, Pal C (2014) Experiments on visual information extraction with the faces of wikipedia. In: AAAI 2014
88. Wolf L, Hassner T, Taigman Y (2011) Effective unconstrained face recognition by combining multiple descriptors and learned background statistics. IEEE Trans Pattern Anal Mach Intell 33:1978–1990
89. Dantcheva A, Cunjian C, Ross A (2012) Can facial cosmetics affect the matching accuracy of face recognition systems? In: 2012 IEEE fifth international conference on Biometrics: Theory, Applications and Systems (BTAS), pp 391–398
90. Bainbridge WA, Isola P, Oliva A (2013) The intrinsic memorability of face photographs. J Exp Psychol Gen 142:1323–1334
91. Chen C, Dantcheva A, Ross A (2013) Automatic facial makeup detection with application in face recognition. In: 2013 International Conference on IEEE Biometrics (ICB), pp 1–8 (2013)
92. Setty S, Husain M, Beham, P, Gudavalli J, Kandasamy M, Vaddi R, Hemadri V, Karure JC, Raju R, Rajan B, Kumar V, Jawahar CV (2013) Indian movie face database: a benchmark for face recognition under wide variations. In: 2013 Fourth National Conference on Computer Vision, Pattern Recognition, Image Processing and Graphics (NCVPRIPG), pp 1–5
93. Hong-Wei N, Winkler S (2014) A data-driven approach to cleaning large face datasets. In: 2014 IEEE International Conference on Image Processing (ICIP), pp 343–347
94. Vieira T, Bottino A, Laurentini A, De Simone M (2014) Detecting siblings in image pairs. Vis Comput 30:1333–1345
95. Grother P, Micheals R, Phillips PJ (2002) Face recognition vendor test performance metrics. In: Kittler J, Nixon M (eds) Audio- and video-based biometric person authentication, vol 2688. Springer, Berlin Heidelberg, pp 937–945

Multi-level Authentication-Based Secure Aware Data Transaction on Cloud Using Cyclic Shift Transposition Algorithm

Prasanta Kumar Bal(✉) and Sateesh Kumar Pradhan

Department of Computer Science & Application, Utkal University, Bhubaneswar, India
sateesh.cs@utkaluniversity.ac.in

1 Introduction

A data "transaction server or transaction processing system (TPS)" is a software system, or combination of hardware and software, that supports transaction processing. In data transaction, computing task is divided into small, indivisible operations, called transactions. Data center approach is used in cloud computing that comprises of thousands of servers to provide services to the end users [1]. Cloud computing has desirable properties, such as elasticity, scalability, and pay-per-use and fault tolerance. As a promising service platform, its applications demand huge data storage. Cloud application also requires data reliability that to be in a highly cost-effective manner [2]. Data storage is most popular and important service. Generally, independent institutions check and certify the security requirements imposed by the users to cloud providers. The cloud provider and cloud end user make formal agreement that includes the security and privacy of the data stored and processed in the cloud. Recently, cloud storage has received huge attractions in personal and business organizations because it is convenient and economical [3].

The main drawback in cloud storage is information integrity and confidentiality. In this paper, these issues are handled by encryption techniques for correctness, confidentiality, and integrity tamper resistance, [4]. Internet is used to connect cloud users having mobile phones, laptops, or fashionable desktops with the cloud platform. Cloud services are usually classified into software as a service (SaaS), platform as a service (PaaS), and infrastructure as a service (IaaS) like raw computing power or cloud storage. A decent cloud security supplier can provide a scalable answer that detects anomalies before they reach the "data center," serving to security [5]. "A broad set of policies, technologies, applications, and controls utilized to protect virtualized IP, data, applications, services, and the associated infrastructure of cloud computing" is referred

M. N. Mohanty and S. Das (eds.), *Advances in Intelligent Computing and Communication*, Lecture Notes in Networks and Systems 109,
https://doi.org/10.1007/978-981-15-2774-6_46

to as cloud security. In recent years, many encryption techniques are used for the above purpose. They have Triple DES, Data Encryption Standard (DES), Advanced Encryption Standard (AES), Twofish encryption, and Rivest–Shamir–Adleman Encrypion (RSA). In this paper, we have used cyclic shift transposition (CST) algorithm. The CST algorithm is basically a transposition algorithm obtained by shifting operations [6].

Lot of researcher has been developing a secure data transaction on cloud. Among them, some of the works are analyzed here; In 2017, Luna et al. [7] developed quantitative reasoning about cloud security using service-level agreement. The technique proposed by them was to help improve the security in cloud data transaction. Adoption framework was used toward achieving data security in cloud by Chang et al. [6] in 2016. They experimented to reduce the injection of Trojans and viruses with success rate of 97.43%. In 2016, Gupta et al. [8] analyzed implementation of cloud ERP with reduced effect of compliance, network, and security. This paper proposed a link between the concerns of cloud vendor and the CSFs that are categorized people, and technological factors. Experimental result showed that the implementation of cloud ERP was moderated. Wei et al. [9] in 2016 developed Revocable Storage Identity Based Encryption (RS-IBE) for secure data sharing in cloud computing. Finally, demonstrated result has advantages in efficiency and functionality. It takes more time to take the conversion of cipher text in cloud computing; its main limitation is security in cloud computing. Awad et al. [10] in 2018 proposed a chaotic searchable encryption for mobile cloud storage in fuzzy keyword. Experimental results showed that the proposed scheme was efficient and suitable for a secure searchable cloud storage system. In 2018, Hu et al. [11] had developed a secure and verifiable access control based on the Nth degree truncated polynomial ring (NTRU) cryptosystem for big data storage in cloud. The design of a secured and privacy preserving for big data storage in a cloud is an extremely challenging problem. It indicates that eligible user can be prevented from cheating and reduce various attacks such as the collusion attack.

2 Proposed Methodology

As the demand for cloud computing is growing across the globe, data security and privacy protection are also two rising issues in cloud computing. To avoid this problem, multi-level authentication-based security aware data transaction on cloud is proposed. The proposed system consists of three modules, namely multi-level authentication, data encryption, and data retrieval. The multi-level authentication process is used to avoid the unauthorized person login process. The proposed methodology is depicted in Fig. 1.

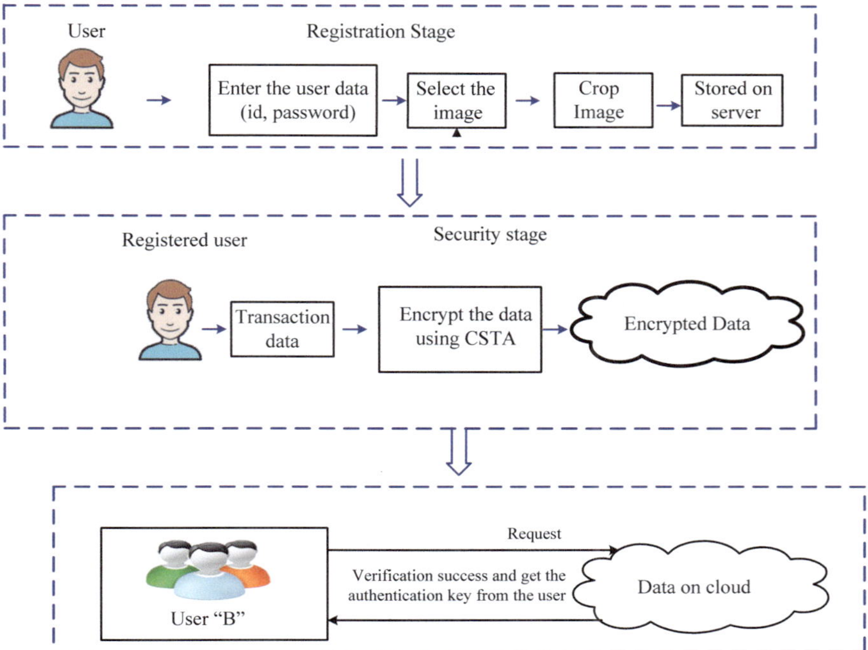

Fig. 1. Overall concept of the proposed methodology

2.1 Multi-level Authentication Process

To avoid the malicious attack and information loss, the authentication process is mandatory. "It becomes easy for unauthorized users to modify the data without the knowledge of the owner in a centralized environment. Hence, it is unavoidable for security breach, thereby making it mandatory to secure the data using multi-level authentication." The multi-level authentication process consists of two stages, namely registration and login.

Registration process:-
During the registration process, initially, the user opens the registration page. Then, the user creates the user id U_{id} and password U_{Pa}. After U_{id} creation, user has to enter all the information about the user. Then, user selects one image S_I from the N number of images available in the database D_b. After the selection process, the selected image S_I is cropped. The cropping is done based on pixel position values. After the cropping process, user information along with cropped image is stored on the cloud server. The registration process is given in Fig. 2.

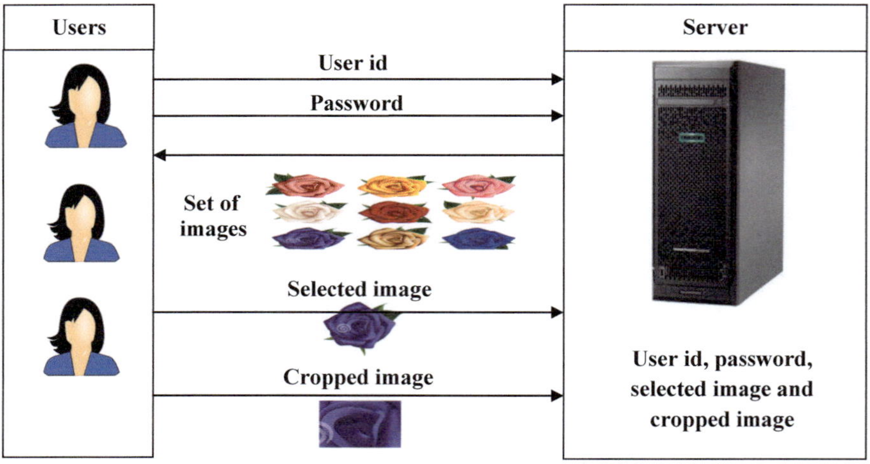

Fig. 2. Registration process

Login process:-
After the registration process, the user is able to upload the data on cloud. A user wants to upload the data means user must login to the server. Using this process, we can avoid the unauthorized person to access the data. In login process, initially, user enters their id U_{id} and password U_{pa}. The information is correct means the server displays n number of images along with registered image. Among the displayed images, the user selects one image S_I which should be same as the registered image. If the selected image is correct, then user crops the image and sends to the server. Finally, server checks the cropped image with registered cropped image. If the information given by the user is correct for the respective user name, the server will allow the user to access the data otherwise, neglect the user request.

2.2 Data Security Using Cyclic Shift Transposition Algorithm (CSTA)

The main concern of cloud data owner is to store the data securely on cloud as data confidentiality is an unavoidable task. Even though different algorithms are used for sensitive data protection from the malicious users, they can easily access the cloud storage framework. To avoid the problem in this paper CSTA method is proposed that does not require any centralized authority such as key management system. In this method, initially, the file is separated into $N \times N$ matrix and so it consists of "n" blocks size and its key size is 2^{2N+1}. The CSTA algorithm consists of two stages, namely encryption and decryption.

Encryption process:-
The encryption process is used to hide the original information. The proposed CSTA performs partition and shifting operation, namely column shift, row shift, primary diagonal shift, and secondary diagonal shift. The step-by-step process of encryption process is explained below;

Step 1 Consider the input file F_i. Initially, we split the file F_i into $N \times N$ matrix format.

Step 2 Then, perform shift column (SC) operation. The SC is performed based on Eq. (1).

$$F'_{r,c} = F_{r + \text{shift}(r,N_b) \bmod N_{b,c}} \tag{1}$$

Here, $\text{shift}(r, N_b)$ only depends upon key value. The key values vary from 0 to 9. The key values represent the cyclic shift of the number of elements, and the modular arithmetic is represented by mod operation.

Step 3 Perform shifting the row (SR) in a particular order. SR transportation can be expressed in Eq. (2).

$$F'_{r,c} = F_{r,c + \text{shift}(r,N_b) \bmod N_b}, \tag{2}$$

Step 4 Perform diagonal shift operation, i.e., shift the diagonal elements from top to right bottom. This can be presented as follows:

$$F'_{r,c} = F_{r + \text{shift}(r,N_b) \bmod N_{b,c + \text{shift}(r,N_b) \bmod N_b}} \tag{3}$$

Step 5 Perform secondary diagonal shift operation in particular order. This can be expressed as follows;

$$F'_{r,c} = F_{(r-1) \bmod N_{b,c}}.$$

Step 6 Then, find the output in a particular order.

$$F'_{r,c} = F_{(c + (N_{b-1})),c}.$$

Step 7 Obtain the encrypted text by converting the output into ASCII format.

Step 8 Finally, we calculate the hash value along with a timestamp and store it in the cloud.

Decryption process The decryption process is a reverse of the encryption process. If any cloud user tries to retrieve the information, the cloud server decrypts the data into original file and sends to the cloud user. The step-by-step process of decryption process is explained below:

Step 1 Evaluate the hash value along with a timestamp and transfer it to the receiver.
Step 2 Consider the output and convert the output into ASCII format to get the encrypted file.
Step 3 After that, shift the row into a specific order.
Step 4 Shifting the column into a specific order
Step 5 Then, shift the matrix diagonally into the corresponding order.
Step 6 Perform secondary diagonal shift operation in corresponding order.
Step 7 Find out the decrypted file.

2.3 Query-Based File Retrieval

After the encryption process, the encrypted data are stored in a cloud. To avoid malicious attack and information loss, the data are securely stored in the cloud. After uploading, any user wants to access the data means they send a request to CSP. After receiving request, the CSP checks the authentication process. If the user is already registered, CSP sends the data; otherwise, it neglects the request.

3 Result and Discussion

The experimental results obtained from the proposed methodology are analyzed in this section. Here, the implementation is done in JAVA with cloud simulator. The effectiveness of the proposed methodology is analyzed in terms of different metrics, namely memory usage, execution time, encryption time, and decryption time.

3.1 Experimental Results

The main objective of proposed methodology is to securely send the data transaction on cloud. Here, two level securities are developed. The first level is in authentication stage, and another one is data security stage. For authentication process, multi-level authentication is developed which is used to avoid the unauthorized user login. For data security, CSTA is developed which is securely transact the data. To prove the effectiveness of the proposed methodology, here, we compare our proposed methodology with different algorithms. The experimental results of proposed against existing is given in this section.

Figure 3 represents the running time of the proposed method. The data size is 10,000 kb, the AES is 5637 s, and the DES is 4632 s. But the proposed CSTA (2) is 3905 is better than the existing methods. Because the "proposed model follows the decentralized architecture, it does not depend on any third-party system."

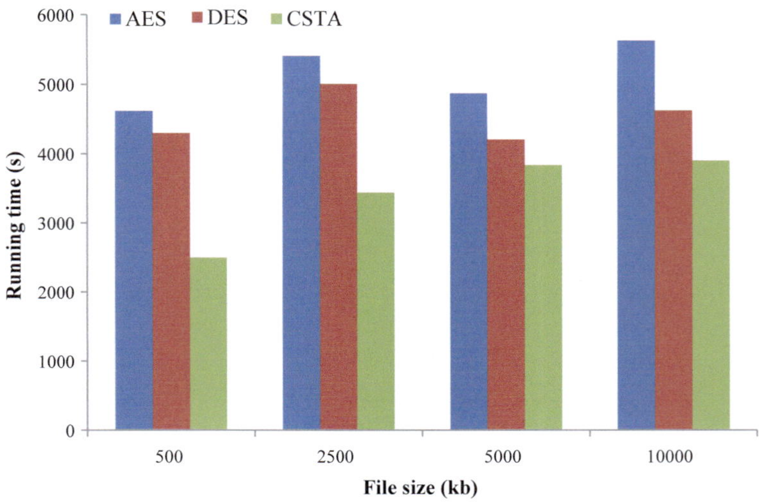

Fig. 3. Running time

To prove the effectiveness of the proposed authentication process, we compare our method with a different authentication process. The performance of the proposed methodology is analyzed in terms of successful authentication is given in Fig. 4. When analyzing Fig. 4, our proposed method attains a better authentication rate. This is because our proposed method has five mandatory fields, namely user id, password, image selection, cropping, and encryption.

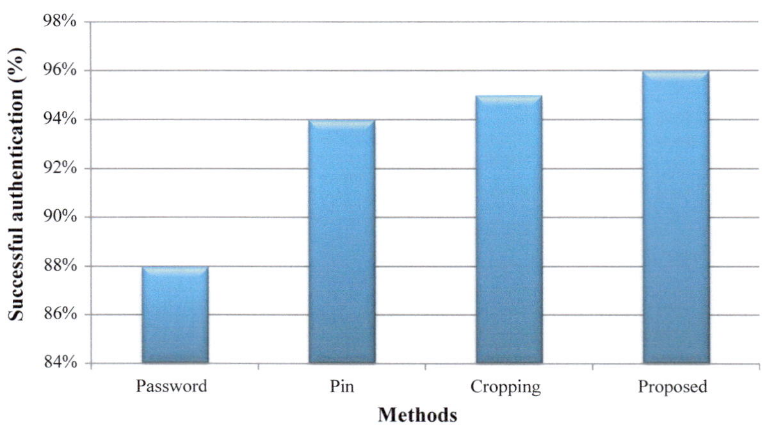

Fig. 4. Performance based on successful authentication

Figure 5 represents the encryption time of the proposed method. The data size is 5000, the AES is 1,911,324s, and the DES is 1,893,721s. But the proposed CSTA 1,857,651s is better than the existing methods. Because the proposed model follows the decentralized architecture, it does not depend on any third-party system.

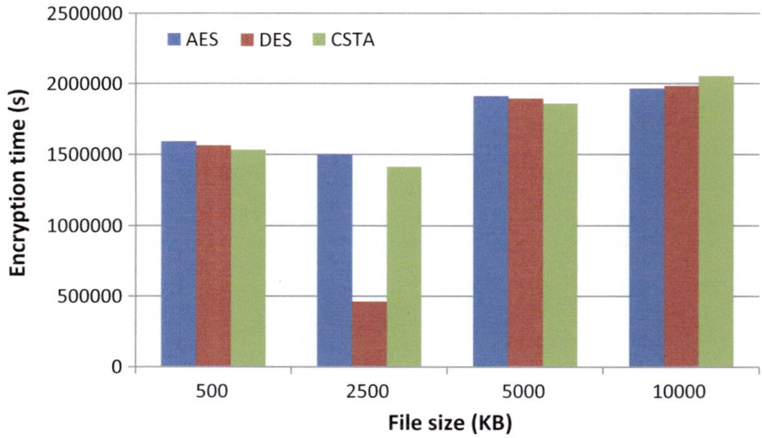

Fig. 5. Encryption time

Figure 6 represents the decryption time of the proposed method. The data size is 10,000, the AES is 16,856,331s, and the DES is 1,620,511s. But the proposed CSTA 1,595,432 is better than the existing methods. Because the proposed model follows the decentralized architecture, it does not depend on any third-party system.

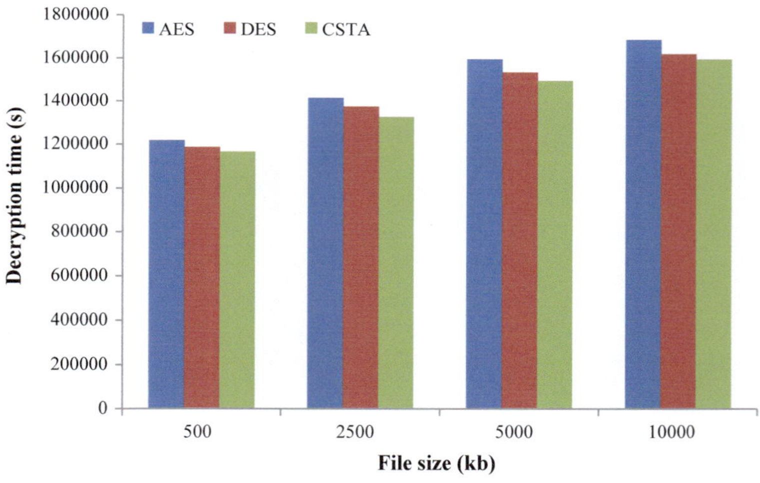

Fig. 6. Decryption time

Figure 7 represents the memory size of the proposed method. The data size is 10,000, the AES is 27,632,481, and the DES is 26,947,263. But the proposed CSTA 26,446,232 is better than the existing methods. Because the proposed model follows the decentralized architecture, it does not depend on any third-party system.

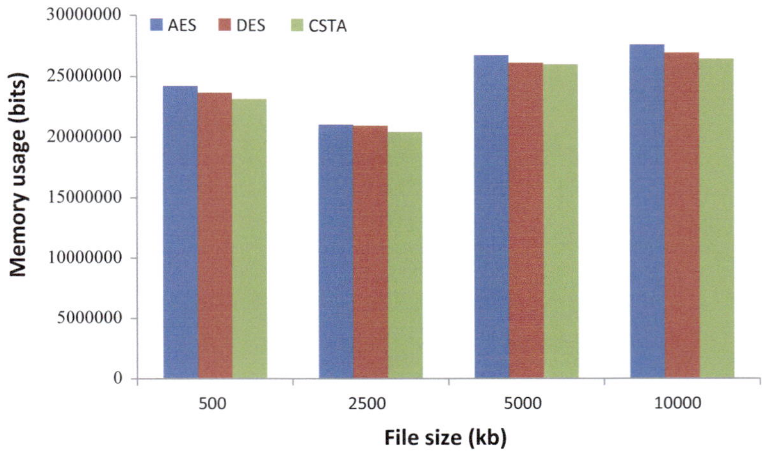

Fig. 7. Memory size

4 Conclusion

In this paper, multi-level authentication-based secure data transaction has been proposed. To avoid the unauthorized user login, multi-level authentication has been developed. To enhance the security of the data, CSTA algorithm has been developed. This method was fully secured; therefore, the third party cannot able to access the data. To prove the performance of the proposed methodology, the proposed algorithm has been compared with different algorithms. The security level of authentication process is also analyzed. The result has clearly shown the effectiveness of the proposed methodology.

References

1. Yan Z, Ding W, Yu X, Zhu H, Deng RH (2016) Deduplication on encrypted big data in cloud. IEEE Trans Big Data 2(2):138–150
2. Li W, Yang Y, Yuan D (2016) Ensuring cloud data reliability with minimum replication by proactive replica checking. IEEE Trans Comput 65(5):1494–1506
3. Jakóbik A, Palmieri F, Kołodziej J (2018) Stackelberg games for modeling defense scenarios against cloud security threats. J Netw Comput Appl 110:99–107
4. Kandil A, El-Deeb H (2016) Exploration of application migration to cloud environment. In 2016 6th international conference-cloud system and big data engineering (Confluence) IEEE, pp 109–114
5. Basu S, Pattnaik PK (2017) A consistency preservation based approach for data-intensive cloud computing environment. In 2017 8th International Conference on Computing, Communication and Networking Technologies (ICCCNT), IEEE, pp 1–5
6. Chang V, Ramachandran M (2016) Towards achieving data security with the cloud computing adoption framework. IEEE Trans Serv Comput 9(1):138–151

7. Luna J, Taha A, Trapero R, Suri N (2017) Quantitative reasoning about cloud security using service level agreements. IEEE Trans Cloud Comput 5(3):457–471
8. Gupta S, Misra SC (2016) Moderating effect of compliance, network, and security on the critical success factors in the implementation of cloud ERP. IEEE Trans Cloud Comput 4 (4):440–451
9. Wei J, Liu W, Hu X (2016) Secure data sharing in cloud computing using revocable-storage identity-based encryption. IEEE Trans Cloud Comput 1–1
10. Awad A, Matthews A, Qiao Y, Lee B (2018) Chaotic searchable encryption for mobile cloud storage. IEEE Trans Cloud Comput 6(2):440–452 (2015), (Apr 2018)
11. Hu C, Li W, Cheng X, Yu J, Wang S, Bie R (2018) A secure and verifiable access control scheme for big data storage in clouds. IEEE Trans Big Data 4(3):341–355

A Novel Graphical User Interface-Based Toolbox for Optimization and Design of Linear Antenna Array

Guru Prasad Mishra, Shibanee Dash, and Saumendra Kumar Mohanty(✉)

Department of ECE, FET, ITER, Siksha 'O' Anusandhan (Deemed to be University), Bhubaneswar, India
saumendramohanty@soa.ac.in

1 Introduction

The performance of single-element antenna is limited. They cannot meet the gain and radiation pattern requirements for any specific application. In many applications, it is also required to design antennas that are suitable for long-distance broadband communication [1]. This demanded antenna should be highly directive and also wideband. A possible solution to obtain a high-gain, high bandwidth antenna is by using the array antenna, which is essentially an assembly of several radiating antennas to form a single antenna system in an electrical and geometrical configuration [2]. The performance of the array increases with the number of elements in the array, but the cost and complexity increase with higher number of elements. Nowadays, antenna arrays are more prominent in wireless terminals, smart antenna, and they are widely used in applications like radar, sonar, and communications [3–5]. Before design, a perfect analysis is required. In this regard, computational electromagnetics is essential for numerical analysis and subsequent computer simulation and physical modeling.

Bossavit [6] proposed computational electromagnetics to model the interaction of electromagnetic fields with the physical structure and environment. In [7], the relationship between field and circuit theory is examined. An extensive study has been carried out on different analytic methods, which are used to analyze antenna structures using computational electromagnetics. Such methods are the method of moments (MoM) and finite-difference time-domain (FDTD). Abd-Alhameed et al. [8] investigated and discussed the design of microstrip patch antennas using MoM. In [9, 10], the multiple sweep method of moments (MSMM) are applied to the three-dimensional (3-D) problems of a long dipole antenna with a reactive load in the center of each arm and a TEM horn antenna with a bend.

Stochastic and metaheuristic optimization methods are now very common and have many advantages over deterministic methods. These methods include genetic algorithm (GA) [11–15], simulated annealing, differential evolution (DE), Tabu search (TS), PSO [16–18], ant colony optimization (ACO), CS optimization, and many others. Further, as global optimizations do not depend on initial conditions, and they handle discontinuous and non-differentiable functions, so they are preferred over local optimization

M. N. Mohanty and S. Das (eds.), *Advances in Intelligent Computing and Communication*, Lecture Notes in Networks and Systems 109,
https://doi.org/10.1007/978-981-15-2774-6_47

techniques. Hence, we have used particle swarm optimization (PSO) and cuckoo search (CS) in optimization of LAA. In [19], the study presents a multi-level optimization (MLO) method applied in particle swarm optimization (PSO) to solve a 3-objective optimization problem. PSO algorithm is introduced in [20] to optimize the element spacing and lengths of a mutually coupled planar array antenna. Baskar has described a method of using particle swarm optimization (PSO) algorithms to optimize the element spacing and lengths of Yagi–Uda antennas in [21]. The radiation pattern and rSLL of a 42 elements linear array, with 20 optimization variables, have been compared using CS and differential evolution [22]. So a comparative study is made between PSO and CS in PAA design using MOM technique.

It is very useful as the actions in a GUI are usually performed through direct manipulation of the graphical elements. Hence, we have gone through the literature or process that involves designing the toolbox. In [23], it gives detail on a quick start with guide, design process, and principles, GUI layout with guide, and some GUI examples.

2 Design Representation of LAA

As MoM considers mutual coupling among the elements of the antenna array, so it is the most prominent method for our purpose. Here, the frequency of operation is in the VHF band, i.e., 30–300 MHz. The antenna array consists of several linear dipole elements and all are energized directly by a discrete feed as shown in Fig. 1.

This array is exclusively designed to operate as an end-fire array. The lengths and diameters of elements, as well as their respective spacing, determine the optimum characteristics. These types of antenna arrays are quite practicable to use because they are lightweight, simple to build, low cost and provide moderately desirable characteristics for many applications.

For this method, Pocklington's integral equation was used to analyze the array structure. So in this chapter, the sections describe the derivation of E and H fields from integral equation. Also, the detailed steps considering in MATLAB to design LAA using MoM will discuss in the next section. As the results from MATLAB need to be validated in a globally accepted simulation software, here LAA is being designed in CST microwave studio. The steps included in design at CST platform will also discuss here in detail. Finally, in the last section, the outcomes from both platforms are matched and analyzed.

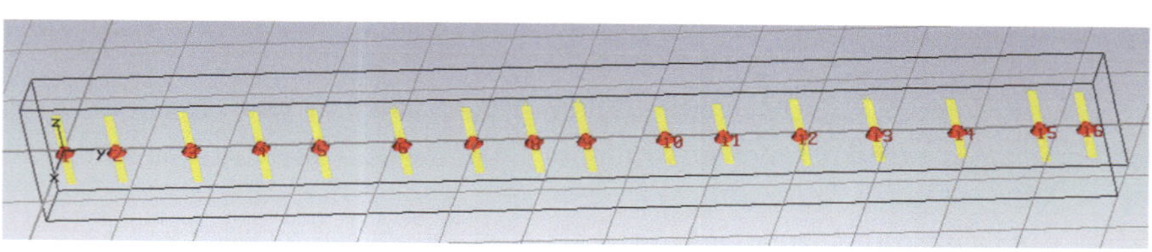

Fig. 1. 16-element dipole LAA

2.1 Integral Equation Moment Method

Here, the design is based on the integral equation for the electric field radiated by the elements in the array, and it will be used to describe the complex current distributions on all the elements and the corresponding radiation patterns. This method uses Pocklington's integral equation to analyze any structure by point matching numerical methods. The number of discrete points must be sufficient in number to allow computed data to match well with experimental results. This equation states that the total field radiating in an unbounded space is generated by an electric current source. Mutual coupling is also considered here, and there is no restriction on the number of elements.

Pocklington's integral equation,

$$\int_{-\frac{l}{2}}^{+\frac{l}{2}} I(z')\left(\frac{\partial^2}{\partial z^2}+k^2\right)\frac{\mathrm{e}^{-jkR}}{R}\mathrm{d}z' = j4\pi\omega\varepsilon_0 E_z^t \tag{1}$$

For small diameter wires, the current on each element can be approximate: where

$$R = \sqrt{(x-x')^2+(y-y')^2+(z-z')^2}$$

$$I_n(z') = \sum_{m=1}^{M} I_{nm}\cos\left[(2m-1)\frac{\pi z'}{l_n}\right] \tag{2}$$

where

I_{nm} complex current coefficient of mode m on element n.

l_n corresponding length of the element.

Once the current distribution is found, the total field of the exciting array is obtained by summing the contribution of each element. The far-zone electric field generated by the m modes of the nth element-oriented parallel to z-axis is given by:

$$E_{\theta_n} \approx -j\omega A_{\theta_n} \tag{3}$$

where

$$A_{\theta_n} \approx -\frac{\mu e^{-jkR}}{4\pi r}\sin\theta\left[\mathrm{e}^{jk(x_n\sin\theta\cos\phi+y_n\sin\theta\sin\phi)} + \int_{-\frac{l_n}{2}}^{+\frac{l_n}{2}} I_n \mathrm{e}^{jkz'_n\cos\theta}\mathrm{d}z'_n\right] \tag{4}$$

So the total field is,

$$E_{\theta} = \sum_{n=1}^{N} E_{\theta_n} \tag{5}$$

These above equations are used in MATLAB to get the total field and their respective parameters of an antenna. MATLAB implementation of the design of LAA using MoM involves the following process. The number of elements, length, and their respective separations is the input parameters to the structure code. Here the frequency of operation is in VHF band (i.e., 30–300 MHz). So here λ is taken as 1 m, and all the dimensions are taken in terms of λ. The radius of dipoles is taken as 0.001 λ, as here they are considered as small wires. After assigning all the above parameters as input, the structure code starts from here. First, the position of the elements in an array is calculated in y-coordinate parallel to z-axis. After placing them, a supply of 1 V is given to the center of each dipole. Then current on each element is found out after placing Eq. (2) in MATLAB code. To operate functions like finite integration and summation, the functions are declared separately to reduce the complexity of the structure code and there call back is used when necessary. Then, electric field is calculated by using Eq. (5), and their respective parameters, like 3 dB beamwidth, front-to-back ratio, directivity (D) and SLL, are found out. Similar steps are also carried out for magnetic field and their respective parameter calculation. Finally, far-field plots are visualized in both polar and Cartesian plots.

3 GUI-Based Toolbox Implementation in MATLAB

In this paper, a GUI has been created where one can design the LAA with and without optimization. This GUI contains six push buttons that perform differently. By clicking the first push-button, one can design an LAA of his own specification of any number of elements, lengths, and their respective spacing. Push-button two and three perform the design of optimized LAA with desired directivity and FSLL for single-objective and multi-objective fitness functions, respectively, using PSO, whereas the fourth and fifth button does the same process as second and third, using CS optimization technique. Finally, a quit button is added to close the GUI.

The input parameters to the different push-button for the above GUI are:

I. The number of elements, lengths of antennas, and their respective spacing.
II. Number of elements, desired directivity, and number of iteration for the PSO optimization process.
III. The number of elements, desired directivity, desired FSLL, and number of iteration for PSO optimization process.
IV. Number of elements, desired directivity, and number of iteration for the CS optimization process.
V. The number of elements, desired directivity, desired FSLL, and number of iteration for the CS optimization process (Fig. 2).

Fig. 2. GUI for optimized LAA design

3.1 Steps for LAA Design in CST Microwave Studio

Before going to start design in CST, one must have to set the units. Units of dimensions, frequency, temperature, and time are the point of concern here. Here, the dimensions, frequency, and temperature are taken in mm, MHz, and ns, respectively. Materials used to design the antenna must be declared then. Here in this work, copper and vacuum materials are used to design the antenna array. Then the structure of the LAA has to be created according to the array. Here 32 numbers of cylinders and 16 numbers of gaps are being designed for a 16-element LAA. According to this design, the frequency range is set in between 0 and 500 MHz as the operating frequency is 300 MHz. The operating frequency must be greater than half of the frequency range. After declaring all the parameters and structure, excitations are set to the gap between the dipoles. Here to match the result with MATLAB, voltage source of 1 V is given to all the elements. Field monitors are an important part of the design process. The user can set E-field, H-field, or far-field with the operating frequency, whichever is needed to analyze the structure. Here only far-field at 300 MHz is set for this purpose. Before going to start the solver, all excitation ports are selected, and simultaneous excitation is set, as a linear array is being designed in CST. There are many solver options like transient solver, frequency solver, eigenmode solver, and TLM solver. Here transient solver is started to solve the structure, and far-field results are obtained. Then the results are analyzed and matched with the MATLAB results by choosing polar, Cartesian plots and 2-D or 3-D plots.

4 Optimization Procedure

4.1 Implementation of PSO

Here, a 16-element LAA is optimized using PSO in MATLAB. Since there are only 16 lengths and 15 spacing along y-axis, total columns of the matrix are taken as 31 and 25 rows for 25 particles in optimization. Here, λ is 1 m and the spacing is fractional multiple of the values of the matrix parameter are taken from 0.6 to 1 for better results. Cases are considered here. Case-I is for a single-objective fitness function, which enhances the directivity only, and case-II considers both directivity and sidelobe level into account. For the time, we have taken only ten iterations (in total 250 times) in the optimization process.

4.2 Implementation of CS

Here we have taken 25 numbers of nests. Then the discovery rate of alien eggs is taken as 0.25. We have restricted the l and d between 0.6l and l. With these limits, 25×21 random matrix is generated, similar to that of PSO.

4.3 Parameters of PSO and CS

To compare both optimizations, we have to keep some variables the same for both the techniques and those are:

- D_{des} = 21 dBi
- $FSLL_{des}$ = −17 dB
- No. of particles or nests = 25
- No. of iterations = 10
- Tolerance limit = 2.

Multi-parameter single-objective fitness function (**case-I**):

First, we have designed a fitness function, whose only aim is to optimize D without considering the other output parameters of the antenna and is given as,

$F = |D_{des} - D(y)|$, where 'D_{des}' stands for the desired directivity and $y = \{l, d\}$.

Multi-parameter multi-objective fitness function (**case-II**):

Here, both D and FSLL are being optimized with antenna lengths and their respective spacing as input parameters to the process. The fitness function is given as,

$F = c_1|D_{des} - D(y)| + c_2|FSLL(y) - FSLL_{des}|$, *where the weighting parameters* c_1 *and* c_2 are taken both as 0.5. 'D_{des},' '$FSLL_{des}$,' stands for the desired directivity and desired FSLL and $y = \{l, d\}$.

5 Result and Discussion

Two different cases, as we discussed earlier, are considered during optimization. case-I directivity optimization of the linear array without taking FSLL into account and case-II a trade-off between D and FSLL. The single-objective (for case-I) results and the multi-objective (for case-II) results are shown in the following table. The lengths and spacing are given in units of λ as we are considering our antenna directivity concerning an isotropic antenna, so the unit is taken in dBi. As the maximum radiation of any field pattern is kept to zero dB, here the values of SLL are expressed in negative. For case-I, the D is 20.673 dBi for PSO and 20.048 dBi for CS. The corresponding FSLL is (−5.27 dB) and (−5.05 dB) for PSO and CS, respectively. The time taken for convergence is found to be 4663.417872 s for PSO and 333.162295 s for CS. For case-II, the D is 19.332 dBi for PSO and 18.138 dBi for CS. The corresponding FSLL is (−17.65 dB) and (−18.71 dB) for PSO and CS, respectively. The time taken for convergence is found to be 3296.936648 s for PSO and 322.260406 s for CS. So if a comparison is made between PSO and CS in terms of convergence time, CS converges approximately 10 times faster than the PSO for both cases having the same number of iterations and same fitness function (Table 1).

6 Simulation Results Obtained from MATLAB and CST Microwave Studio

All the simulations are carried out using Intel(R) Pentium(R) CPU B960@ 2.20 GHz processor with 4 GB RAM. The optimized performance parameters are estimated, and the radiation patterns are verified in CST microwave studio. It is observed from the figures and the table that radiation pattern obtained in MATLAB best approaches with the pattern obtained from CST microwave studio.

The simulation results of the PSO optimized linear array are as shown below (Fig. 3) (Table 2).

The simulation results of CS optimized linear array are as shown below (Fig. 4).

Table 1. Optimized parameters of 16-element linear array (case-I and case-II)

Case-I					Case-II				
PSO			CS		PSO			CS	
Element	Length (in λ)	Spacing (in λ)	Length (in λ)	Spacing (in λ)	Element	Length (in λ)	Spacing (in λ)	Length (in λ)	Spacing (in λ)
1	0.9314	–	0.9782	–	1	0.9961	–	0.9828	–
2	0.9311	0.9021	0.9242	0.7841	2	0.9958	0.8779	0.9881	0.9737
3	0.9404	0.8282	0.9584	0.7593	3	0.9614	0.8923	0.9070	0.7285
4	0.9752	0.9448	0.9202	0.8225	4	0.9461	0.8664	0.9478	0.9553
5	0.9766	0.9178	0.9127	0.8088	5	0.9805	0.8931	0.9739	0.8567
6	0.9506	0.9164	0.9686	0.7486	6	0.9354	0.8818	0.9734	0.7049
7	0.9679	0.8637	0.9911	0.9681	7	0.9691	0.7280	0.9816	0.7388
8	0.9522	0.9104	0.9751	0.8321	8	0.9040	0.7895	0.9533	0.8780
9	0.9560	0.8371	0.9823	0.9011	9	0.9788	0.8522	0.9561	0.9165
10	0.9698	0.9096	0.9225	0.7631	10	0.9186	0.9134	0.9347	0.7111
11	0.9558	0.8115	0.9387	0.8413	11	0.9308	0.9981	0.9551	0.7529
12	0.9793	0.8747	0.9419	0.7150	12	0.9522	0.7504	0.9021	0.7697
13	0.9582	0.8559	0.9936	0.8488	13	0.9125	0.9069	0.9737	0.9145
14	0.9682	0.9229	0.9106	0.7809	14	0.9047	0.9554	0.9834	0.9312
15	0.9894	0.9248	0.9531	0.9053	15	0.9617	0.7209	0.9836	0.8768
16	0.9881	0.8404	0.9621	0.9244	16	0.9494	0.7708	0.9033	0.9619
D (**dBi**)	**20.673**		**20.048**		*D* (**dBi**)	**19.332**		**18.138**	
FSLL (**dB**)	**−5.27**		**−5.05**		*FSLL* (**dB**)	**−17.65**		**−18.71**	
Time (**s**)	**4603.41**		**333.16**		*Time* (**s**)	**3296.93**		**322.26**	

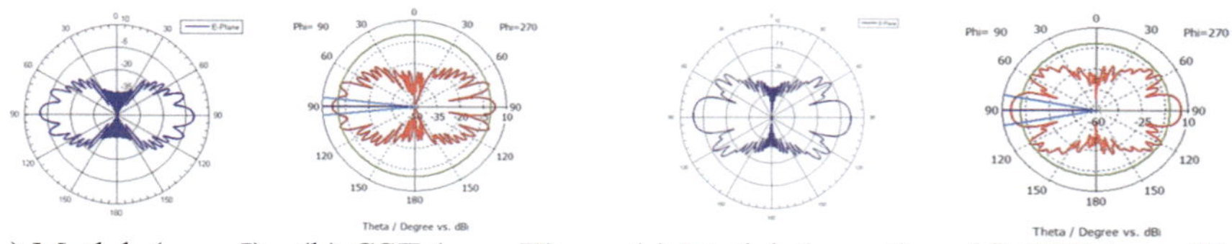

(a) Matlab (case-I) (b) CST (case-II) (c) Matlab (case-I) (d) CST (case-II)

Fig. 3. E-theta versus θ (at Φ = 90°) for free-space environment using MATLAB and CST microwave studio for case-I and case-II, respectively

Table 2. Comparison of results from MATLAB and CST microwave studio

	Case-I				Case-II			
	PSO		CS		PSO		CS	
	D (dBi)	FSLL (dB)	D (dBi)	FSLL (dB)	D (dBi)	FSLL (dB)	D (dBi)	FSLL (dB)
MATLAB	20.67	−5.27	20.04	−5.05	19.33	−17.65	18.13	−18.71
CST	19.93	−4.92	19.15	−4.87	19.23	−16.52	18.62	−16.91

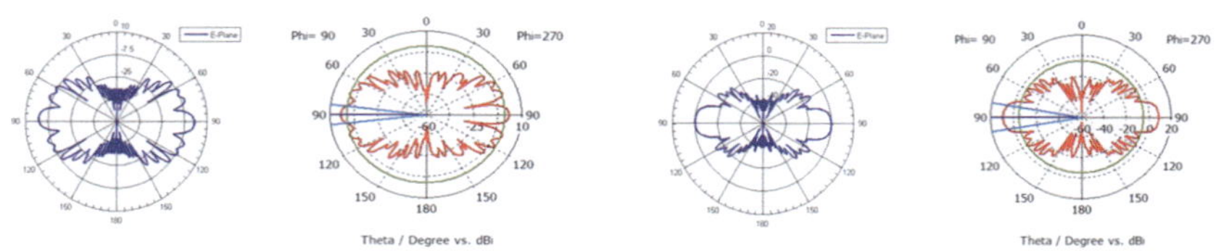

(a) Matlab (case-I) (b) CST (case-II) (c) Matlab (case-I) (d) CST (case-II)

Fig. 4. E-theta versus θ (at Φ = 90°) for free-space environment using MATLAB and CST microwave studio for case-I and case-II, respectively

7 Conclusion

For this work, the structure code is developed in MATLAB for LAA, and the structures are optimized by PSO and CS optimization techniques. The performances of optimized and un-optimized results of LAA are compared. The optimized proved to have better performance than un-optimized in both design and performance parameters. The comparative study shows that CS is better than PSO in terms of convergence time for all cases. Then by taking these optimized parameters such as lengths and their respective spacing of elements, the LAA is designed in CST simulation software. Both outcomes are seemed to be fairly agreeing with each other for non-uniform spacing, non-uniform LAA. The novelty of the above discussion is that a GUI-based toolbox is developed here to design and to optimize LAA in MATLAB, which reduces the complexity for an amateur. The GUI having both options to design LAA with or without process optimization made it completely from all aspects.

References

1. Wang X, Gao F, Liu Q (2004) Design of antenna array used as smart antenna for TD-SCDMA systems. Int Conf Comm Circuits Syst 1:176–180
2. Balanis CA (2005) Antenna theory: analysis and design, 3rd edn. Wiley, London
3. Kummer WH (1992) Basic array theory. Proc IEEE 9(3):127–139
4. Hejres JA (2004) Adaptive sidelobe using the positions of selected elements of the phased antenna array. IEEE Antennas Propag Soc Symp 3:2655–2658
5. Lee A, Chen L, Wei J, Hwang HK (2005) Simulation study of wideband interference rejection using adaptive array antenna. In: IEEE aerospace conference, pp 1–6
6. Bossavit A (1998) Computational electromagnetism. Eng Sci Educ J 7(6):275–281
7. Kettunen L (2001) Fields and circuits in computational electromagnetism. IEEE Trans Electromagn 37(5):3393–3396
8. Abd-Alhameed RA, McEwan NJ, Excell PS, Ibrahim MM, Hejazi ZM, Musa M (1997) New procedure for design of microstrip patch antennas using method of moments. In: Tenth International Conference on Antennas and Propagation, vol 1, no. 436, pp 178–181
9. Colak D, Newman EH (1998) The multiple sweep method of moments (MSMM) design of wide-band antennas. IEEE Trans Antenna Propag 46(9):1365–1371
10. Mishra GP, Jena MR, Mangaraj BB (2016) Investigation on design and performance of linear cantor array using strip dipole and V-dipole for UHWF band application. In: IEEE International Conference on Wireless Communication, Signal Processing and Networking, pp 1810–1814
11. Wang W, Gong S, Wang X, Guan Y, Jiang W (2010) Differential evolution algorithm and method of moments for the design of Low-RCS antenna. IEEE Antennas Wirel Propag Lett 9:295–298
12. Sivanandam SN, Deepa SN (2008) Introduction to genetic algorithms. Springer, Berlin, Heidelberg
13. Johnson J, Rahmat-Samii Y (1997) Genetic algorithms in engineering electromagnetics. IEEE Trans Antennas Propag Magazine 39(2):7–21
14. Yan KK, Lu Y (1997) Sidelobe reduction in array pattern synthesis using genetic algorithm. IEEE Trans Antennas Propag 45:1117–1122
15. Robinson J, Rahmat-Samii Y (2004) Particle swarm optimization in electromagnetics. IEEE Trans Antennas Propag 52:397–407
16. Eberhart RC, Shi Y (2001) Particle swarm optimization: developments, application, and resources. In: Proceedings of IEEE conference on evolutionary computation, pp 81–86
17. Khodier M, Al-Aqeel M (2009) Linear and circular array optimization: a study using particle swarm intelligence. Prog Electromagn Res 15:347–373
18. Nguyen TH, Morishita H, Koyanagi Y, Izui K, Nishiwaki A (2013) A multi-level optimization method using PSO for the optimal design of an L-Shaped folded monopole antenna array. IEEE Trans Ant Propag 62(1):209–215
19. Mohanty SK, Sahoo AB, Pradhan H, Mangaraj BB (2017) Comparative study of PSO and CS for optimization of 3 x 5 planar antenna array using MOM. Int J Inf Commun Technol 11(1):128–149

20. Baskar S, Alphones A, Suganthan PN, Liang JJ (2005) Design of Yagi-Uda antennas using comprehensive learning particle swarm optimization. IEEE Proc Microwaves Antennas Propag 152(5):340–346
21. Mohanty SK, Mishra GP, Mangaraj BB (2014) Implementing Taguchi and Cuckoo Search to Optimize LAA. In: Annual IEEE India Conference (INDICON), 11–13 December 2014, pp 1–5
22. Building GUIs with MATLAB (1997) The math works, 5th edn.
23. Moshe Y (2004) GUI with Matlab. In: Signal and Image Processing Laboratory (SIPL)

Wideband Patch Antenna Element for Broadcast Applications in Lower 5G Bands

Sonagara Abhishek M(✉) and Rahul Vishwakarma

Department of Electronics and Communication Engineering, ITER, Siksha 'O' Anusandhan (Deemed to be University), Bhubaneswar, Odisha 751030, India

1 Introduction

The telecom industries have currently been preparing themselves to embrace new 5G cellular technologies. The commercial products supporting 5G are supposed to be available by 2020 [1]. Increasing demands of 5G mobile technology has given a boost to ongoing research in 5G supporting antennas. In accordance with demand, the frequency bands between 3 GHz and 5 GHz have been allocated for 5G services in many countries such as 3.1–3.55 GHz and 3.7–4.2 GHz in USA, 3.4–3.8 GHz in Europe, and 3.3–3.6 GHz and 4.8–4.99 GHz in China [2]. To cover all 5G bands, base station antenna having wide bandwidth with overall high efficiency and high gain is required. For that, many new antenna designs are proposed in 5G bands [3]. In the evolution of modern 5G services, compared to other types of antennas, patch antenna can play vital role due to its characteristics like small size, low profile, easy integration, and low cost [4]. But at the other side, patch antenna also has limitations in the form of low bandwidth and lower efficiency, especially in wideband [5]. So microstrip patch antenna can be a good candidate for 5G base station antenna design due to its mentioned advantages, if such design can overcome its limitations [6, 7]. Wideband operation of simple microstrip patch antenna can be achieved by applying partial ground and stacking technique [8, 9]. Considering demand on 5G, the proposed antenna design provides good results for one single antenna element which can be arrayed to produce good results for base station application [10].

2 Antenna Design

The proposed antenna consists of two grounds one of which is partial and another one is having length and width same as of substrate as shown in Figs. 1 and 2.

Patch antenna with partial ground plane is suspended over full ground plane. The substrate used is FR-4 (lossy) dielectric having thickness of 1.6 mm with loss tangent 0.025 and dielectric constant 4.3. The procedure to find length and width for

M. N. Mohanty and S. Das (eds.), *Advances in Intelligent Computing and Communication*, Lecture Notes in Networks and Systems 109,
https://doi.org/10.1007/978-981-15-2774-6_48

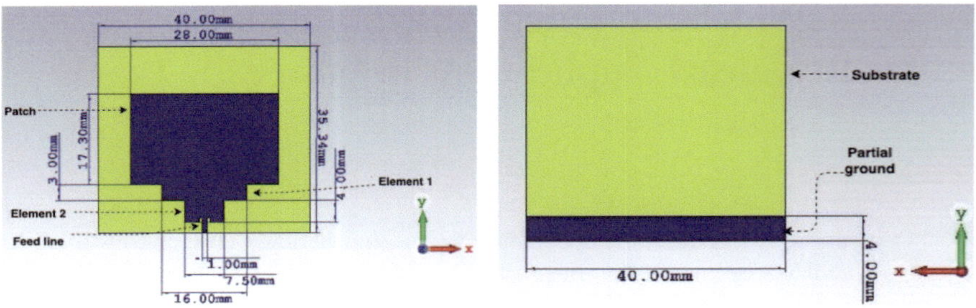

(a) Top view (b) Back view

Fig. 1. Patch antenna with partial ground

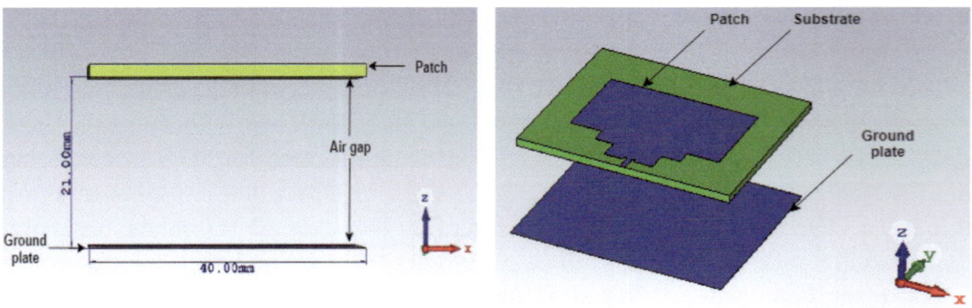

(a) Side view (b) Perspective view

Fig. 2. Patch antenna with partial ground suspended over full ground plate

rectangular patch is carried out using the below-mentioned equations. The specific parameters required are dielectric constant ($\in_r$), height of substrate (h), and frequency of resonance (f_r) [5].

$$W = \frac{v_0}{2f_r}\sqrt{\frac{2}{\in_r + 1}} \tag{1}$$

$$\in_{\text{reff}} = \frac{\in_r + 1}{2} + \frac{\in_r - 1}{2}\frac{1}{\sqrt{1 + 12\frac{h}{W}}} \tag{2}$$

$$\Delta L = 0.412 * h\frac{(\in_{\text{reff}} + 0.3)}{(\in_{\text{reff}} - 0.258)}\frac{\left(\frac{W}{h} + 0.264\right)}{\left(\frac{W}{h} + 0.8\right)} \tag{3}$$

$$L = \frac{v_0}{2f_r\sqrt{\in_{\text{reff}}}} - 2\Delta L \tag{4}$$

The metal patch over the substrate is extended at one end by loading rectangular metal strips with successively reduced widths and lengths which is shown in Fig. 1a. Partial ground can also be seen in Fig. 1b. Metal thickness used here is 0.020 mm.

Inset feeding technique is used to achieve better impedance matching. The air gap between patch antenna with partial ground and full ground plate is also shown in Fig. 2a. The perspective view is also shown in Fig. 2b for better understanding.

The given lengths and widths of the metal strips and partial ground are obtained to achieve higher required bandwidth of 2 GHz. For the simulation of the proposed design CST studio software is used.

3 Results and Discussion

Results are obtained for two designs, (1) patch antenna with partial ground and (2) patch antenna with partial ground suspended over full ground plate. S11 parameter for patch antenna with partial ground remains less than −10 dB for 3.26–4.23 GHz with bandwidth of 1.16 GHz (31% of center frequency 3.745 GHz) and minimum return loss reaching at −14.10 dB at 3.68 GHz. These wideband characteristics are due to partial ground and extended patch elements of different widths and lengths. S11 parameter and VSWR results for both designs are given in Figs. 3 and 4, respectively.

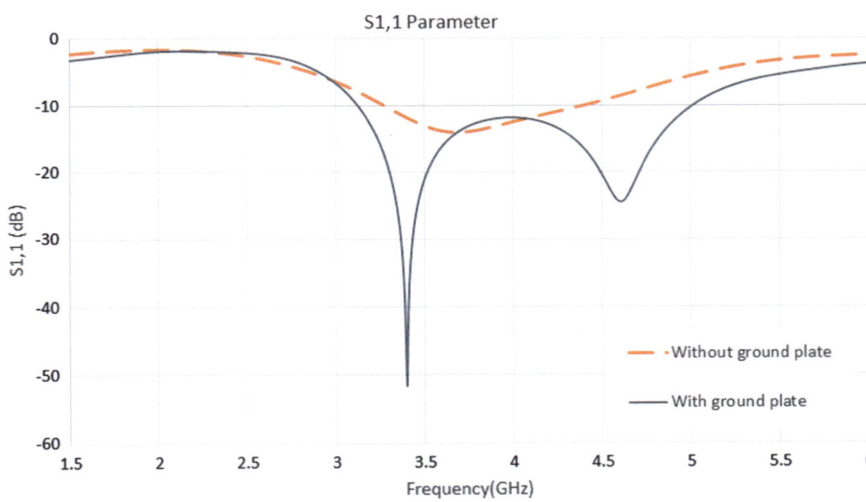

Fig. 3. S11 parameter comparison

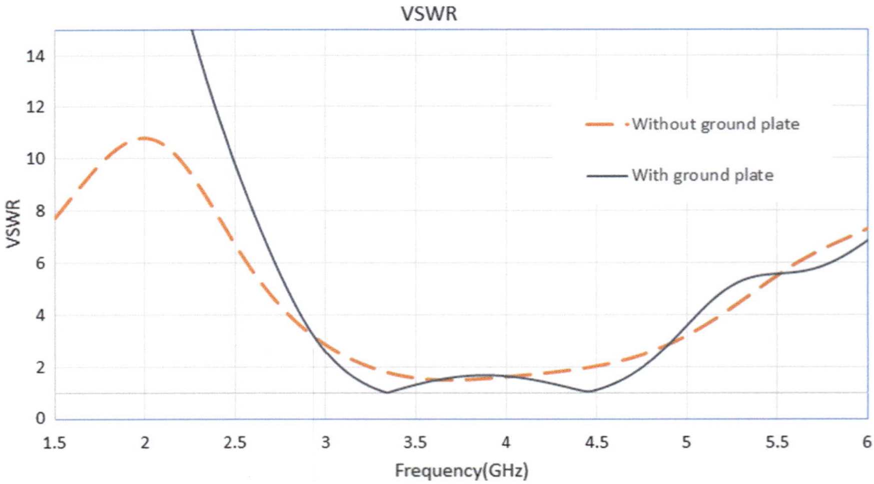

Fig. 4. VSWR comparison

S11 parameter for patch antenna with full ground plate shows wideband characteristics for frequency band from 3.1–5 GHz having 10 dB impedance bandwidth of 1.9 GHz (47% of center frequency 4.05 GHz), which covers entire lower 5G bands with minimum return loss of −51.5 dB at 3.4 GHz and −24.5 dB at 4.6 GHz. VSWR value for the proposed antenna is also below 2 for most of the covered bands. As shown in figures, both return loss and VSWR is improved in patch antenna with partial ground suspended over full ground plate.

From realized gain plot in Fig. 5, role of ground plate below patch antenna can be understood. For most of the interested frequency bands, realized gain is more than 5 dBi with reaching maximum gain of 6.77 dBi at 3.4 GHz frequency. But the design of antenna without ground plate has maximum gain of only 2.3 dBi, and it remains less than the antenna with full ground plate for whole bandwidth.

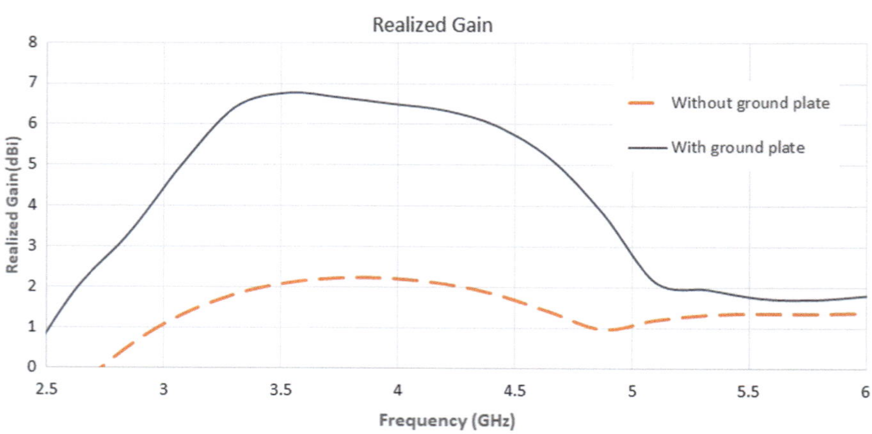

Fig. 5. Realized gain comparison

Figure 6 shows that total efficiency is also improved in the antenna with full ground plate compared to antenna without ground plate. Impedance smith chart for both the designs has been plotted in Fig. 7a, b, and we can observe that smith chart is also in good agreement with the return loss graph showing that more impedance matching occurs when impedance curve moves nearer to the normalized 1 value of impedance.

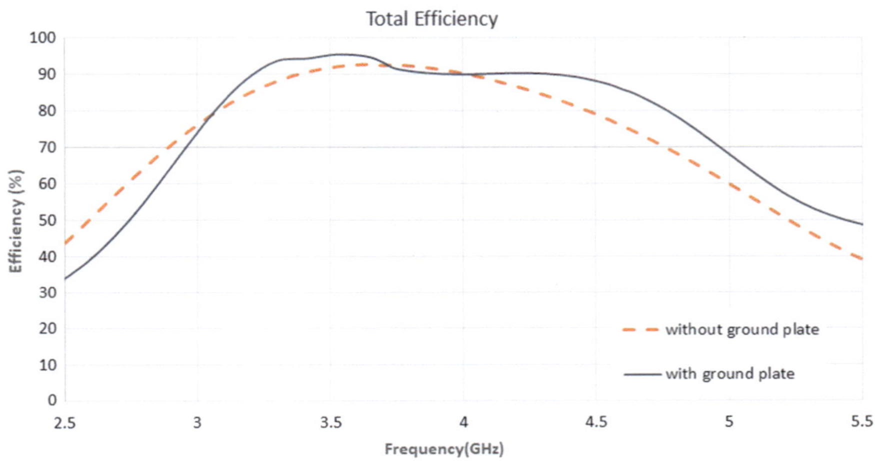

Fig. 6. Total efficiency comparison

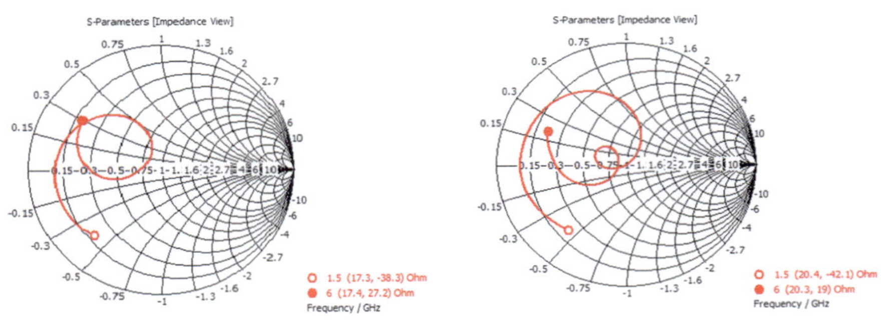

(a) Without ground plate

(b) With ground plate

Fig. 7. Impedance Smith chart

In Figs. 8 and 9, E plane and H plane directivity polar plots indicate that after applying ground plate, back radiation is eliminated and giving more radiation in desired direction, which is desired characteristic for base station antenna (Table 1).

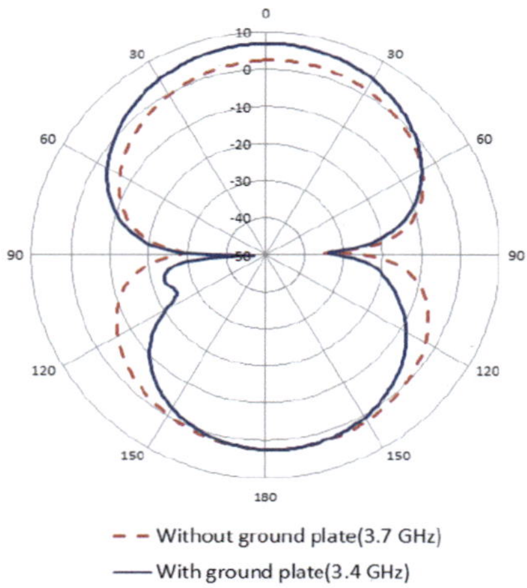

(a) E plane

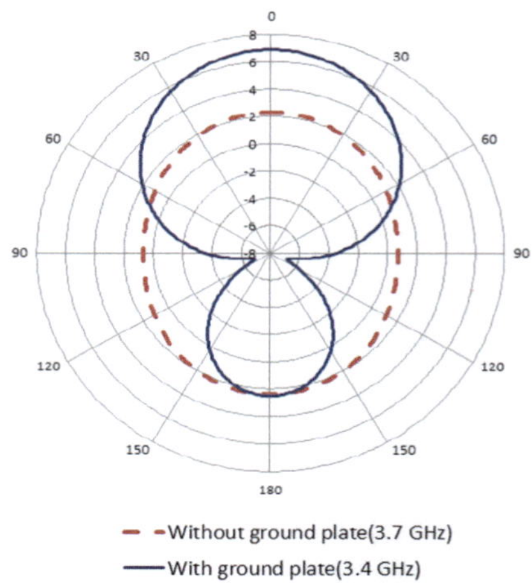

(b) H plane

Fig. 8. Directivity at 3.4 GHz and 3.7 GHz

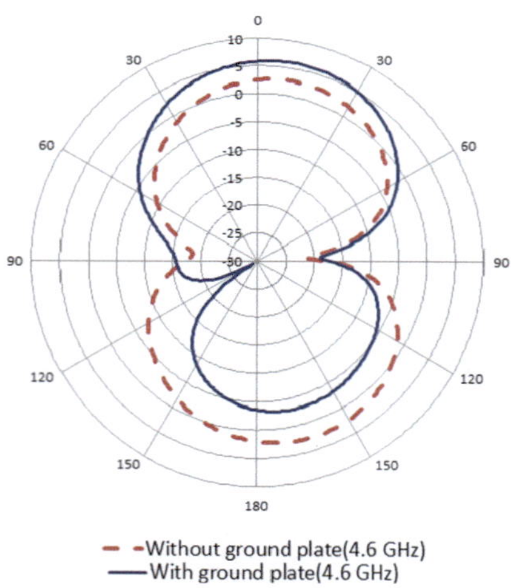

(a) E plane

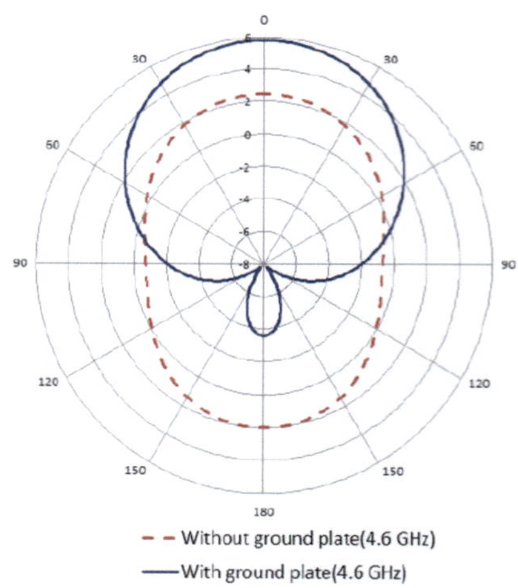

(b) H plane

Fig. 9. Directivity at 4.6 GHz

Table 1. Summary of results

Parameter	Proposed designs	
	Without ground plate	With ground plate
S11	−14.1 dB @ 3.7 GHz	−51.5 dB @ 3.4 GHz −24.5 dB @ 4.6 GHz
VSWR	1.5 @ 3.7 GHz	1.005 @ 3.4 GHz 1.13 @ 4.6 GHz
10 dB Impedance bandwidth	1.145 GHz (31% of center freq. 3.745 GHz)	1.9 GHz (47% of center freq. 4.05 GHz)
Total efficiency	92.5% @ 3.7 GHz	95.35% @ 3.4 GHz 86% @ 4.6 GHz
Realized gain	2.23 dBi @ 3.7 GHz	6.77 dBi @ 3.4 GHz 5.36 dBi @ 4.6 GHz

4 Conclusion

In this simulation work, microstrip patch antenna with partial ground backed by full ground plate is proposed for entire lower 5G bands for broadcast purpose. The simulated S11 parameter of the proposed structure shows that return loss remains under −10 dB for all the required 5G frequency bands between 3 and 5 GHz. The proposed design has overall good efficiency which is more than 80% for almost all required frequencies. Good performance for gain and radiation patterns is obtained. Therefore, the proposed antenna can be a good candidate for antenna element of antenna array for broadcasting purpose in 5G communication technologies.

References

1. Shafi M, Molisch AF, Smith PJ, Haustein T, Zhu P, De Silva P, Tufvesson F, Benjebbour A, Wunder G (2017) 5G: a tutorial overview of standards, trials, challenges, deployment, and practice. IEEE J Sel Areas Commun 35(6):1201–1221
2. Ancans G, Bobrovs V, Ancans A, Kalibatiene D (2017) Spectrum considerations for 5G mobile communication systems. Procedia Comput Sci 104:509–516
3. Ali MMM, Sebak AR (2016) Dual band (28/38 GHz) CPW slot directive antenna for future 5G cellular applications. In: 2016 IEEE International Symposium on Antennas and Propagation (APSURSI), pp 399–400
4. Chandana RS, Sai Deepthi P, Sriram Teja D, Veera JayaKrishna N, Sujatha M (2018) Design of a single band microstrip patch antenna for 5G applications. Int J Eng Technol 7:532
5. Balanis C (2005) Antenna theory analysis and design, 3rd edn. Wiley, Publication
6. Yassine J, Gharnati F, Said AO (2017) Design of a compact dual bands patch antenna for 5G applications. In: 2017 International Conference on Wireless Technologies, Embedded and Intelligent Systems (WITS), pp 1–4
7. Tarpara N, Rathwa R, Kotak N (2018) Design of slotted microstrip patch antenna for 5G application. Int Res J Eng Technol (IRJET) 05(04)

8. Marzudi WNNW, Abidin ZZ, Dahlan SH, Ramli KN, Majid HA, Kamarudin MR (2015) Rectangular patch with partial ground wearable antenna for 2.4 GHz applications. In: 2015 IEEE international RF and microwave conference (RFM), pp 104–9
9. Ooi B-L, Qin S, Leong M-S (2002) Novel design of broadband stacked patch antenna. IEEE Trans Antennas Propag 50(10):1391–1395
10. Khattak MI, Sohail A, Khan U, Barki Z, Witjaksono G (2019) Elliptical slot circular patch antenna array with dual band behaviour for future 5G mobile communication networks. Prog Electromagnet Res C 89:133–147

Anonymized Credit Card Transaction Using Machine Learning Techniques

B. K. Padhi[1(✉)], S. Chakravarty[1], and B. N. Biswal[2]

[1] Department of Computer Science & Engineering, Centurion University of Technology & Management, Bhubaneswar, Odisha, India

[2] Department of Computer Science & Engineering, Bhubaneswar Engineering Colleges, Bhubaneswar, Odisha, India

1 Introduction

Today, due to the rapid growth of digitalization that has increased online transactions in the mobile phone drastically created mobile transaction fraud. Mobile phones have become the best devices for the attacker to gain lots of money within less time with less effort. Due to the fraudulent transaction, the business and bank are facing massive financial loss. For years, it has been considered that the information or data collected from credit card transactions are anonymized or genuine.

1.1 Fraud

Fraud can be known as acquiring someone's property illegally. Credit card fraud (CAF) means the use of a credit card without authorization and even stealing the sensitive data by the fraudsters such as card number, security pin, and one-time password [1]. CAF can be classified into two categories, such as internal fraud and external fraud [2], while other classification categories, such as, are conventional-related card frauds, tradesman-related frauds, and network-related frauds [3].

1.2 Fraud Detection

Fraud detection (FD) is the most effective way to identify a successful transaction as a fraudulent or non-fraudulent [4]. Credit card fraud detection is a challenging job due to the following; firstly, information on normal and fraudulent behaviors changes from time to time, and secondly, due to the skewed and imbalanced credit card transactional datasets. FD is considered as a data mining classification problem, where the main objective should be to accurately classify the credit card transactions as fraud or non-fraud [3].

1.3 Machine Learning

ML techniques can be highly recommended for FD, because of its correctness rate and detection rate [5]. In ML, we provide data to the particular generic algorithm without

M. N. Mohanty and S. Das (eds.), *Advances in Intelligent Computing and Communication*, Lecture Notes in Networks and Systems 109,
https://doi.org/10.1007/978-981-15-2774-6_49

writing any programming code, and thus, it builds logic based on the data provided. ML algorithms use computational methods to "learn" information directly from data without relying on a predetermined equation as a model.

2 Literature Survey

In paper [1], a bagging classifier is applied to the credit card transaction dataset to explain the benefits of the bagging ensemble techniques. In paper [2], support vector machine (SVM)-based method with multiple kernels is used to analyze the user profile. In paper [3], the analysis and judgment on the performance of naive Bayes, k-nearest neighbor, and logistic regression on highly skewed and imbalanced credit card fraud datasets are evaluated based on an undersampling and oversampling. In paper [4], ten different machine learning algorithms in the field of credit card fraud detection have been used with time features in the dataset which is considered as one of the important features for the performance evaluation. In paper [5], the objective is to create secure credit card transactions based on deep learning algorithms such as logistic regression, naive Bayesian, support vector machine, neural network, artificial immune system, k-nearest neighbor. In paper [6], artificial neural network techniques have been adopted to detect fraud. In paper [7], different types of supervised machine learning algorithms are applied in credit card transactions to detect fraud transactions. Undersampling is used for handling the imbalanced dataset. In paper [8, 9], genetic algorithm has been used to detect fraud transactions. In paper [10], support vector machine techniques are used for credit card fraud detection. In paper [11], the mining techniques such as a frequent itemset mining algorithm have been used for credit card fraud detection. In paper [12], the decision tree algorithm has been used for detecting fraud. In paper [13], migrating birds optimization techniques has been used for fraud detection in credit cards. In paper [14], naive Bayes techniques are been used for fraud detection in credit cards. In paper [15], for credit card fraud detection, good approaches such as logistic regression and naive Bayes are carried out. In paper [16], the performance of Bayesian and neural network has been calculated in credit card fraud detection. In paper [10], decision tree, neural networks, and logistic regression have been performed and tested for their applicability in fraud detection.

3 Methodology

In this phase, we have followed the working flow diagram for detecting fraud. In this section, we have focused on two approaches:

- Classification algorithm
- Ensemble learning algorithm (Fig. 1).

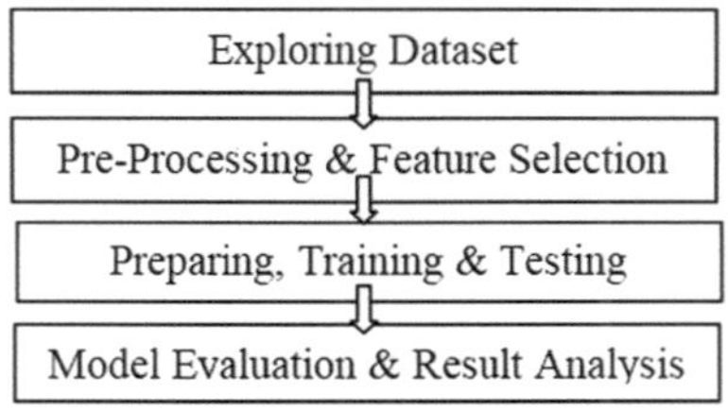

Fig. 1. Working flow diagram

3.1 Classification Algorithms

3.1.1 SVM, QDA, LDA, and CART

In 1995, the SVM algorithm was introduced by Vapnik and his team members [4]. The main idea behind the SVM algorithm is to find the optimal hyperplane that has the maximum width between the vectors in any dimensional space. An optimal hyperplane can be considered as one of the linear decision functions with maximum width between the vectors [5, 7, 17].

Quadratic discriminant analysis (QDA) is similar to LDA, where QDA considers that estimation from each class in the dataset is normally distributed. As comparing to LDA, QDA does not consider the covariance of each of the classes is identical [18].

LDA is one most preferred supervised learning techniques where the input region is divided into regions known as decision regions, and its boundaries are called decision surfaces or decision boundaries [19]. It is used for solving and providing a better result for the classification predictive modeling problems.

Classification and regression trees (CART) are implemented by using a binary decision tree which was developed by Breiman et al. (1984). This technique is used for continuous and non-parameter data. The condition for division of tree is depended on attributes, quantity, and Gini index. Two subsets of the data are generated based on the condition, and the same process continues again for each subset to find out the next condition for the division.

3.2 Ensemble Learning Algorithm

Ensemble learning is one of the meta-classifier approaches which help to get better results by the mixture of multiple machine learning classifiers to achieve well the predictive results [20, 21]. This algorithm includes two types of learning technique:

- Sequential learning (boosting)
- Parallel learning (bagging).

Our main focuses will be on the boosting algorithm.

3.2.1 Boosting Algorithm

Boosting machine learning methods are used to solve complex and convoluted. In this case, all the features are taken into consideration to make the classification and prediction accuracy. Boosting is a process that combines many machine learning algorithms to convert weak learners (WL) to strong learners so that the accuracy of the model increases. In the above figure, the output result (Out) of all the WL are grouped together in order to produce strong learner (SL) [21].

XGBoost, CatBoost, LGBoost, and Random Forest

The advanced version of gradient boosting is XGBoost, which improves the efficiency of the models to get better output results. It extremely boosts the performance of the models [22].

In CatBoost, the categorical columns are given with the indices so that it can be encoded as one-hot encoding using one_hot_max_size for all the features with no of different value less than or equal to the given parameter [22].

LGBoost (also known as light gradient boosting machine), from Microsoft and published in 2017, is used to increase the performance as compared to XGBoost. LGBoost uses XGBoost as a base and performs better than it in training dataset and handling capacity sizes of the dataset [22].

RF method is commonly known as the random forest method used for solving classification and regression problems. RF uses the collection of the decision tree values; each one is slightly different from each other [7, 20].

4 Performance Evaluation

4.1 Measuring Algorithm Performance

To evaluate the performance of the model, we have considered the measuring factors such as false positive (FP) and positive predictions that are considered as negative; true positive (TP) and positive predictions that are considered as positive; false negative (FN) and negative predictions that are considered as positive; true negative (TN) and negative predictions that are considered as negative.

4.2 Confusion Matrix

This is a graphical table representation for evaluating the performance of any classification model [4, 23, 24]. "Y" is yes and "N" is no, whereas "P Class" is predicted class and "A Class" is an actual class. Confusion matrix looks like (Fig. 2).

	A Class= Y	A Class =N
P Class =Y	TP	FP
P Class =N	FN	TN

Fig. 2. Confusion matrix

4.2.1 Precision, Recall/Sensitivity/True-Positive Rate (TPR), Specificity/True-Negative Rate (TNR), and F1 score

$$\text{Precision} = \frac{\text{TP}}{(\text{TP} + \text{FP})} \tag{1}$$

$$\text{TPR} = \frac{\text{TP}}{(\text{TP} + \text{FN})} \tag{2}$$

$$\text{TNR} = \frac{\text{TP}}{(\text{TP} + \text{FP})} \tag{3}$$

$$\text{F1 Score} = 2 * \text{Pr} * \frac{\text{Rc}}{(\text{Pr} + \text{Rc})} \tag{4}$$

4.2.2 Area Under ROC Curve

A receiver operating characteristic curve (ROC curve) is a graph representation that shows the performance of a classification algorithm as compared to all classification thresholds [23, 24]. This curve plot is obtained from TPR versus FPR.

4.2.3 Matthews Correlation Coefficient Score (MCCS)

It is a performance score obtained from true and false positives, as well as true and false negatives [24].

$$\text{MCCS} = \frac{\text{TP} * \text{TN} - \text{FP} * \text{FN}}{\sqrt{(\text{TP} + \text{FP}) * (\text{FN} + \text{TN}) * (\text{TP} + \text{FN}) + (\text{FP} + \text{TN})}} \tag{5}$$

4.2.4 Area Under Precision–Recall Curve

A precision–recall curve is the graph representation curve that shows performance between precision against recall [25].

4.2.5 Cross-validation Score Report

Cross-validation is one of the statistical methods used to estimate the accuracy of machine learning models [25].

5 Result Analysis

5.1 Exploring the Dataset

We have taken a dataset that is available in the ULB Machine Learning Group [6] collected in two days made through credit cards by European cardholders in September 2013. This dataset is transformed by the PCA transformation method, and it provides

28 principal components or features from V1 to V28, but 30 components are taken for the evaluation process. It is found that there are 284,807 transactions, out of which 492 are fraud transactions and only 0.172% are of fraud transactions from all transactions; thus, this dataset is highly unbalanced and skewed toward fraud transactions.

5.2 Pre-processing and Features Selection

Pre-processing and selection of features from the above dataset can be done in two ways, such as:

- Data normalization
- Outlier removal and dimensional reduction.

5.2.1 Data Normalization

Due to the highly unbalanced dataset which is skewed toward the positive class [26], we have used a StandardScaler method, which will transform the data such that its distribution will have a mean value 0 and a standard deviation of 1. To create a subsample dataset with balanced class distributions, we have taken 446 fraudulent transactions in the training data and 446 non-fraudulent transactions randomly from normal samples. We got 892 transactions as a subsample dataset for analyzing by concatenation (Fig. 3).

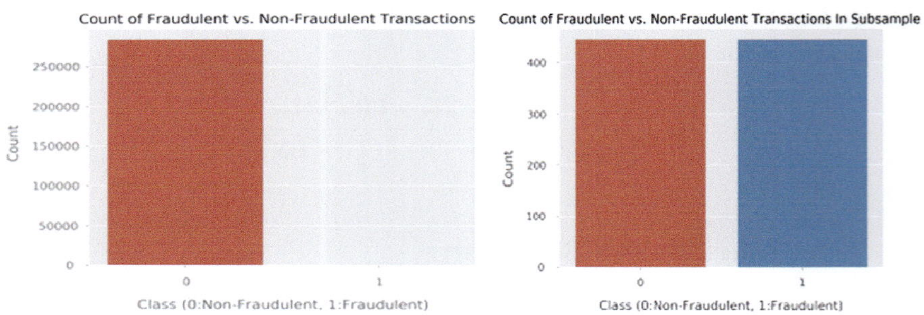

Fig. 3. Before sampling and after resampling

Then, we shuffled the subsampled dataset. Now after resampling, we found out of 892, non-fraudulent transactions are 446 and fraudulent transactions are 446.

5.2.2 Outlier Removal and Dimensional Reduction

To remove extreme outliers, we have used the quantile technique which reduces our data size from 892 transactions by 264 transactions to 628 transactions. For dimensional reduction, we have t-Distributed Stochastic Neighbor Embedding (t-SNE) to reduce the number of random variables.

5.3 Model Evaluation and Result Analysis

5.3.1 Cross-validation Score Report

In the cross-validation technique, we have used $k = 10$ and random state = 42. We got the CatBoost model with better results (Tables 1 and 2).

Table 1 Comparison of cross-validation score

Classification algorithms		Ensemble learning algorithms	
Mean value	Standard deviation value	Mean value	Standard deviation value
SVM: 0.972,559	0.021100	CatBoost: 0.977419	0.016286
LDA: 0.975472	0.018522	XGBoost: 0.972844	0.017558
QDA: 0.967962	0.021107	LGBoost: 0.968263	0.018790
CART: 0.909202	0.044487	RF: 0.961604	0.026427

Table 2 Confusion matrix

Classification algorithms		Ensemble learning algorithms	
SVM: [[83 2] [10 31]]	QDA: [[81 4] [6 35]]	CatBoost [[84 1] [10 31]]	XGBoost: [[84 1] [9 32]]
LDA: [[85 0] [12 29]]	CART: [[76 2] [8 33]]	LGBoost: [[85 2] [8 33]]	RF: [[85 0] [9 32]]

5.3.2 Confusion Matrix Report

From the confusion matrix, we state that LGBM gives better outcomes. It predicts high positive values and less negative values.

5.3.3 ROC Curve

From Fig. 4, we found that the LGBM algorithm gives better results (Figs. 4, 5 and 6).

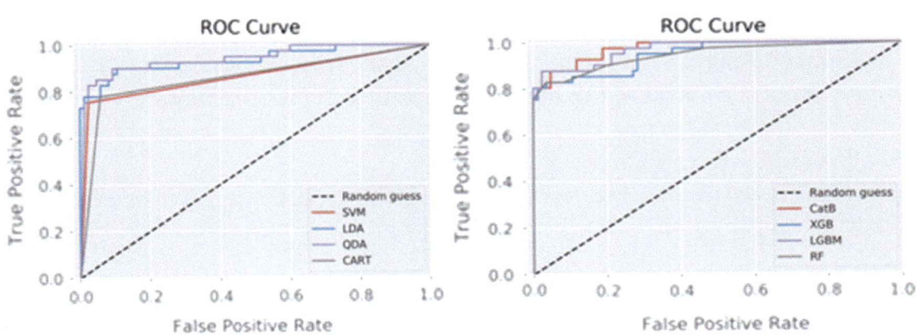

Fig. 4. ROC curve

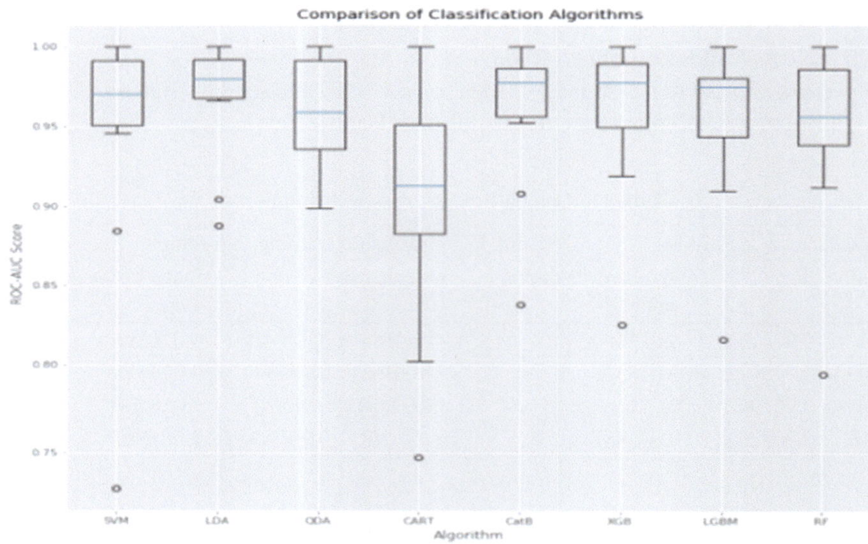

Fig. 5. Comparison of algorithms based on ROC-AUC curve

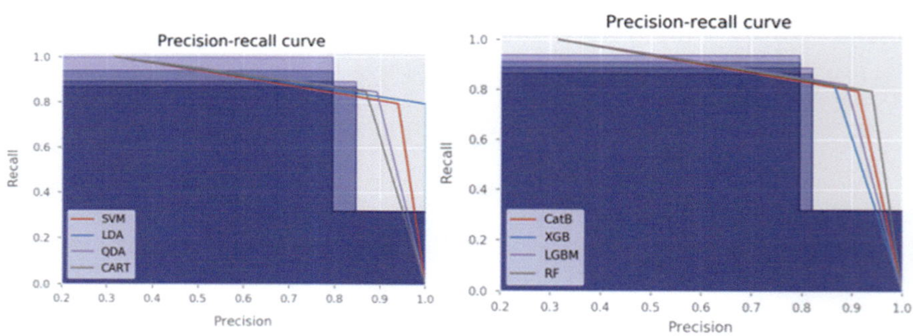

Fig. 6. Comparison of algorithms based on precision–recall curve

5.3.4 ROC-AUC Curve

From the comparison plot, we focus not only on achieving the highest accuracy but also to extract business values and information. Therefore, we found that LGBM takes over the CatBoost to achieve the highest degree of comprehensiveness.

5.3.5 Precision–Recall Curve

From the precision–recall curve, we found LGBM gives better result than other models (Table 3).

Table 3 Comparison of precision, recall, F1 score, and **MCC score**

Algorithms				
Selected model	Precision value	Recall value	F1 score value	MCC score value
SVM	0.94	0.76	0.84	0.78
LDA	1.00	0.71	0.83	0.79
QDA	0.90	0.85	0.88	0.82
CART	0.86	0.78	0.82	0.78
CatBoost	0.97	0.76	0.85	0.80
XGBoost	0.97	0.78	0.86	0.82
LGBoost	1.00	0.80	0.89	0.86
RF	0.94	0.81	0.87	0.80

To evaluate imbalanced datasets, it is better to consider the four-parameters, in order to achieve high comprehensiveness and if the dataset is highly imbalanced, i.e., Matthew Coefficient Correlation Score value, F1 score value, and the area under the precision–recall curve value. Therefore, we found that the LGBM algorithm gives better outcomes among all models.

6 Conclusion and Future Work

In order to detect fraud is a tremendous and complex issue that requires better active planning before applying a machine learning algorithm on it. Fraud detection is an application of data science, and machine learning can give a better result, to keep the customer's money safe without losing integrity. We propose our future work to include a comprehensive tuning of the LGBoost algorithm and apply grid search techniques, with a non-anonymized dataset in which we can find important features for detecting fraudulent transactions.

References

1. Zareapoor M, Shamsolmoali P (2015) Application of credit card fraud detection: based on bagging ensemble classifier. In: International Conference on Computer, Communication and Convergence (The authors published by Elsevier B.V)
2. Shen A, Tong R, Deng Y (2007) Application of classification models on credit card fraud detection. In: 2007 International conference on service systems and service management, IEEE, pp 1–4
3. Awoyemi JO, Adetunmbi AO, Oluwadare SA (2017) Credit card fraud detection using machine learning techniques: a comparative analysis. In: 2017 International Conference on Computer Networking and Informatics (ICCNI), Lagos, pp 1–9
4. Rajora S, Li D-L, Jha C, Bharill N, Patel OP, Joshi S, Puthal D, Prasad M (2018) A comparative study of machine learning techniques for credit card fraud detection based on time variance, 978–1-5386-9276-9/18

5. Popat RR, Chaudhary J (2018) A survey on credit card fraud detection using machine learning. In: Proceedings of the 2nd International Conference on Trends in Electronics and Informatics (ICOEI 2018) IEEE Conference Record:# 42666; IEEE Xplore ISBN:978-1-5386-3570-4
6. Mittal S, Tyagi S (2019) Performance evaluation of machine learning algorithms for credit card fraud detection, 978-1-5386-5933-5/19/$31.00
7. Dhankhad S, Mohammed EA, Far B (2018) Supervised machine learning algorithms for credit card fraudulent transaction detection: a comparative study. In: 2018 IEEE international conference on information reuse and integration for data science
8. Rama Kalyani K, Uma Devi D (2012) Fraud detection of credit card payment system by genetic algorithm. Int J Sci Eng Res 3(7):1–6 (ISSN 2229-5518)
9. Meshram PL, Bhanarkar P (2012) Credit and ATM card fraud detection using genetic approach. Int J Eng Res Technol 1(10):1–5 (ISSN: 2278-0181)
10. Singh G, Gupta R, Rastogi A, Chandel MDS, Riyaz A (2012) A machine learning approach for detection of fraud based on SVM. Int J Sci Eng Technol 1(3):194–198 (ISSN : 2277-1581)
11. Seeja KR, Zareapoor M (2014) Fraud miner: a novel credit card fraud detection model based on frequent itemset mining. Sci World J 2014:1–10. https://doi.org/10.1155/2014/252797 (Hindawi Publishing Corporation)
12. Patil S, Somavanshi H, Gaikwad J, Deshmane A, Badgujar R (2015) Credit card fraud detection using decision tree induction algorithm. Int J Comput Sci Mobile Comput 4(4): 92–95 (ISSN: 2320-088X)
13. Duman E, Buyukkaya A, Elikucuk I (2013) A novel and successful credit card fraud detection system implemented in a turkish bank. In: 2013 IEEE 13th International Conference on Data Mining Workshops (ICDMW), pp 162–171
14. Bahnsen AC, Stojanovic A, Aouada D, Ottersten B (2014) Improving credit card fraud detection with calibrated probabilities. In: Proceedings of the 2014 SIAM international conference on data mining, Society for Industrial and Applied Mathematics, pp 677–685
15. Ng AY, Jordan MI (2002) On discriminative vs. generative classifiers: a comparison of logistic regression and naive bayes. Adv Neural Inf Process Syst 2:841–848
16. Maes S, Tuyls K, Vanschoenwinkel B, Manderick B (2002) Credit card fraud detection using Bayesian and neural networks. In: Proceedings of the 1st international naiso congress on neuro fuzzy technologies, pp 261–270
17. Hsu C, Chang C, Lin C (2003) A practical guide to support vector classification. Technical report, Department of Computer Science and Information Engineering, National Taiwan University
18. Ogwueleka FN (2011) Data mining application in credit card fraud detection system. J Eng Sci Technol 6(3):311–322
19. Mahmoudi N, Duman E (2015) Detecting credit card fraud by modified Fisher discriminant analysis. Expert Syst Appl 42:2510–2516
20. Sorournejad S, Zojaji Z, Atani RE, Monadjemi AH (2016) A survey of credit card fraud detection techniques: data and technique oriented perspective. arXiv:1611.06439
21. Herawan T et al (eds) An application of oversampling, undersampling, bagging and boosting in handling imbalanced datasets. In: Proceedings of the first international conference on advanced data and information engineering (DaEng-2013). Lecture notes in electrical engineering vol 285. Springer Science+Business Media Singapore. https://doi.org/10.1007/978-981-4585-18-7_2
22. CatBoost, LGBoost, XGBoost: https://towardsdatascience.com/catboost-vs-light-gbm-vs-xgboost-5f93620723db

23. Syaripudin A, Khodra ML (2014) A comparison for handling imbalanced datsets. In: 2014 International Conference of Advanced Informatics: Concepts, Theory and Aplications (ICAICTA), 978-1-4799-5100-0/14
24. Thennakoon A, Bhagyani C, Premadasa S, Mihiranga S, Kuruwitaarachchi N (2019) Real-time credit card fraud detection using machine learning. 978-1-5386-5933-5/19
25. Nuno Abrunhosa Carneiro (2017) A data mining approach to fraud detection in e-tail. Published in Decision Support Systems. https://doi.org/10.1016/j.dss.2017.01.002
26. Lee S (2000) Noisy replication in skewed binary classification. Comput Stat Data Anal 34:165–191

Performance Comparison of Adaptive Algorithms for Smart Antenna of Dipole Array

Barsa Samantaray[1(✉)], Kunal Kumar Das[1], and Jibendu Sekhar Roy[2]

[1] ECE Department, Institute of Technical Education and Research, Siksha 'O' Anusandhan Deemed to be University, Bhubaneswar, Odisha, India
{barsasamantaray, kunaldas}@soa.ac.in

[2] School of Electronics Engineering, Kalinga Institute of Industrial Technology (KIIT), Deemed to be University, Bhubaneswar, Odisha, India

1 Introduction

Smart antenna is smart with reference to smart signal processing algorithm. In adaptive smart antenna, by using signal processing algorithm the direction of arrival can estimate and beam direction toward the desired user and null toward the undesired direction can be generated [1, 2]. Various beam-forming Adaptive signal processing algorithms with its advantages and disadvantages are discussed in [1–4]. Most of the algorithms used for adaptive beamforming in smart antenna have fixed step-size parameters. Some literatures are available where variable step-size parameters are used [5–8], where mainly variable step-size least mean square (LMS) algorithm is used.

Here, sign LMS (SLMS) and variable step SLMS (VS-SLMS) algorithms are used for the design of smart antenna. SLMS is a variant of LMS algorithm. Here, a comparison is done between the performances of these two algorithms. The VS-SLMS algorithm has better performance than SLMS algorithm in adaptive beam generation of smart antenna.

2 Adaptive Beamforming Algorithms

For the design of smart antenna, adaptive signal processing algorithms, sign LMS (SLMS) algorithm and variable step sign LMS (VS-SLMS) algorithm, are used. Sign (also called signed regressor) LMS (SLMS) is used for the adaptation process to be very fast.

The equation used for updating the weight in sign LMS is [9].

$$\omega(n+1) = \omega(n) + \mu(n) \cdot e^{*}(n) \cdot \mathrm{sgn}[x(n)] \quad (1)$$

M. N. Mohanty and S. Das (eds.), *Advances in Intelligent Computing and Communication*, Lecture Notes in Networks and Systems 109,
https://doi.org/10.1007/978-981-15-2774-6_50

The sign function can be defined as
where

$$\mathbf{sgn}[\boldsymbol{x}(\boldsymbol{n})] = \mathbf{1}; \quad \mathbf{for}\,\boldsymbol{x}(\boldsymbol{n}) > \mathbf{0}$$
$$= 0; \quad \text{for}\,x(n) = 0 \tag{2}$$
$$= -1; \quad \text{for}\,x(n) < 0$$

$\mu(n)$ is the step size, and $e^*(n)$ is the error signal between input and output signals.

In VS-SLMS, the step size $\boldsymbol{\mu}$ is also variable [5, 6] with iteration number and in every iteration, and during weight updating, step-size parameter varies according to the formula [5]

$$\begin{aligned}\mu_{n+1} &= \alpha\mu_n + \delta\varepsilon_n, \quad \text{if}\,0 < \mu_{n+1} < \mu_{\max} \\ &= \mu_{\max}, \quad \text{otherwise}\end{aligned} \tag{3}$$

where maximum value of convergence parameter, $\mu_{\max}$, is defined as $\mu_{\max} < 2/\lambda_{\max}$ and is the largest $\lambda_{\max}$ which is the maximum eigen value of the correlation matrix of the signal. In (3), 'α' and 'δ' are constant parameters, and in this paper, in simulation, the values chosen for those are α = 0.95 and δ = 0.0003.

The factor ε_n in (3) is related to the weight vector as

$$\varepsilon_n = \|w(n+1) - w(n)\|/\|w(n+1)\| \tag{4}$$

The weight updated VS-SLMS algorithm equation is

$$\boldsymbol{w}(\boldsymbol{n}+\mathbf{1}) = \boldsymbol{w}(\boldsymbol{n}) + \boldsymbol{\mu}_{\boldsymbol{n}+\mathbf{1}}\boldsymbol{e}^*(\boldsymbol{n})\mathbf{sgn}[\boldsymbol{x}(\boldsymbol{n})] \tag{5}$$

3 Simulated Design of Smart Antenna of Dipole Array

The diagram of a linear dipole array is shown in Fig. 1, where the dipole antennas are separated by a distance of 'd'.

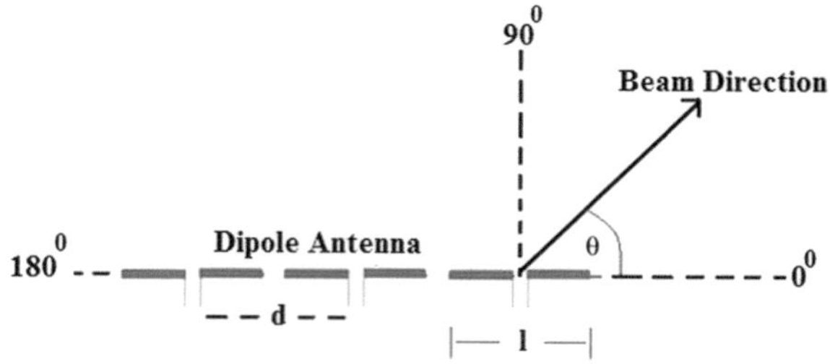

Fig. 1. Linear dipole array

For a dipole antenna of length 'l', the radiated electric field is expressed as [10]

$$E(\theta) = j\eta \frac{I_0 e^{-j\beta r}}{2\pi r} \left[\frac{\cos\left(\frac{\beta l}{2}\cos\theta\right) - \cos\left(\frac{\beta l}{2}\right)}{\sin\theta} \right] \quad (6)$$

where $\beta = 2\pi/\lambda$ is the propagation constant, I_0 is current amplitude fed to the dipole antenna, $\eta = 120\pi\Omega$, and r is the distance from the radiator array.

For a linear array of N number of dipole antennas, the total radiated field, with mutual coupling, is given by

$$E_{\text{total}}(\theta) = \sum_{n=1}^{N} I_0 E(\theta) e^{j(n-1)\left(\frac{2\pi d}{\lambda}\cos\theta + \alpha\right)} \quad (7)$$

where λ is the wavelength, α is the progressive phase shift of the array, and I_0 is the current fed to the antennas, and it is assumed that all the antennas are fed by equal current I_0. Array factor of (7) is used for dipole array using adaptive algorithms SLMS and variable step-size VS-SLMS. Adaptive beam generation algorithm makes possible to generate beam in the desired direction and null in the unwanted direction.

In simulated design of smart antenna of dipole array, a number of dipole antennas in arrays are N = 16 and N = 20, and in all the cases, inter-element spacing is $d = 0.5\lambda$. The results for different beam directions (BD) and null directions (ND) for N = 16, obtained using SLMS and VS-SLMS, are plotted in Figs. 2 and 3, respectively (BD = 50°, ND = 60°).

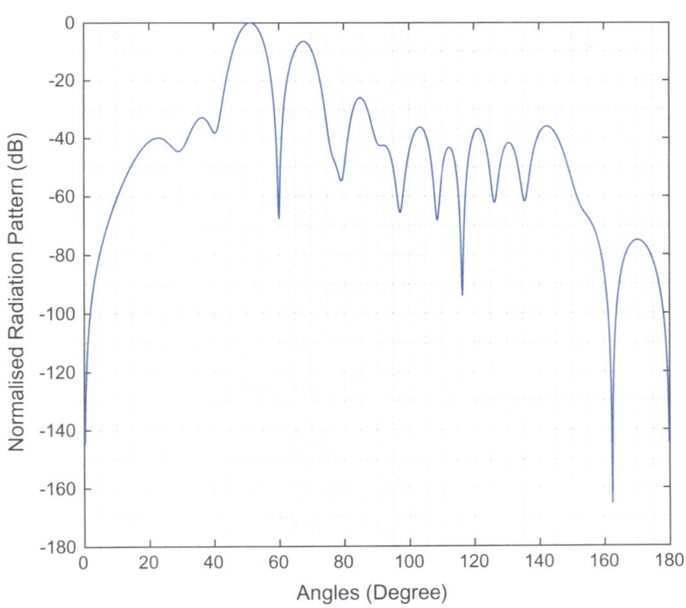

Fig. 2. Beamforming using SLMS algorithm

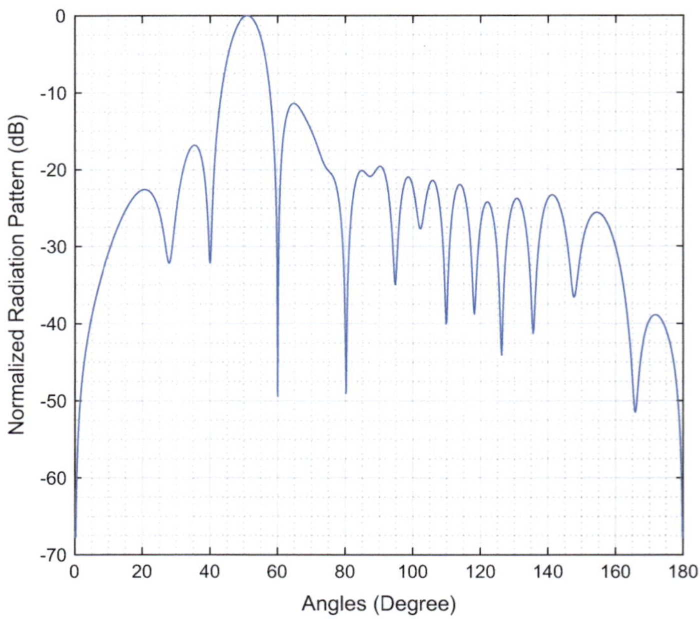

Fig. 3. Beamforming using VS-SLMS algorithm

Similarly, the results for different beam directions (BD) and null directions (ND) for $N = 20$, obtained using SLMS and VS-SLMS, are plotted in Figs. 4 and 5, respectively (BD = 70°, ND = 75°).

The variation for variable step-size parameter for VS-SLMS algorithm for $N = 16$ is shown in Fig. 6.

Mean square error (MSE) plots for SLMS and VS-SLMS are plotted in Fig. 7 and in 8, respectively, for $N = 20$.

From the graphs of Figs. 7 and Fig. 8, it is evident that the convergence of VS-SLMS algorithm is better than SLMS algorithm. The results for beamforming are tabulated in Table 1.

From Table 1, it is evident that the performance of VS-SLMS algorithm is better to obtain lower side lobe level (SLL). Using SLMS for $N = 16$ and $N = 20$, maximum SLLs (SLL_{max}) are −6.5 dB and −6 dB, respectively, whereas using VS-SLMS $N = 16$ and $N = 20$, maximum SLLs (SLL_{max}) are −11.5 dB and −8.2 dB, respectively. Also, convergence of VS-SLMS is better than SLMS algorithm.

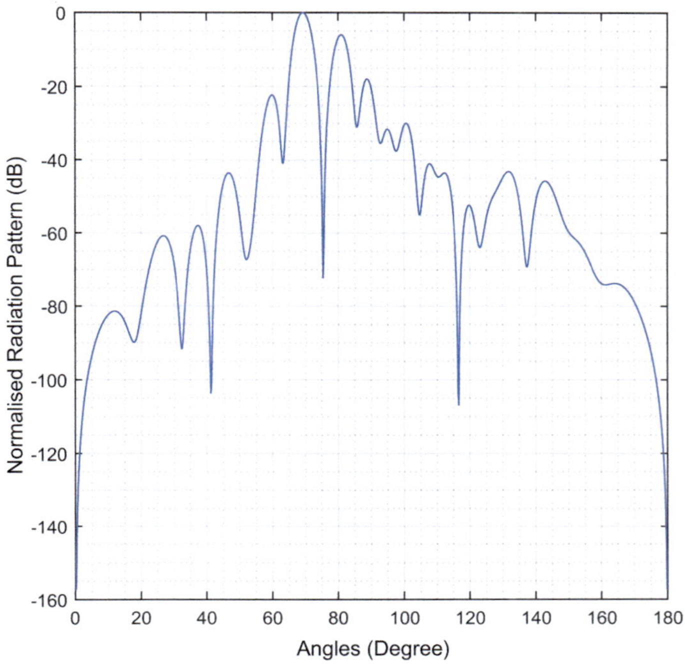

Fig. 4. Beamforming of SLMS algorithm

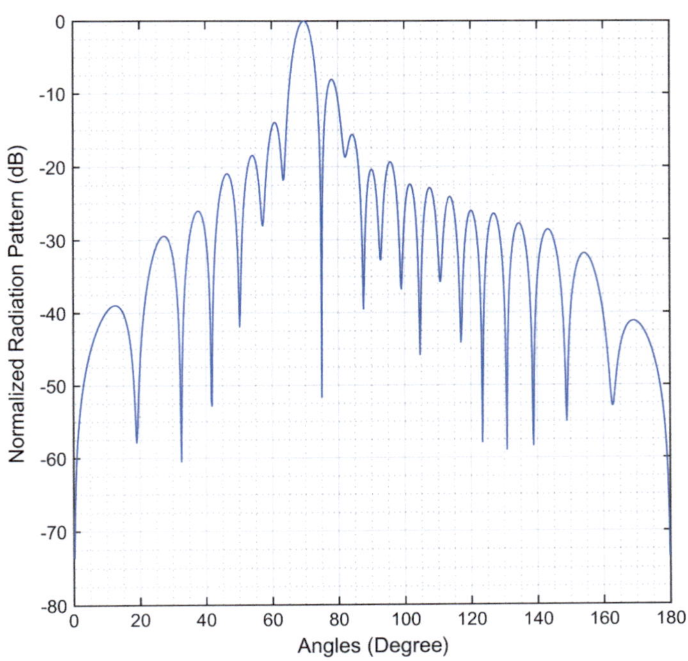

Fig. 5. Beamforming using VS-SLMS algorithm

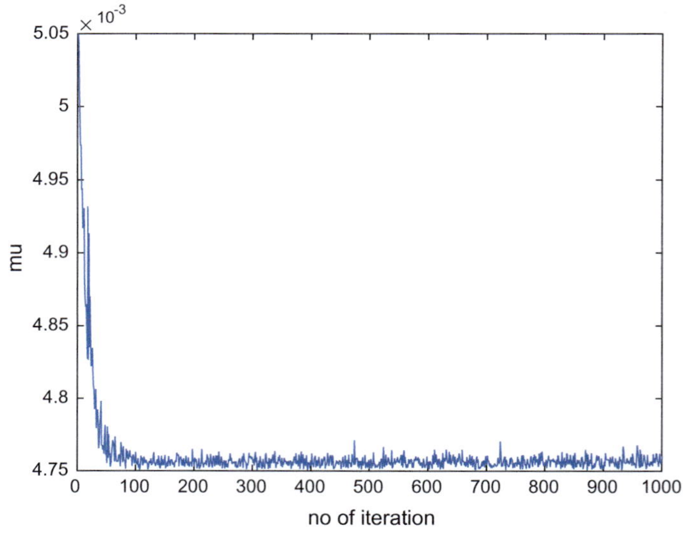

Fig. 6. Variation of step-size parameter of VS-SLMS algorithm

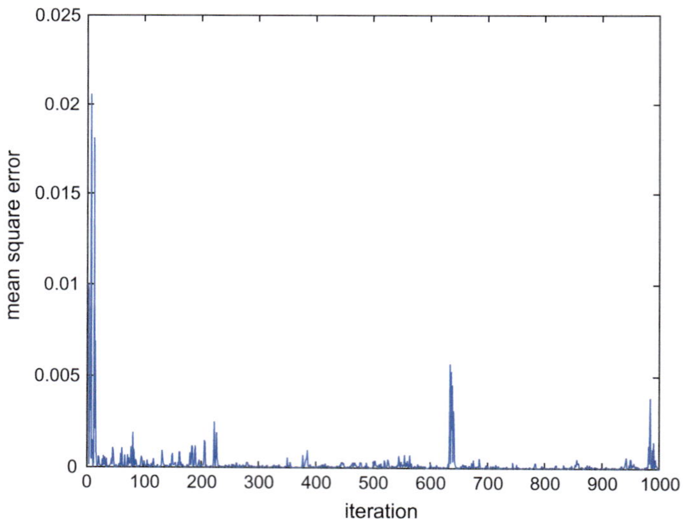

Fig. 7. MSE plot for SLMS (N = 20)

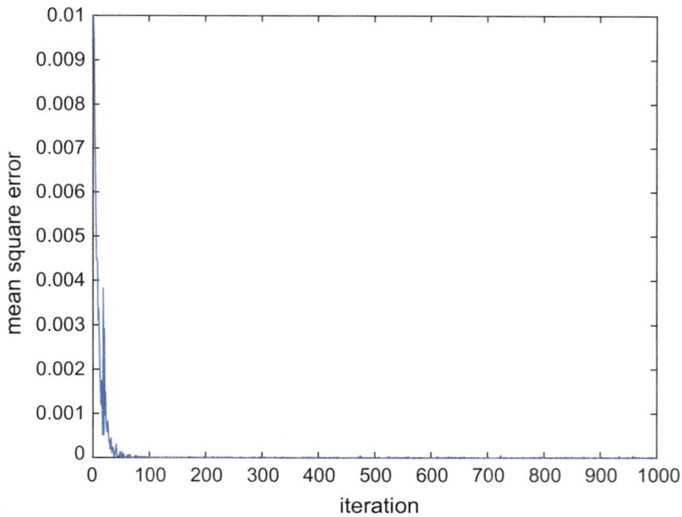

Fig. 8. MSE plot for VS-SLMS

Table 1 Comparison of beamforming performances of SLMS and VS-SLMS algorithms

Adaptive algorithms	No. of dipoles	Desired angle (°)	Desired null (°)	Obtained angle (°)	Obtained null (°)	SLL_{max} (dB)
VS-SLMS	$N = 16$	50	60	51	60	−11.5
	$N = 20$	70	75	69.6	75	−8.2
SLMS	$N = 16$	50	60	51.6	60	−6.5
	$N = 20$	70	75	69	75.4	−6

4 Conclusion

Smart antennas of dipole arrays are generally used for cellular communication from UHF to microwave frequencies, typically up to 6 GHz. A performance comparison between SLMS and VS-SLMS algorithms is presented here which will help to design adaptive smart antenna of dipole array. In both the algorithms, the number of iteration is 1000. In SLMS, the step-size parameter, $\mu = 0.05$. Convergence of variable step-size algorithm is better, and using variable step size algorithm, more accurately desired beam direction and null direction can be achieved. Side lobe level reduction using VS-SLMS algorithm is more than SLMS algorithm. For a smart antenna array of 16 dipoles, 5 dB lower side lobe level is achieved using VS-SLMS algorithm than SLMS algorithm.

References

1. Godara LC (1997) Application of antenna arrays to mobile communications, part II: beamforming and direction-of-arrival considerations. Proc IEEE 85(8):1195–1245
2. Bellofiore S, Balanis CA, Foutz J, Spanias AS (2002) Smart antenna systems for mobile communication network, part 1, overview and antenna design. IEEE Antennas Propag Mag 44(3):145–154
3. Halder A, Senapati A, Roy JS (2018) Smart antenna of microstrip array for sector beamforming for 28 GHz millimeter wave mobile communication. Int J Microwave Opt Technol (IJMOT) 13(3):203–208
4. Sharma R, Senapati A, Roy JS (2018) Beamforming of smart antenna in cellular network using leaky LMS algorithm and its variants. Int J Microwave Opt Technol (IJMOT) 13(4): 263–268
5. Lau YS, Hussain ZM, Harris RJ (2004) A weight-vector LMS algorithm for adaptive beamforming. In: IEEE TENCON Conference, New Jersey, USA, pp 94–498
6. Luo XD, Jia ZH, Wang Q (2006) A new variable step size LMS adaptive filtering algorithm. Acta Electronica Sinica 34(6):1123–1126
7. Zhang Y, Li N, Chambers JA, Hao Y (2008) New gradient-based variable step size LMS algorithms. EURASIP J Adv Signal Process 2008(7):1–9
8. Huang H, Lee J (2012) A new variable step size NLMS algorithm and its performance analysis. IEEE Trans Signal Process 60(4):2055–2060
9. Senapati A, Ghatak K, Roy JS (2015) A comparative study of adaptive beamforming techniques in smart antenna using LMS algorithm and its variants. In: International Conference on Computational Intelligence & Networks (CINE-2015), KIIT University, India, 12–13 Jan 2015. IEEE Xplore, pp 58–62
10. Balanis CA (2005) Antenna theory—analysis and design, 3rd edn. Wiley

Effect of Buffer Size on Performance of Wireless Sensor Network

Manoranjan Das(✉) and Benudhar Sahu

Institute of Technical Education and Research, SOA (Deemed to be University), Bhubaneswar, India

1 Introduction

Wireless communication technology though an old concept for information transmission between a pair of transmitter and receiver is still treated as one of the advanced communication technologies so far as the data sharing among users is concerned. More specifically when the users belong to a communication network where the network is limited to distance between users and cost of installation, the wireless is preferred as an ultimate among various types of communication technology. Thus, wireless communication network has reserved a worthy place in the current research scenario. The cellular network, WLAN, wireless ad hoc network are the examples of wireless communication network. These wireless communication networks are further can be categorized as either infrastructure-based or infrastructure-less networks. The infrastructure-based wireless communication networks are those where the presence of a fixed infrastructure (like a fixed base station) plays an important role for functioning of the network. Example of such a kind wireless network is the cellular network. In the other side, an important feature of an infrastructure-less wireless communication network is that the users (here onwards named as nodes) in these networks communicate among themselves without the help of any fixed base station (or access point). Wireless ad hoc networks where the nodes are either static or mobile by nature come under this category. The popularly known wireless sensor network (WSN) [1, 2] is the noted example to describe the operation of an infrastructure-less wireless communication network.

The performance of such wireless communication networks in compared to wired communication networks suffers from various factors such as network size, traffic density, and routing. To be more explainable, in a wireless sensor network the amount of data collected at the end node (i.e., sink node) directly reflects the network performance. Further, in order to have a good collection of data at the end node it is required that the sensor nodes (which are the only constituents of a WSN) must have good efficiency not only in data sensing but also data handling (i.e., both data storing and data forwarding). It is worthy enough to mention here that in a wireless sensor network the job of a sensor node is not limited to data sensing and transmitting its data to the end node only. Rather a sensor node is also required to carry forward the data sensed by other sensing node which is not in the transmission range of sink node. Alternatively, the sensor nodes also take the responsible of handling the data traffic efficiently. In order to handle this job, the sensor node uses its buffer which is of a fixed

M. N. Mohanty and S. Das (eds.), *Advances in Intelligent Computing and Communication*, Lecture Notes in Networks and Systems 109,
https://doi.org/10.1007/978-981-15-2774-6_51

size. However, many a times it may so happen that the network gets congested in route when the traffic in the network is heavy enough. Further, when the sensor node is loaded with heavy traffic, the buffer size is not enough to handle. In this condition, the data packets may get dropped because of buffer overflow [3]. Thus, the buffer size of the sensor node is also treated as one strong factor that affects the performance of a wireless sensor network. Hence, it is important to study the effect of buffer size on the performance of wireless sensor network [3, 4].

As explained above, it is important in a wireless sensor network to have a good amount of collection of data packets from the sensing nodes that show a good network performance. Thus in this paper, we study the effect of buffer size on the performance of a wireless sensor network that assuming the nodes are static by nature [5, 6].

The rest of the paper is organized as follows. In Sect. 2, a brief overview about wireless sensor network is given. Section 3 defines the performance metrics used for analyzing the performance of the network. The list of parameters being chosen for simulation is also given in Sect. 3. In Sect. 4, analysis toward the performance of network under different environments with the help simulation plots is presented. Conclusion of the paper is presented in Sect. 5.

2 Overview of Wireless Sensor Network

Among the different infrastructure-less wireless network, the WSN is quite popular so far as the application is considered. The WSN is preferably suitable for application environment such as battlefield, habitat monitoring, and mines area. The installation of a wireless sensor network is done through the deployment of large number of tiny sensor nodes in a given geographical area. The distribution of the nodes in the specified geographical area may follow a random (mostly) or regular fashion depending on the network design. Further, it is also noticed that in almost all WSN applications the sensor nodes have two major responsibilities (1) data acquisition and (2) data transmission to an end node (normally called as sink node) as and when desired. Thus, as the name suggests the nodes belonging to a WSN are usually equipped with both sensing devices and communication devices along with limited resources such as power supply (i.e., battery-powered) and storage. Usually, all the nodes present in WSN shares their information through wireless mode of communication. However, since the nodes are small in size, and associatively, they are battery-powered the communication range of some sensor node is not enough to directly communicate the information with the sink node. This problem is solved by allowing a sensor node to carry forward the data packet of another sensor node which is neighbor of each other. Thus, the sensor nodes in WSN also bear the responsibility of routing the data packets in a route between source and destination node pair.

3 Simulation Environment and Performance Metrics

In this work to study the effect of buffer size on the network performance, we preferred a simulation methodology where the simulation environment replicates the specified real-time scenario. The network considered for the simulation is a static one which means the nodes present inside network area are static by nature. Also, the distribution of nodes in the network follows random topology. The network simulator used for the said purpose is Global Mobile Information System Simulator (GloMoSim) [7], and the associated parameters assigned with the values are as given in Table 1.

Table 1. Parameters used for simulation

Parameters	Value/Specification
Terrain area	1500 M × 300 M
Number of nodes	100
Node mobility	None
Number of sources	25
Simulation time	5 M
Transmission power	20 dBm
Mac protocol	802.11
Routing protocol	AODV
Packet size	512 bytes
Type of data traffic	Constant bitrate (CBR)
Packet transmission rate	1 packet/s, 5 packet/s
Buffer size	2, 5, 10, 50, 100

Performance Metrics. In this study, the performance analysis is carried with the help of the following performance metrics as defined below.

(a) *Packet Delivery Fraction.* This metric gives a value that defines the ratio of the amount of data packet delivered at the destination node to the total data packets transmitted from the source nodes during the simulation period.

(b) *Packet Drop Fraction.* This metric presents the fraction of amount of data packets lost (dropped) at the routing nodes due to buffer overflow for a given of data packets being transmitted by the source nodes in the network during the simulation period.

4 Result Analysis

In order to get the plots for different performance metrics, the network simulator runs for 300 s against five different seeds, where the seed differs from each other based on spatial coordinates of nodes in the network area. This means for each seed, the spatial coordinates of nodes are different. Accordingly, the average of five output values corresponding to five different seeds is computed to give a point on the plot.

Figure 1 presents the variation in packet delivery fraction versus buffer size for different packet transmission rate. It is observed from Fig. 1 that the amount of packets delivered successfully at the destination increases with increase in buffer size. This is due to the fact that, the increase in buffer capacity at the sensor node enables it to store more incoming data packets which are to be forwarded later. It is also noticed from the plots in Fig. 1 that with increase in packet transmission rates the packet delivery fraction decreases for a fixed value of buffer size. This happens so because the sensor nodes have a fixed data receiving rate. Thus, when packet transmission rate increases, the receiving sensor node gets less time to receive all the data packets and to put it in the buffer. However at the same time, it can be seen in Fig. 1 that for higher packet transmission rate the percentage increase in packet delivery fraction is more in compared to that for lower packet transmission rate. The specific reason behind this is nothing but the number of data packets that get a successful route to be transmitted in-between the source and destination nodes. Lastly, it is also depicted from Fig. 1 that for increase in buffer size at the lower range (i.e., from 2 to 10) the percentage change in packet delivery fraction is more in compared to the increase in buffer size in the higher range (i.e., from 10 to 100). Also for low packet transmission rate, it seems that the change in buffer size at a higher range has almost nil effect on packet delivery fraction. However for high packet transmission rate, there is a small change in packet delivery fraction with increase in buffer size at the higher range. This is because when packet transmission rate is low, a précised portion of the buffer area remains empty most of the time.

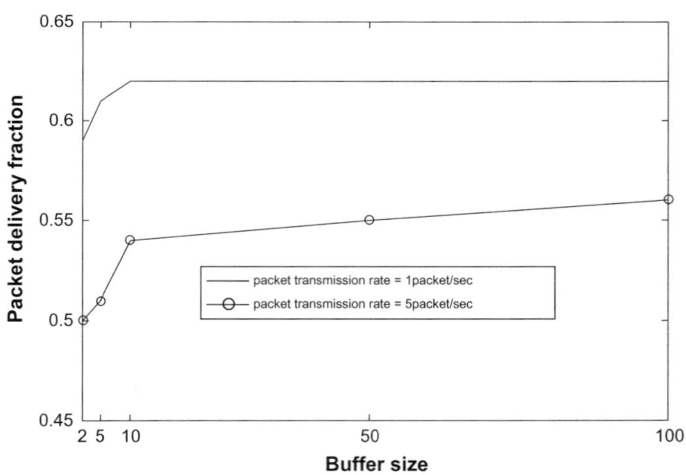

Fig. 1. Packet delivery fraction versus buffer size for different packet transmission rate in a WSN of 100 nodes with 25 sources

Figure 2 given indicates the effect of buffer size on performance metric (i.e., packet drop fraction) for different packet transmission rate. It is observed from Fig. 2 that with increase in buffer size, the packet drop fraction due to buffer overflow decreases. It is quite correct since the increase in buffer size at the sensor node allows the node to store more incoming data packets. It is also noticed in Fig. 2 that there is no packet drop take place for high range of buffer size (i.e., 10 and more than 10) when the packet transmission rate is low. However for high packet transmission rate, the trend of decrease in packet drop fraction continues even in high buffer size range. Further, it is clear from Fig. 2 that with increase in packet transmission rate the amount of data packet dropped due to buffer flow is more in compared to that for lower packet transmission rate for a fixed buffer size. This is because when the packet transmission rate is made to increase, the number of data packets that arrive at a sensor node also increases. This increase in incoming data packets at the sensor node creates an overload burden to the sensor node since the node has also some other responsibility (i.e., data acquisition) at the same time and collected data also has to be stored in the same buffer. Hence, congestion at the intermediate sensor node occurs and so the incoming data packets cannot be accommodated in the buffer and dropped.

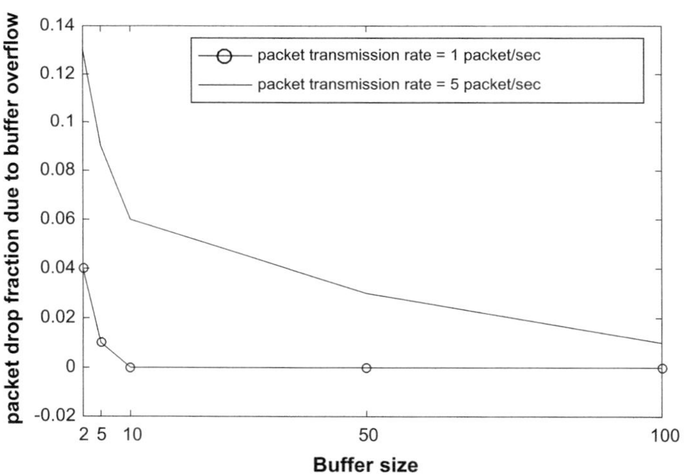

Fig. 2. Packet drop fraction due to buffer overflow versus buffer size for different packet transmission rate in a WSN of 100 nodes with 25 sources

5 Conclusions

The study on different performance metrics for different buffer sizes in a static WSN shows that selection of suitable buffer size is important so as to maintain a good quality performance. Further, it is also observed from the plots obtained from simulation results that the effect of buffer size is more precious when packet transmission rate in the network increases. It is also seen from the simulation plots that with increase in buffer size to a higher range, the performance of the network improves. However, since the sensor nodes are very small in size and the number of sensor nodes being deployed in the geographical area is large, the increase in buffer size to support high data transmission rate is not economical. Hence, there must be a tradeoff between selecting the buffer size with respect to the rate of transmission of data packet in between nodes.

References

1. Sohraby K, Minoli D, Znati T (2010) Wireless sensor networks, 2nd edn. Wiley, India
2. Fahmy HMA (2016) Wireless sensor networks: concepts, applications, experimentation and analysis. Springer
3. Kempa WM (2017) On buffer overflow duration in WSN with a vacation-type power saving mechanism. In: Proceedings of the 24th international conference on systems, signals and image processing (IWSSIP)
4. Shakir ZD, Yoshigoe K, Lenin RB (2012) Adaptive buffering scheme to reduce packet loss on densely connected WSN with mobile sink. In: Proceedings of IEEE consumer communications and networking conference (CCNC)
5. Pathak AA, Deshpande VS (2015) Buffer management for improving QoS in WSN. In: Proceedings of ICPC
6. Rault T, Bouabdallah A, Challal Y (2013) WSN lifetime optimization through controlled sink mobility and packet buffering. In: Proceedings of the global information infrastructure symposium (GIIS)
7. Zeng X, Bagrodia R, Gerla M (1998) GloMoSim: a library for the parallel simulation of large-scale wireless networks. In: Proceedings of 12th workshop on parallel and distributed simulation

Noise Cancellation Using a Novel Self-adaptive Neuro-fuzzy Inference System (SANFIS)

Laxmipriya Samal, Debashisa Samal(✉), and Badrinarayan Sahu

Department of ECE, ITER, S'O'A Deemed to be University, Bhubaneswar, India
debashishamal@soa.ac.in

1 Introduction

In order to remove noise contained in signals, the noise cancellation is a unique method [1, 2]. It is applied in systems like transformers, hands-free phones, machineries, speech enhancement, echo cancellation, biomedical signal and image processing [3–7]. The concept of noise cancellation using adaptive filter was first proposed by Widrow [8, 9]. Many methods are suggested in order to overcome this problem. More recently least mean square algorithm (LMS), normalized least mean square algorithm (NLMS), recursive least square (RLS) algorithms are used [10–12]. In this work, self-adaptive neuro-fuzzy inference system (SANFIS) is proposed for noise cancellation. It is being extended to the nonlinear area by using nonlinear adaptive system which uses the concept of neural network and fuzzy logic. The neural network is used to adapt the parameter of the fuzzy system of the noise source into an interfering component for the detection of signal. The performance of SANFIS technique gives better results with previously proposed LMS and NLMS algorithm.

2 Adaptive Noise Cancellation (ANC)

In the above block diagram [13], the information-bearing signal $s(k)$ was mixed with noise which was originated from another noise by a dynamic procedure. The objective of this is to produce a signal which is exactly the same as that of source signal (Fig. 1).

3 Adaptive Algorithms for Noise Cancelling

In the past, many algorithms have been widely studied, and some are discussed below.

3.1 The Least Mean Square (LMS) Adaptive Algorithm

This algorithm suggested by Widrow and Hoff [14] uses the method of gradient descent approach. To reduce the error, it calculates minima adjusting the filter coefficients. LMS algorithm is discussed below

M. N. Mohanty and S. Das (eds.), *Advances in Intelligent Computing and Communication*, Lecture Notes in Networks and Systems 109,
https://doi.org/10.1007/978-981-15-2774-6_52

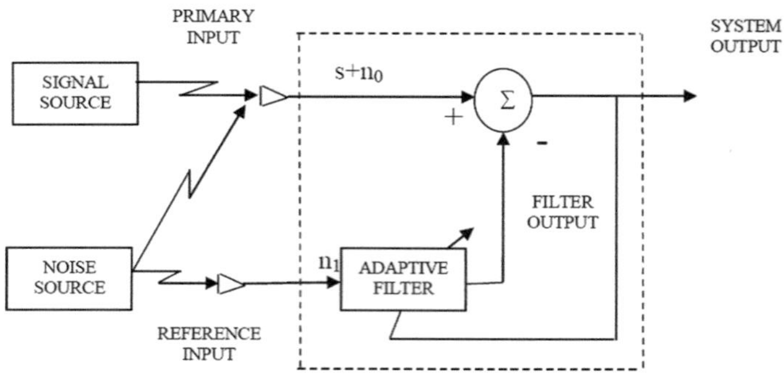

Fig. 1. Generalized block diagram of adaptive noise cancelling

$$w_l(p+1) = w_l(p) + \eta * e(p) * s(p-l) \tag{1}$$

$$y(p) = \sum_{l=0}^{k-1} w_l(p) * s(p-l) \tag{2}$$

$$e(p) = d(p) - y(p) \tag{3}$$

where

$y(p)$ filter output,
$e(p)$ signal error,
$s(p)$ input signal,
η *step size*,
$w(p)$ filter weights.

3.2 Normalized Least Mean Square (NLMS) Adaptive Algorithm

In order to overcome the drawback of LMS algorithm, NLMS algorithm [15, 16] was suggested which uses variable step size. By doing so, it converges faster than the LMS. The steps of NLMS algorithm are given below

$$\eta(p) = \frac{\beta}{d + |s(p)|^2} \tag{4}$$

$$w(p+1) = w(k) + \frac{\beta}{d + |s(p)|^2} e(p) * s(p) \tag{5}$$

where

β is the NLMS adaption parameter
d normalization constant.

4 The Self-adaptive Neuro-fuzzy Inference (SANFIS) System

The self-adaptive neuro-fuzzy inference system creates its name from adaptive neural fuzzy inference system where the parameters are updated itself using back-propagation algorithm. The combination fuzzy systems with neural networks possess good the learning ability in optimizing the coefficients (Fig. 2).

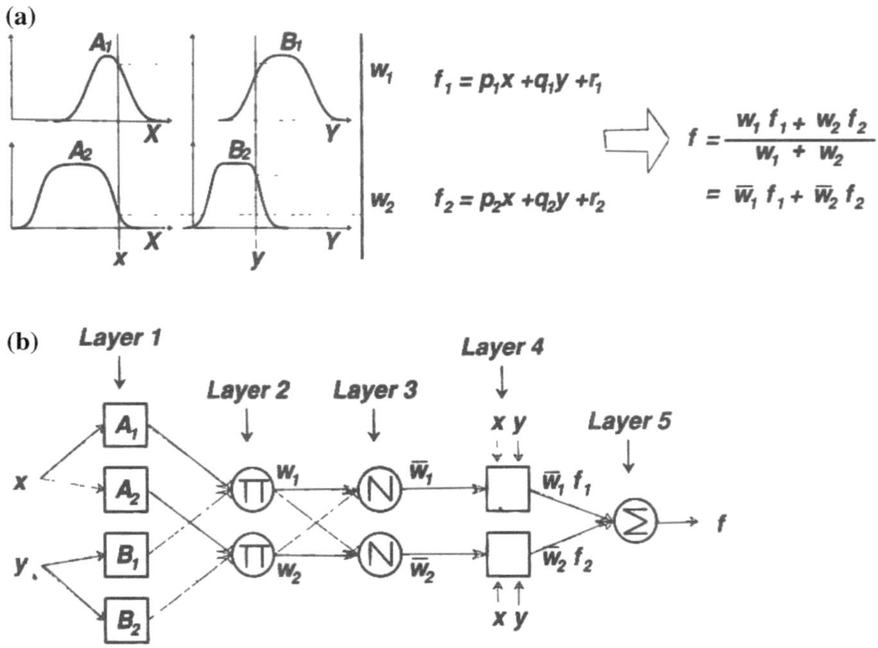

Fig. 2. SANFIS architecture

The architecture consist of [17–19] described as below

$$\text{axiom 1:let}\{x\,\text{is}\,A_1\;\text{and}\,y\,\text{is}\,B_1\}\;\text{so}\,f_1 = \{p_1x + q_1y + r_1\} \tag{6}$$

$$\text{axiom 2:let}\{x\,\text{is}\,A_2\;\text{and}\,y\,\text{is}\,B_2\}\;\text{so}\,f_2 = \{p_2x + q_2y + r_2\} \tag{7}$$

First layer: each node with activation function is given by

$$R_{1,k} = \mu_{A_k}(x), \quad k = 1, 2 \tag{8}$$

$$R_{2,k} = \mu_{B_{k-2}}(x), \quad k = 3, 4 \tag{9}$$

Second layer: every node in this level is given by

$$R_{2,k} = w_j = \mu A_k(x)\mu B_k(y), \quad k = 1, 2 \tag{10}$$

Third layer: every node in this is given by

$$O_{3,k} = \bar{w}_k = \frac{w_k}{w_1 + w_2}, \quad k = 1, 2 \tag{11}$$

Fourth layer: every node in this level is given by

$$O_{4,k} = \bar{w}_k f_k = \bar{w}_k (p_k x + q_k y + r_k) \tag{12}$$

Fifth layer: every node in this level is given by overall response

$$O_{5,k} = \sum_k \bar{w}_k f_k = \frac{\sum_k w_k f_k}{\sum_k w_k} \tag{13}$$

The suggested methodology of noise cancellation with SANFIS approach is expressed below.

In Fig. 3, $X(k)$ is the information-bearing signal which is to be reconstructed from the primary noise signal n(k). This is passed through nonlinear dynamics (f) and generates signal $Z_1(k)$, which summed to $X(k)$ to form the measured output signal $Y(k)$. The purpose is to retrieve $X_1(k)$ from the measured signal $Y(k)$ and delayed version of primary noise $Z_2(k)$ which should be the same version of information signal $X(k)$.

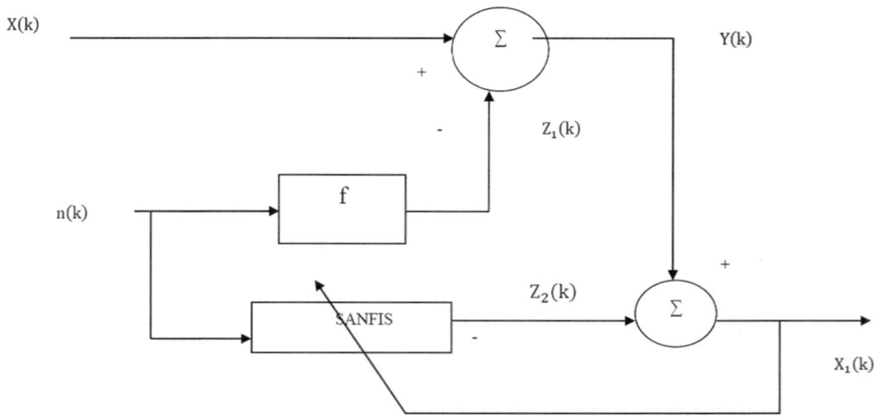

Fig. 3. Block diagram of the proposed SANFIS method

5 Results and Discussion

The results of simulation are analysed using MATLAB 2018. Figure 4 shows the information signal which is a sinusoidal signal. Figure 5 shows the noise signal generated from a primary noise source by a nonlinear process. Figure 6 is the measured signal obtained by sum of information with noise signal. Figure 7 is the output of SANFIS system, and Fig. 8 is the recovered signal which is the exact replica of

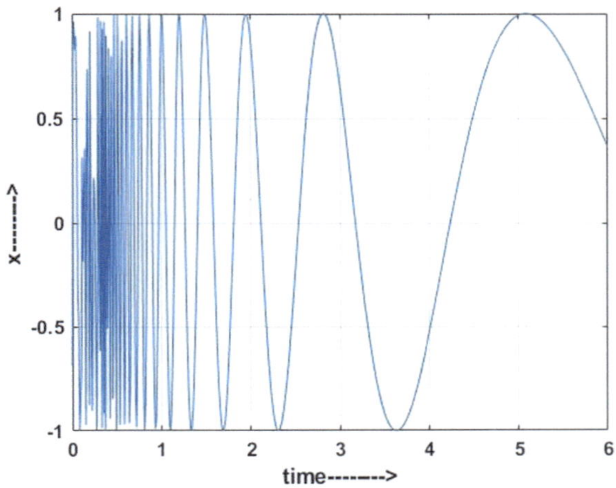

Fig. 4. Information signal

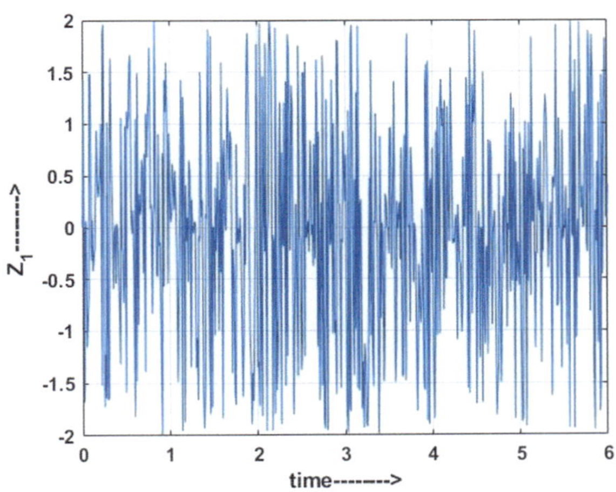

Fig. 5. Noise signal Z_1

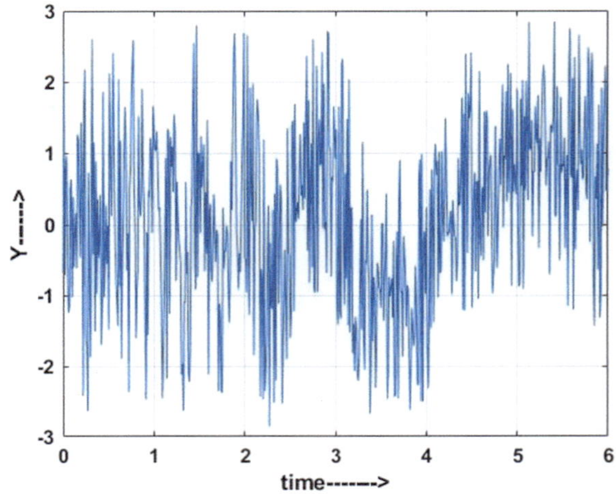

Fig. 6. Measured signal *Y*

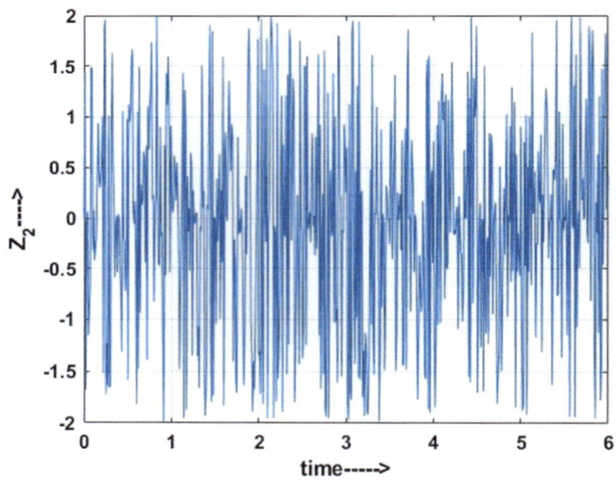

Fig. 7. Output of SANFIS Z_2

information signal. In order to find the performance measure, we plot mean squared error (MSE) curves of LMS, NLMS and SANFIS algorithms shown in Fig. 9. We observe that LMS algorithm converges around 40 iterations whereas NLMS algorithm around 30 iterations. Similarly, for the same data, SANFIS system converges around 20 iterations. From these results, we conclude that NLMS algorithm converges faster compared to LMS algorithm. Similarly, the converge rate of SANFIS system algorithm is more faster compared to NLMS. Thus, we conclude SANFIS system is better than LMS and NLMS algorithms in terms of quality performances.

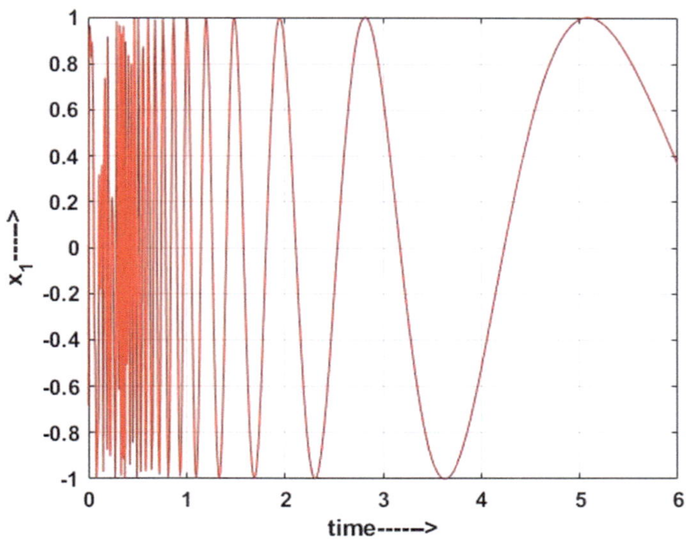

Fig. 8. Recovered signal X_1

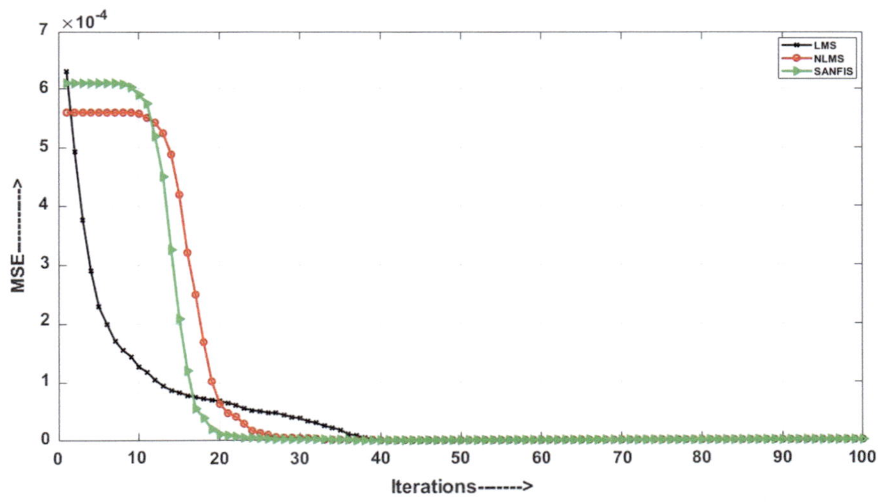

Fig. 9. MSE curve for comparisons of LMS, NLMS and SANFIS system

6 Conclusion

In this work, the idea of fuzzy logic and neural network are merged to develop SANFIS. This SANFIS technique is used for noise cancellation. We extend the method of linear adaptive noise cancellation into the nonlinear by using nonlinear adaptive systems. The performance of SANFIS technique for adaptive noise cancellation, compared with the LMS, NLMS algorithm. The future work may include the optimization algorithms for different kinds of noises and to implement with DSP microcontroller and LABVIEW.

References

1. Sambur M (1978) Adaptive noise canceling for speech signals. IEEE Trans Acoust Speech Signal Process 26(5):419–423
2. Burgess JC (1986) Adaptive signal processing. In: Widrow B, Stearns SD (eds)
3. Wu JD, Lin SL (2010) Audio quality improvement of vehicular hands-free communication using variable step-size affine-projection algorithm. Int J Wavelets Multiresolut Inf Process 8 (06):875–894
4. Sasaoka N, Shimada K, Sonobe S, Itoh Y, Fujii K (2009) Speech enhancement based on adaptive filter with variable step size for wideband and periodic noise. In: IEEE international midwest symposium on circuits and systems, August 2009, pp 648–652
5. Ahmad MS, Kukrer O, Hocanin A (2013) A 2-D recursive inverse adaptive algorithm. SIViP 7(2):221–226
6. Kim PU, Lee Y, Cho JH, Kim MN (2011) Modified adaptive noise canceller with an electrocardiogram to enhance heart sounds in the auscultation sounds. Biomed Eng Lett 1 (3):194
7. Widrow B, Glover JR, McCool JM, Kaunitz J, Williams CS, Hearn RH, Goodlin RC (1975) Adaptive noise cancelling: principles and applications. Proc IEEE 63(12):1692–1716
8. Haykin S (2008) Adaptive filter theory. In: 27th annual international conference of the engineering in medicine and biology society. IEEE Press, Pearson Education India, pp 1212–1215
9. Albert TR, Abusalem H, Juniper MD (1991) Experimental results: detection and tracking of low SNR sinusoids using real-time LMS and RLS lattice adaptive line enhancers. Naval Ocean Systems Center, San Diego, CA
10. Kazemi R, Farsi A, Ghaed MH, Karimi-Ghartemani M (2008) Detection and extraction of periodic noises in audio and biomedical signals using Kalman filter. Sig Process 88(8):2114–2121
11. Akingbade KF (2014) Separation of digital audio signals using least-mean-square (LMS) adaptive algorithm. Int J Electr Comput Eng 4(4):2088–8708
12. Dhiman J, Ahmad S, Gulia K (2013) Comparison between adaptive filter algorithms (LMS, NLMS and RLS). Int J Sci Eng Technol Res (IJSETR) 2(5):1100–1103
13. Dixit S, Nagaria D (2017) LMS adaptive filters for noise cancellation: a review. Int J Electr Comput Eng 7(5):2088–8708
14. Haykin S (2008) Adaptive filter theory. In: 27th Annual international conference of the engineering in medicine and biology society. IEEE Press, Pearson Education India, pp 1212–1215
15. Paulo SD (2008) Adaptive filtering algorithms and practical implementation. Int Series Eng Comput Sci 23–50
16. Mohammed J (2012) A study on the suitability of genetic algorithm for adaptive channel equalization. Int J Electr comput Eng 2(3):285
17. Juang CF, Chiou CT, Lai CL (2007) Hierarchical singleton-type recurrent neural fuzzy networks for noisy speech recognition. IEEE Trans Neural Netw 18(3):833–843

18. Theocharis JB (2006) A high-order recurrent neuro-fuzzy system with internal dynamics: Application to the adaptive noise cancellation. Fuzzy Sets Syst 157(4):471–500
19. Qin H, Yang SX (2007) Adaptive neuro-fuzzy inference systems based approach to nonlinear noise cancellation for images. Fuzzy Sets Syst 158(10):1036–1063
20. Chandrakar C, Kowar MK (2012) Denoising ECG signals using adaptive filter algorithm. Int J Soft Comput Eng (IJSCE) 2(1):120–123, 2088–8708
21. Dixit S, Nagaria D (2017) Design and analysis of cascaded LMS adaptive filters for noise cancellation. Circuits Syst Signal Process 36(2):742–766
22. Prasetyowati SAD, Susanto A (2015) Multiple processes for least mean square adaptive algorithm on roadway noise cancelling. Int J Electr Comput Eng 5(2):355

Frequency Regulation in an Islanded Microgrid with Optimal Fractional Order PID Controller

Narendra Kumar Jena[1], Subhadra Sahoo[1], Amar Bijaya Nanda[1], Binod Kumar Sahu[1(✉)], and Kanungo B. Mohanty[2]

[1] Department of Electrical Engineering, Siksha 'O' Anusandhan Deemed to be University, Bhubaneswar, Odisha, India
{narendrajena, subhadrasahoo, amarnanda}@soa.ac.in
[2] Department of Electrical Engineering, NIT, Rourkela, Rourkela, Odisha, India
kbmohanty@nitrkl.ac.in

1 Introduction

Microgrid (MG) paves a solution for electrification in an isolated area. Apart from this, incessant rise in demand, limited conventional resources, and environment issues entail to develop MGs. MGs in off-grid condition bring the voltage and frequency regulation problem, which seldom faces such problem in the grid-connected mode. The problems in the isolated MG are (a) low inertia due to the microturbine DGs, (b) sporadic nature of renewable energy sources, and (c) no inertia with complex structure of energy-storing elements. So, by the robust secondary controller, LFC problem tackled to bring the frequency deviation within a predefined boundary [1, 2].

Different researchers delved with LFC study on MGs endorsing different kinds of controllers such as PID controller [2], adaptive PI controller [3], fuzzy PI controller [4], type-2 fuzzy PI controller [5–7], *H*-infinite controller [8], $H_{\infty,}$ and μ-synthesis-based controller [9], tilted integral derivative (TID) controller [10], and virtual inertia control of microgrid [11]. In fuzzy-type controller, rule formulation is a difficult task by which with improper input–output scaling parameters leads the system into an unstable region or the response carries high steady-state error. In *H*-infinite type controller, it is difficult to know all the internal states, and in practice, it is too difficult to implement. But PID is a conventional one which is used profusely in industries due to its simple structure and can be easily implemented. From this angle in this paper, a fractional order-based PID controller is adopted which is a maiden attempt to use as a secondary controller in the LFC of a microgrid. This FOPID controller can reach out to produce an improved dynamic response in comparison with PID controller because it carries two numbers of extra adjusting knobs. Adjusting these extra two knobs, undershoot, overshoot, and settling time can be improved than the result produced by PID controller. To get the optimal gains of controllers, various researchers employed different computational

M. N. Mohanty and S. Das (eds.), *Advances in Intelligent Computing and Communication*, Lecture Notes in Networks and Systems 109,
https://doi.org/10.1007/978-981-15-2774-6_53

techniques like PSO [2], Big Bang–Big Crunch optimization [5], salp swarm optimization (SSA) [7], and grasshopper optimization [12]. In this paper, to have a better transient response of the microgrid, the gains of PID and FOPID controllers are optimized by a SOS algorithm.

The objective of this paper is assimilated as follows:

a. A fractional order PID controller is designed by SOS algorithm to employ as a secondary controller in the proposed isolated MG.
b. Performance of this proposed controller (FOPID) is validated by comparing with PID controller.
c. Robustness of the controller is examined by subjecting power generation variation as well as load variation.

2 Systems, Proposed Controller, and Algorithm

2.1 System Considered

The MG shown in Fig. 1 [9] is configured by its linearized model as depicted in Fig. 2. The microgrid contains PV source, WTG, diesel generator, FC, MTG, FH storage unit, and BES unit whose specifications are tabled in Appendix and are taken from [9]. The total power required to meet the demand is the sum of power generated by renewable sources, diesel generator, FC, MTG, and power exchange of storage elements. The power balance equation is expressed in Eq. (1).

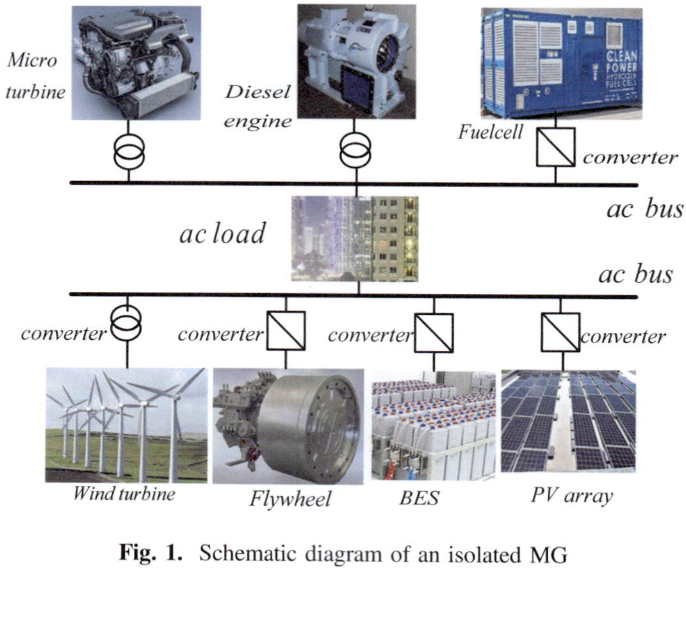

Fig. 1. Schematic diagram of an isolated MG

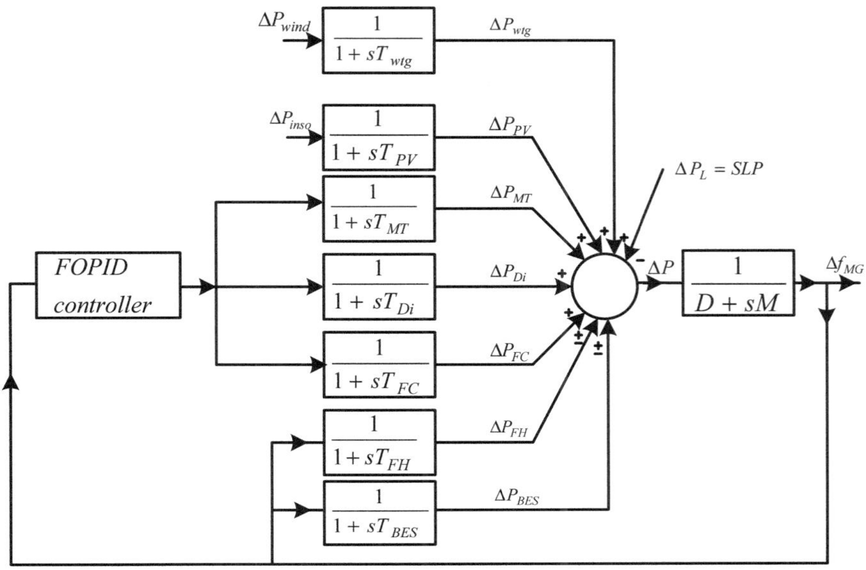

Fig. 2. Linearized model of MG

$$P_L = P_{\text{PV}} + P_{\text{wtg}} + P_{\text{MT}} + P_{\text{Di}} + P_{\text{FC}} \pm P_{\text{FH}} \pm P_{\text{BES}} \tag{1}$$

With an incremental change of load, the power balance equation becomes as given in Eq. (2).

$$\Delta P_L + \Delta P_{\text{PV}} + \Delta P_{\text{wtg}} + \Delta P_{\text{MT}} + \Delta P_{\text{Di}} + \Delta P_{\text{FC}} + \Delta P_{\text{FH}} + \Delta P_{\text{BES}} = 0 \tag{2}$$

The frequency deviation from Fig. 2 is given in Eq. (3).

$$\Delta f = \frac{\Delta p}{D + sM} \tag{3}$$

where

$$\Delta p = \Delta P_{\text{PV}} + \Delta P_{\text{wtg}} + \Delta P_{\text{MT}} + \Delta P_{\text{Di}} + \Delta P_{\text{FC}} + \Delta P_{\text{FH}} + \Delta P_{\text{BES}} - \Delta P_L,$$

D = damping coefficient, and M = inertia constant.

Mathematical modeling of different sources is elaborated below.

$$\Delta \dot{P}_{\text{wtg}} = \frac{\Delta P_{\text{wind}} - \Delta P_{\text{wtg}}}{T_{\text{wtg}}} \tag{4}$$

$$\Delta \dot{P}_{\text{PV}} = \frac{\Delta P_{\text{inso}} - \Delta P_{\text{PV}}}{T_{\text{PV}}} \tag{5}$$

$$\dot{\Delta P}_{MT} = \frac{u - \Delta P_{MT}}{T_{MT}} \tag{6}$$

$$\dot{\Delta P}_{\text{Di}} = \frac{u - \Delta P_{\text{Di}}}{T_{\text{Di}}} \tag{7}$$

$$\dot{\Delta P}_{\text{FC}} = \frac{u - \Delta P_{\text{FC}}}{T_{\text{FC}}} \tag{8}$$

$$\dot{\Delta P}_{\text{FH}} = \frac{\Delta f_{\text{MG}} - \Delta P_{\text{FH}}}{T_{\text{FH}}} \tag{9}$$

$$\dot{\Delta P}_{\text{BES}} = \frac{\Delta f_{\text{MG}} - \Delta P_{\text{BES}}}{T_{\text{BES}}} \tag{10}$$

$$\dot{\Delta f} = \frac{\Delta p - D\Delta f}{M} \tag{11}$$

The state-space model of the microgrid as in [9] taking all the differential equations from Eqs. 4–11 is expressed in Eq. 12.

$$\begin{pmatrix} \dot{\Delta P}_{\text{wtg}} \\ \dot{\Delta P}_{\text{PV}} \\ \dot{\Delta P}_{\text{MT}} \\ \dot{\Delta P}_{\text{Di}} \\ \dot{\Delta P}_{\text{FC}} \\ \dot{\Delta P}_{\text{FH}} \\ \dot{\Delta P}_{\text{BES}} \\ \dot{\Delta f} \end{pmatrix} = \begin{pmatrix} \frac{-1}{T_{\text{wtg}}} & 0 & 0 & 0 & 0 & 0 & 0 & 0 \\ 0 & \frac{-1}{T_{\text{PV}}} & 0 & 0 & 0 & 0 & 0 & 0 \\ 0 & 0 & \frac{-1}{T_{\text{MT}}} & 0 & 0 & 0 & 0 & 0 \\ 0 & 0 & 0 & \frac{-1}{T_{\text{Di}}} & 0 & 0 & 0 & 0 \\ 0 & 0 & 0 & 0 & \frac{-1}{T_{\text{FC}}} & 0 & 0 & 0 \\ 0 & 0 & 0 & 0 & 0 & \frac{-1}{T_{\text{FH}}} & 0 & \frac{-1}{T_{\text{FH}}} \\ 0 & 0 & 0 & 0 & 0 & 0 & \frac{-1}{T_{\text{BES}}} & \frac{-1}{T_{\text{BES}}} \\ 0 & 0 & 0 & 0 & 0 & 0 & 0 & \frac{-2D}{M} \end{pmatrix} \begin{pmatrix} \Delta P_{\text{wtg}} \\ \Delta P_{\text{PV}} \\ \Delta P_{\text{MT}} \\ \Delta P_{\text{Di}} \\ \Delta P_{\text{FC}} \\ \Delta P_{\text{FH}} \\ \Delta P_{\text{BES}} \\ \Delta f \end{pmatrix}$$
$$+ \begin{pmatrix} \frac{1}{T_{\text{wtg}}} & 0 & 0 \\ 0 & \frac{1}{T_{\text{PV}}} & 0 \\ 0 & 0 & 0 \\ 0 & 0 & 0 \\ 0 & 0 & 0 \\ 0 & 0 & 0 \\ 0 & 0 & 0 \\ 0 & 0 & \frac{2}{M} \end{pmatrix} \begin{pmatrix} \Delta P_{\text{wind}} \\ \Delta P_{\text{inso}} \\ \Delta P_L \end{pmatrix} + \begin{pmatrix} 0 \\ 0 \\ \frac{1}{T_{MT}} \\ \frac{1}{T_{Di}} \\ \frac{1}{T_{FC}} \\ 0 \\ 0 \\ 0 \end{pmatrix} u, \text{ and} \tag{12}$$

$$y = \Delta f$$

To evaluate a prompt and smooth dynamic response of the proposed microgrid, FOPID and PID controllers are optimized by SOS algorithm which is explained in the next section.

2.2 Fractional Order PID Controller

Podlubny [13] explored non-integer-based controller $\left(PI^{\phi}D^{\lambda}\right)$ in which ϕ and λ are two fractional operators. The operational plane of fractional controller from Fig. 3a shows that PID only works at a single point but $PI^{\phi}D^{\lambda}$ controller works on the whole quadrant. So, from this whole area suitable points are available for controlling the system. The control signal of $PI^{\phi}D^{\lambda}$ controller is given in Eq. (13).

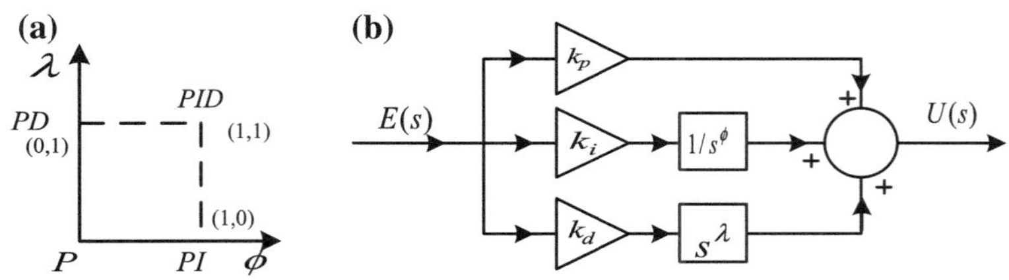

Fig. 3. **a** Operational quadrant of PID, and fractional PID. **b** Structure of FOPID controller

$$u_{\text{FOPID}}(t) = k_p e(t) + k_i D_t^{\phi} e(t) + k_d D_t^{\lambda} e(t) \tag{13}$$

where $e(t)$ is the error signal injecting into the controller. In the Laplace domain, the transfer function of $PI^{\phi}D^{\lambda}$ is expressed in Eq. (14).

$$G_{\text{FOPID}}(s) = k_p + \frac{k_i}{s^{\phi}} + k_d s^{\lambda} \tag{14}$$

2.3 Symbiotic Organisms Search (SOS)

The SOS algorithm [14] imitates the symbiotic collaboration procedures for the most part received by different organisms to endure and scatter in the biological system. Organisms never isolate themselves to lead their life because of dependence on different species for sustenance and even endurance. Symbiosis epitomizes the bonding between at least two kinds of organisms for their feeding association. Often, living beings get advantage one another or one gets an advantage from the other without doing any damage to the other living being. Therefore, according to their dependency with each other, the SOS algorithm technique contains three stages in particular mutualism stage, commensalism stage, and parasitism stage which are represented in the flowchart in Fig. 4 eloquently. SOS has only few parameters which are fixed and does not require any parameter tuning like other algorithms.

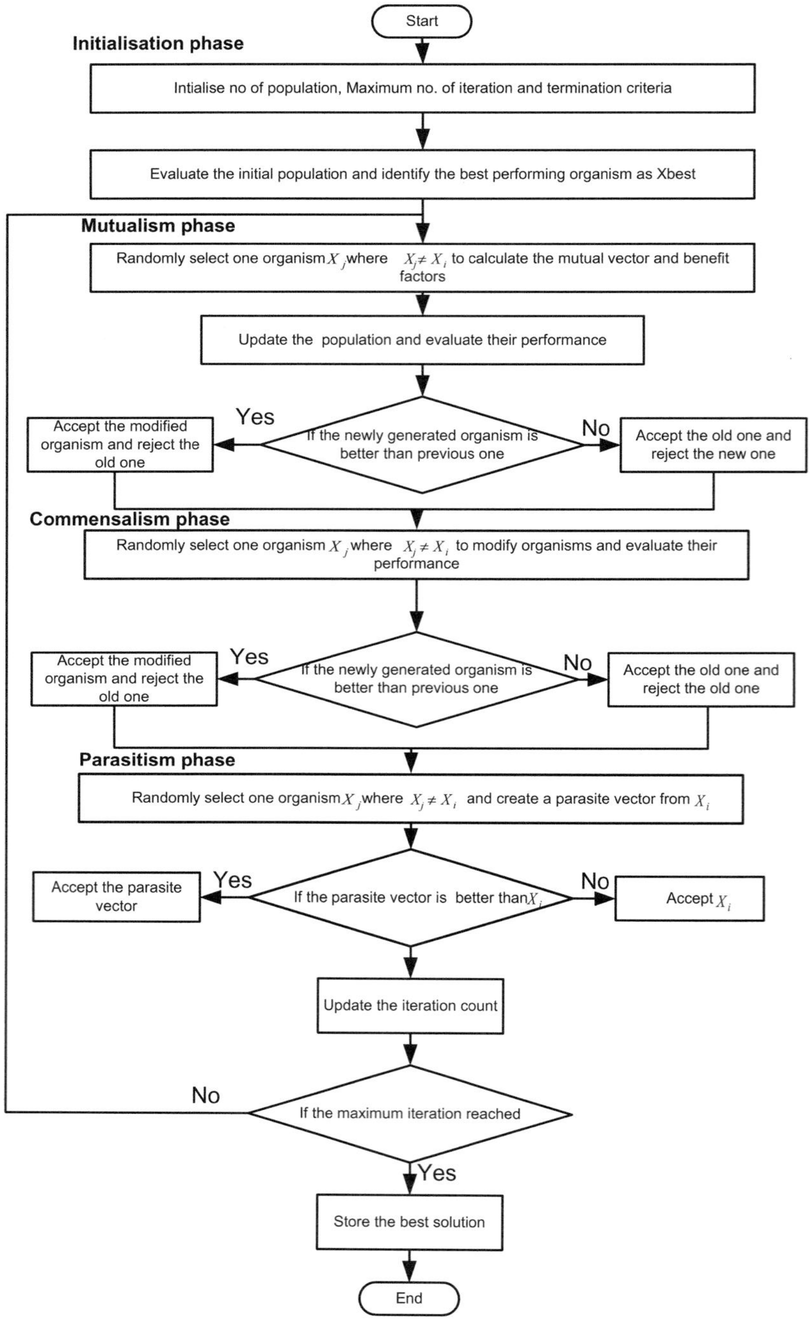

Fig. 4. Flowchart for SOS computational technique

3 Result and Discussion

The proposed model in Fig. 1 is deployed in MATLAB/Simulink to study the frequency control under five different cases which are based on different combinations among the change of solar power, wind power, and load. The dynamic response is assessed by employing PID and FOPID controller designed by SOS algorithm. The enumerated gain parameters are given in Table 1. In this paper, with these same designed gain values all the rest cases have been studied which is different from other cited papers.

Table 1. Gain parameters of controllers

Controller	k_p	k_i	k_d	ϕ	λ
PID	−14.3624	−2.7633	−16.1832	–	–
FOPID	−19.6294	−15.4307	−8.4892	0.9321	0.8007

Case-1 Dynamic behavior under variable load and constant solar, and wind power
In this case, a load deviation of 0.01 is imposed injecting constant solar power and wind power of each 0.2 pu. The frequency response is depicted in Fig. 5a. The overshoots and settling times (2% band) of frequency excursions by PID and FOPID controllers are 13.876 MHz, and 5.985 s, and 6.797 MHz and 3.875 s, respectively. The FOPID controller gives a prompt response curtailing both the overshoot and the settling time.

Case-2 Dynamic behavior under variable load and constant solar, and wind power
After getting the success in the first case, frequency control is studied in the same system by subjecting a wider load variation (−0.035 to 0.1 pu) endorsing FOPID controller. The wind power and solar power are 0.2 pu each. The load variation curve and frequency deviation curve are captured in Fig. 5b. The time curve in Fig. 5b shows a minor deviation in frequency which shows the design of the controller and is optimal.

Case-3 Dynamic behavior under constant load, constant solar power, and variable wind power.
Under this condition, frequency deviation dynamics is studied keeping solar power at 0.2 pu, load at 0.01, and variable wind power ranging from −0.1 to 0.4 pu. This excess increment or decrement of power injected into the system is balanced by the energy-storing element by which the frequency deviation dynamics are stabilized. The excursion of frequency portrayed in Fig. 5c shows a skimpy overshoot than the response observed in case 1.

Case-4 Dynamic behavior under constant load, constant wind power, and variable solar power
In this context, a variable solar power varied from −0.1 to 0.3 pu along with a constant wind power of 0.2 pu and a constant load of 0.01 pu. The time response of frequency

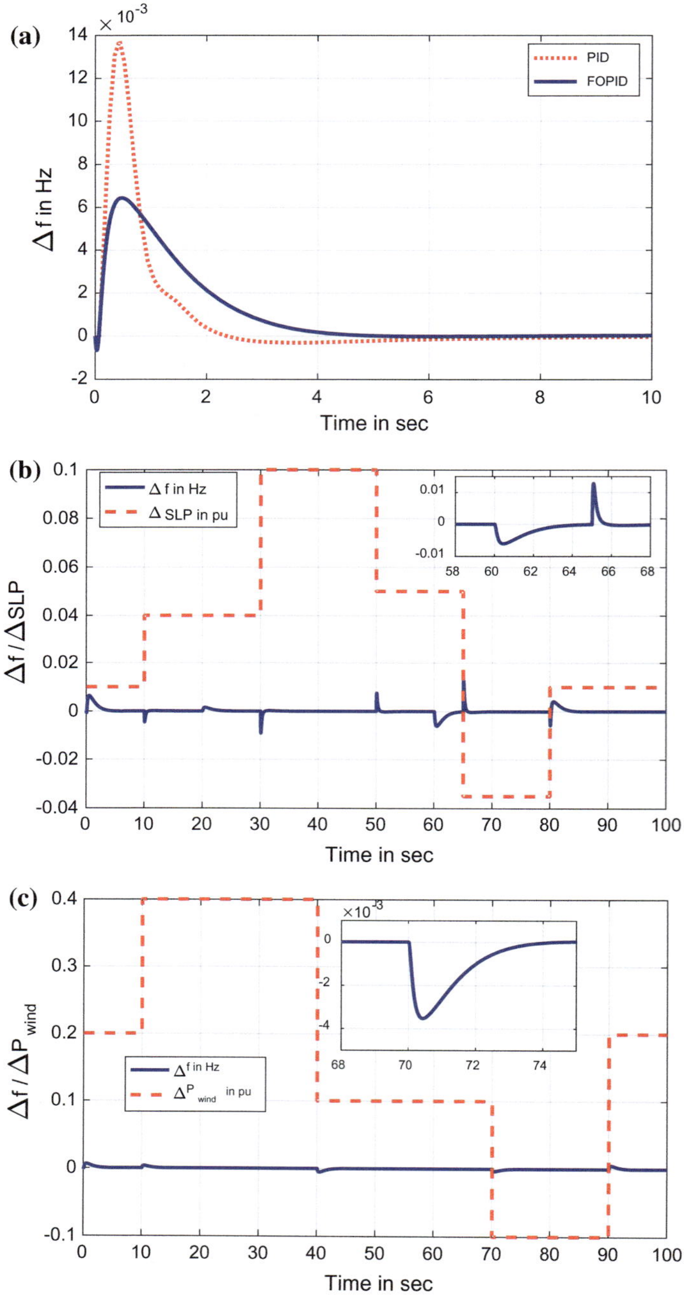

Fig. 5. **a** Frequency deviation curve. **b** Frequency deviation curve against variable load. **c** Frequency deviation curve against variable wind power. **d** Frequency deviation curve against variable solar power. **e** Frequency deviation curve against variable load, solar power, and wind power

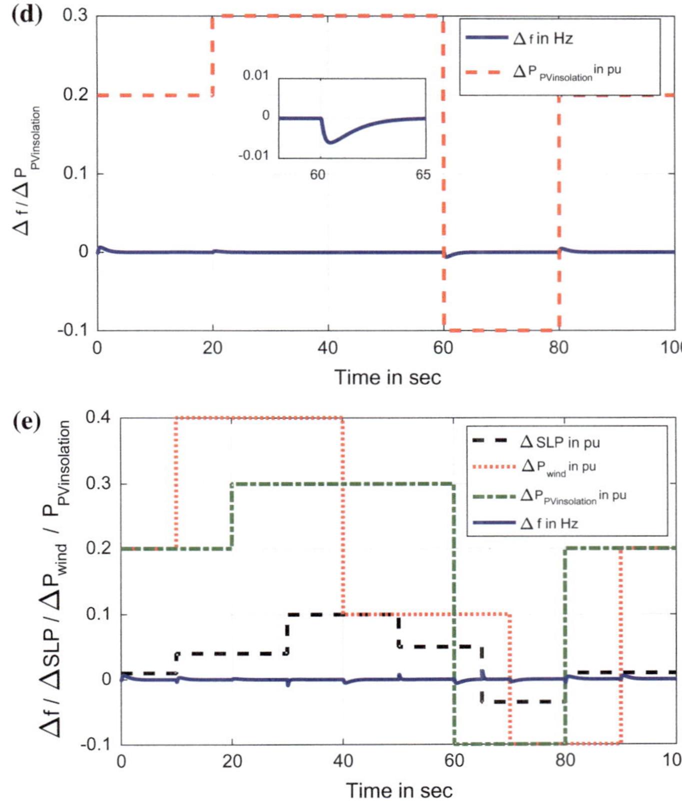

Fig. 5. *(continued)*

deviation depicted in Fig. 5d shows an improved transient response in comparison with case 2 because of the power balance correction is carried out by the energy-storing components.

Case-5 Dynamic behavior under variable load, wind power, and solar power The last case deals with the perturbation of load in the range of −0.035 to 0.1 pu, wind power variation from −0.1 to 04 pu, and solar power variation from −0.1 to 0.4 pu. The transient behavior frequency along with variable load, wind power, and solar power time diagram as shown in Fig. 5e shows the stability of the system and is increased due to the power exchanged by the energy-storing element into the system. The transient response is quite faster with exhibiting minimal overshoot.

4 Conclusion

The simulation of islanded MG enticing FOPID as a secondary controller optimized by SOS algorithm is carried out successfully. The FOPID controller ameliorates the transient response produced by the PID controller efficiently. The proposed controller tackles the response proficiently and promptly in the wake of load variation, solar, and wind power variation independently as well as combining. The appreciable excursion of frequency deviation in all cases and in all forms (undershoot, overshoot, and settling time) confirms the robustness of the proposed controller that is perceived.

Appendix

D = Damping coefficient = 0.012	T_{MT} = Time constant of MT = 2 s	T_{FH} = Time constant of FH = 0.1 s
M = Inertia constant = 0.2	T_{Di} = Time constant of diesel engine = 2 s	T_{BES} = Time constant of BES = 0.1 s
T_{PV} = Time constant of PV = 1.8 s	T_{FC} = Time constant of FC = 4 s	T_{wtg} = Time constant of WTG = 1.5 s

References

1. Yang J, Zeng Z, Tang Y, Yan J, He H, Wu Y (2015) Load frequency control in isolated micro-grids with electrical vehicles based on multivariable generalized predictive theory. Energies 8(3):2145–2164
2. Bevrani H, Habibi F, Babahajyani P, Watanabe M, Mitani Y (2012) Intelligent frequency control in an ac microgrid: online PSO based fuzzy tuning approach. IEEE Trans Smart Grid 3(4):1935–1944
3. Zheng S, Tang X, Song B, Lu S, Ye B (2013) Stable adaptive PI control for permanent magnet synchronous motor drive based on improved JITL technique. ISA Trans 52(4): 539–549
4. Li X, Song YJ, Han SB (2008) Frequency control in micro-grid power system combined with electrolyzer system and fuzzy PI controller. J Power Sources 180(1):468–475
5. Engin Y (2014) Interval type-2 fuzzy PID load frequency controller using big bang-big crunch optimization. Appl Soft Comput 15:100–112
6. Khooban MH, Niknam T, Blaabjerg F, Davari P, Dragicevic T (2016) A robust adaptive load frequency control for micro-grids. ISA Trans 65:220–229
7. Sahu PC, Mishra S, Prusty RC, Panda S (2018) Improved-salp swarm optimized type-II fuzzy controller in load frequency control of multi area islanded AC microgrid. Sustain Energy Grids Netw 16:380–392
8. Singh VP, Mohanty SR, Kishor N, Ray PK (2013) Robust H-infinity load frequency control in hybrid distributed generation system. Int J Elect Power Energy Syst 46:294–305

9. Bevrani H, Feizi MR, Ataee S (2015) Robust frequency control in an islanded microgrid: H_∞, and μ-synthesis approaches. IEEE Trans Smart Grid 7(2):706–717
10. Khadanga RK, Padhy S, Panda S, Kumar A (2018) Design and analysis of tilt integral derivative controller for frequency control in an islanded microgrid: a novel hybrid dragonfly and pattern search algorithm approach. Arab J Sci Eng 43(6):3103–3114
11. Kerdphol T, Rahman FS, Watanabe M, Mitani Y, Turschner D, Beck HP (2019) Enhanced virtual inertia control based on derivative technique to emulate simultaneous inertia and damping properties for microgrid frequency regulation. IEEE Access 7:14422–14433
12. Barik AK, Das DC (2018) Expeditious frequency control of solar photovoltaic/biogas/biodiesel generator based isolated renewable microgrid using grasshopper optimisation algorithm. IET Renew Power Gener 12(14):1659–1667
13. Podlubny I (1999) Fractional-order systems and PI/sup/spl/lambda//D/sup/spl/mu//-controllers. IEEE Trans Autom Control 44(1):208–214
14. Min-Yuan C, Prayogo D (2014) Symbiotic organisms search: a new metaheuristic optimization algorithm. Comput Struct 139:98–112

Biometric Detection Using Stroke Dynamics

Abhishek Das, Saumendra Kumar Mohapatra(✉),
and Laxmi Prasad Mishra

ITER, Siksha 'O' Anusandhan (Deemed to be University), Bhubaneswar, India
laxmimishra@soa.ac.in

1 Introduction

Alphonse Bertillon, a famous French Police Officer, was the first person who developed an identification system based on body measurements for criminals in the 1870s. 'Anthropometry' developed by him was the first scientific biometric system used by police to identify criminals. The reason behind this development was impersonation. Impersonation is a crime associated with pretending to be someone else. Earlier criminals changed their appearances to evade capture, so he used body measures to identify and capture criminals. That means life measures can be used to authenticate a person.

Biometrics means life measures is a method to identify persons based on their characteristics. It depends on unique biological characteristics to verify an individual. These biometric authentication systems compare currently extracted data with the stored data. If both data match, then authentication is confirmed; otherwise, the person is blocked to use the system. The applications for biometrics can be found in government centers, military applications, commercial applications, banking, etc. Biometrics can be divided into two kinds: physiological and behavioral.

The system requires no dedicated action from the user, so it avoids the problems of need for physical contact. And also, for capturing biometric data, no external hardware is required. The system discriminates the users by using the combination of human motion analysis and typing rhythm biometrics. Continuous authentication (CA) is the method in which we have to check the identity of the user continuously, based on each user's action performed. This method is to authenticate persons both statistically and continuously to provide continuous security without much interrupting the user.

Our work is ordered as follows: Section 1 gives the brief introduction of this project with its objective and motivation. Section 2 describes the existing system. Section 3 gives the design of the proposed system and the description of the method. Implementation is organized in Sect. 4. Result analysis is included in Sect. 5. Section 6 concludes the work.

M. N. Mohanty and S. Das (eds.), *Advances in Intelligent Computing and Communication*, Lecture Notes in Networks and Systems 109,
https://doi.org/10.1007/978-981-15-2774-6_54

2 Related Works

(A) Face Recognition

The principal completely mechanized personification-based face acknowledgment approach fit for working with information procured in the wild is presented. 3D face structure is produced from a solitary 2D picture and contrasts it and a reference model for acquiring a minimal portrayal of face highlights deviations. But it requires expensive camera, and construction of the 3D model is difficult and takes more time [1]. Face recognition is done using linear discriminant analysis method. SVD-LDA method is used to reduce the time complexity. LDA is taking mean as the discriminating factor not taking the covariance, and so it does not contain as much varied data. Also, this method is supervised so cannot work efficiently with new features [2].

Then, user's face detection is done using the principal component analysis method. By using PCA, the researchers have reduced the amount of variables required. This identification was carried out by providing a new image in the eigenface subspace and by comparing the spot in eigenface subspace with the spot of already stored images. But the method has worked with the still images and is not useful for real-time face recognition. And also, the method did not work for different size face image recognition [3]. The fiducial points were determined by the system in one matching process, and thus, it eliminates the requirement for analyzing the equality of all the models with corresponding graphs individually. But the method is responsive to illuminating conditions. A lot of graphs have to be stored manually, and it will use huge storage [4].

(B) Keystroke dynamics

In literature, we can find a number of works were done related to key stroke estimation. They used amount of disarray of the array to differentiate among different typing data. They computed the mean distance of the provided sample by using all the data of different users. If the corresponding mean distance is small to the mean distance of the samples, then the user is considered as a legal user. So authentication depends on all legal users. That is, for the development of a user profile, it relies on other users [5].

In the approach by Dowland, calculating the mean and also the standard deviation of digraphs were used for determining a user profile. Typing example was collected by monitoring the users during their regular behavior. But it showed high error rates [6].

In authentication via keystroke, dynamics clustered the data with the help of a variation in the maxi-min distance algorithm. Another variable is the typing speed of a user which is used as the criteria for clustering. But there is a need to recluster the data as the system is used [7].

(C) Mouse Dynamics

A power transformation method is used to analyze the holistic attributes of mouse trajectories. A schematic representation is made using these holistic attributes. But the method listening carefully on the user verification in a static scenario. And also, it is using fixed sequence of mouse behavior [8].

The users have to perform a precise job, and the job consisted of moving the mouse between two lines, i.e., from a point where it was started and the ending point. The participants are instructed to perform this task 5 times per session, 6 sessions in total resulting in total 30 runs. But the problem is that, instruction to do a specific task did not represent the natural user's behavior [9].

Active authentication by mouse movements developed a random look like environment for revealing user's heuristic mouse movements than just a fixed path for every user. User's one sample movement of all procedure is recorded as a reference data. Training and verification phases are there. Features from 9 paths between 7 squares are displayed consequently [10].

User's behaviors are extracted from interaction with a Web page in Web biometrics. It uses the normal login process of a conventional online security system. User's verification is by Web interaction [11].

(D) Combining Keystroke and Mouse Dynamics

A work uses MD and KD in multi-model decision fusion for continuous authentication. Architecture with 11 number of sensors are used. A dataset having data of 67 users while working in an office for a week is used. A mixture of naive Bayes and SVM is used to analyze the performance [12].

A new reporting technique was used to evaluate the system performance. They have obtained a good result, i.e., 50 out of 53 unblocked users out by their system while 3 users were occasionally wrongly locked out by the system [13].

3 Methodology

3.1 Problem Statement

The problem is to develop a mechanism for providing static and continuous authentication for humans. Static authentication is based on face recognition, and continuous authentication is provided by mouse and keyboard interaction behavior. On the off chance that the present conduct of a client digresses altogether from the ordinary conduct, the framework hails this conduct as irregular.

3.2 Design

Biometric authentication systems compare confirmed authentic data in a database with captured biometric data. Authentication is confirmed, if the two examples of biometric information coordinate. Ceaseless confirmation is a type of dynamic validation which changes the point of view of verification from an occasion to a procedure. Design of the proposed system is displayed in Fig. 1.

The proposed method starts with static authentication, and then, the continuous analysis is performed by the combination of mouse and keystroke dynamics. Both the static and continuous authentication methods consist of two different phases:

1. Training state:

During which the user's data is collected and a final model is produced.

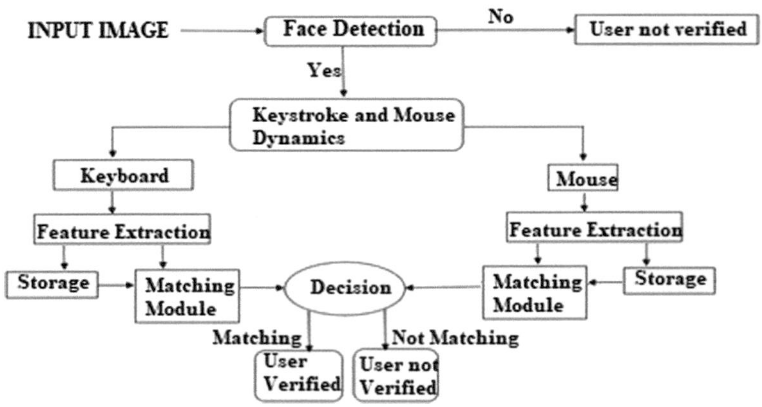

Fig. 1. System architecture

2. Authentication phase:

The model is then used to constantly screen they believed clients account. In the event that the present conduct of a client strays altogether from typical conduct, at that point the framework hails this conduct as atypical.

3.3 Overall System Design

Building of face recognition is by principal element study. Eigenface approach transforms faces into a small set of essential characteristics, eigenfaces. Recognition is performed by extracting the face features and comparing them with the measurements stored in the database. Face recognition allows the user to continue to use the system, if both face features are matched. It can use the existing hardware infrastructure and existing camera. Then, the user is continually monitored by the combination of keystroke and mouse dynamics to avoid the session hijacking. Keystroke dynamics is a form of behavioral biometrics. The aim behind this keystroke dynamics is to identify humans and it is based on the analysis of their typing rhythms on the keyboard. By estimating the stay time (the period of time a key is held down) and fly time (the time term from a key discharged to the following key squeezed) for every console activity, elements can be removed.

In mouse dynamics, for each user, a schematic characteristic representation is made by using the holistic attributes (travelled distance, time, etc.). Then, procedural characteristic representation is made using the dynamic process of mouse movements by performing behavior segmentation.

4 Implementation

4.1 Face Recognition

Face identification is a biometric method of identifying an individual by comparing live capture with the stored record for that person. It is an "Easy-to-collect" biometrics. The whole recognition process involves two steps:

A. Initialization process.
B. Recognition process.

When a user starts using a system, then his face image is captured. After preprocessing the face images, the features are extracted.

PCA is used to overcome the expensive computation and the need for great amounts of storage of older face recognition technique such as correlation methods. The features extracted are stored.

In authentication phase also, the features are extracted in the same way as in the training phase and classification method of nearest neighbor is used for classification. When the authorized user uses the system after his training phase, then the user identified as genuine and if a user, whose features are not stored starts using the system he is identified as an unauthorized user and system is blocked to prevent further movements of the unauthorized user in the system.

If an authorized user is using the system, he is continued to use the system and keystroke, and mouse dynamics is used to avoid session hijacking. So that the user can be continually monitored (Fig. 2).

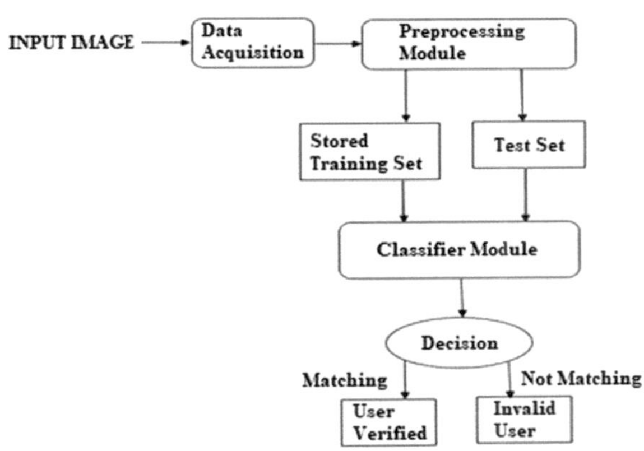

Fig. 2. Face recognition

PCA is the method for compressing a lot of data into something that captures the essence of original data. It is done by the decomposition of covariance matrix. PCA finds the principal components of data, and it does what it says on the name. The eigenvector with the highest eigenvalue is the principal component. Every image in the training set is represented as eigenfaces. Eigenface maximizes the scatter between all face images in the training set.

Keystroke dynamics aims to identify humans based on the analysis of their typing rhythms on the keyboard. And it is a behavioral biometric. The time and value of all key press and key release events are recorded. The method has two phases: enrollment and verification. Learning phase extracts the dynamics, and the verification phase compares the current user with the dataset. K-nearest neighbor is used for the classification at authentication phase. By using KNN, the genuine user and malicious user can be found out. Each user exhibits distinctive hand motion gestures and habits, and the mouse interaction behavior appears to be unique. Here, users are discriminated by the way they interact with the mouse.

4.2 Procedural Feature Construction

The dynamic process of a mouse movement is captured by partitioning each mouse movement into several segments. By partitioning the mouse movement, movement segments are obtained. Dividing each mouse movement into four movement segments (four phases).

1. First phase: Speed increases from 0 and user starts moving the mouse from starting position.
2. Second phase: Speed reaches the maximum user accelerates.
3. Third phase: Speed reduces and user decelerates the mouse moving in the target direction.
4. Fourth phase: Speed reduces to 0. User adjusts mouse to approach the target.

Features are extracted from segments, and the extracted features then combined to form a vector. It represents the mouse segments of each mouse operation.

The verification system mainly consists of four main components:

1. Recorder
2. Feature construction
3. User verification model
4. Fusion and decision maker.

Fsch = Schematic feature vector classification score.

Fpro = Procedural feature vector classification score.

The combination rules used are:

1. Product rule: Fsch * Fpro.
2. Sum rule: Fsch + Fpro.
3. Max rule: max (Fsch, Fpro).
4. Min rule: min (Fsch, Fpro).
5. Weighted sum rule: Wsch*Fsch + Wpro*Fpro.

Recognize the legitimate user by calculating the scores. The classification scores from both the schematic features and procedural features are combined using the combination rules. The weighted sum rule gives the best performance while the product rule showed the worst performance.

The result from both the verification models (keyboard and mouse) is compared to make a decision. The result from the keyboard authentication and mouse authentication is compared to decide whether the user is genuine or not. So a continuous authentication of the user is performed by comparing both the results (Fig. 3).

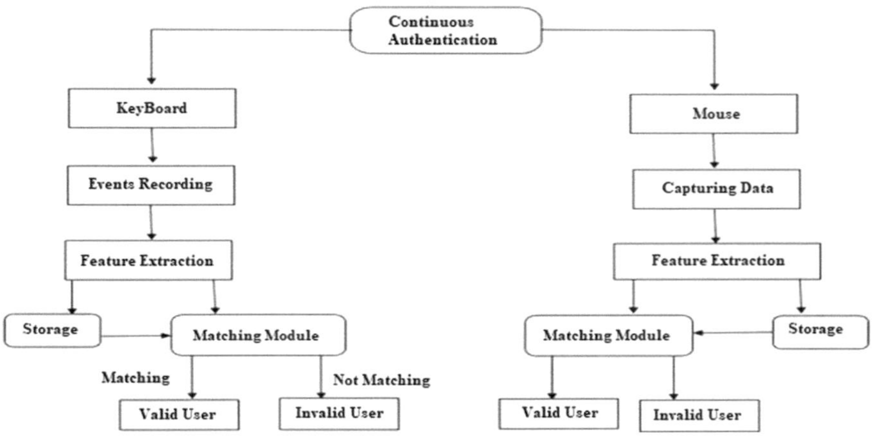

Fig. 3. Combination of keyboard and mouse

5 Results and Discussions

In order to evaluate the proposed recognition system, experiments were conducted. These total 25 users were tested using the method. Out of 25, 13 are genuine users and 12 are unauthorized users. We do test in initial learning and also in behavioral learning. The results of test done after behavioral learning are shown in Fig. 4. And the result of test done after initial learning is shown in Fig. 6. The performance measures found out during the test are shown in Figs. 5 and 7. From the figures, it is clear that the results obtained by the proposed method are better than doing single biometric authentication.

BEHAVIORAL LEARNING					
	NO. OF TESTS	TP	TN	FP	FN
FACE	25	9	11	2	1
KEYBOARD	25	10	9	2	2
MOUSE	25	11	10	1	4
COMBINATION	25	10	10	3	2

Fig. 4. Behavioral learning test result

METHOD	ACCURACY	PRECISION	RECALL	F SCORE
FACE	81%	0.81	0.75	0.78
KEYBOARD	83%	0.32	0.85	0.85
MOUSE	76%	0.74	0.67	0.70
COMBINATION	91%	0.95	0.85	0.86

Fig. 5. Performance measures in behavioral learning

INITIAL LEARNING					
	No. of Tests	TP	TN	FP	FN
FACE	25	9	9	1	4
KEYBOARD	25	10	8	2	4
MOUSE	25	9	9	2	3
COMBINATION	25	11	11	1	3

Fig. 6. Initial learning test result

METHOD	ACCURACY	PRECISION	RECALL	F SCORE
FACE	79%	0.88	0.68	0.77
KEYBOARD	80%	0.78	0.79	0.76
MOUSE	78%	0.68	0.69	0.74
COMBINATION	86%	0.91	0.79	0.83

Fig. 7. Performance measure in initial learning

Both in initial learning and behavioral learning, the combination of face, keyboard, and mouse is showing more accuracy, precision, etc., than the single biometric system.

On comparing both the performance measures results, it is shown that the results got from the behavioral learning test are best than the initial learning test.

6 Conclusion

Biometric-based recognition of humans based on keystroke and mouse dynamics with face recognition is a simple and efficient static and continuous authentication system based on human–device interaction behavior. Continuous authentication is a form of dynamic and risk-based authentication. It changes the perspective of an event to a process. Face recognition is appropriate for static authentication because it cannot be fooled by any other methods. After the successful face authentication, the user is continually monitored by the keyboard (by using the typing pattern) and mouse (by using the user's mouse behavior). By behavioral learning, more accuracy can be achieved. Without right security measures, critical data would get into the hands of cybercriminals and also impersonation is a huge challenge. So utilizing biometrics for personal authentication is convenient and considerably more accurate than the existing methods.

References

1. "A Leopard Cannot Change Its Spots": Improving Face Recognition Using 3D-Based Caricatures. Joao Neves. IEEE Trans Inform Forensics Secur 14(1), Jan 2019
2. Deswal M, Kumar N, Rathi N (2014) Face recognition based on singular value decomposition linear discriminant analysis method. Int J Eng Innov Technol 3(12), June 2014
3. Chandrapaul L, Al Sumam A (2012) Face recognition using principal component analysis. Int J Adv Res Comput Eng Technol 1(9), Nov 2012
4. Wiskott L, Fellous JM, Kruger N, Von Der Malsburg C (1999) Face recognition by elastic bunch graph matching. IEEE Trans Pattern Anal Mach Intell 19(07), July 1999
5. Gunetti D, Picardi C (2005) Keystroke analysis of free text. ACM Trans Inform Syst Secur 8(3)
6. Dowland PS, Fvrnell SM, Papadaki M (2002) Keystroke analysis as a method of advanced user authentication and response. In: Proceedings of the IFIP TC11 17th information security
7. Monrose F, Rubin A (1997) Authentication via keystroke dynamics. In: Fourth ACM conference on computer and communication security, April 1997
8. Shen C, Cai Z, Liu X, Guan X, Maxion RA (2016) Mouse identity: modelling mouse interaction behavior for a user verification system. IEEE Trans Human Mach Syst 46, October 2016
9. Bours P, Fullu CJ (2009) A login system using mouse dynamics. In: Proceedings of the 5th international conference on intelligent information hiding and multimedia signal processing, pp 1072–1077
10. Aksari Y, Artuner H (2009) Active authentication by mouse movements. In: Proceedings of the international symposium on computer and information sciences
11. Gamboa H, Fred ALN, Jain AK (2007) Web biometrics: user verification via web interaction. In: Proceedings of the biometrics symposium, p 16
12. Fridman L, Stolerman A, Acharya S, Brennan P, Juola P, Kam M (2015) Multi model decision fusion for continuous authentication. Comput Electr Eng
13. Bours P, Mondal S (2015) Performance evaluation of continuous authentication system. In: IET Biometrics, pp 1–7

Hardware Partitioning Using Parallel Genetic Algorithm to Improve the Performance of Multi-core CPU

Suryakanta Nayak(✉) and Mrutyunjaya Panda

Department of Computer Science and Application, Utkal University, Bhubaneswar, India
mrutyunjaya74@outlook.com

1 Introduction

Hardware platform comprises the outstanding digital accessories that perform software application programming packages. Owing to the need of limiting the development time of digital products, advancement in CAD tools for hardware architecture is a major challenging matter [1–3]. VLSI circuit partitioning is a key technique owing to fact that it regulates the components which to be applied on hardware circuit, which works on the software program. Over the last decades, several researches had been implemented for HW partitioning [4, 5]. Further, HW partitioning can confirm that faster throughput, less cost and efficient performance. At the commencement and procedure stage, HW partitioning was taken as a graph partitioning representation. The partitioning procedure is demonstrated as an undirected graph partitioning [6]. Moreover, specified model is marked into two unique kinds of HW partitioning issue, out of which first one may be determined ideally by polygon time complex in nature, whereas the second one is basically used to NP-hard problem [7]. Moreover, it is pertaining to mathematical aspects of partitioning, and two types of classifications for calculations are used, specific accurate techniques and approximate approaches which is applied with heuristic calculations. The predetermined calculation is utilized to obtain exceptional goals for the small size. The common exceptional ways are involved in dynamic programming branch and bound [8, 9] and linear programming [10]. In the partitioning process, the deficiency estimation and the arrangement region of HW partitioning are incremented exponentially. Heuristic calculations are used in mode simulation choices because of their modern ability to procure the best choices inside the restricted processing time. Aside this, the commencement stage of hardware circuit partitioning where space is required in heuristics, including tools arranged for heuristic calculation, had been proposed [11, 12]. The very beginnings complete challenges in product goals and continuously moves from one to other the technique, while the last arrangement begins with a product application converting segments to equipment. The equipment placed in the system is treated as a presentation requirement process. The software application technique is treated as a period limitation process. After that point, many general heuristics and meta-heuristics methods had been additionally evolved, like hereditary calculation [7, 13], simulated annealing [14, 15], particle swarm

M. N. Mohanty and S. Das (eds.), *Advances in Intelligent Computing and Communication*, Lecture Notes in Networks and Systems 109,
https://doi.org/10.1007/978-981-15-2774-6_55

optimization (PSO) [16]. In the present work, Sect. 2 discusses related work, Sect. 3 discusses proposed methodology, Sect. 3.1 discusses multi-core correction by genetic algorithm, Sect. 3.2 discusses operations of hardware partitioning, and Sect. 5 narrates conclusion.

2 Related Work

The application of segmentation to an undirected graph is nicely demonstrated in [7], where it is described the segmentation with the guide of measurements are, communication price (CP), hardware partitioning (HP), which might be developed. So, it deals with the insufficiencies of the above procedure, an inclining framework fixated based upon the two-dimensional search was suggested to take care of the arrangement space of problem (*P*). Further, the strategy for searching of *P* was governed with the guide of the goals of boundary *P*0 [7]. Summing up the advantages of direct method and indirect method, at first, the solution of the problem is investigated by genetic algorithm, and next, the development idea is adopted from the second indirect ways for problem-specific constraint matter. More often, our improvement strategy is unlike that of existing methods [12, 13], and simply one impracticable solution is corrected. Here, a number of impracticable solutions are corrected. In this way, plan can improve the search concentration and ultimately improve the solution quality and address the constraint concern. Moreover, techniques are projected to equal the architecture of CPU hardware and reduce the run time, otherwise it would be time taking. In our framework, various infeasible arrangements are adjusted. In this strategy, our procedure can expand this pursue. A novel parallel example for our revision methods is proposed to analyze great the structure of CPU and GPU equipment. The analyses confirm the adequacy of our framework. In multi-core CPU and GPU has emerged as a well-liked parallel computing platform with low energy absorbs and a high ratio of performance to cost. This is a vital role in the science and engineering domains [17, 18]. For that reason, we determine the process of use of how to use a parallel multi-core CPU and GPU on individual computers for hardware partitioning is an attractive matter. The present research proposes a brand name—new CPU and GPU by employing parallel genetic algorithm scattering revision for HW apportioning. The difference in structure among CPU and GPU information said that switch among CPU and GPU is through PCI-E bus. The design of the CPU is limpness arranged. At the point when there is complex decision-making ability in programming applications, the structure of CPU shows enormous information over that of GPU. The significant matters are that CPU is throughput oriented, and it measures CPU performance, the primary part supervises the nodes and keeps a large portion of the region, to minimize the circuit even as in a introverted chip of CPU, number reduction of time and space majority of the circuit. There are more than a couple of parallel programming dialects helping multi-core CPU and GPU. This paper represents that how to compute unified device architecture.

3 The Proposed Method

In a circuit partitioning, meta-heuristic technique is utilized for calculation of throughput by using genetic selection. The basic components inside the genetic calculation integrate replacement, amalgam and change. In our framework, these center components are adjusted for HW partitioning. We stop our calculation when the same type of result will be done, in minimization techniques. In this proposed technique, calculation is done with scattering rectification for the HW partitioning issue. It introduces a basic and simple to utilize a component of multi-stringing on more than one CPU when parallelizing a sequential application. Hence, a feasible calculation must rationalize the exchange of fixed cost among GPU and CPU. In this original copy, we planned an innovative CPU and GPU parallel example to assist genetic calculation through dispersion revision. The interior elements in the genetic search contain crossover, reproduction and mutation (Fig. 1).

3.1 Multi-Core Correction by Genetic Algorithm

In normal generic calculation, an underlying factor is the initial element to the subsequent stages. Let us consider for hardware partitioning, x represents (x_1, x_2, x_n). Here, x means an HW partitioning of concern P. It proposes that the node V_i is allocated to hardware partitioning. Subsequently, the length of communication is equivalent to the number of nodes in the task graph. It recommends an outline of randomized members. The rest of the graph are produced to utilize arbitrarily arrangement of their properties, two qualities are inside the circuit first, and second is distinct and arbitrary calculation said that population cost is high. The full-adder circuit will take as our information structure. Figure 2 demonstrates the 17 hubs, and the underlying allocation is finished by the hereditary calculation depends on the guideline of typical inherited qualities and predictable assurance (Fig. 3).

3.2 Genetic Operations for Hardware Partitioning

At every single cycle, common genetic calculation chooses two groups in the present. For registering the product expense and the equipment cost, they consent to the identical technique because of the calculation of the equipment charge. In Fig. 4 illustrate the computing cost, generally position of hardware partitioning initialization are required for computation. For the processing, the discussion rate relies upon the representation of the endeavor diagram on CPU. The weight on every single region must indicate the discussion rate between couples of hubs. The consecutive method of correspondence rate is $O\ (n^2)$. In view that two hubs of each characteristic are taken by methods for their strings, the information situation of the response is the vector set on mutual memory. In this manner, the summary of every single feature is completed strings.

Algorithm 1
1. residual 1 = R − (S(y) + C(y))
2. at the same time residual $\geq$ 0 three.
3. for each and every $y_i = 0$

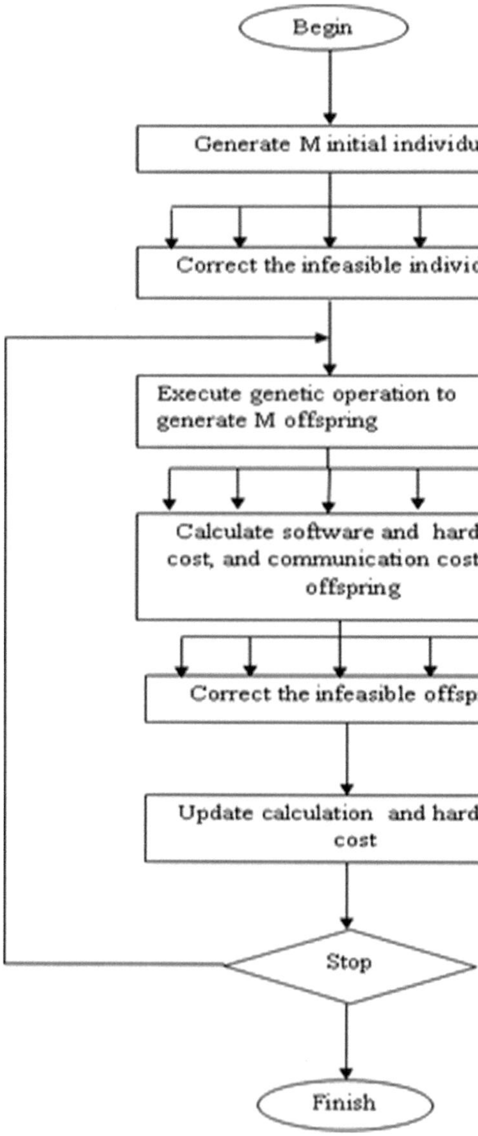

Fig. 1. Proposed model of hardware partitioning

4. do
5. assume $y_i = 1$
6. calculate Oci and hello/(s_i + Oci)

The goal is our to minimizing the hardware cost, for HW partitioning, the smaller the hardware fee, the more the saved hardware cost.

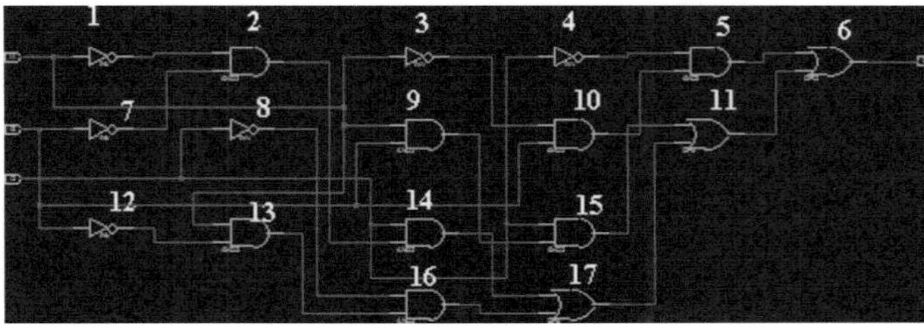

Fig. 2. A full-adder circuit

Individual 1	1	0	1	0	1	0	1	
Individual 2	1	0	1	0	1	0	1	0
Individual n	1	0	1	0	1	1	1	0

Fig. 3. An initialized population for HW partitioning

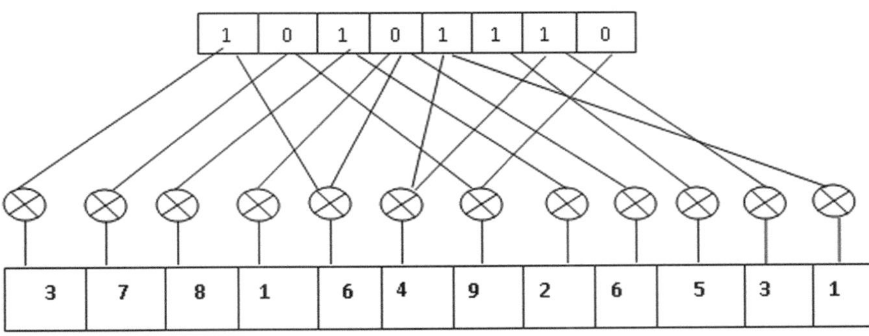

Fig. 4. Computing cost on CPU

Algorithm 2

1. residual1 = R − (S(y) + C(y))
2. while residual $\geq$ 0
3. for each $y_i = 0$
4. do
5. assume $y_i = 1$
6. Compute Oci and $h_i/(s_i + Oci)$
7. decide node k with maximum residual
8. end for
9. if exists node k then

10. if exists node k then
11. update H (y),S(y), C(y)
12. residual
13.update H (y), S(y),C(y)
14.else
15.else
16.break
17.end if
18.end while.

4 Results and Discussion

The proposed methodology is consolidated the advantages of both the immediate techniques and roundabout strategies. The current deviating strategies are outperformed by the listed methods [6, 7], which is a realistic method. In the present research calculating data transfer to the CPU and GPU, minimization of the equipment cost is our objective. It is calculated in the first output to represent data transfer send to GPU. The technique is the high caliber and straightforward. The figure four represents the performance analysis in between array size and speed. It recommends that the appropriate response quality by utilizing our system is the high caliber in general cases. In the figure five represents the data transfer to CPU and GPU simultaneously, *x*-axis represents the array size in byte, *y*-axis represents transfer speed GB/s, and in the read write data transfer, CPU is better than GPU. Aside from, taking the data transfer technique, the current ways are better than the proposed hereditary calculation in [7].

4.1 Physical Realization of the System

See Figs. 5 and 6.

5 Conclusion

The present investigation accomplishes a CPU and GPU parallel hereditary calculation with inclusion of scattering rectification for the hardware dividing issue. The work is portrayed as per the following: at first, a more grounded inherited calculation with distribution adjustment is exhibited. Further, including preparing costs calculation and dispersion revision are kept running in parallel. At last, we have displayed a novel parallel technique by method for utilizing the corresponding intensity of multi-center CPU as well as many centers GPU. Finally, the suggested method reveals that at the appointed time an innovative substitute design for CPU and GPU.

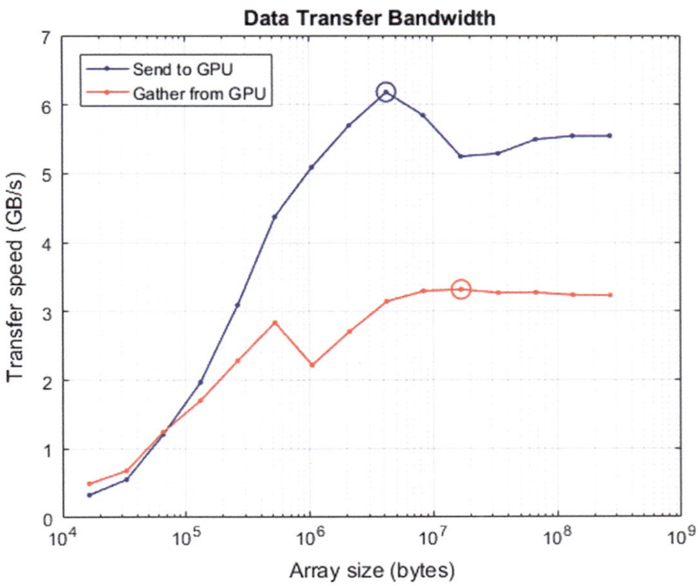

Fig. 5. Relationship between array size and data transfer speed

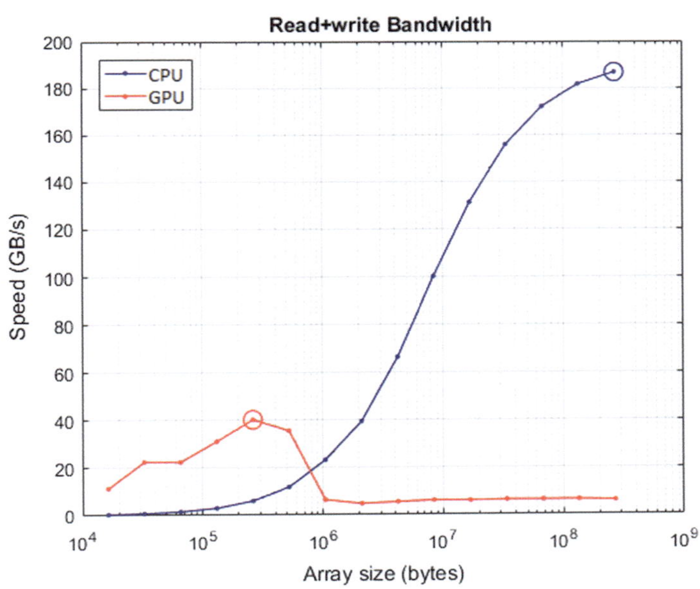

Fig. 6. Relationship between read write bandwidth and data transfer speed

References

1. Wolf WH (2011) Hardware-software co-design of embedded systems. Proc IEEE 82:968–989
2. Wolf WH (2017) A decade of hardware/software co-design. Computer 36(4):38–43
3. Teich J (2018) Hardware/software co-design: the past, the present, and predicting the future. Proc IEEE 100:1411–1430
4. Mhadhbi I, Ben Othman S (2016) A comprehensive survey on hardware/software partitioning process in co-design. Int J Comput Sci Inf Secur 14(3):263
5. Wang R, Hung WN (2016) Uncertainty model for configurable hardware/software and resource partitioning. IEEE Trans Comput 65(10):3217–3223
6. Arató P, Juhász S (2003) Hardware-software partitioning in embedded system design. In: Presented at the IEEE international symposium intelligent signal process., Budapest, Hungary, Sept 2003, pp 197–202
7. Arató P, Mann ZA, Orbán A (2015) Algorithmic aspects of hard-ware/software partitioning. ACM Trans Des Autom Electron Syst 10:136–156
8. Shi W, Wu J (2016) Algorithms for bi-objective multiple-choice hardware/software partitioning. Comput Electr Eng 50:127–142
9. Trindade AB, Cordeiro LC (2016) Applying SMT-based verification to hardware/software partitioning in embedded systems. Des Autom Embedded Syst 20(1):1–19
10. Mann ZA, Orbán A (2012) Finding optimal hardware/software partitions. Formal Methods Syst Des 31(3):241–263
11. Gupta RK, Micheli GD (1993) Hardware-software co synthesis for digital systems. IEEE Des Test Comput 10(3):29–41
12. Ernst R, Henkel J, Benner T (1993) Hardware-software co synthesis for microcontroller. IEEE Des Test Comput 10(4):64–75
13. Janakiraman N, Kumar PN (2014) Multi-objective module partitioning design for dynamic and partial reconfigurable system-on-chip using genetic algorithm. J Syst Archit 60(1):119–139
14. Abdelhalim MB, Habib SED (2011) An integrated high-level hard-ware/software partitioning methodology. Des Autom Embedded Syst 15: 19–50
15. Yan X, He F, Hou N, Ai H (2017) An efficient particle swarm optimization for large-scale hardware/co-design system. Int J Cooperat Inf Syst 9:59–64
16. Henkel J, Ernst R (2009) An approach to automated hardware/software partitioning using a flexible granularity that is driven by high-level estimation techniques. IEEE Trans Very Large Scale Int (VLSI) Syst 9(2): 273–28
17. Wang G, Gong W, Kastner R (2006) Application partitioning on programmable platforms using the ant colony optimization. J Embedded Comput 2(1):119–136
18. Ferrandi F, Lanzi PL, Pilato C, Sciuto D (2013) Ant colony optimization for mapping, scheduling and placing in reconfigurable systems. In: Presented at the IEEE NASA/ESA conference adaptive hardware system, Turin, Italy, (2013), pp 47–54
19. Koudil M, Benatchba K, Tarabet A, Sahraoui EIB (2007) Using artificial bees to solve partitioning and scheduling problems in codesign. Appl Math Comput 186:1710–1722
20. Wang N, Liu C, Yuhong L, Shen J (2017) A visible light communication (VLC) based intelligent transportation system for lorry fleet. IEEE, ICOCN, pp 1–3

An Internet of Healthcare Things (IoHT)-Based Healthcare Monitoring System

Harleen Kaur[1(✉)], Mohd. Atif[1], and Ritu Chauhan[2]

[1] Department of Computer Science and Engineering, School of Engineering Sciences and Technology, JamiaHamdard, New Delhi, India
[2] Centre for Computational Biology and Bioinformatics, Amity University, Noida, U.P, India

1 Introduction

Traditional strategies of supplying protection cannot be directly carried out in IoT because of extraordinary requirements and verbal exchange stacks involved. Information and communication technologies (ICTs) deployed as part of medical records systems must assure various sizable safety requirements together with integrity, confidentiality, availability, authorization, non-repudiation, and authentication duty with a purpose to comfortable medical facts without impacting the efficiency of services and privacy of sufferers' records.

The primary problem that each affected person mainly residing in faraway places observed turned into unavailability of medical doctors and remedy on crucial situations. This had very dreadful effects on human's mind about the hospitals and medical doctors' offerings. Nowadays, with the implementations of new technology by using making use of IoT devices for healthcare monitoring machine, these problems have been taken care of to massive quantity. IoT is now not able to keep patients safe and healthy most efficiently but to enhance the proper way doctors provide care. Healthcare IoT can also enhance patient engagement and pride by allowing patients to spend more time with their doctors. The Internet of things (IoT) is a widespread environment in health care. More incorporated blessings and tactics with a position for the so-called Internet of healthcare things (IoHT) or the Internet of medical things (IoMT) are sought within the general context of medical care and eHealth.

Kevin Ashton coined the time net of factors for the first time in 1999. The RFID organization, based upon fashionable conversation protocols, defines the Internet of factors as the worldwide network of interconnected devices. It contains conventional areas such as embedded systems, controlling and automation systems, and wireless sensor networks to facilitate network conversation between devices (D2D). The idea was first used at MIT's Auto-ID center. RFID has become visible as a precondition for implementing structures classified as IoTs. RFID is an IoT. Today, it has both private and corporate customer programs.

The development of miniature devices has been brought about by improvements and the convergence of micro-electromechanical systems (MEMS), wireless

M. N. Mohanty and S. Das (eds.), *Advances in Intelligent Computing and Communication*, Lecture Notes in Networks and Systems 109,
https://doi.org/10.1007/978-981-15-2774-6_56

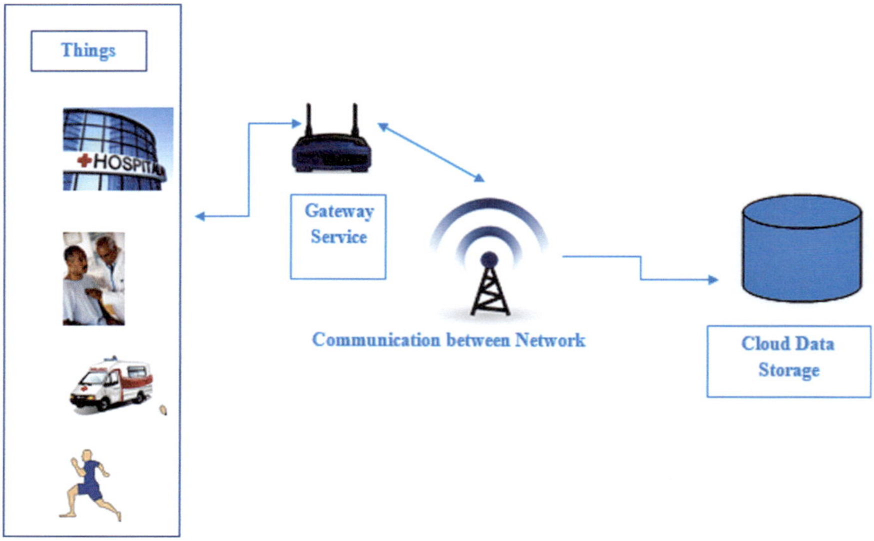

Fig. 1. Internet of things

communications, and digital electronics. They are capable of feeling, calculating, and talking wirelessly. These small devices are referred to as wireless sensor networks (WSNs) nodes [1].

In the Internet for things (IoT), it is easy to collect, file, and analyze new streams of information faster and more precisely by making gadgets quickly collect and share statistics with each other and the cloud. Numerous application domains take place on the Internet of things. The packages can be classified mainly based on Web access, insurance, scale, diversity, lack of originality, participation with individuals, and implications [2].

2 Related Works

Firstly, in 1999 Kevin Ashton suggested an IoT idea and described IoT as a unique identification of the RFID generation, loosely coupled linked accessories. Luigi et al. addressed on the IoT and the combination of several technology and communication solutions that make this promising paradigm possible. Identification and tracking technology, wired and wireless networks for sensors and actuators, improved communication protocols, and intelligence for smart products maximum applicability. The fundamentals of IoT as the aggregate of net and the rising technology have been mentioned [3].

Shen has studied that the eHealthcare gadget contains three domain names in particular: frame region, conversation and connectivity, and provider. A number of wireless placement (WAAN) networks, each similar to a consumer, describe the framework location domain. The key function of the communication and networking domain is to connect frame locations and provider domains. WAAN gateways link to the Web and allow for the efficient communication of mutual data between WAANs by advanced Wi-fi technologies. In the service field, an authority reliance maintains an online server receiving personal health and fitness information and records [4].

The IoT platform's structure and problems in the design of IoT's hardware and software additives have been addressed [5]. The IoT platform's structure and problems in the design of IoT's hardware and software additives have been addressed. They have worked out the various IoT utilities, together with intelligent cities, healthcare, farming, and nanoscale programs. In their paper, the country of IoT artwork was studied and numerous key drivers of technology were provided. The possibility of lowering the DTLS overhead by means of compression of 6LoWPAN headers was investigated, and the first DTLS compression specifier for 6LoWPAN was presented. The updated hardware, communication and computing requirements for uHealth's next technology systems have been fully assessed. Instead, they talked about new technological and technological trends and how to deal with health predictions.

Current strategies have been studied to design efficient and comfortable eHealth monitoring. They provided a comprehensive framework for superior eHealth monitoring by describing the entire life cycle in detail. They also highlighted the critical supplier additives, accurately recognizing the data series in the aspect of the individual. In order to make the proposed framework more efficient, we have offered and discussed the major challenges to solve, in order to develop a green and relaxed person-oriented tracking machine [6].

First, there is a description of the safety and privacy issues of the use of the **frame sensor community (ASN)** in health applications. Later, although most of the popular ASN studies have fully recognized the security challenges, they found that they still fail to incorporate robust safety offers that could preserve the privacy of individuals affected. Finally, it was proposed that the ASN, called ASNCare, could efficiently fulfill many safety needs of the ASN-based healthcare system, using mainly aligned IoT device [7].

The first vulnerabilities, including the latest TMIS protocol proposed in the literature, were explored. The attacks proposed mainly on the basis of stamping weaknesses, reader request calculation, and tag answer messages were not monitored in detail using a one-way hash function. Second, they proposed an efficient protocol to protect health and high performance for cell authentication dual RFID-TMIS. Its inspiration was to develop and expand the previous protocol in which TMIS was designed to use both promising technologies with the same authentication device to support the RFID era. The evaluation showed that the progressive protocol should address the weaknesses in protocol protection studied and provide mobility and efficiency [8].

In order to eliminate these vulnerabilities, a new authentication protocol on radio-frequency identification has been proposed primarily on the basis of elliptic curve encryption (ECE). In order to generate a shared short key for encrypting subsequent messages, they have also used an elliptical curve Diffie–Hellman (ECDH) key settlement protocol. Their protocol, carried out in a set of safety homes, likes mutual authentication, anonymousness, confidentiality, forward safety, privacy, middle class attack resistance, resistance to replay and impersonation attack resistance. The consequences of implementation suggest that our proposed protocol exceeds complexity over time compared with other similar protocols and calls for significantly less operational volatility [9].

There is a comfortable IoT framework that ensures end-to-end protection against IoT devices from an IoT utility is proposed and is composed of IoT, IoT and IoT devices. The IoT broker manages its personal devices and collects sensing information. The IoT app offers IoT offers to customers. It wants access to sensing information to utilize IoT services. In particular, intermediate protection problems should be taken into account when offering real-time healthcare services since patient's medical facts are surely a very sensitive data on privacy. The maximum IoT protocols in terms of end-to-end security, including CoAP and MQTT, however, were of greatest importance to DTLS safety. Consequently, we proposed a new IoT framework which would comply with the safety feature of the end-to-end CoAP communication. A symmetrical encryption and mainly based encoding of attributes are encoded for oral exchange and computing efficiencies in this context. In addition, every IoT tool has as one of its attributes a completely unique identity. Thus, while the IoT broker is a medium node, only decrypt and display facts when all the attributes are met [10].

A secure user profiling structure [11] was provided which includes personal information on the individuals involved. At the same time, an affected person and a clinic are keeping the information up-to-date. Data can be leaked while they are being percented and communicated. To solve the problems of security, the safe communications channel must be mounted and an OTP input price generated using a two-way hash feature, with a single password between a boss and health center. This work provides a dual hash functional method for creating a one-time password to ensure that the secured key is relaxed on the channel. As a result, attackers cannot decrypt the information leaked due to the secured key and the proposed method exceeds existing strategies in calculated value phrases. The Internet of things is being used; a concept within the health field is not now without further data and therefore overhead transfers of facts. In an attempt to maximize capabilities and broaden reputation/use of things that have been provided by the Internet, new metrics and methods are added to cope with these overheads. The above technique is evaluated experimentally within the paradigm of health without losing its generality. The focus is on an overview that contains statistical codecs and transmission methods and the choice of the highest combination, which can result in cost reduction/minimization. An analytical method is submitted and experimentally evaluated using theoretical metrics.

Various famous ICT paradigms, including cloud, IoT, and big data, have been mentioned [12]. It provided a thorough evaluation of the works of art and their convergence. Then, they proposed a M2M gadget for electronic health programs, based on the decentralized cloud structure, fashionable systems, and RTUs. The device has been designed in such a way that data can be aggregated to create virtual sensors for big data processing of sensor records, and some dimension implications were supplied.

The data protection problems of the patient have been diagnosed [13], and their corresponding mechanisms have been found in the literature selected. The evaluation found that IAE, AAE, and its variations are the most used way to deal with the privacy concerns of the affected person in health clouds. Other strategies based on theoretical modes and frameworks which do not use any encryption strategy are not therefore implemented in the current world situation.

A light smash glass has been suggested to manipulate (LiAAC), which will help to provide access to encrypted medical files by entering and accessing damage glass attribution-based files [14]. In daily situations, a scientist with an attribute set can decipher and get admission to the information, which please the medical file policy. Under emergency conditions, the access mechanism for break-glasses bypasses the right of entry to clinical report coverage to allow a timely transfer of data via emergency hospital therapy or rescue. LiAAC is lightweight because the overheads for garage and transmission are low and only a few calculations are made through devices within the healthcare network. The LiAAC version is officially proven comfortable, and extensive experiments in the famous version are carried out to show its performance.

A light smash glass has been suggested to manipulate (LiAAC), which will help to provide access to encrypted medical files by entering and accessing damage glass attribution-based files [15]. In daily situations, a scientist with an attribute set can decipher and get admission to the information, which please the medical file policy. Under emergency conditions, the access mechanism for break-glasses bypasses the right of entry to clinical report coverage to allow a timely transfer of data via emergency hospital therapy or rescue. LiAAC is lightweight because the overheads for garage and transmission are low and only a few calculations are made through devices within the healthcare network. The LiAAC version is officially proven comfortable and extensive experiments in the famous version are carried out to show its performance.

A review of and analysis of the weakness of an authentication protocol for WMSNs is mainly based on the medical services [6]. They found that the protocol within the authentication and session key agreements phase was not always correct so that Ui and Sn could not authenticate each other correctly and could not agree on a consultation key. The biometrics are provided as the third aspect of authentication [11] and in comparison with preceding protocols the new consumer authentication protocol, which is based primarily on WMSNs is designed in an attempt to eliminate drawbacks of (He 2015) protocol.

The proposed [10] scheme was analyzed, and its scheme was indicated to suffer from a repeat attack and to be deficient. Crypto et al. proposed a reliable power-efficient control system to overcome the weaknesses of the system for wireless sensor networks. In addition, their new scheme has proven itself easy to cope with many kinds of attacks. The framework for coalesce storing personal and oral exchange has been proposed instead of developing uneven safety answers for garage and verbal trade, [9]. Fusion has been suggested. Fusion utilizes existing IoT relaxed conversation protocols consisting of Internet Protection Protocol (IPsec) and DTLS, and re-uses defined storage conversation protection mechanisms. Thus, advanced security frameworks for communication extend to the storage area. With this mechanism, the records requested can be transmitted from the recording device directly without decrypting statistical blocks. As a consequence, Fusion offers speed and performance advantages that are essential for useful IoT devices with limited resources.

An upheaval [16] method was derived called GSD technique for the detection of malicious attacks in IoTs which include intrusion detection, based entirely on mathematical morphology. They have successfully produced GSD clusters to show the number of energy nodes in a wireless sensor community at once because the GSD curves are comparable while the amount of power nodes in the wireless sensor network is consistent. The facts of each node are used because the network parameters are used in this method.

Various aspects of IoT technology have been examined, and diverse architectures and structures have provided for the healthcare network to facilitate the access of the IoT backbone and the transmission and reception of scientific records. In the IoT-driven fitness services and applications, substantial R&D efforts were made. The paper also provides research on how the IoT can address pediatric and old people's treatment, persistent health surveillance, personal health, and health management. The paper presents a broad overview of the way that recent and ongoing progress has inspired cheap healthcare gadgets and linked health services to unlimitedly increase the capacity of IoT-based mainly healthcare services to further developments for deeper insights into business developments and technological enabling solutions. The paper considers various safety requirements and challenges to increase the understanding of IoT healthcare safety and reveals extraordinary research problems in the vicinity that could suggest a version that mixes the related safety dangers. The dialog on many critical issues, including standardization, the community type, business model, the quality of the carrier, and protection of health facts, will help to provide the basis for similar IoT-based studies on entirely healthcare services. This paper provides rules and rules on eHealth and IoT for the advantage of various stakeholders interested in evaluating IoT healthcare technology.

3 Need and Challenges of IoHT in Healthcare

A. Need of IoT in Healthcare

(i) In order to make better patient decisions, IoT ensures that all information is taken into account. Networked health alternatives and fully interactive technology strengthen and improve treatment effectiveness.

(ii) Sophisticated IoT technology enriches emergency treatment and makes emergency assistance more efficient.

(iii) Sophisticated medical devices call on patients to use prescribed medication on time and other essential health initiatives. IoT tends to improve patient comfort.

(iv) IoT helps physicians to discuss urgent cases with specialists across the world.

(v) Various sophisticated sensors and different equipments help doctors to examine the patient's health and his/her internal feelings with ease.

B. Challenges of IoT in Healthcare
   (i) As the daily data generated is huge, therefore to maintain security and privacy of data is the biggest challenge.
   (ii) Number of medical devices can create network congestion due to overload and can lower the speed of data transfer. This is due to the lack of sophisticated protocols.
   (iii) The cost of affording the IoT devices is yet a big challenge.

4 Conclusion

The long projected IoHT health revolution is already under way in this paper. As new applications increase, the urgent demand for cheap, hand care continues to be addressed. In the meantime, the automatic IoHT building blocks and verbal exchange of device to machine are still hooked up are added bureaucracy to the service layer throughout the IoT infrastructure. This revolution is characterized by providing IoT-driven healthcare solutions for end-to-end treatment and connectivity. However, a machine would like to be designed according to this assessment, where doctors can analyze the critical parameters of the affected person in real time using comfortable mechanisms elsewhere. A device must be designed to provide a whole security against a certain attack, including access to the IoT-based healthcare control and authentication protocol. IoHT demands are enormous and can massively support the healthcare sector. Use of IOHT technology can reach every patient globally and communicate physicians with patients. In future, it will keep growing and serve the purpose of treating patient at ease. In coming year, hospitals will acquire the IoHT technology to strengthen their medical facility and patient will get treated in less amount of time. IoHT helps to enhance the interaction between patients and doctors.

References

1. Bagci IE, Raza S, Roedig U, Voigt T (2016) Fusion: coalesced confidential storage and communication framework for the IoT. Sec Commun Netw, pp 2656–2673
2. Bandyopadhyay D, Sen J (2011) Internet of things: applications and challenges in technology and standardization. Wireless Personal Commun, pp 49–69
3. Korteum G, Kawsar F, Fitton D, Sundramoorthy V (2010) Smart objects as building blocks for the internet of things, pp 44–51
4. Liang X, Xu i, Shen Q, Lu R, Lin X, Shen XS, Zhuang W (2012) Exploiting prediction to enable secure and reliable routing in wireless body area networks, pp 388–396
5. Benssalah M, Djeddou M, Drouiche K (2016) Dual cooperative RFID-telecare medicine information system authentication protocol for healthcare environments. Sec Commun Netw, pp 4924–4948
6. He D, Kumar N, Chen J, Lee C, Chilamkurti N, Yeo SS (2015) Robust anonymous authentication protocol for health-care applications using wireless medical sensor networks. Multimedia Syst, pp 49–60

7. Gope P, Hwang T (2016) BSN-care: a secure IoT-based modernhealthcare system using body sensor network, pp 1368–1376
8. Li X, Niu J, Kumari S, Liao J, Liang W, Khan MK (2016) Anew authentication protocol for healthcare applications using wirelessmedical sensor networks with user anonymity. Sec Commun Netw, pp 2643–2655
9. Alamr AA, Kausar F, Kim J, Seo C (2016) A secure ECC based RFID mutual authentication protocol for internet of things. J Supercomput, pp 1–14
10. Garkoti G, Peddoju SK, Balasubramanian R (2014) Detection of insider attacks in cloud based e-healthcare environment. Int Conf Inf Technol, pp 195–200
11. Ko H, Song MB (2015) A study on the secure user profiling structure and procedure for home healthcare systems. J Med Syst, pp 1–9
12. Cvitic I, Vujic M, Husnjak S (2016) Classification of security risks in the IoT environment. In: Proceedings of the 26th DAAAM international symposium on intelligent manufacturing and automation, pp 0731–0740
13. Choi J, In Y, Park C, Seok S, Seo H, Kim H (2016) Secure IoT framework and 2D architecture for End-To-End security, pp 1–15
14. Bruce N, Sain M, Lee HJ (2014) A support middleware solutionfor e-healthcare system security. In: 16th international conference on advanced communication technology
15. Chi L, Hu L, Li H, Sun Y, Yuan W, Chu J (2013) Improvedenergy-efficient access control scheme for wireless sensor networks basedon elliptic curve cryptography. Sensor Lett, pp 953–957
16. Islam SMR, Kwak D, Kabir MH, Hossain M, Kyung-Sup Kwak, KS (2015) The internet of things for health care: a comprehensive survey. IEEE Access, pp 678–708

Botnet Detection Technology Based on DNS-Based Approach

Bhavya Alankar(✉)

Department of Computer Science and Engineering, School of Engineering Sciences and Technology, JamiaHamdard, New Delhi, India

1 Introduction

All the businesses and services are dependent on the Internet and the servers that store the data in the cloud and provide easy access across the globe. It gives several services that are extremely helpful to use. The Internet may be helpful positively, but the cyber crimes are prevailing across it as well. The confidentiality and integrity of the data can be harmed by doing breaches in the information security, theft of identity, and several other attacks. The person who attacks also known as botmaster spread Trojan or malwares or both which increases the bots number present in the network. Bot in the botnet is referred to robot and net to the Internet which states that botnet is a robot network or computer/servers where attacker controls and gains the access of the systems of the network, while the end is not even known to it.

The bots are used remotely through command and controlling server which is controlled by the attacker. The attacker uses one of the bots of the botnet which is the command and control server to control and communicate with the other bots through instructions. The bots can be singular as well as in a group at the same time through commands. Botnet's size can be increased compromising the devices or servers in the network. Botnet sustains the property of propagation which helps it in spreading all over the Internet [1]. The attacks that can be done through botnets are the fraud of phishing, fraud of clicks, stealing of password, spamming, fraud of bitcoin, theft of mass identity, traffic sniffing, new malware spreading, and logging of the key.

Internet Relay Chat (IRC) and Hypertext Transfer Protocol (HTTP) are the communication protocols for botnets [1]. The widely used botnet protocol is IRC as they are extremely popular and it can be easily found for use by a botmaster. IRC server is easily found and taken down if the botnet is detected, which makes a disadvantage of IRC. A zombie computer is a name given to the compromised systems; hence, botnet is also known as a zombie network. The most crucial time in the making of the botnet was the first five years. The first botnet is created by the attackers which was called as "Eggdrop" in 1993. More advanced botnets are created after that which have new features and functions until 2002. During these years, the use of botnets was started by most of the attackers which resulted in a rapid increase of cyber attacks. The world has not got sufficient data and a convincing procedure for detecting a botnet. The hazard of botnet is not strange to anyone as they are existing from a long time; there is a lot of

M. N. Mohanty and S. Das (eds.), *Advances in Intelligent Computing and Communication*, Lecture Notes in Networks and Systems 109,
https://doi.org/10.1007/978-981-15-2774-6_57

research that needs to be done to put a check on them [1]. Nowadays, the things are adding on to check the botnet in the server.

A lot of research work has been done on botnet detecting ways. However, researchers have advised ways that are distant from each other on botnets. It has been examined that to notice and audit botnets, two major ways are used [3, 6]. The major first way is using honeynets by installing them into the servers. This honeynet will be set up to focus on collecting reports of botnets and know the ways they behave. They may not be able to find botnet, but it collects the information which is further used to create a protection for botnets [7]. A setup with intentional vulnerabilities is known as honeynet. It main motive is to get attacked by the botnets; this enables us to study the activities of the attacker and several methods can be developed to secure the network. The second approach for botnet detection is established on quietly checking the network movement and analysing it. The approaches which follow the ways for detecting a botnet are the anomaly-based and Domain Name System (DNS)-based detection techniques which we have studied further in the paper. The movement of the DNS server and IDS server is been tracked, and any abrupt behaviour is saved to further study the attacker and develop methods to secure the network. This approach is more efficient as compared to other approaches.

2 Background

A lot of research has been done in botnet detection. We focus on different techniques of botnet detection and botnet suppression in this section [1–3] gave the mechanism of various botnet detection techniques. A study also gave network-based botnet detection techniques in which network traffic is inspected related to IRC protocol which may sense the presence of a botnet. The method of detection of a botnet which related to alarms when an intruder from a different network.

IRC botnet deployment is done all around in past years. The use of IRC protocol as a C&C method spreads because of its versatility, redundancy, and scalability. There are a large knowledge and code base for IRC-based bots, which enables botnet creators to reuse the source code and create a new botnet, for example, Agobot variants. A well-designed and modular Agobot's code is online available, which makes it easy to create their own botnet for botnet author.

The fundamental weaknesses of an IRC botnet are that stem from its C&C servers which are centralized. The source of command is easily identified and a single point is possessed by these servers, which make it very easy to disrupt a botnet which is based on IRC and get the IP addresses of bots present in a botnet. Botnets which use IRC as communication means for C&C are easy to detect relatively to others. The hostility is adapted by the botmasters towards IRC bots by somehow optimizing the IRC botnet's C&C architecture or by creating a new botnet. The shift currently is to decentralized P2P C&C architectures from centralized client–server. The IRC-based bots are being changed by the authors to make them resilient because of which IRC bots are still seeing widespread use P2P botnets are documented and discovered in the wild, e.g., Phatbot, Peacomm, Sinit, Slapper, and Nugache. The shift is already made to a new botnet C&C control methodology, i.e., P2P C&C. Following are the approaches which follow the ways for detecting a botnet:

1. Detection based on signature

For detecting botnets, this approach tries to find a signature. It gives us the important instruction around the form of botnet intrusion, and it is easy to apply. This approach detects the botnet at very huge per cent, and most of the detection detected by this is real. No new botnet can be detected by this approach as it has signature saved for only already familiar botnet which can be the only defect in this approach. The correlation produces signature for IDS system. The strategy driven by IDS to detect botnet has problem. The signature produced by the IDS system sometimes causes problem in detection as they can be in encrypted form and it can be difficult to decrypt the signature. The signature or sign provided by a botnet after detection is saved into the database on this technique, and the resistance is created for that particular botnet which is the main drawback as it is related to a particular botnet only [19, 20]. If another botnet of same type attacks the system again, then it is not able to detect it as it has signature saved of only one botnet of that type, the botnet can attack again and again, and this system will not be able to detect the botnet.

2. Detection based on anomaly

The botnet is detected in this approach by movement irregularity, also known as anomaly on the net [6]. This approach works by analysing the whole grid and finds each irregular act of the movement all over the grid. The defect in detection based on signs is also removed by this approach as it has the ability to find botnets that are never used in any intrusion. It costs more as compared to detection based on signature and also superior to it in detecting a botnet. The main two proposed anomaly detection methods are based on resulting large deviations for packet level and flow data. Anomaly represents a set of records of interaction in both types of anomaly detecting method.

3. Detection based on DNS

The approach is based on DNS, where botnet is detected by getting the data of the DNS and checking if a botnet connects to the DNS or not which is not like above-mentioned approaches. Detection-based DNS and anomaly are similar as procedure practiced in detection through anomaly is practiced in DNS movement too. The group of hosts which behave abruptly are taken into account by this method which is like botnets. DNS TTL is not honoured in these group of hosts, and DNS queries are carried out to servers which are not local. This method looks into large number of response from DNS with NXDOMAIN as code of error. This technique is able to detect botnet with high efficiency.

3 Life Cycle of Botnet Technology

The botnet life cycle is a procedure for adding a new bot in a botnet. The different phases for making a botnet or creating a new bot in a botnet are contained in a life cycle. There are different researchers who state differently about the life cycle of botnet. The distinct steps in a lifecycle of botnet are infection and injection, control and command and application of botnet [1, 2]. Figure 1 describes about a botnet life cycle.

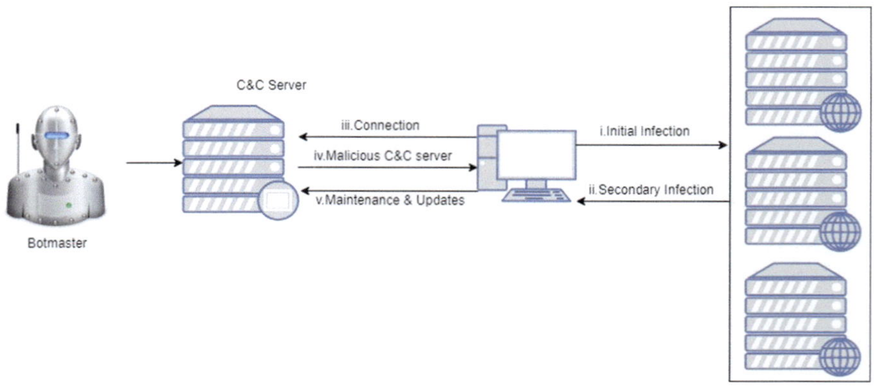

Fig. 1. Life cycle of a botnet

Other researchers state that the life cycle of the botnet is in five phases: infecting the server, injecting to the server, connecting to the server, sending virus to the server and maintaining and updating the botnet [3]. When the botnet infection is inserted into a new device which is connected to the Internet, then the injection of virus into device by HTTP protocol and P2P protocol connects to the C&C server of botnet. It starts working as a zombie computer for botnet after malicious code injection into the system. Now, the new victim device is controlled by the attacker through command and control server. The masters keep the updating and maintenance work [1, 4].

Network Topologies in Botnets

When talking about botnets network topologies needed to be discussed because it creates a huge difference in botnet performing its functions. Different topologies and architecture have been used by attackers [5]. The main topologies botnets uses are:

A. *Star Topology*

A very good bot managing and communicating system is provided between the bots by this topology. The big disadvantage of this topology is that it has only C&C server, which creates the problem of failing at one place; i.e., the whole system breaks if C&C gets blocked. The connection from botnet can be blocked by even a legitimate user themselves [5]. The Main C&C contains all the information that is needed to run the botnet. If the main C&C fails, then the botnet is of no harm. One does not need to remove that if the main C&C is not working for a botnet in star topology. A number of systems that are connected to this type of topology may vary; it can be as few as one or can be in large numbers as hundreds. The risk of connecting several systems to this botnet is high as it becomes dead if the main C&C server breaks.

B. Multi-server Topology

In this topology, multiple servers or C&C control and manage the network as a whole which makes the communication system better between all C&Cs and bots in the network. For some reasons, if a C&C fails, all other servers remain working and also make a decision about removing the C&C which has failed. The risk failing of the network at a point is removed in this topology [5]. This topology is highly efficient as compared to the star topology. The C&C servers in this topology can be many which can create a big network as each C&C server can connect to several other systems. If the C&C server which has many systems connected to it fails, then it is replace by other C&C servers which also connect to the other systems which were previously connected to the failed C&C server and the operation continues without smoothly. Figure 2 describes above topology (Fig. 3).

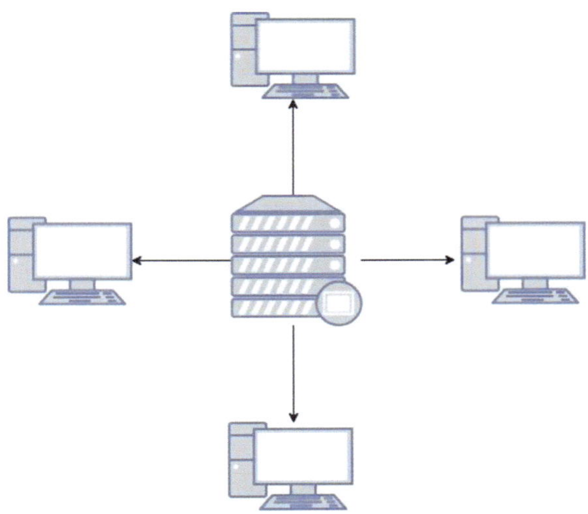

Fig. 2 Star topology

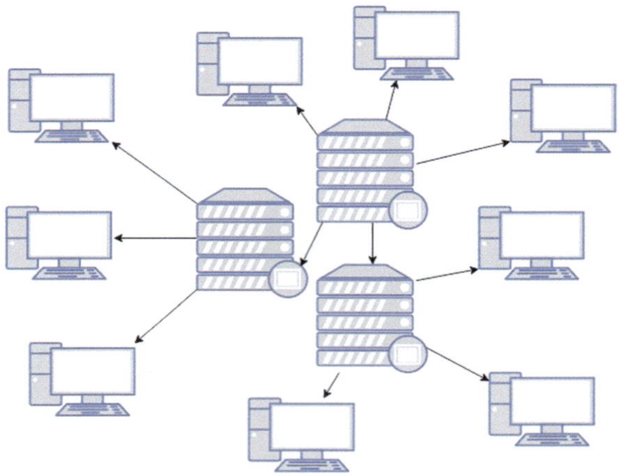

Fig. 3 Multi-server topology

C. *Hierarchical Topology*

In this topology, botnet contains multiple C&C servers are set in a way in cluster for providing a reliable network. It is easier for a botmaster in renting a bot due to hierarchical topology. Using hierarchical topology, botnets have many benefits as they are not easily detectable. In botnets based on hierarchical topology, the location of the bot agent is not known to the other bot agent which makes it tough to find a botnet and it also does not reveal the botnet size [5]. The botnets using hierarchical topology are very hard to detect as the C&C server location is not known if one C&C server gets caught, then another C&C server cannot be detected. Each server has many systems connected to it in a hierarchical way so that each level contains a C&C server and makes the work faster and easier. The same C&C server can be rented out to many clients and one server can be used to carry out functions for different clients at the same time. It reduces the number of machines used in the botnet which makes it more efficient than the star topology and multi-server topology. The cost factor also rises in this type of topology as the server increases so it takes space to put data of several clients into a C&C server and use it according to the functions and commands. Hierarchical topology can be very useful and efficient as its servers are distributed all over the network at different levels which reduce the risk of failure in the system (Fig. 4).

D. *Random Topology*

Botnet agents spread the malware through the same procedures although this topology does not contain any centralized C&C infrastructure. As per the condition, any zombie computer can act as a C&C server. Bots can communicate through many paths and ways, and in this topology, botnets are very difficult to detect and control as C&C infrastructure is not centralized; i.e., a new C&C server will be replaced into the network if a working C&C server is hijacked and is removed by new server. Bots respond to the command given by C&C in more time as compared to other

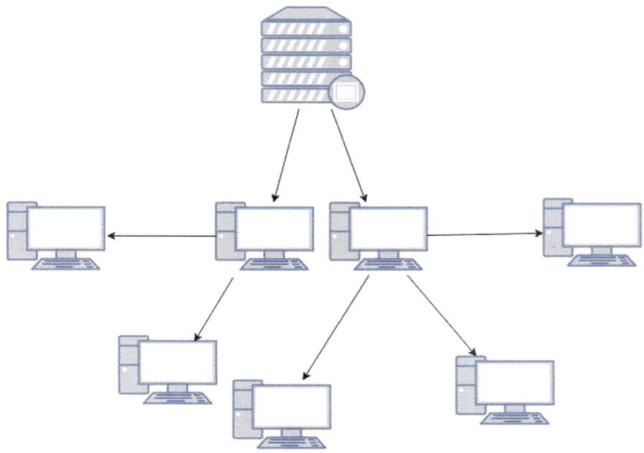

Fig. 4 Hierarchical topology

topologies. It is still better to use as it provides many better functions and features than other topologies of botnet [5]. This can also be known as dynamic master–slave or peer-to-peer relationship topology. In the dynamic master–slave, master refers to the C&C servers, while the slave is the system which is dynamic means the master–slave keeps on changing. Any system can be a master or a slave at a particular instance of time. The peer-to-peer relationship is that the system and server both are treated as peers which can be related to each other in the same way at any given time. The peers remain same as it can act as server and the system as well (Fig. 5).

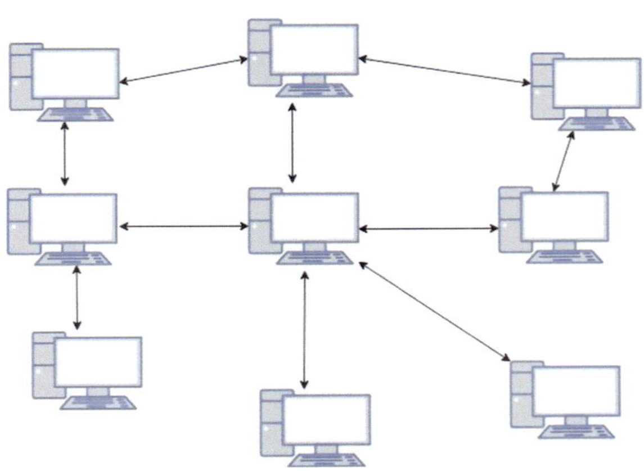

Fig. 5 Random topology

4 Problem Statement

The vital elements of our daily lives are the Internet and World Wide Web which are critical to business and personal activities of the society, but they have also became a home to cyber crimes. Several means have been developed which protect the network against the attacks which include computer worms, viruses, scanning, and denial of service, but more stealthy techniques have been developed by the attackers. The attackers are motivated for cyber crimes by the inclusion of money in network. A significant security threat on the network is botnet, and its operator is growing for their own profit. One can attack a country or enterprise for political gain through botnets by getting the secret data. Botnets exist not only in old well-known network but also in a new created application states [15–17]. Signature-based techniques on selected network or honeypots are used to detect botnet in the current state.

5 Proposed Solution

The proposed solution mainly consists of:

(i) **C&C servers**: The command and control servers are the servers that control the botnet. All the instructions of the botnets like attacking a server, infecting a server, and overtaking a server are passed through these servers only. These are the headquarters of the botnets, and the botnet can be destroyed if its C&C servers stop working.

(ii) **Positive Bot**: The bot that will be introduced by us that will detect the botnet if any of it is present in the system. It then intrudes in the system and kills the main C&C servers that are working with the botnet to infect the system.

(iii) **Killed Servers**: The C&C servers that are infected by our bot and will stop working are known as killed servers. Our bot attacks those servers and kills them by infecting virus inside them which make stop its working.

Procedure: Currently, the solution to remove the botnet is to reset the whole server which will remove all the data from the server. This process may result in the loss of some important data; to overcome this problem, the proposed solution is that we can create a bot which will first detect the presence of botnets in the network and then intrude into the network if it is present. It will then kill and replace a C&C server by first infecting it with anti-bot code.

After intruding in the network, the positive will detect and kill the C&C servers in the network of the botnet. In the botnets using star topology, it will be easy to remove the whole network by detecting botnet, intruding into a that network, and killing a server as there is only one server which will be removed by us, while in the botnets using multi-server topology and hierarchical topology it will detect the botnet first, intrude into the network by killing and replacing a C&C server, and then kill the remaining servers one by one.

This will be a lot harder in the botnets using random topology as each and every computer can act as a C&C server; for this topology, it can be a long task as it will have to check and kill each and every system connected to the network for a botnet (Fig. 6).

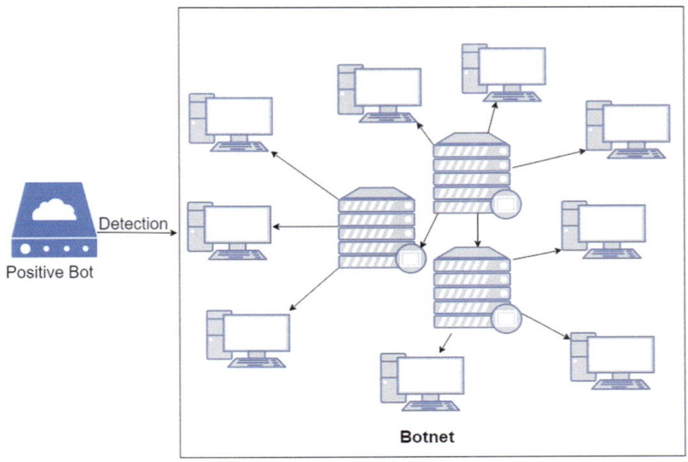

Fig. 6 Detection of botnet

In this figure, we can see that our bot is detecting the presence of bot in the server. This is mainly done through three techniques detection based on signature, detection based on anomaly, and detection based on DNS [17] (Fig. 7).

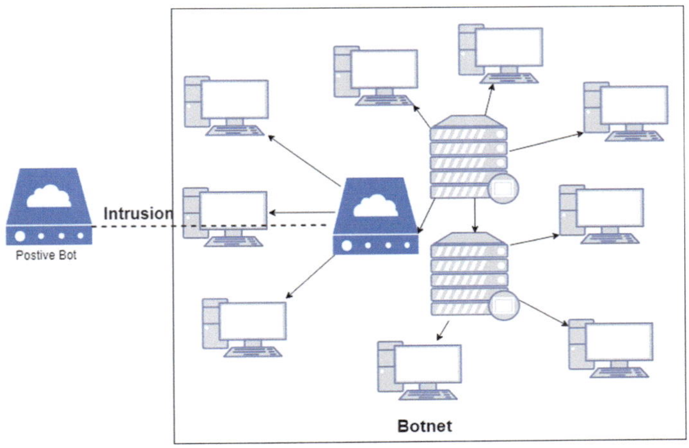

Fig. 7 Intrusion of positive bot into botnet

In this figure, we can see that out bot it trying to intrude into the network of the bots. The positive bot intrudes into the system and infects the main C&C servers by the virus so that they stop working and the botnet is totally malfunctioned (Fig. 8).

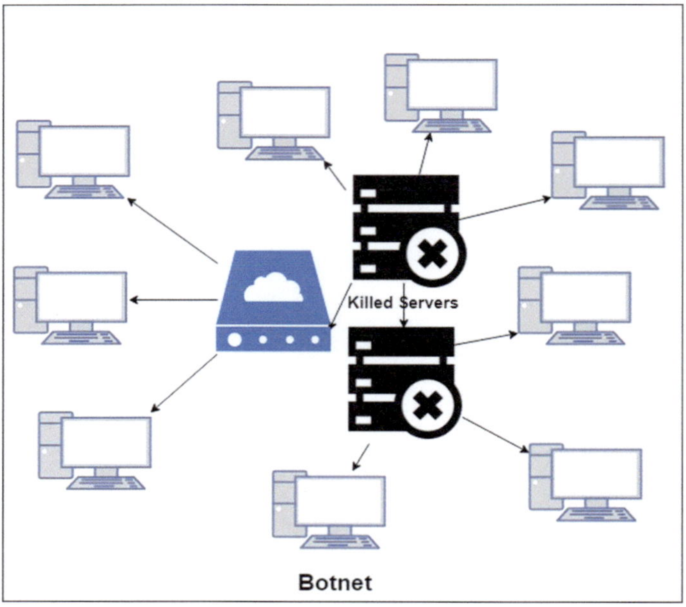

Fig. 8 Positive bot killing servers in the botnet

The positive bot finally kills the botnet by injecting the virus into the system of the botnet and disabling the C&C server of the botnet which stops the functioning of the botnet and hence make our system bot-free.

6 Conclusion

In this paper, we have analysed that what are the botnets and what threats they can cause tous. This study discussed some techniques through which the botnets can be detected and monitored and the network topologies used by the botnets in attacking a server. In the end, we can say that the Internet has become a basic need to everyone around the world nowadays and there are several works running across it. The cyber thieves use new techniques every time to introduce and steal the data through botnets. We need create new techniques to detect and remove the botnet from the server without affecting our data and to safeguard our server.

Millions of users use the Internet globally as it is becoming more common in the upcoming days. New threats are coming on Internet for attacking users. Botnets are into Internet for quite a long time; still we did not see emergence of best detection techniques. We need a technique to detect and kill the botnet in the network.

References

1. Rossow C, Andriesse D, Werner T, Stone-Gross B, Plohmann D, Dietrich CJ, Bos H (2013) SoK: P2PWNED—modeling and evaluating the resilience of peer-to-peer botnets. In: Proceedings of the 2013 IEEE symposium on Security and Privacy (SP), San Francisco, CA, USA, 19–22 May 2013, pp 97–111
2. Hu X, Knysz A, Shin KG (2011) Measurement and analysis of global IP-usage patterns of fast-flux botnets. In: Proceedings of the IEEE INFOCOM, Shanghai, China, 10–15 Apr 2011
3. Ji Y, He Y, Jiang X, Cao J, Li Q (2016) Combating the evasion mechanisms of social bots. Comput Secur 58:230–249
4. Jerkins JA (2017) Motivating a market or regulatory solution to IoT insecurity with the Mirai botnet code. In: Proceedings of the 2017 IEEE 7th annual Computing And Communication Workshop and Conference(CCWC), Las Vegas, NV, USA, 9–11 Jan 2017, pp 1–5
5. Zhao G, Xu K, Xu L, Wu B (2015) Detecting APT malware infections based on malicious DNS and traffic analysis. IEEE Access 3:1132–1142
6. Asha S, Harsha T, Soniya B (2005) Analysis on botnet detection techniques. In: Proceedings of the international conference on Research Advances in Integrated Navigation Systems (RAINS), Karnataka, India, 6–7 May 2016, pp 1–4. Saha B, A Gairola (2005) Botnet: an overview. CERT-In White Paper CIWP-2005-05, 2005
7. Alomari E (2012) Botnet-based Distributed Denial of Service (DDoS) attacks on web servers: classification and art. 49(7): 24–32
8. Barford P, Yegneswaran V (2006) An inside look at botnets. Adv Inf Sec, Springer
9. Plohmann D, Gerhards-Padilla E, Leder F (2011) Botnets: detection, measurement, disinfection and defence. Eur Netw Inf Secur Agency Tech Rep
10. Gu G, Perdisci R, Zhang J, Lee W (2011) Botminer: clustering analysis of network traffic for protocol and structure-independent botnet detection. Usenix Security Symposium, 2008. BotHunter, Bothunter: a network-based botnet diagnostic system
11. http://www.bothunter.net/. (Online). Accessed 12 Dec 2011
12. Mody N, O'Reirdan M, Masiello S, Zebek J (2009) Common best practices for mitigating large scale bot infections in residential networks. MAAWG, July 2009
13. Sakib MN, Huang C (2016) Using anomaly detection based techniques to detect HTTP-based botnet C&C traffic. In: Proceedings of the IEEE international Conference on Communications (ICC), Kuala Lumpur, Malaysia, 23–27 May 2016, pp 1–6
14. Stone-Gross B, Cova M, Gilbert B, Kemmerer R, Kruegel C, Vigna G (2011) Analysis of a botnet takeover. IEEE Secur Priv 9:64–72
15. The Honeynet Project (2016) Know your enemy: fast-flux service networks. Available online http://www.honeynet.org/papers/ff/. Accessed on 19 May 2016
16. Passerini E, Paleari R, Martignoni L, Bruschi D (2008) FluXOR: detecting and monitoring fast flux service networks. In: Proceedings of the 5th international conference on Detection of Intrusions and Malware, and Vulnerability Assessment (DIMVA '08), Paris, France, 10–11 July 2008, pp 186–206

17. Wang T, Lin H, Cheng W, Chen C (2017) DBod: clustering and detecting DGA-based botnets using DNS trafficanalysis. Comput Secur 64:1–15
18. Kolias C, Kambourakis G, Stavrou A, Voas J (2017) DDoS in the IoT: mirai and other Botnets. Computer 50:80–84
19. Kwon J, Lee J, Lee H, Perrig A (2016) PsyBoG: A scalable botnet detection method for large-scale DNS traffic. Comput Netw 97:48–73

Development of an IoT-Based Tourism Guide System

Vedanta Prusty[1(✉)], Abhisek Rath[1], Kshirod Kumar Rout[2(✉)], and Sivkumar Mishra[3]

[1] Department of ETC, IIIT, Bhubaneswar, Odisha, India
{b214059,b214003}@iiit-bh.ac.in
[2] Department of EE, IIIT, Bhubaneswar, Odisha, India
kshirod@iiit-bh.ac.in
[3] Department of EE, CAPGS, BPUT, Rourkela, Odisha, India
sivkumar@iiit-bh.ac.in

1 Introduction

Tourism is one of the promising areas over which the fulcrum of economic prosperity of many countries, including India, depends. With an annual average GDP of 6–7% since 2014, India's tourism story is no alien when it comes to contribution in the finances of the country. According to the World Travel and Tourism Council, tourism not only produced US$ 210 billion, i.e., contributing 9.4% of country's GDP, but also created 41.622 million jobs thereby contributing 8% of the total employment. In real-time scenario, people are investing more on traveling places, seeking adventure and aspiring to bridge or at least understand the mysticism and achievements of the past and present-day utility-based life, thereby making India as the most viable option. India not only provides charm of heavenly, picturesque and pristine hill stations, adrenaline pumping travel destination but is also considered as the bedrock of culture and traditions and legacy of physical artifacts, inherited from the past generations.

The concept of tourism is not a new phenomenon in India. Since 1948, with the setting up of a Traffic Tourism Committee (TTC), different agencies like India Tourism Development Corporation (1966) and campaigns, namely "Incredible India" (2002) and followed by "Incredible India-2.0" (2017), were launched. These agencies and campaigns have played a significant role in boosting tourism prospects of the country. Different types of variations have been bought into tourism as a concept, and it has resulted into diversification of the idea into the following forms, namely—Leisure Tourism, Business Tourism, Ecological Tourism, Pilgrimage Tourism, Historical Tourism, Medical Tourism, Ayurveda and Yoga Tourism, Adventure Tourism, Sports Tourism and Wildlife Tourism, hereby making India as the one-stop destination for vast number of tourists.

Figure 1 demonstrates how Tourism impacts other dominant sectors of the Indian Economy. It not only can boost investment in Transportation Sector and Hospitality Sector but also can significantly impact progress of small and medium enterprises (SMEs). Following are some factors contributing to the prosperity of the Tourism Sector in India:

M. N. Mohanty and S. Das (eds.), *Advances in Intelligent Computing and Communication*, Lecture Notes in Networks and Systems 109,
https://doi.org/10.1007/978-981-15-2774-6_58

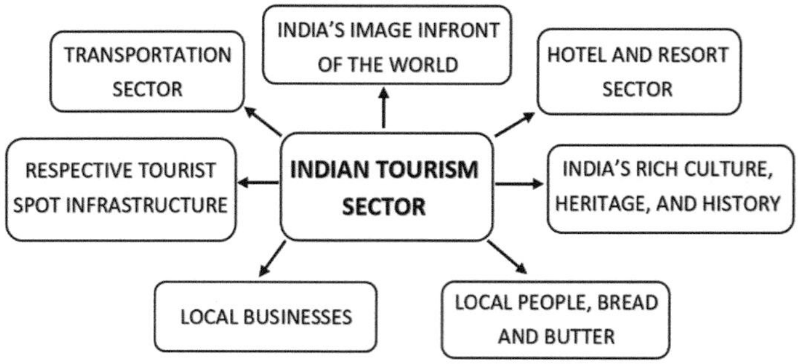

Fig. 1. Impact of Indian Tourism sector

(a) Price competitiveness of Tourism.
(b) Good freight/air transport facilities.
(c) Ground transport infrastructure.
(d) Availability of vast number of natural and cultural resources.

With the above-mentioned positives, there are certain oddities in terms of low penetration of technological advancements which if achieved, can act in favor of India's growth story. The most versatile and one of the futuristic technical advancement which can harness tourism sector to its fullest is Internet of things, abbreviated as IoT [1]. The concept of IoT has shrunk the world into a smartphone, which facilitates the interaction between the physical entities via Internet. The main application of IoT consists of controlling and monitoring functions. This innovative and simple to understand and easy-to-use technology has penetrated into both social, economic, academic and energy sectors, where it facilitates efficient and optimal usage of resources. In the present-day scenario, the multi-faceted nature of technology raises various possible transient and random failures which are vulnerabilities that can be exploited for illicit means by cyber attackers. The various possible attacks on IoT applications include—denial-of-service (DoS), controlling, eavesdropping, physical damage and node capture. These threats can be handled by different features of IoT, namely: reliability, responsibility, privacy which include data sharing and management, data collection and data security, safety and identification and authentication, being the crucial features [2].

Research in the field of IoT has paved way to application oriented domains like home automation [2], smart irrigation system [3], smart tourism and many other fields which form a part of our day-to-day life. Many previous pioneering works focused on development results of IoT studies and discuss ideas on how to apply them to business [4], IoT platform for supporting a real-time recommender system [5] and signature-based authenticated key establishment scheme for IoT environment [6], which stressed upon the security aspect of the IoT. The area of research was extended to the big data analytics where many works on behavioral pattern analysis [7] using the existing database which made further research possible in the study of many e-commerce models [8]. Big data analytics study enabled synthesis of millions of data points into a meaningful pattern, thereby resulting in automation of a physical entity for which the analytics was performed. The IoT and big data together has resulted in some of research opportunities like design of the intelligent monitoring system for scenic spot [9], providing an IoT information technology public platform for regional revival [10], concept of smart and connected communities (SCC) [11], innovation approach of mobile dynamic trip planner [12], application of Geo2Tag LBS Platform [13], implementation of GPS and Electronic Map as a multi-mode [14] and numerous other techniques and technologies. The ideation for the above-mentioned researches is the result of various case studies in the field of tourism conducted world-wide, namely on Liaoning Tourism Resources Management Information System Based on GIS [15], Guided Tour [16], IoT application of South African Tourism [17] and i-Tour model [18]. The sustainability of the project work depends on the flexibility of the idea and focusing on this various research opportunities in the field of rural folk tourism [19] and eco-tourism has gained momentum [20–23].

In this paper, we have tried to bring in an entirely new concept of transforming Indian Tourism sector digitally by automating the tourism sector with the right use of technology such as embedded systems and Internet of things. And by this we think we can really capitalize our place on world's tourism fraternity and also help in improving tourist's satisfaction level, meet to the expectations of tourists and hence safeguard our country's image and precious amazing history.

1.1 Problem Statement

(A) **Problems in Present Scenario**:

The present scenario presents us with many opportunities when the below-mentioned problems are catered with essential and efficient solutions:

1. Indian history, culture and heritage at stake.
2. Lack of effective tourist management and support.
3. Tourist's level of expectation versus satisfaction rate is very low.
4. Lack of knowledge of guides and lack of authentic resources to know in-depth about the rich history and heritage of India.
5. Guides lacking English speaking ability along with cheating and theft being the most critical issues to be looked at to safeguard our country's image globally.
6. Every tourist, visiting various tourists' places, is not capable enough to afford a guide which keeps them away from getting to know the actual authentic facts behind the incredible Indian History.
7. Old people, physically impaired, blind people, etc. also have equal rights to know about their motherland, but sadly in the present scenario they are deprived of that.
8. The real fact of such magnificent history is being shadowed and will remain shadowed if proper steps are not taken to mitigate the issue.

So, it is high time to take action to save our country's pride.

B. **Where and Who is Undergoing the Problem**:

- The problem exists in the tourist places all over India.
- Problem is faced by all potential tourists who are a part of India's Tourism.
- People who are interested in knowing about the rich history, culture and heritage of India.

2 Bharat Darshan—A Concept of Digitalizing Indian Tourism

Bharat Darshan is entirely a new concept of transforming Indian Tourism digitally with the help of embedded systems and IoT (Fig. 2). This concept mainly

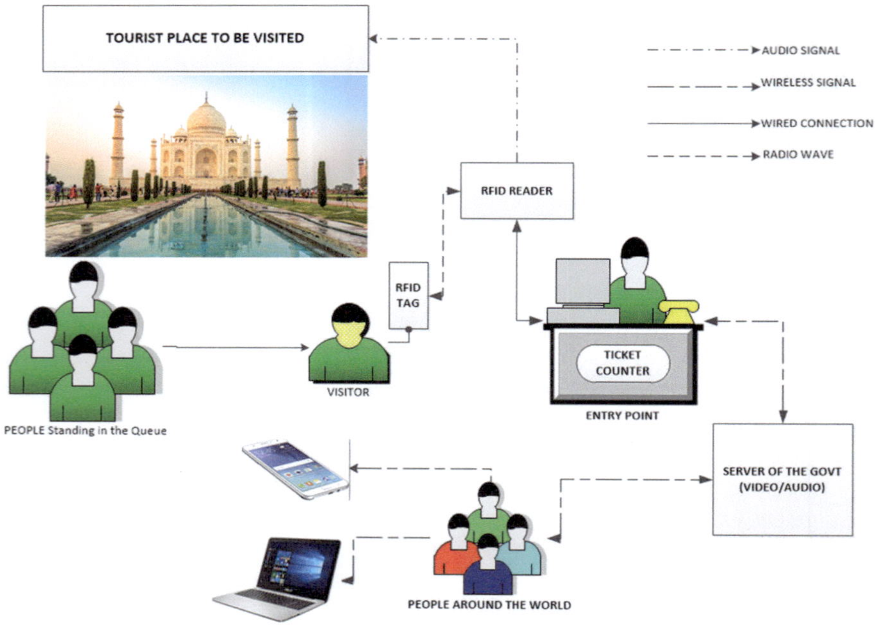

Fig. 2. Flow diagram describing the basic schema of the Bharat Darshan prototype

aims at ensuring that the true and authentic facts relating to the incredible history, magnificent heritage, marvelous culture and tradition of India. Below described are the motivating factors behind the creation of prototype along with the description of the module:

(A) **Technology/Process Solution**:
1. Complete tourist management and support using RFID and IoT implementation.
2. Authentic resource management using audio recordings and videos about the rich culture, history, heritage and tradition of India.
3. Authentic guide resources directly to smart phones.

(B) **What is Unique/Innovative**:
1. Authentic resources and facts to guide tourists.
2. RFID and IoT in Tourism Sector.
3. Provision for old, pregnant women, physically disabled people, etc. to equally get the opportunity of enjoying the beauty of the place at their own convenience.
4. Language preference option for people to enjoy the pre-recorded resources.

(C) **How is it Different from Existing Solutions**:
1. Makes authentic knowledge enriching experience and fun-filled space for the tourists.
2. Now everybody can afford a guide and get to know the real facts from the pre-loaded authentic resources.
3. People's problem (disorders, age, etc.) is not a matter with our solution. Everyone gets an equal opportunity to enjoy India's beauty at their own convenience.

(D) ***Detailing Of the Concept—Bharat Darshan***:

The descriptive content of the module "**Bharat Darshan**" has been divided into three parts:

1. **Section 1**: Represents the first phase which categorizes the available data according to timescale.
2. **Section 2**: Represents the event occurred in the respective timescale according to the Section 1. The event is categorized into two sub-categories, namely
   (a) Event value/consequence.
   (b) Size of the scenic spot.
3. **Section 3**: Represents the sub-sections of the Section 2 into the below-mentioned categories:

(a) Event value/consequence:
   (i) Popularity and influence
   (ii) Scientific significance
   (iii) Historical significance.

(b) Size of the scenic spot:
   (i) Scenic region combination
   (ii) Ornamental significance
   (iii) Tourism environmental capacity.

Bharat Darshan unit consists of two independent sub-systems:

1. Embedded Division
2. Internet Of things (IoT) Division.

Tourists get the option of choosing either of the two systems or even both as per their convenience.

(I) **Embedded Division**:

1. During ticket collection by the visitors, they will be asked about their preferred language.
2. Govt. will be recording pre-recorded voices as well as interesting videos with enriching and authentic facts about the particular historical place where the system is to be used.
3. The voices and videos will be recorded in different languages such as English, Hindi and the local language of that area.
4. During ticket collection, visitors will be provided with the preferred language RFID tags (passive).
5. When the visitor scans the card of his/her preferred language, the voice gets played in the same preferred language as scanned by the user.
6. When one visitor is listening or once the card is scanned and the voice is being played and at that point of time someone else tries to interfere within that period then automatically the access is not granted to the other user as currently one user is using the device. (Illegal scanning is disabled in the system, i.e., if a person is listening and another tries to interfere by scanning the card in between then that scanning is not considered as authentic and access is denied to the user at that point of time.)
7. Secondary check is done whether the person after scanning the card is present or not using an ir sensor.
8. If in case a person scans the card and goes away without listening the setup or system automatically detects the same and shuts the voice just after few seconds, hence allowing other visitors to access the device.
9. The voice is played every time according to the scanned card (preferred language card).
10. Another check is also provided in the setup that if a person while listening lives in between and no one else is also present over there to listen then at that point of time the system detects it and shuts the voice automatically within few seconds, hence making the system ready to be accessed by another user.

(II) **IoT Division**:

(Providing visitors with another option to get to know the facts directly on their own respective smart phones).

1. A local server is to be installed at the tourist spot.
2. A webpage is developed in that local server.
3. That Web site can only be accessed in that particular tourist's spot by getting connected with the local network that the server is connected with.
4. By accessing that website, one gets access to interesting audio—visual contents that are pre-recorded videos which are developed by Govt. for that particular tourist's spot.
5. Now people get to know the authentic resources and facts about the rich culture and heritage of India through interesting videos and authentic facts which enrich their knowledge with true and real facts and hence with all these above steps helps in increasing India's image in world tourist arena.

Value Proposition

1. Safeguarding the rich culture, heritage and history of India by transferring the traditional values and cultural knowledge via IoT and RFID implementation
2. Digital transformation of our Tourism Sector which gives a completely new and beautiful perspective to Indian Tourism.
3. Growth of Tourism Sector directly contributing to India's gross economy.
4. Tourist management and support (TMS) via IoT interface which is user-friendly.
5. Generate employment opportunity both in rural as well as heritage tourism which can result in better new India ideas of tourism village, bettering opportunities of country-side tourism.

3 Conclusion

Tourism Sector of India being one of the biggest game changers in India's total economic growth which directly leads to country's development must not be neglected at any cost (Fig. 3).

Digital Transformation of Tourism sector will not only encourage a greater number of tourists to visit India to get enlightened with the real amazing Indian history, culture and tradition but also will have a direct impact on country's growth.

What more proud moment can it be for India when outsider's as well as our country people will get to know about the original authentic facts related to India's glorious history, heritage and start appreciating them in the world tourist arena?

It is high time to think and act upon this issue and we believe the concept of Bharat Darshan truly has great potential to make Indian Tourism sector the best in the world and not only that but also will surely help India generate a good amount of revenue which further defines the growth of India in the upcoming future.

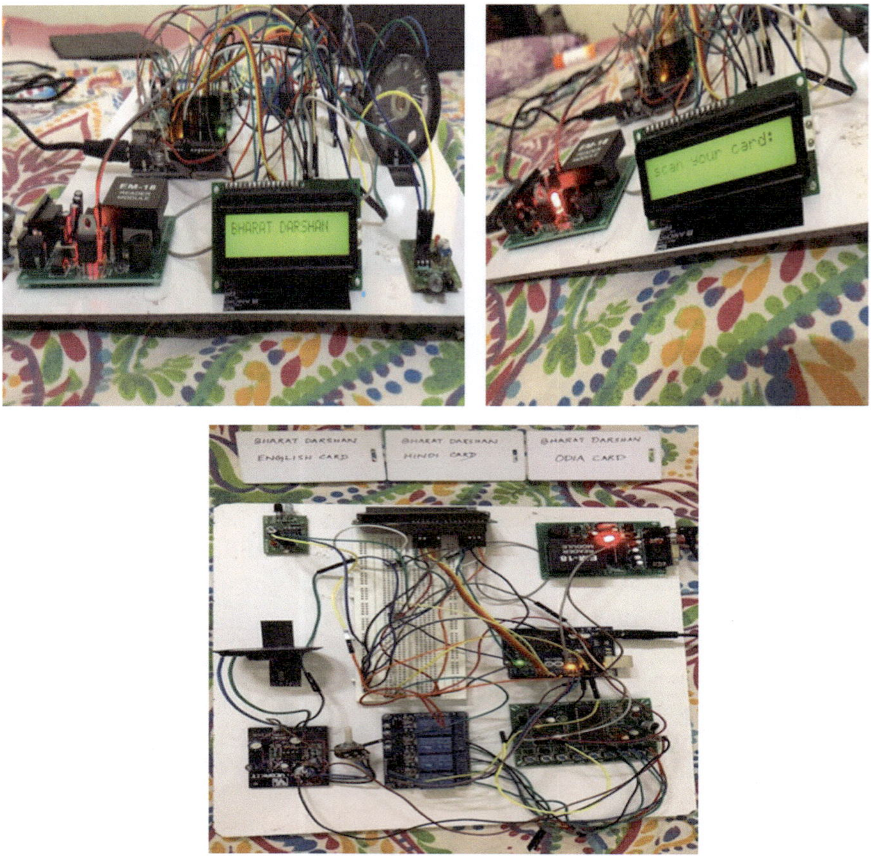

Fig. 3. Working prototype of Bharat Darshan

References

1. Rout KK, Mishra S, Routray A (2017) Development of an IoT based Introductory Laboratory for Under graduate engineering students. In: 16th International Conference on Information Technology (ICIT-2017), Bhubaneshwar, India
2. Rout K, Mallick S, Mishra SK (2018) Design and implementation of an Internet of Things based prototype for smart home automation network. ICRIEECE, Bhubaneswar, India
3. Rout K, Mallick S, Mishra SK (2018) Solar powered smart irrigation system using Internet of Things. In: 2nd International Conference on Data Science and Business Analytics (ICDSBA 2018), ChangSha, Hunan, China
4. Balandina E, Balandin S, Koucheryavy Y, Mouromtsev D (2015) IoT use cases in healthcare and tourism. In: 2015 IEEE 17th conference on business informatics
5. Cha S, Ruiz MP, Wachowicz M, Tran LH, Cao H, Maduako I (2016) The role of an IoT platform in the design of real-time recommender systems. In: 2016 IEEE 3rd World Forum on Internet of Things (WF-IoT)

6. Challa S, Wazid M, Das AS, Kumar N, Reddy AG, Yoon E-J, Yoo K-Y (2016) Secure signature-based authenticated key establishment scheme for future IoT applications. IEEE Access 2016
7. Shafiee S, Ghatari AR (2010) Big data in tourism industry. In: 10th international conference on e-commerce with focus on e-tourism
8. Fang M, Zheng Z (2011) Innovation pattern of China tourism e-commerce on the post-crisis era. In: 2011 2nd international conference on Artificial Intelligence, Management Science and Electronic Commerce (AIMSEC)
9. Lin Y (2011) The application of the internet of things in Hainan tourism scenic spot. In: 2011 seventh international conference on computational intelligence and security
10. Gautam BP, Asami H, Batajoo A, Fujisaki T (2016) Regional revival through IoT enabled Smart Tourism Process Framework (STPF): a proposal. In: 2016 joint 8th international conference on soft computing and intelligent systems and 2016 17th international symposium on advanced intelligent systems
11. Sun Y, Song H, Jara AJ, Bie R (2015) Internet of things and big data analytics for smart and connected communities. IEEE Access, vol 14, no 8, August 2015
12. Alghamdi H, Zhu S, El Saddik A (2016) E-tourism: mobile dynamic trip planner. In: IEEE 2016, international symposium on multimedia
13. Balandina E, Balandin S, Koucheryavy Y, Mouromtsev D (2015) Innovative e-tourism services on top of Geo2Tag LBS platform. In: 2015 11th international conference on signal-image technology & internet-based systems
14. Hu Z, Wang Z, Liu H, Bie H (2010) Self-service folk tourism guiding technology on mobile terminal with multi-mode: application of GPS and Electronic Map. In: IGARSS 2010, IEEE 2010
15. Chen L, Wang H, Yang Z (2010) Study on Liaoning tourism resources management information system based on GIS. In: 2010 international conference on networking and digital society
16. Wu S-T, Lee B-W (2017) An innovative way of guided tour: a virtual experience of dark tourism. In: 2017 IEEE international conference on information, communication and engineering
17. Gcaba O, Dlodlo N (2016) The internet of things for South African tourism. In: IST-Africa 2016 conference proceedings
18. Tripathy AK, Tripathy PK, Ray NK, Mohanty SP (2018) iTour: the future of smart tourism. In: IEEE consumer electronics magazine, May 2018
19. Ma P, Zhao W, Hu Z, Duan F (2010) Three-dimensional display component for rural folk-custom tourism. In: 2010 18th international conference on geoinformatics
20. Shafiee MM, Najafabadi SI (2010) The interaction of technological progress and tourism industry development in the developing countries: the case of Iran's Tourism Industry. In: 10th international conference on e-commerce with focus on e-tourism
21. Wang J, Dong S, Wan Y (2014) Abundance of tourism resources and regional tourism development. In: 2014 22nd international conference on geoinformatics
22. Xingqi W (2009) Study on core capabilities of tourism networks. In: 2009 International conference on environmental science and information application technology
23. Yi W, Zhao J (2011) Research on systematic coupling symbiosis of low-carbon tourism and eco-tourism. IEEE

Design and Simulation of a Standalone Photovoltaic System Using Synchronous Boost Converter and Reduced Switch Five-Level Inverter

Kshirod Kumar Rout[1](✉), Sivkumar Mishra[2], and Pravat Kumar Biswal[3]

[1] Department of EEE, IIIT, Bhubaneswar, Odisha, India
[2] Department of EEE, CAPGS, BPUT, Rourkela, Odisha, India
[3] Department of EEE, KIIT, Bhubaneswar, Odisha, India

1 Introduction

The advancement of new energy sources is consistently improved in the recent decades. But, the need of sustainable, renewable, and clean electric energy sources has become indispensible in the present energy scenario of the world, and it seems the solar energy has emerged as the viable alternative [1]. In the recent decades, humungous amount of research has been done by the scientists on the PV cells and the ways to increase their efficiency and converting it into usable energy form [2]. The market for PV-based systems is expanding around the globe, and presently, solar PV gives around 4800 GW of electric energy. During the years from 2004 to 2009, network-associated PV limits achieved 21 GW and were expanding at a yearly normal rate of 60% [3]. Usually, each polycrystalline PV cell generates voltage ranging from 0.5 to 0.8 volts depending upon the sort of semiconductor material utilized and the developed innovation of highly efficient and commercial PV cells. This voltage is sufficiently low as it cannot be useful. In order to get advantage from this innovation, many PV cells ranging from 36 to 72 numbers of cells are affixed into a matrix. Thus, obtained modules are arranged in series and parallel lines to obtain required voltage and current. In the event that these modules are associated in arrangement, their voltages are included with a similar current. In any case, when they are associated in parallel, their currents are added while the voltage is the same. Due to various reasons, solar panel's output voltage may not be constant. Thus, the requirement of a boost converter system arises. A synchronous boost converter's (SBC) [4, 5] output voltage is always higher than the given input voltage depending upon switching frequency and the reference voltage level set. The boost converters operated at higher frequencies produces a hike in switching losses during turn ON/OFF. For the reduction of the switching losses occurred in the adopted synchronous converter, soft switching techniques also known as zero-voltage switching (ZVS) and zero-current switching (ZCS) [6] are proposed involving an added auxiliary circuit which consists of a capacitor, a 1N4007 diode,

M. N. Mohanty and S. Das (eds.), *Advances in Intelligent Computing and Communication*, Lecture Notes in Networks and Systems 109,
https://doi.org/10.1007/978-981-15-2774-6_59

a helper switch, and an inductor. All the switches in the proposed converter system operate under ZVS and ZCS conditions to reduce the losses occurred during switching and to improve the efficiency compared to a conventional boost converter system. As the efficiency of a conventional chopper is not satisfactory, it can be improved by replacing a diode with MOSFET operated in synchronization mode. This reduces the power losses caused by diode as well as MOSFET has much lesser internal resistance as compared to a diode. MPPT [7, 8] systems provide a good control over maintaining the required voltage and current levels obtained at the output of the SBC. This controls the gate pulse to be delivered to the SBC arrangement. In order to serve the domestic loads, a five-level inverter [9, 10] requires two different DC supplies ideally each of same voltage output, which in turn is converted into AC voltage. Multi-level inverters (MLIs) with reduced number of switches are preferred to conventional cascaded MLI as the former has lesser harmonic distortion [11].

In light of above developments, in this paper a standalone PV system is designed with a SBC and a reduced switch five-level inverter for driving AC loads. The SBC is simulated in MATLAB/Simulink, and the simulation observations are shown to support theoretical analysis. Section 2 elaborates the proposed model and its sub-systems, Sect. 3 explains the simulations performed for the design purpose, and Sect. 4 shows the analysis and results of the work. The use of SBC with MPPT and PI controller and RSFL inverter makes the overall design quite efficient.

2 Proposed Model

Photovoltaic panels are usually of 36 cells or 72 cells. This work utilizes two 72 cell solar panels with 0.5 V per each cell affixed in a matrix form in the panel, whose output ranges within 30–50 V, 5 A. As the PV curve (Fig. 4) gives the knowledge about the variations in the voltage and current outputs of the PV panel at different temperatures. Output of the solar panels is connected to respective synchronous boost converters to step up the DC voltage to 115 V. MPPT system controls the pulses to be sent to the SBC by measuring the voltage and current output of the installed polycrystalline solar panels. The outputs of the two SBCs are given to RSFL converter as input so that it can generate 230 V AC. The outline of the proposed design is shown in Fig. 1.

A. *Solar panel*

Each polycrystalline silicon solar cell has an internal circuit with a current source, P-N junction, shunt, and series resistances as shown in Fig. 2. The output voltage is nearly 0.8 volts. Matrix formation of several such cells is known as a module. Collection of such modules is known as panel (Figs. 3 and 4).

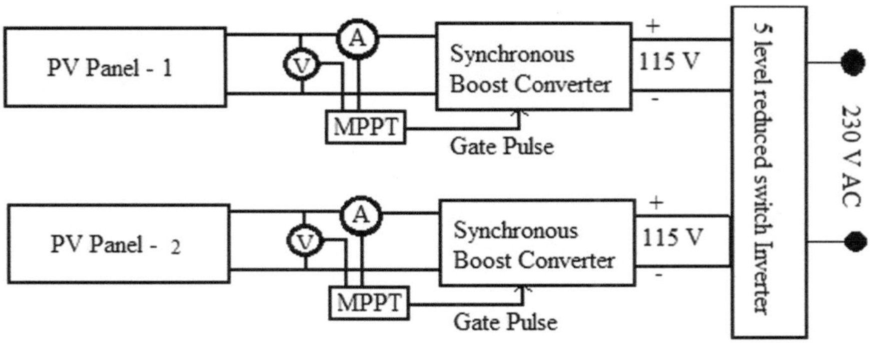

Fig. 1. Overview of the proposed design

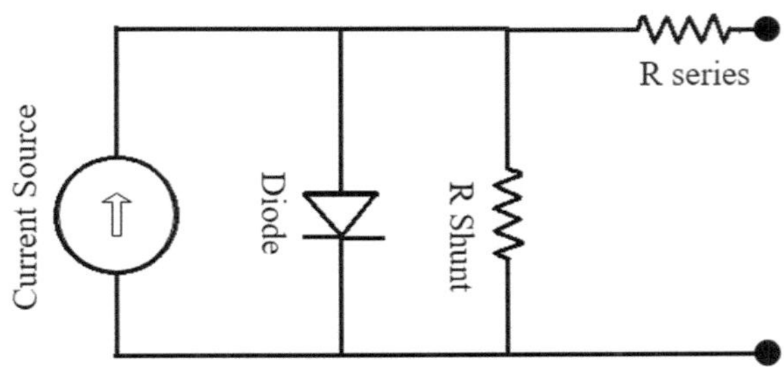

Fig. 2. Internal circuit of a PV cell

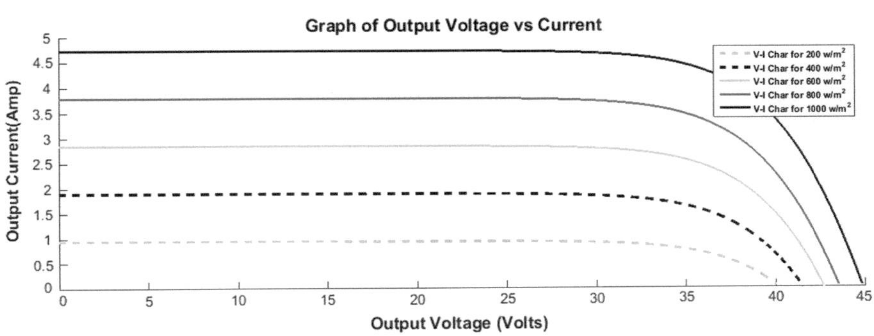

Fig. 3. IV characteristics of PV panel

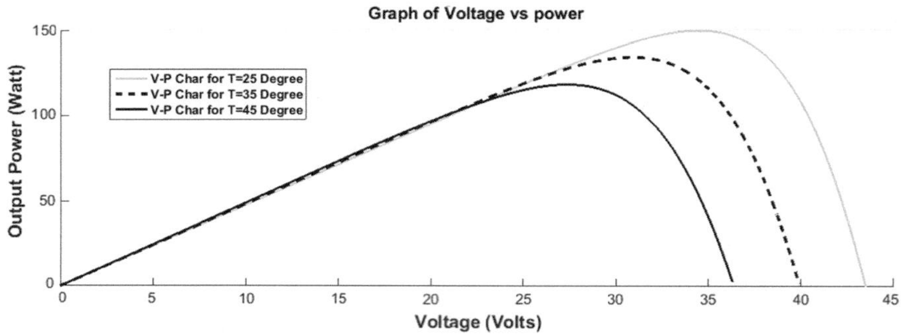

Fig. 4. PV characteristics of solar panel

B. *Measurement of electrical parameters*

Measuring the DC voltage and current can be achieved by V-A method or A-V method. V-A method is much accurate than the other. Voltage divider circuit is used for measuring the voltage, and Hall effect sensor is used for current. Analog outputs of the sensors are connected to the MPPT system.

C. *MPPT*

Maximum power point tracking is used to extract maximum possible power from the solar panel. Perturbation and observation (P&O) method is adopted as it is less complex to implement [12]. The algorithm for the chosen MPPT scheme is shown in Fig. 5.

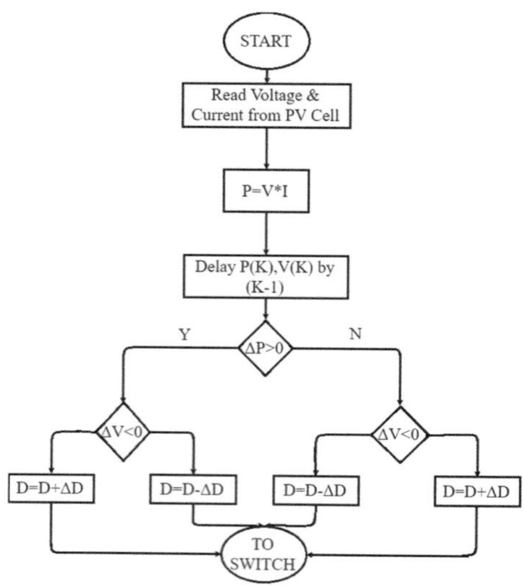

Fig. 5. Algorithm for P&O method

D. *Synchronous boost converter*

Proportional and integral control system has been adopted for the proposed SBC as a negative feedback input to control the output potential difference according to the base voltage set. The assistant circuit is made out of primary switch (S1), a helper switch (S2), a full capacitor (Cr), a resounding inductor (Lr), and two diodes (D1 and D2), as appeared in Fig. 2. PWM [11] duty cycle is varied to operate the switches at the desired frequency (Fig. 6).

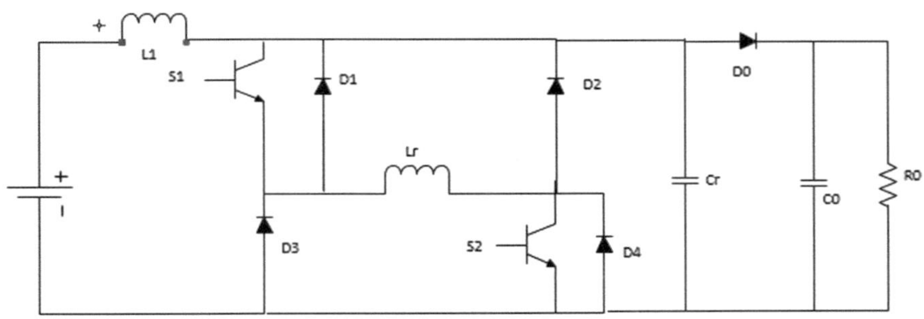

Fig. 6. Circuit outline of SBC

E. *Reduced switch inverter*

A five-level reduced switch inverter (Fig. 7) utilizes two DC sources. In the place of the DC sources, we replace it with the output of the SBCs. This configuration uses only five switches which lead to lesser switching loss. Required PWM is simulated.

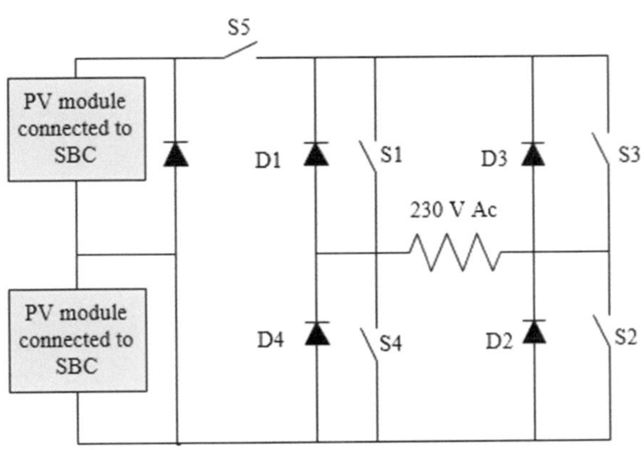

Fig. 7. Five-level reduced switch (FLRS) inverter

3 Simulation

A. *Overall view*

The SIMULINK model of the proposed design is shown in Fig. 8. The parameters like irradiance and temperature to the panel's model are given as inputs. As the solar panel acts as a current source, it is simulated using a controlled current source in Simulink toolbox. Required circuitry is made for diagnosis of the designed system. MPPT with the algorithm specified in Fig. 5 is integrated in the above simulation shown.

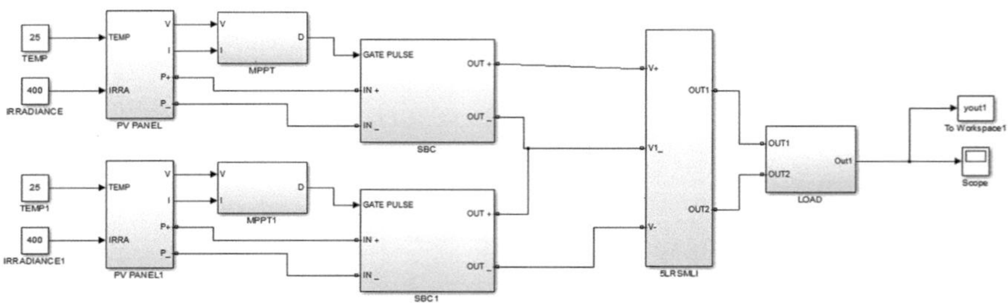

Fig. 8. Overall view of the design

B. *SBC simulation*

The Simulink design of the SBC is as shown in Fig. 9. It uses high-frequency switching devices for the converter action. PWM has been utilized for their operation. The input voltage has been set to 50 V which can vary up to 100 V DC. The output is taken as negative feedback loop with a PI controller [13, 14] in it. The reference voltage

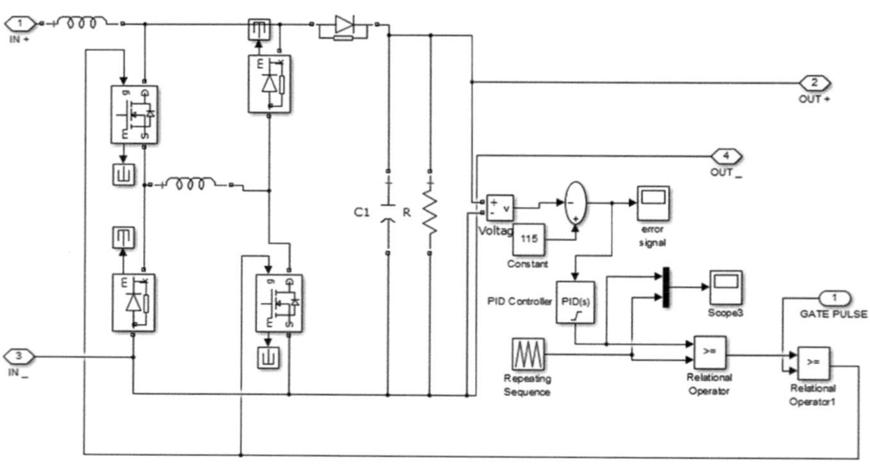

Fig. 9. Modeling of synchronous boost converter sub-system

level is set as 115 V as it is necessary for the reduced switch inverter to generate AC output voltage. Scope is introduced into the system to monitor the PWM at the switching devices, the error signal, and the output voltage waveform.

C. *RSFL converter simulation*

The reduced five-level inverter has been simulated as shown in Fig. 10. Required PWM is supplied for the MOSFET to perform the switching operation at desired frequency. This kind of inverter is easy to implement and analyze. It requires two different DC voltage sources to work as a five-level inverter. Delivering output voltage is 230 V AC.

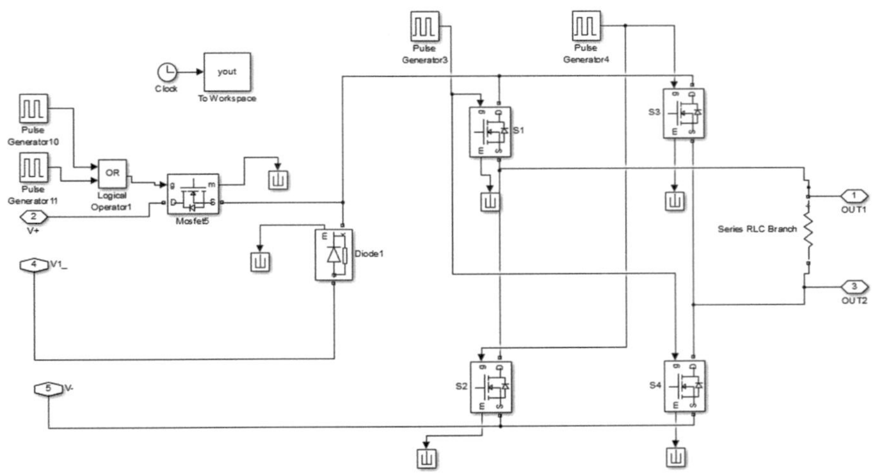

Fig. 10. Reduced switch five-level inverter

D. *Simulation of load sub-system*

The type of load utilized in this design process is resistive type and is connected to a voltmeter for measuring the voltage across the load.

4 Results

From the simulations performed, the output waveforms of the SBC, MPPT, and RSFL inverters are as follows. The pulse train of the MPPT at 10 kHz is also shown in the results.

A. *Synchronous boost converter*

SBC takes 50 V DC as input and delivers 115 V DC as output as shown in Fig. 11. Output voltage settles at 115 V DC within 0.06 s due to the PI controller used from the observations. Figure 12 shows the corresponding MPPT pulses.

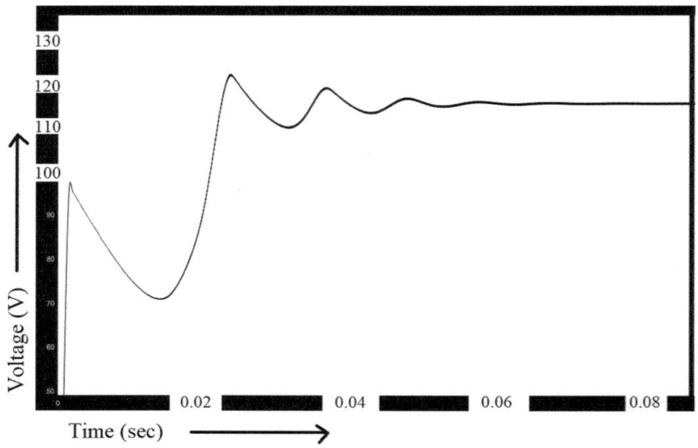

Fig. 11. Controlled output voltage of SBC

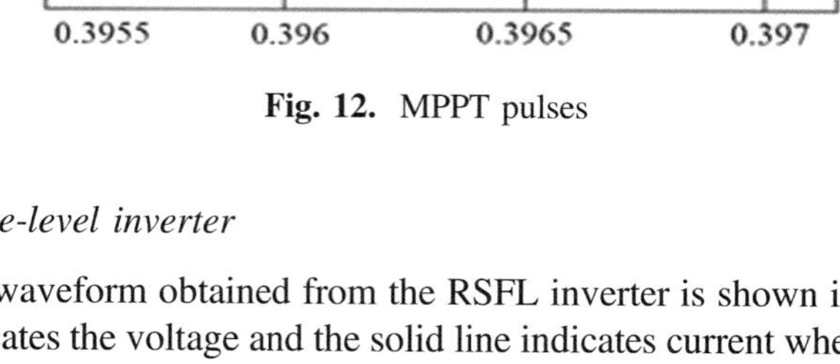

Fig. 12. MPPT pulses

B. *Reduced five-level inverter*

The output waveform obtained from the RSFL inverter is shown in Fig. 13, where dotted line indicates the voltage and the solid line indicates current when connected to a resistive load. The voltage obtained is purely a five-level inverted wave. As the load is resistive, the voltage and current are in phase with each other. The total harmonic distortion (THD) of the voltage is found to be 26.49% and is shown in Fig. 14.

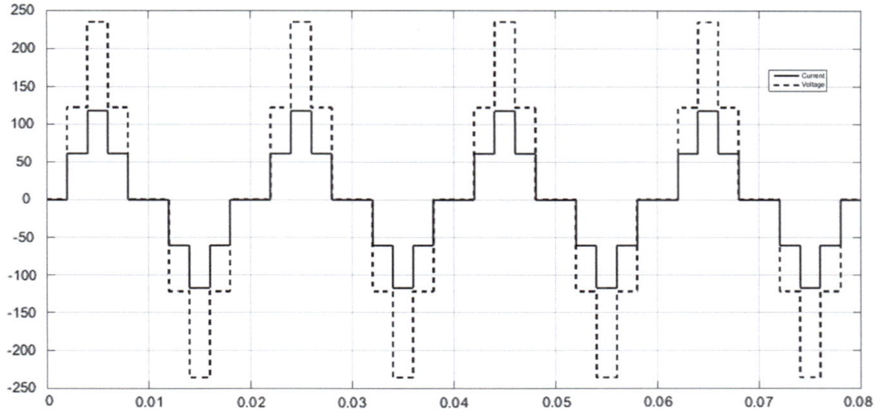

Fig. 13. Output of the RSFL inverter

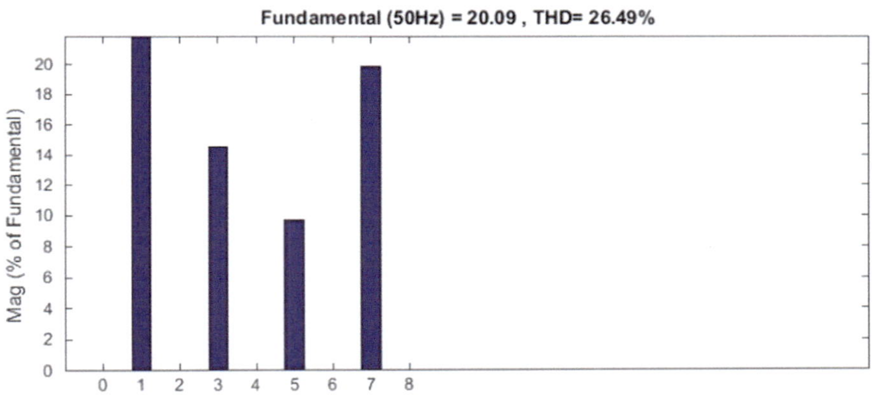

Fig. 14. THD of output voltage (RSFL)

5 Conclusion

In this paper, the conversion of unstable DC voltage obtained from the solar panels has been first converted to two constant DC115V sources with help of SBCs with a PI feedback control and a P&O-based MPPT scheme. A 230 V AC output has been generated from a RSFL inverter with the two 115 V constant DC as its inputs. The entire PV system has been simulated to drive a resistive load. The theoretical analysis has been supported by the MATLAB/Simulink models of all the components and waveforms of the corresponding sub-systems. The hardware implementation of the proposed PV system is included in future work.

References

1. Rosyid OA (2017) Comparative performance testing of solar panels for smart city micro-grids. In: 2017 international conference on smart cities, automation and intelligent computing systems (ICON-SONICS), Yogyakarta, pp 81–86
2. Goel S, Sharma R (2017) Performance evaluation of stand alone, grid connected and hybrid renewable energy systems for rural application: a comparative review. Renew Sustain Energy Rev 78:1378–1389
3. Gul M, Kotak Y, Muneer T (2016) Review on recent trend of solar photovoltaic technology. Energy Explor Exploit 34(4):485–526
4. Qi L, Xiao M (2017) Robust control of boost converter. In: 2nd international conference on cybernetics, robotics and control (CRC), Chengdu, China, pp 61–65
5. Kim J-H, Jung D-Y, Park S-H, Won C-Y, Jung Y-C, Lee S-W (2009) High efficiency soft-switching boost converter using a single switch. J Power Electron 9(6):929–939
6. Jabbari M, Farzanehfard H (2009) A new soft switching step-down/up converter with inherent PFC performance. J Power Electron 9(6):835
7. Xiao S, Balog RS (2018) An improved adaptive perturb & observe maximum power point tracking technique. In: IEEE Texas Power and Energy Conference (TPEC), College Station, TX, pp 1–6
8. Karabetsky D (2017) Conceptual design of maximum power point tracking for solar rechargable airplane. In: IEEE 4th International Conference Actual Problems of Unmanned Aerial Vehicles Developments (APUAVD), Kiev, pp 58–60
9. Garg V, Sinha SK, Dave MP (2016) Phase shifted PWM cascaded multilevel inverter for solar PV standalone systems. In: IEEE 1st International Conference on Power Electronics, Intelligent Control and Energy Systems (ICPEICES), Delhi, pp 1–5
10. Kabalci E, Kabalci Y, Canbaz R, Gokkus G (2015) Single phase multilevel string inverter for solar applications. In: International Conference on Renewable Energy Research and Applications (ICRERA), Palermo, pp 109–114
11. Rout KK, Mishra SA (2018) A modified 7-level reduced switch symmetrical inverter. In: National Power Engineering Conference(NPEC),Madurai, India
12. Mohapatra A, Nayak B, Das P, Mohanty KB (2017) A review on MPPT techniques of PV system under partial shading condition. Renew Sustain Energy Rev 80:854–867
13. Qian K, Xiong H, Lei D (2017) Design of PWM rectifier based on fractional order PID control. In: International Conference on Industrial Informatics - Computing Technology, Intelligent Technology, Industrial Information Integration (ICIICII), Wuhan, China, pp 215–218
14. Karuppiah M, Karthikumar K, Arunbalj A (2017) A transformerless buck-boost converter with PID controller. In: IEEE International Conference on Intelligent Techniques in Control, Optimization and Signal Processing (INCOS), Srivilliputhur, India, pp 1–7

Designing and Sizing of a Stand-alone Photovoltaic System: A Case Study

Kshirod Kumar Rout[1(✉)], Sivkumar Mishra[2], and Pravat Kumar Biswal[3]

[1] Department of EE, IIIT, Bhubaneswar, Odisha, India
kshirod@iiit-bh.ac.in
[2] Department of EE, CAPGS, BPUT, Rourkela, Odisha, India
[3] Department of EE, KIIT, Bhubaneswar, Odisha, India

1 Introduction

With the current transition that this world is going through, electricity has already become another basic necessity apart from food, clothing and shelter. Since the beginning, the major source of electricity is fossil fuel, and on its combustion, it releases huge amount of harmful gases to our surrounding. To prevent it, many nations have considered this a serious global agenda and are committed to eradicate gases that cause the greenhouse effect from our atmosphere by producing cleaner energy through renewable sources. Also, the production of electricity as a source of energy is quite costly in terms of exploitation of earth's resources such as coal and oil which are limited and are going to be scarce over the next 50 years. Moreover, UN has also taken initiatives to encourage nations for generating clean and green energy. As a result, different countries have initiated different strategic measures to reduce their greenhouse gases and eventually their carbon footprints. Hence, different countries are exploring other sources of energy which are renewable in nature such as hydro (water), wind and solar. This exploration toward alternate clean energy resulted in a steady increase in utilization of nonconventional sources of energy to resolve acute energy shortage and reduce the effects of global warming [1, 2]. In the present day scenario, renewable sources of energy are world's fastest growing energy tank, and its consumption is expected to increase from 9% to 16% till the year 2040 [3]. Extensive research has already been done in the field of exploring more unique ways of increasing the accuracy and efficiency of the PV cells and optimal sizing conditions.

Now, focusing on the energy produced from sunlight, it has been found that the use of solar energy is quite attractive for an alternative source of energy because it is silent, safe, nonpolluting and can be adjusted for any increase in load, and most importantly, it is fail proof with an average working life of 20–30 years [1]. Moreover, it contains no special software or any sort of moving parts and is virtually maintenance free.

A PV system is a system that uses solar energy as its input and converts it into electricity. These systems are usually classified based upon the connection of the system components with the power sources, i.e., stand-alone (SA) and utility-interactive (UI) systems. Taking the "stand-alone" factor into consideration, as the name suggests the SAPV systems are usually designed to operate independently and

M. N. Mohanty and S. Das (eds.), *Advances in Intelligent Computing and Communication*, Lecture Notes in Networks and Systems 109,
https://doi.org/10.1007/978-981-15-2774-6_60

sized to supply a specific amount or average amount of DC and/or AC electrical loads [4]. Now, taking the solar radiation into consideration, the sunlight irradiation intensity that reaches our planet earth varies with respect to the time, seasons, geographical location and different weather conditions. The term "irradiation" is defined as the net energy that we obtain on a daily basis, and it represents the strength of the sunlight intensity. The unit for irradiation is Wh m^{-2} per day [1, 2, 5].

The other physical factors that affect the working as well as the efficiency of the solar PV system are the different geographic regions that experience different weather patterns, number of sun days in a year and best tilt angle for the solar panels. The PV cells would be exposed to more sunlight if a tracker is installed that changes the angle according to the movement of the sun; however, that method is way too expensive to meet the daily needs [6]. The phenomenon or the process of evaluating or calculating the adequate voltage, current and power ratings for each component of the PV system to meet the electric load demand and at the same time calculating the total expenses for the entire system from design to implementation including transportation cost and other costs is called as system sizing.

In this paper, we are primarily considering the solar energy as the conventional source of energy and focusing on designing a stand-alone photovoltaic (SAPV) system by assuming some standard loads present in a house. The sizing scenario presented in this paper aims at fulfilling the daily electrical needs for a residential housing in the city of Bhubaneswar, Odisha, India.

2 Methodology of Sizing of SAPVs

2.1 Components and System Requirements

a. **PV Module**: It is a semiconductor containing *p–n* junctions that convert sunlight to electricity which is DC in nature. Commonly, a PV module includes single poly-crystalline silicon and amorphous silicon [1].
b. **Battery**: The battery stores energy for meeting the peak load demands and is mostly useful during dark days or no-sun days. The commonly used types are lead–acid type batteries, that are designed to gradually discharge and recharge 80% of their capacity hundreds of times [1].
c. **Solar Charge Controller**: The solar charge controller is mainly used to maintain the battery levels at optimal state alongside protecting it from overcharge of the PV panels and excess discharge from the connected loads and therefore elongates the battery life [6].
d. **Inverter**: It primarily converts the DC output from the PV cells to AC form and feeds it to the AC load. Inverters are further classified based upon the waveform output and on the basis of installation. The inverter is known as power conditioner because of its ability to change the form of the electric power. The efficiency of inverters reaches its maximum limit when the load demand is greater than half the rated load [7].
e. **Load**: These are the end user connected electronic appliances.

The complete schematics of the SAPV are shown in the following Fig. 1.

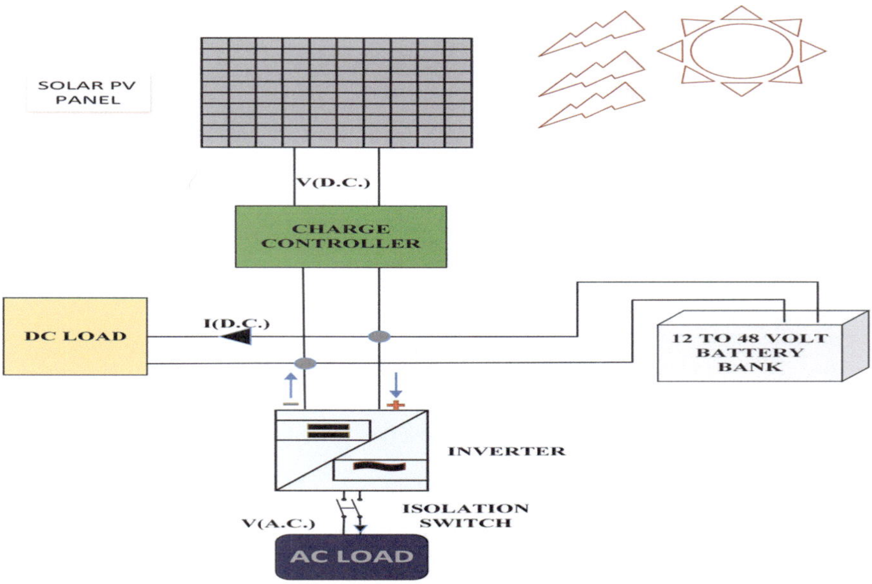

Fig. 1. Schematics of a stand-alone photovoltaic system (SAPV)

2.2 System Sizing

The phenomenon or the method of anticipating the required voltage and current ratings for each component of the PV system to fulfill the electric demand and at the same time calculating the total expenses for the entire system from design to implementation including transportation cost and other costs is called as system sizing. The following sizing is implemented based on intuitive methods [8–12].

a. **Home Appliance Sizing**:

The voltage and current requirements for the connected loads are first calculated along with their time of operation, and an average value of required power is estimated on a daily basis [1].

b. **Sizing of Solar Modules**:

It deals with the number of solar arrays required for the desired power output, and it also includes the manner in which the arrays needed to be connected, i.e., in series or parallel manner to gain higher efficiency. It completely depends on the daily energy in watt-hour (E), the DC voltage (V_{DC}) and the average sun-hour per day (T_{min}) of the system, and these [9] parameters must be calculated at first. Moreover, to avoid the undersizing of the system, losses are taken into consideration.

Therefore, to avoid undersizing, the daily average energy consumption is divided by the system efficiencies to deduce the required output from the solar panel array. Mathematically:

$$E_{\mathrm{r}} = \frac{\text{Daily average Energy Consumption}}{\text{Product of system component's efficiency}} = \frac{E}{\eta_{\text{overall}}} \tag{1}$$

The peak power is calculated by dividing the value of E_{r} by average sun-hours per day ($T_{\min}$) for the given area. Mathematically:

$$P_{\mathrm{p}} \frac{\text{Energy required on daily basis}}{\text{Minimum sun time per day}} = \frac{E_{\mathrm{r}}}{T_{\min}} \tag{2}$$

For calculating, the net current required:

$$I_{\mathrm{dc}} = \frac{\text{Peak Power}}{\text{System Voltage(DC)}} = \frac{P_{\mathrm{p}}}{V_{\mathrm{Ddc}}} \tag{3}$$

The number of modules connected in parallel can be calculated by:

$$N_{\mathrm{p}} = \frac{\text{Total current of the PV Module}}{\text{Rated value of current of one PV Module}} = \frac{I_{\mathrm{dc}}}{I_{\mathrm{r}}} \tag{4}$$

Modules to be connected in series can be calculated by:

$$N_{\mathrm{s}} \frac{\text{System Voltage(DC)}}{\text{Rated value of Voltage of one PV Module}} = \frac{V_{\mathrm{dc}}}{V_{\mathrm{r}}} \tag{5}$$

Lastly, the total number of modules that needed to be connected are:

$$N_{\mathrm{m}} = N_{\mathrm{s}} * N_{\mathrm{p}} \tag{6}$$

c. **Sizing of Battery**:

The approximated amount of energy needed is calculated by multiplying the number of no-sun days or autonomy days and the total power demand. Mathematically:

$$E_{\text{rough}} = \text{Total Power Demand(E)} * \text{Days of Autonomy(D)} \tag{7}$$

Considering safety factor, the result is divided by the maximum allowable level of discharge (MDOD):

$$E_{\text{safe}} = \frac{\text{Required energy for storage}}{\text{Maximum depth of Discharge}} = \frac{E_{\text{rough}}}{\text{MDOD}} \tag{8}$$

Now, at this step, the battery bank capacity (in A-h) can be evaluated by dividing the safe energy storage required by the DC voltage of one of the batteries:

$$C = \frac{E_{\text{safe}}}{V_{\text{b}}} \tag{9}$$

Therefore, the number of batteries needed is obtained by:

$$N_{\text{batteries}} = \frac{C}{C_{\text{b}}}; \text{ where } C_{\text{b}} \text{ refers to the capacity.} \tag{10}$$

Batteries to be cascaded in series are given by:

$$N_{\text{s}} = \frac{V_{\text{DC}}}{V_{\text{b}}} \tag{11}$$

and the number of parallel paths are given by:

$$N_{\text{p}} = \frac{N_{\text{Batteries}}}{N_{\text{s}}} \tag{12}$$

d. **Sizing of the Voltage Controller**:

Even though the name is voltage controller, the parameter that it controls is current. For a voltage controller, its main aspect is to withstand the maximum load current along with the maximum array current. The product of the short-circuit current and the safety factor results in deducing the ratings for the voltage controller. Mathematically:

$$I = I_{\text{sc}} * N_{\text{p}} * F_{\text{safe}} \tag{13}$$

We have considered the safety factor mainly because to ensure that the controller withstands and handles the maximum array currents produced by the arrays, which could be higher than the calculated value [10]. The number of controllers required is calculated by:

$$N_{\text{controller}} = \frac{I}{\text{Amps of Each Controller}} \tag{14}$$

e. **Inverter Sizing**:

The inverter sizing is determined from the first step as the actual power drawn will determine the rating of the inverter

f. **Sizing of the wiring system**:

The required type of wire to be used can be actuated based upon the above result.

3 Example

The location of the application is considered at Bhubaneswar, Odisha, India. The geographical location of Bhubaneswar is at 85.82° longitude and 20.29° latitude [13]. It receives an annual irradiance of about 5.32 kWh/m^2/day [14].

The home appliance sizing is considered as follows (Table 1).

Table 1. Daily consumption table (based upon approximated values)

Individual load	Quantity	Volt.	Amp.	AC wattage	Usage in h/day	Usage in days/week	÷7 days	Watt-hours AC
Domestic fan	1	220	0.54545	120	8	**7**	**7**	960
Grinder	1	220	2.72	600	0.3	**7**	**7**	180
Electric iron	1	220	4.54	1000	0.8	**4**	**7**	457.143
Desktop	1	220	0.56	120	2	**7**	**7**	240
Bulb	4 * 15	220	0.272727	60	5	**7**	**7**	300
Radio	1	220	0.363636	80	4	**7**	**7**	320
Fridge	1	220	0.909091	200	12	**7**	**7**	2400
Television	1	220	0.568182	125	6	**7**	**7**	750
Electric washing machine	1	220	1.136364	250	0.5	**5**	**7**	89.28571
Total AC connected Watts				2555	Average AC daily load			5696.43

Total average energy consumption 5696.426 approximated to 5700

The selected panel for the above mentioned requirements is luminous solar panel, 250 W, 24 V, 8.31 A. It is of polycrystalline silicon type having 60 number of cells with P_{max} as 250 W, V_{OC} as 37.42 V, I_{sc} as 8.85 A, V_{mp} as 30.2 V and I_{mp} as 8.31 A, cell efficiency as 0.176.

From the above formulas,

a. **Sizing of Arrays**:

$$E_r = \frac{E}{\eta_{overall}} = 5700/0.8 = 7125\,\text{Wh/day or } 7.125\,\text{kWh/day}$$

$P_p = \frac{E_r}{T_{min}} = 7.125/6 = 1.18\,\text{kWhp}$ [15]

$$I_{DC} = \frac{P_p}{V_{DC}} = 1187.5/24 = 49.47\,\text{A}$$

For modules,

$$N_p = \frac{I_{DC}}{I_r} = 49.47/8.31 = 5.95\,\text{Panel} = 6\,\text{Panels}$$

$$N_s = \frac{V_{DC}}{V_r} = 24/24 = 1$$

$$N_m = 6 * 1 = 6\,\text{Panels}$$

Hence, the PV array consists of six panels connected in parallel.

b. **Sizing of Battery**:

$$\text{Net Average Energy} = 5700\,\text{Wh}$$
$$\text{No Sun - days} = 3(\text{say})$$
$$E_{rough} = 5700 * 3 = 17.1\ \text{kWh}$$
$$E_{safe} = \frac{E_{rough}}{\text{MDOD}} = 17100/0.75 = 22800\,\text{Wh}$$
$$C = \frac{E_{safe}}{V_b} = 22800/12 = 1900\,\text{A-h}$$
$$N_{batteries} = \frac{C}{C_b} = 1900/250 = 7.6\,\text{Batteries} = 8\,\text{Batteries}$$
$$\text{Batteries in series} = N_s = \frac{V_{DC}}{V_b} = 24/12 = 2$$
$$\text{Number of parallel paths} = N_p = 8/2 = 4$$

c. **Sizing of Voltage Controller**:

The rated current of the voltage controller is calculated

$$I = I_{sc} * N_p * F_{safe} = 8.85 * 6 * 1.25 = 66.375\ \text{A}$$

$$N_{controller} = \frac{I}{\text{Amps of Each Controller}} = 66.375/60 = 1.10$$

Hence, we need two regulators connected in parallel.

d. **Sizing of Inverter**:

The total power of the devices when connected together:

P_{total} = 2555 W. This means the inverter that is needed must be capable of handling 2555 W at 220 V AC.

4 Conclusion

The location that is considered (Bhubaneswar, Odisha, India) qualifies itself to be famed as a sun-rich region having annual irradiance of more than 5.32 kWh/m^2 annually [14]. The solar stand-alone photovoltaic system is fully designed from scratch and is implemented based upon the calculated values. Several factors that affect the process directly or indirectly are explained in this report. The mentioned processes for the implementation of the solar stand-alone photovoltaic system can be used in any region having a tropical, subtropical or dry climate.

References

1. Al-Shammani AN, Othman MYH, Mat S, Ruslan MH, Abed AM, Sopian K (2015) Design & sizing of stand-alone solar power systems a house Iraq. In: Recent advances in renewable energy sources, April 2015
2. Goyal S, Sharma R (2017) Performance evaluation of stand alone, grid connected and hybrid renewable energy system for rural application: a comparative review. Renew Sustain Energy Rev 78:1378–1389
3. Energy Data. https://www.iea.org/weo2017/. Accessed on April 2018
4. Ritchie RW (1999) Using sunlight for your own solar electricity: build your own system, become independent of the grid, domestic photo voltaics. Ritchie Unlimited Publications
5. AL-Rousan N, Mat Isa NA, Desa MKM (2018) Advances in solar photovoltaic tracking systems: a review. In: Renewable and sustainable energy reviews, vol 82, Part 3, pp 2548–2569, Feb 2018
6. Messenger RA, Ventre J (2003) Photovoltaic systems engineering. CRC Press
7. Sonnenenergie DGF (2007) Planning and installing photovoltaic systems: a guide for installers, architects and engineers. Earthscan
8. Faraa L, Craciunescua D (2017) Output analysis of stand-alone PV systems: modeling, simulation and control, vol 112, pp 595–605, Mar 2017
9. Jadin MS, Nasiri IZM, Sabri SE, Ishak R (2015) A sizing tool for PV standalone system. ARPN J Eng Appl Sci 10(22), Dec 2015
10. Nordin ND, Rahman HA (2015) An optimization method for designing stand alone photovoltaic system using iterative method. In: IEEE International Conference on Smart Energy Grid Engineering (SEGE), 12 Nov 2015
11. Bataineh K, Dalalah D (2012) Optimal configuration for design of stand-alone PV system. Smart Grid Renew Energy 3:139–147
12. Srinivasarao M, Sudha KR, Bhanu CVK (2016) A simple and reliable method of design for standalone photovoltaic systems. Journal IE (India), Series B, 19 Aug 2016
13. Location Data. http://www.synergyenviron.com/tools/solar-irradiance/india/odisha/bhubaneswar. Accessed on April 2018
14. Irradiation Data. http://www.synergyenviron.com/tools/solar-irradiance/india/odisha/bhubaneswar. Accessed on April 2018
15. Average Sun-Hours Data. https://www.currentresults.com/Weather/India/annual-sunshine.php. Accessed on April 2018
16. Aziz NIA, Sulaiman SI, Shaari S, Musirin I, Sopian K (2017) Optimal sizing of stand-alone photovoltaic system by minimizing the loss of power supply probability. Solar Energy 150:220–228

17. Ibrahim IA, Khatib T, Mohamed A (2016) Optimal sizing of a standalone photovoltaic system for remote housing electrification using numerical algorithm and improved system models. Energy 126:392–403
18. Khatib T, Mohamed A, Sopian K (2013) A review of photovoltaic systems size optimization techniques. Renew Sustain Energy Rev 22:454–465
19. Rajesh R, Mabel MC (2015) A comprehensive review of photovoltaic systems. Renew Sustain Energy Rev 51:231–248
20. Sumathia V, Jayapragash R, Bakshi A, Akella PK (2017) Solar tracking methods to maximize PV system output—a review of the methods adopted in recent decade. Renew Sustain Energy Rev 74:130–138

A Model for Optimizing Cost of Energy and Dissatisfaction for Household Consumers in Smart Home

Nilima R. Das[1(✉)], Satyananda C. Rai[2], and Ajit Nayak[1]

[1] Faculty of Engineering & Technology, Siksha 'O' Anusandhan Deemed to Be University, Bhubaneswar, Odisha, India
nilimadas@soa.ac.in

[2] Department of IT, Silicon Institute of Technology, Bhubaneswar, Odisha, India

1 Introduction

Smart grid is an uprising concept in electrical grid systems. It claims the advancement of technology and information system with highly advanced electrical infrastructures that can increase the reliability, security, and efficiency. There are many challenges in the implementation of smart grids and a lot of research is going on in this area from the last few years. The implementation of smart grid now involves various demand-side management (DSM) techniques. The DSM programs comprise of load management and energy conservation activities at the user side that tries to maintain a balance between the demand and supply of electricity to facilitate the use of the existing energy resources avoiding investment in additional sources. It results in reduced electricity payments for the consumers and increased availability of the electricity. The time-varying prices also help a lot to encourage the consumers to reduce their demand during peak hours by shifting a fraction of their consumption from peak hours to off-peak hours as the peak hours are charged with high prices. The method proposed in this work follows a similar approach. It uses a time-varying pricing technique to optimize the energy consumption behavior of the user by producing an operation time schedule for the user's appliances which reduces the peak hour demand and electricity cost of the user. The next section provides a brief literature survey in this area. Section 3 elaborates the mathematical structure for the proposed method. Section 4 provides a precise description about the simulation process and its results. Finally, Sect. 5 provides the conclusion.

2 Related Works

The users in a smart grid can use various energy management techniques to lower their peak hour power consumption which can also help them to reduce their daily electricity bill. A lot of research works are being done to implement DSM for the consumers so that they can reduce their consumption during peak hours. The authors have used an energy management technique that can predict upcoming prices in a real-time

M. N. Mohanty and S. Das (eds.), *Advances in Intelligent Computing and Communication*, Lecture Notes in Networks and Systems 109,
https://doi.org/10.1007/978-981-15-2774-6_61

environment that helps to control and manage the peak hour consumption of the users [1]. The authors [2] have used an integer linear programming (ILP) method to generate the time schedule for the appliances which can minimize the peak time load and generate optimal operation time for user appliances. They have also proposed a consumption scheduling mechanism using integer linear programming (ILP) and game theory approach to minimize the peak time load [3]. The authors [4], follow a game theory approach for generating an optimized schedule for the consumers. In this work, the game is played between the users and each user tries to minimize its cost. The authors have proposed a quadratic convex programming method to determine the optimal operation time for the appliances for cost minimization in [5]. The authors [6, 7] have described a game theory-based optimization procedure where the game is played between the users and the utility provider. In the game, the user's strategy is to minimize its electricity payment and the strategy of utility provider is to adjust the energy price parameter to reduce users' consumption. Ye et al. [8] provided a real-time information-based energy management technique based on game theory optimization method to minimize the peak-to-average ratio in demand.

All the works in the literature try to reduce the peak time load and electricity payment for the users. However, in [1], the authors have tried to minimize the waiting time for each appliance along with cost reduction. Every appliance has to operate in a particular time interval which is decided by the user. After optimization, the appliances have to follow the time schedule generated by the optimization process. If the starting time of an appliance is postponed, then the waiting cost is increased and the optimization process also makes an effort to reduce the waiting cost. The waiting cost depends on the gap between the scheduled starting time and the beginning of the time interval previously set by user for the appliance. The waiting cost increases with the percentage increase in the waiting time.

There are some other scenarios possible where the user has a preferred starting time for the execution of an appliance from the allowed time of operation set for it. The optimization process may generate a starting time before or after the preferred starting time of the user and if the optimization process does not generate the time schedule according to the user preference then the user may get some dissatisfaction. In this work, a method has been proposed that reduces the dissatisfaction caused due to the mismatch of scheduled starting time and user's preferred starting time of an appliance as an outcome of cost minimization. The procedure is described in detail in the next sections.

3 Mathematical Formulation of the System Model

The method proposed in this work gives a solution to a multi-objective optimization problem. The objective of the system is to minimize the daily energy cost for the user as well as reduce the level of dissatisfaction caused due to the cost minimization. It generates an optimal time schedule for the operation of the user appliances for a whole day (12 a.m. to 11 p.m.) in order to achieve the objective. The total time horizon taken for simulation is one day. One day is divided into 24 time slots and one slot is equivalent to one hour.

3.1 Formulation of the Cost Function

The cost of the energy during an hour is defined in terms of total energy consumed during that hour. The cost for an hour is proportional to the total hourly energy usage of the consumer. In literature, several cost functions have been used. Since a logarithmic cost function follows a linear growth, it can reduce the inconvenience of the user caused due to increase in cost. The cost function for an hour h is represented as $p(l_{\mathrm{h}})$ which is defined as follows:

$$p(l_{\mathrm{h}}) = \kappa \times l_{\mathrm{h}} \times \log(l_{\mathrm{h}} + 1) \tag{1}$$

where κ is the cost parameter. The value set for κ is more during daytime and less during nighttime that can control the consumption of the users during daytime, because generally the daytime is considered as the peak time in a day. l_{h} is the total load consumed during hour h which is the sum of the energy consumption of all the appliances during that hour described by the following equality.

$$l_{\mathrm{h}} = \sum_{a \in A} e_a^{\mathrm{h}} \tag{2}$$

A is the set of all appliances the user possesses. The energy consumption of an appliance a during hour h is denoted as e_a^{h}. The cost during an hour is increased when load l_{h} is increased during that hour. By representing $p(l_{\mathrm{h}})$ as P_{h}, the total daily electricity cost termed as C can be defined as:

$$C = \sum_{\mathrm{h}=1}^{24} P_{\mathrm{h}} \tag{3}$$

The optimizer tries to minimize the total cost to be paid for the day which is being considered for optimization.

3.2 Formulation of the Dissatisfaction Factor

Every appliance has a specific energy requirement and time of operation. The allowed time period during which an appliance can be operated is decided by the user which is based on its energy requirement. An appliance can start at any hour within this allowed time of operation and has to finish operating on or before the end of this period. The optimizer generates the starting time of every appliance which lies between the allowed time intervals for the appliance with a purpose to minimize the cost. However, the user can have a preferable starting time for every appliance. When the scheduled starting time does not match with the preferred starting time, the user gets some displeasure. The scheduled start time may fall before or after the preferred start time. When the difference between the scheduled starting time and preferred starting time increases the dissatisfaction is also increased which has to be reduced. The description of the dissatisfaction factor in the optimization process is given below.

The rate of dissatisfaction caused due to cost optimization is estimated using the following function.

$$f_a = x/x_{\max} \tag{4}$$

where $x = |p_a - T_a|$

p_a is the user's preferred starting hour for appliance a and T_a is the starting hour calculated by the scheduler. It is to be noted that the values of p_a and T_a are considered in hours. $x_{\max}$ is the maximum possible value of x. The value of x lies between [0, 23]. The minimum value for x occurs when scheduled starting time becomes equal to preferred starting time. Whereas the maximum value for x occurs when for an appliance the possible time of operation is whole day, that means the preset time interval for the operation of the appliance is from 1st hour of the day to the last hour of the day and value of preferred starting time slot is 1(1st hour of the day) and scheduled starting time slot is 24 (last hour of the day). So, in that case, the gap between the preferred starting time and scheduled starting time is 23. In the worst case, the time gap is 23, and in the best case, it is 0. For simplicity of calculation, fraction of a time slot (hour) has not been considered here. f_a is increasing with every increase in x. The range of f_a lies between [0, 1]. The function used for calculating the unhappiness or dissatisfaction denoted as U_a is defined as follows:

$$U_a = (v_a)^{f_a} \tag{5}$$

where $v_a > 1$. The value of v_a can be different for different appliances which is set by the user. The appliance for which the user wants minimum dissatisfaction the value of v_a for that appliance will be higher than the values set for other appliances. Thus, for higher values of v_a, the value of U_a will be higher which means the dissatisfaction term will have higher value in the optimization function. The value for U_a is increased when the rate of dissatisfaction f_a is increased. When x is 0, the value for U_a is 1 as f_a is 0 and when x is maximum, U_a attains its maximum value for a specific value of v_a. It means when the scheduled starting time is equal to the preferred starting time, the dissatisfaction is zero. However, when there is a difference between the two, there is dissatisfaction and when the difference is increased, the dissatisfaction is increased. The level of dissatisfaction for all the appliances is to be minimized through optimization.

3.3 Formulation of the Optimization Function

With the above generalizations, the optimization function can be defined as:

$$\text{minimize} \quad \theta_1 \times C + \theta_2 \times D \tag{6}$$

Subject to the following constraints

$$e_a^{\mathrm{h}} = 0, \quad \text{if } h \notin [S_a, F_a]$$

$$L_a = \sum_{h=1}^{24} e_a^{\mathrm{h}}$$

$$p_{\min}^a(a) \le e_a^{\mathrm{h}} \le p_{\max}^a(a)$$

where $D = \sum_{a \in A} U_a$

The parameters θ_1 and θ_2 represent the significance of each objective of the proposed optimization problem. θ_1 describes the importance of energy cost reduction, and θ_2 describes the importance of dissatisfaction level minimization. The sum of θ_1 and θ_2 is 1. S_a is the allowed starting hour of appliance a, and F_a is the allowed finishing hour of a. The interval $[S_a, F_a]$ for an appliance a is decided by the user. So for every appliance, the allowed time of operation is determined by the user and beyond which the energy consumption of that appliance is 0 as described by the first constraint. The second constraint says that the total energy consumption for an appliance a in a whole day denoted as L_a is predefined based on the user's consumption behavior or user's choice. The third constraint says every appliance has a minimum power level and maximum power level denoted as $p_{\min}^a$ and $p_{\max}^a$, respectively. C is the total daily cost and D represents the total dissatisfaction for all the appliances operated in the whole day.

4 Results of Optimization

In this work particle, swarm optimization (PSO) technique has been used for optimization process. First optimization is done for one user only and then it is applied for more than one user.

The optimization process minimizes the objective function and generates the operation time schedule for each appliance. That means it generates the optimal e_a^{h} value for every appliance for all the hours in a whole day. The following figures confirm the effectiveness of the algorithm.

Figure 1 shows the convergence of the proposed optimization algorithm. Figure 2 shows that when the optimization is not done the energy consumption of the user is more during peak time. The forecasted load curve (general residential consumption pattern without optimization) used here is similar to the consumption pattern used in [9]. It is clear from the figure that when optimization is followed a part of the energy consumption is shifted to non-peak time making the demand low during peak hours which is very much required to create a proper balance between demand and supply of electricity. Figure 3 shows the comparison of the energy costs with and without optimization for 25 days. It shows that the cost for all these days remains lower with optimization. Figure 4 shows that the energy consumption pattern is changed when more importance is given satisfaction rather than cost minimization. There is a trade-off between the cost minimization and dissatisfaction minimization. Figure 5 shows a comparison between the monthly electricity payment made without optimization and with optimization for 15 different users. It is obvious that the cost calculated after

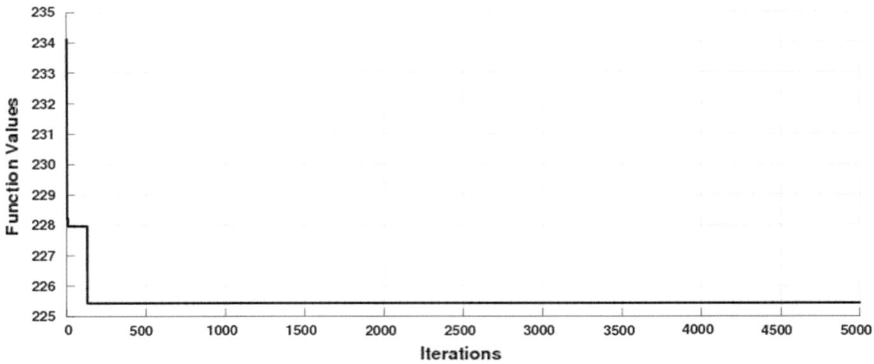

Fig. 1. Convergence of the algorithm

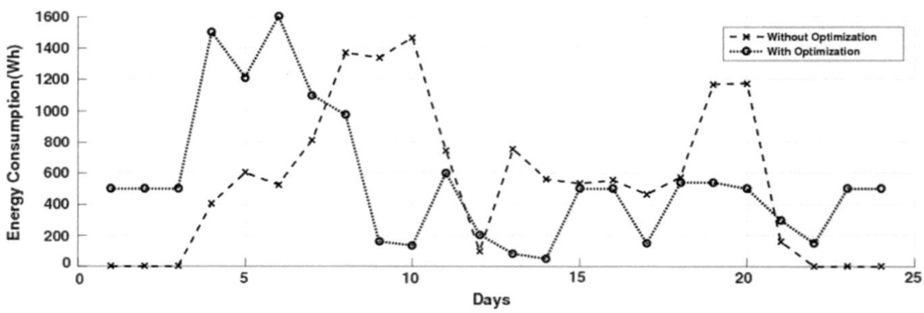

Fig. 2. Comparison of energy usage in a day before and after optimization

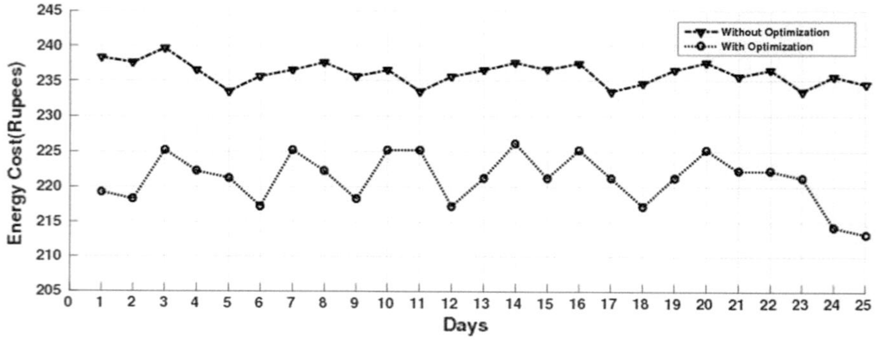

Fig. 3. Cost of energy before and after optimization for a single user for 25 days

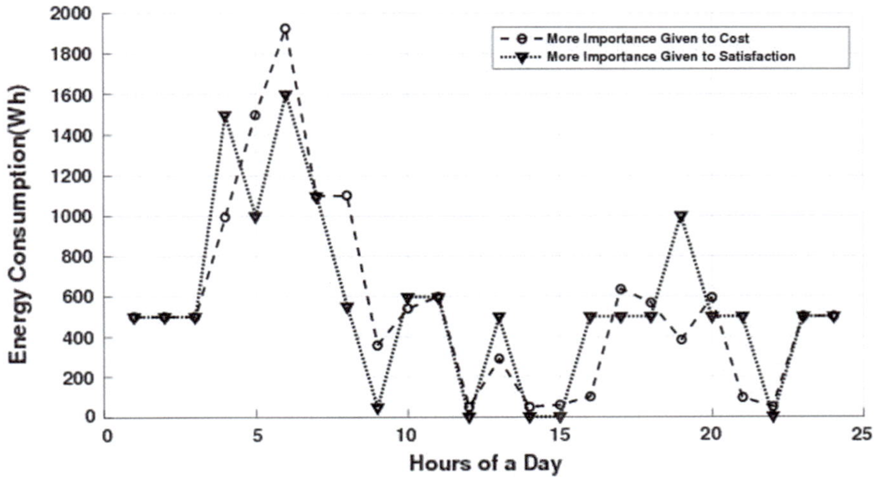

Fig. 4. Difference in energy consumption pattern with a different objective

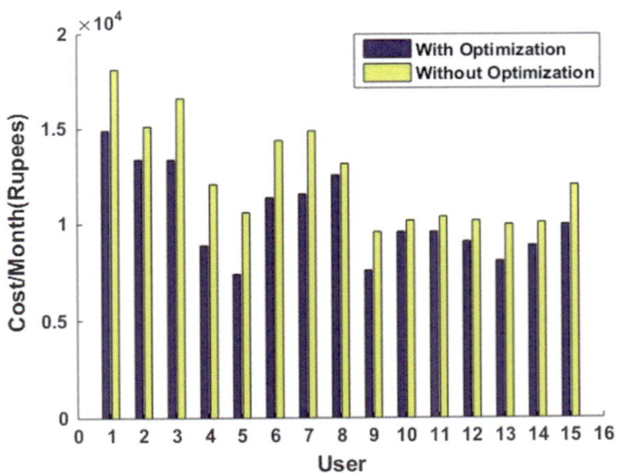

Fig. 5. Comparison of cost before and after optimization for 15 users

optimization is lower than that of before optimization. Figure 6 shows the fact that for more satisfaction the user has to pay more. On the other hand, the user can save some money if it can tolerate a little discomfort.

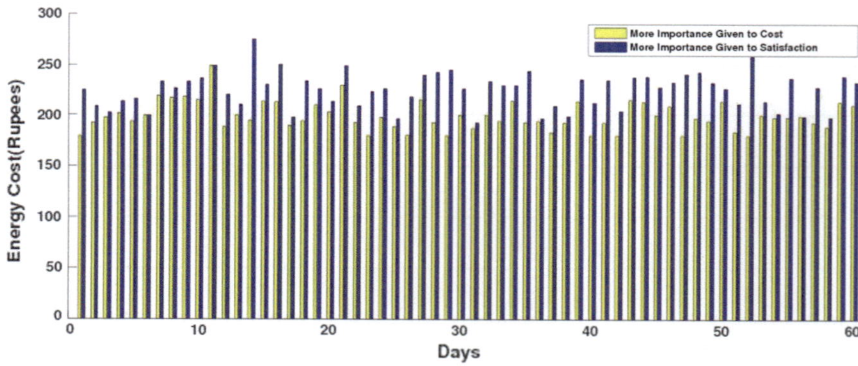

Fig. 6. Higher cost for higher satisfaction

5 Conclusion

The proposed method reduces the user demand for electricity during peak hours by producing a schedule for the operation of the electrical appliances for the whole day. Without proper scheduling, the user may try to operate its appliances mainly during peak hours which increases the peak hour consumption and the electricity costs. But when the user follows the schedule generated by the optimizer, the electricity payments is minimized. The optimizer process used here not only minimizes the electricity payments but also reduces the dissatisfaction level of the user while using the appliances maintaining a balance between the two. If the user wants less dissatisfaction, then the electricity payment made will be more and if a reduced payment is preferred, then a little discomfort has to be tolerated. However, in both the cases, the cost will be less than the scenario with no optimization.

References

1. Mohsenian-Rad H, Leon-Garcia A (2010) Optimal residential load control with price prediction in real-time electricity pricing environments. IEEE Trans Smart Grid 1(2):120–133
2. Zhu Z, Tang J, Lambotharan S, Chin WH, Fan Z (2011) An integer linear programming and game theory based optimization for demand-side management in smart grid. In: IEEE international workshop on smart grid communications and networks
3. Zhu Z, Tang J, Lambotharan S, Chin WH, Fan Z (2012) An integer linear programming based optimization for home demand-side management in smart grid. IEEE PES Innov Smart Grid Technol (ISGT) (2012)
4. Nguyen HK,Songl JB, Hanl Z (2012) Demand side management to reduce peak-to-average ratio using game theory in smart grid. In: Proceedings of IEEE INFOCOM (2012)
5. Liu Y, Yuen C, Huang S, Hassan NU, Wang X, Xie S (2014) Peak-to-average ratio constrained demand-side management with consumer's preference in residential smart grid. IEEE PES Innov Smart Grid Technol 8(6):1084–1097

6. Soliman HM, Leon-Garcia A (2014) Game-theoretic demand-side management with storage devices for the future smart grid. IEEE Trans Smart Grid 5:1475–1485
7. Fadlullah ZMd, Quan DM, Kato N, Stojmenovic I (2014) GTES: an optimized game-theoretic demand-side management scheme for smart grid. IEEE Syst J 8:588–597
8. Ye F, Qian Y, Hu RQ (2016) A real-time information based demand-side management system in smart grid. IEEE Trans Parallel Distrib Syst 27(2):329–339
9. Labeeuw W, Deconinck G (2013) Residential electrical load model based on mixture model clustering and markov models. IEEE Trans Industr Inf 9(3):1561–1568

A Circular Ultra-Wideband Antenna for Wearable Applications

Shaktijeet Mahapatra(✉), Sarmistha Satrusallya, and Mihir Narayan Mohanty

Department of ECE, ITER, Siksha 'O' Anusandhan (Deemed to be University), Bhubaneswar, India
shaktijeetmahapatra@soa.ac.in

1 Introduction

With the development of wearable antennas, the wearable devices and sensors are being increasingly used by the medical practitioners, the military and the scientific community. The reason is that the antennas with smaller profile and low weight are being developed. The UWB band for wearable antennas (3–10.6 GHz) offers more freedom to the designers in terms of low-power, frequency and bandwidth [1]. The band generally is used for ubiquitous computing but nowadays it is being used for wearable applications as well owing to less congestion in this band.

Klemm et al. [2] designed two ultra-wideband antennas—coplanar waveguide fed disc monopole antenna and microstrip-fed annular slot antenna—for body area network applications. These antennas were designed entirely using textile only. Almpanis et al. [3] presented an inverted truncated annular conical dielectric resonator antenna for body area network applications. The operating range of antenna was 3.0–5 GHz. See et al. [4] presented three antennas for UWB applications, namely directional suspended plate antenna, omnidirectional monopole antenna and printed diversity antenna for on-body communications. Shin et al. [5] presented a CPW-fed stair-shaped antenna having ultra-wide bandwidth for wireless body area network (WBAN). Kang et al. [6] presented a folded antenna working in UWB range aiming to reduce backward radiation and other effect arising out of proximity to human body.

In this paper, we present a circular patch antenna with four square slots cut on the edges. The advantages of this antenna are that it is small, very lightweight and has a thin profile and is easy to fabricate. The ground plane has been kept full to cut down backward radiation.

The paper is organized into the following sections: In Sect. 2, the design of the antenna is described; in Sect. 3, simulation results are presented; in the last section, conclusions are drawn.

M. N. Mohanty and S. Das (eds.), *Advances in Intelligent Computing and Communication*, Lecture Notes in Networks and Systems 109,
https://doi.org/10.1007/978-981-15-2774-6_62

2 Design Considerations

The proposed antenna is a circular patch antenna of 13 mm radius on a cylindrical substrate of radius 20 mm and height of 1.6 mm, fed by a microstrip line. FR4-epoxy, with a relative permittivity of 4.4 and a dielectric loss tangent of 0.02, was chosen as material for the substrate. Four square slots of 5 mm were cut from the edges. The feed was recessed by 1 mm and a quarter-wave line was inserted for matching input impedance with the 50 Ω port impedance. Figure 1 shows the proposed antenna.

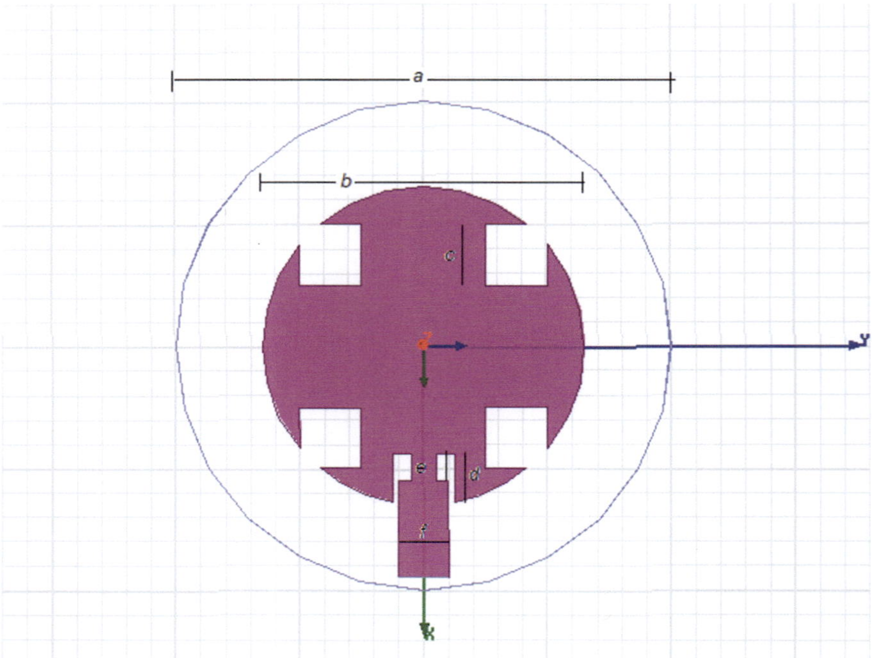

Fig. 1. Design of the antenna (proposed)

The design equation is as follows [7]:

$$b = \frac{F}{\left\{1 + \frac{2h}{\pi \epsilon_r F}\left[\ln\frac{(\pi F)}{2h} + 1.7726\right]\right\}^{1/2}} \quad (1)$$

$$F = \frac{8.791 \times 10^9}{f_r\sqrt{\epsilon_r}} \quad (2)$$

where h is in cm (Table 1).

Table 1. Design parameters

Parameters	Values (in mm)
a	40
b	26
c	5
d	3.9
e	2.2
f	4

3 Simulation Results and Discussions

As can be clearly seen from Fig. 2, the antenna resonates at 3.8 and 9.74 GHz. The impedance bandwidths obtained are 1.87 GHz and 0.38 GHz, respectively. The S11 or the return loss obtained is −19 dB and −22 dB, respectively.

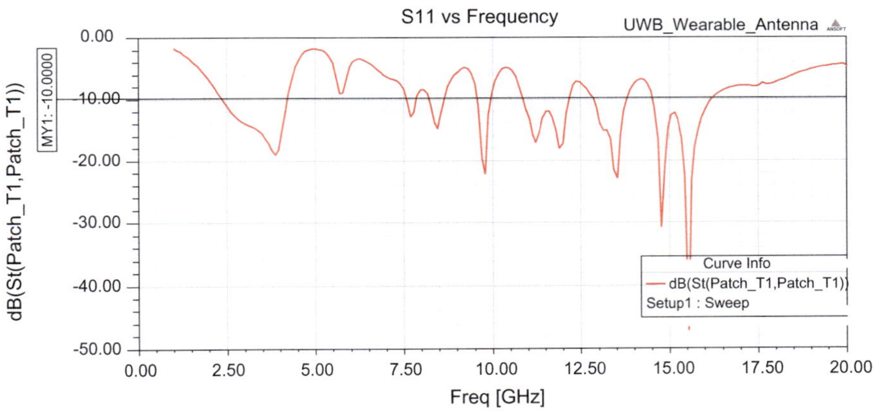

Fig. 2. S11 versus frequency

Figure 3 shows the electric field distribution on the patch at 3.8 GHz. As can be clearly seen that due to the slots, the value of the electric field at the edges range from 654.4 to 915 V/m. This results in almost directional beam as can be seen from Figs. 4 and 5. A closer inspection also reveals that due to full ground structure there is no backward radiation, making this antenna an excellent candidate for wearable applications and on-body communications.

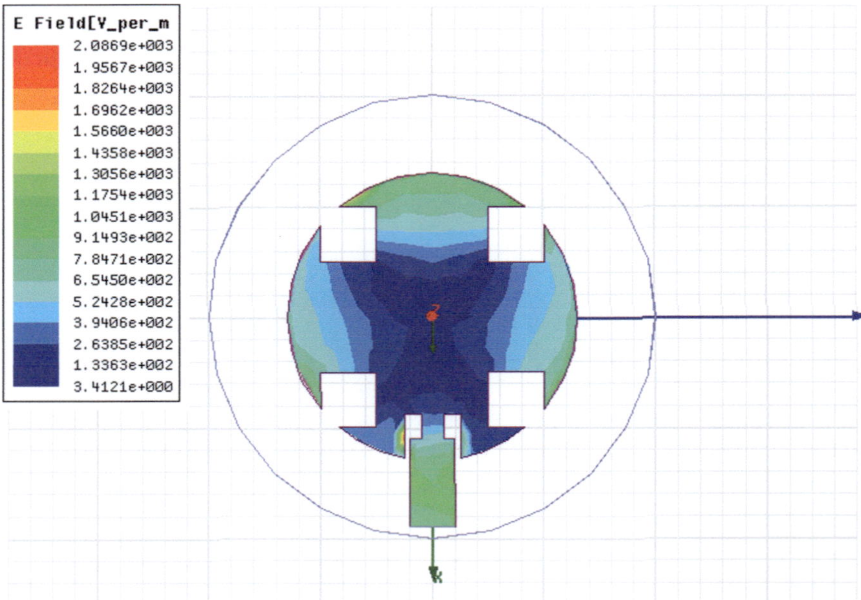

Fig. 3. Electric field at 3.8 GHz

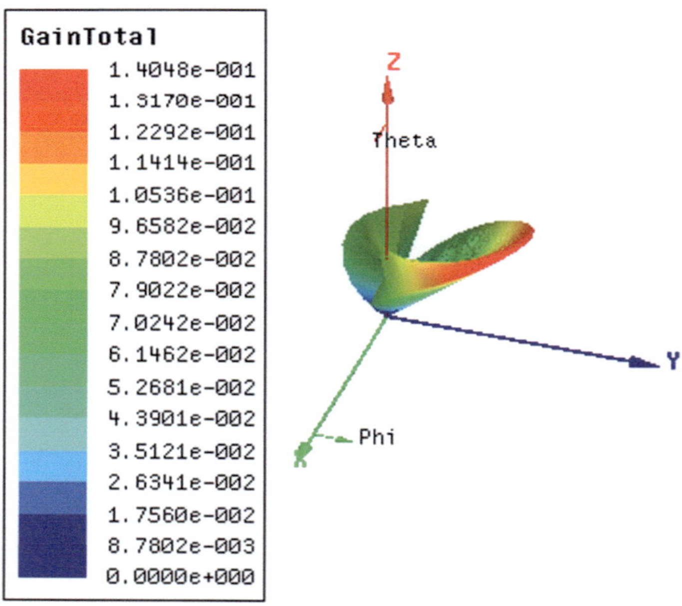

Fig. 4. 3D polar plot of gain at 3.8 GHz

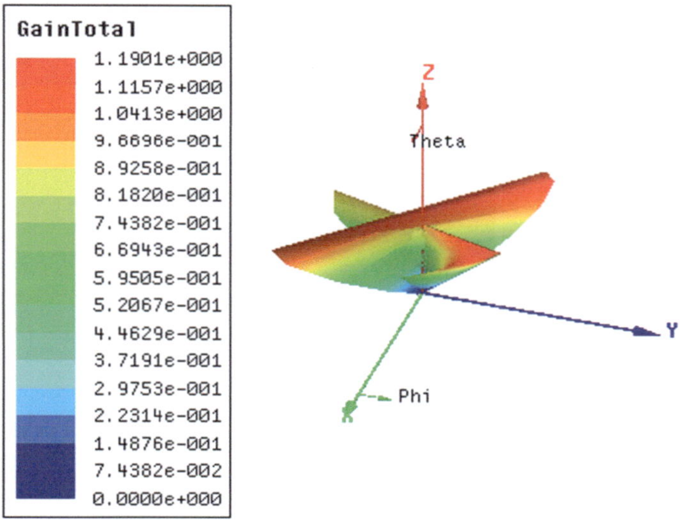

Fig. 5. 3D polar plot of gain at 9.74 GHz

4 Conclusions

A microstrip-fed slotted circular patch antenna has been designed for UWB wearable applications. The antenna was found to be resonating at 3.8 and 9.74 GHz. The advantages of this design lie in the simplicity of design, and no backward radiation. Owing to the above-mentioned advantages, the proposed antenna can be easily used for body area network applications and on-body communication applications easily.

References

1. Hall PS, Hao Y (2012) Antennas and propagation for body-centric wireless communications. Artech house
2. Klemm M, Troester G (2006) Textile UWB antennas for wireless body area networks. IEEE Trans Antennas Propag 54(11):3192–3197
3. Almpanis G et al (2008) A truncated conical dielectric resonator antenna for body-area network applications. IEEE Antennas Wireless Propag Lett 8:279–282
4. See TS, Chen ZN (2009) Experimental characterization of UWB antennas for on-body communications. IEEE Trans Antennas Propag 57(4):866–874
5. Shin H, Kim J, Choi J (2009) A stair-shaped CPW-fed printed UWB antenna for wireless body area network. In: 2009 Asia Pacific Microwave Conference, IEEE
6. Kang C-H, Wu S-J, Tarng J-H (2011) A novel folded UWB antenna for wireless body area network. IEEE Trans Antennas Propag 60(2):1139–1142
7. Balanis CA (2016) Antenna theory: analysis and design, wiley

Author Index

M. N. Mohanty and S. Das (eds.), *Advances in Intelligent Computing and Communication*, Lecture Notes in Networks and Systems 109,
https://doi.org/10.1007/978-981-15-2774-6

MIX
Papier aus verantwortungsvollen Quellen
Paper from responsible sources
FSC® C105338

If you have any concerns about our products,
you can contact us on
ProductSafety@springernature.com

In case Publisher is established outside the EU,
the EU authorized representative is:
Springer Nature Customer Service Center GmbH
Europaplatz 3, 69115 Heidelberg, Germany

Printed by Libri Plureos GmbH
in Hamburg, Germany